Foundations of Arithmetic Differential Geometry

Mathematical
Surveys
and
Monographs

Volume 222

Foundations of Arithmetic Differential Geometry

Alexandru Buium

AMERICAN MATHEMATICAL SOCIETY
Providence, Rhode Island

2010 *Mathematics Subject Classification.* Primary 11E57, 11F85, 14G20, 53C21.

For additional information and updates on this book, visit
www.ams.org/bookpages/surv-222

Library of Congress Cataloging-in-Publication Data

Buium, Alexandru, 1955-

Foundations of arithmetic differential geometry / Alexandru Buium.

Description: Providence, Rhode Island : American Mathematical Society, [2017] | Series: Mathematical surveys and monographs ; volume 222 | Includes bibliographical references and index.

Identifiers: LCCN 2016056302 | ISBN 9781470436230 (alk. paper)| ISBN 9781470475772 (Softcover)

Subjects: LCSH: Geometry, Differential. | AMS: Number theory – Forms and linear algebraic groups – Classical groups. msc | Number theory – Discontinuous groups and automorphic forms – p-adic theory, local fields. msc | Algebraic geometry – Arithmetic problems. Diophantine geometry – Local ground fields. msc | Differential geometry – Global differential geometry – Methods of Riemannian geometry, including PDE methods; curvature restrictions. msc

Classification: LCC QA641 .B774 2017 | DDC 516.3/6–dc23

LC record available at https://lccn.loc.gov/2016056302

Printed in the United States of America.
Reprinted by the American Mathematical Society, 2023

♾ The paper used in this book is acid-free and falls within the guidelines established to ensure permanence and durability.
Visit the AMS home page at `http://www.ams.org/`

11 10 9 8 7 6 5 4 3 2 28 27 26 25 24 23

Contents

Preface

Aim of the book. The aim of this book is to introduce and develop an arithmetic analogue of *classical differential geometry*; this analogue will be referred to as *arithmetic differential geometry*. In this new geometry the ring of integers $\mathbb{Z}$ will play the role of a ring of functions on an infinite dimensional manifold. The role of coordinate functions on this manifold will be played by the prime numbers $p \in \mathbb{Z}$. The role of partial derivatives of functions with respect to the coordinates will be played by the *Fermat quotients*, $\delta_p n := \frac{n-n^p}{p} \in \mathbb{Z}$, of integers $n \in \mathbb{Z}$ with respect to the primes p. The role of metrics (respectively 2-forms) will be played by symmetric (respectively anti-symmetric) matrices with coefficients in $\mathbb{Z}$. The role of connections (respectively curvature) attached to metrics or 2-forms will be played by certain adelic (respectively global) objects attached to matrices as above. One of the main conclusions of our theory will be that (the "manifold" corresponding to) $\mathbb{Z}$ is "intrinsically curved;" this curvature of $\mathbb{Z}$ (and higher versions of it) will be encoded into a $\mathbb{Q}$-Lie algebra $\mathfrak{hol}_{\mathbb{Q}}$, which we refer to as the *holonomy algebra*, and the study of this algebra is, essentially, the main task of the theory.

Needless to say, *arithmetic differential geometry* is still in its infancy. However, its foundations, which we present here, seem to form a solid platform upon which one could further build. Indeed, the main differential geometric concepts of this theory turn out to be related to classical number theoretic concepts (e.g., Christoffel symbols are related to Legendre symbols); existence and uniqueness results for the main objects (such as the arithmetic analogues of Ehresmann, Chern, Levi-Cività, and Lax connections) are being proved; the problem of defining curvature (which in arithmetic turns out to be non-trivial) is solved in some important cases (via our method of analytic continuation between primes and, alternatively, via algebraization by correspondences); and some basic vanishing/non-vanishing theorems are being proved for various types of curvature. It is hoped that all of the above will create a momentum for further investigation and further discovery.

Immediate context. A starting point for this circle of ideas can be found in our paper [**23**] where we showed how to construct arithmetic analogues of the classical jet spaces of Lie and Cartan; these new spaces were referred to as *arithmetic jet spaces*. The main idea, in this construction, was to replace classical derivation operators acting on functions with *Fermat quotient operators* acting on numbers and to develop an *arithmetic differential calculus* that parallels classical calculus. There were two directions of further development: one towards a theory of *arithmetic differential equations* and another one towards an *arithmetic differential geometry*. A theory of *arithmetic differential equations* was developed in a series of papers [**23**]-[**42**], [**7**] and was partly summarized in our monograph [**35**] (cf. also the survey papers [**43, 102**]); this theory led to a series of applications to invariant theory [**27, 7, 28, 35**], congruences between modular forms [**27, 37, 38**], and Diophantine geometry of Abelian and Shimura varieties [**24, 36**]. On the other hand an *arithmetic differential geometry* was developed in a series of papers [**42**]-[**47**], [**8**]; the present book follows, and further develops, the theory in this latter series of papers.

We should note that our book [**35**] on *arithmetic differential equations* and the present book on *arithmetic differential geometry*, although both based on the same conceptual framework introduced in [**23**], are concerned with rather different objects. In particular the two books are independent of each other and the overlap between them is minimal. Indeed the book [**35**] was mainly concerned with arithmetic differential calculus on Abelian and Shimura varieties. By contrast, the present book is concerned with arithmetic differential calculus on the classical groups

$$GL_n, SL_n, SO_n, Sp_n,$$

and their corresponding symmetric spaces.

Of course, the world of Abelian/Shimura varieties and the world of classical groups, although not directly related within *abstract algebraic geometry*, are closely related through *analytic* concepts such as *uniformization* and *representation theory.* The prototypical example of this relation is the identification

$$M_1(\mathbb{C}) \simeq SL_2(\mathbb{Z})\backslash SL_2(\mathbb{R})/SO_2(\mathbb{R})$$

where M_1 is the coarse moduli space of Abelian varieties

$$\mathbb{C}/\text{lattice}$$

of dimension one; the curve M_1 is, of course, one of the simplest Shimura varieties. It is conceivable that the analytic relation between the two worlds referred to above could be implemented via certain *arithmetic differential correspondences* (correspondences between arithmetic jet spaces); in particular it is conceivable that the subject matter of the present book and that of our previous book [**35**] might be related in ways that go beyond what can be seen at this point. A suggestion for such a possible relation comes from the theory of *δ-Hodge structures* developed in [**22**] and from a possible arithmetic analogue of the latter that could emerge from [**7, 35**].

There are at least two major differential geometric themes that are missing from the present book: *geodesics* and the *Laplacian.* It is unclear at this point what the arithmetic analogues of geodesics could be. On the other hand, for first steps towards an arithmetic Laplacian (and, in particular, for a concept of curvature based on it), we refer to [**34**]. But note that the flavor of [**34**] is quite different from that of the present book. Again, it is conceivable that the arithmetic Laplacian theory in [**34**] could be related, in ways not visible at this point, to our theory here.

Larger context. By what has been said so far this book is devoted to unveiling a new type of "geometric" structures on $\mathbb{Z}$. This is, of course, in line with the classical effort of using the analogy between numbers and functions to the advantage of number theory; this effort played a key role in the development of number theory throughout the 20th century up to the present day, as reflected in the work of Dedekind, Kronecker, Hilbert, Artin, Weil, Iwasawa, Grothendieck, Manin, Parshin, Arakelov (to name just a few of the early contributors). However, as we shall explain in the Introduction, this classical theory (which views number fields as analogous to Riemann surfaces) differs in essential ways from the theory of the present book (which views number fields as analogous to infinite dimensional manifolds). Our theory should also be contrasted with more recent geometric theories of "the discrete" such as: arithmetic topology [**104**]; the theory around the p-curvature conjecture of Grothendieck [**87**]; the Ihara differential [**78**]; the Fontaine-Colmez calculus [**64**]; discrete differential geometry [**10**]; and the geometries over the field with one element, $\mathbb{F}_1$, proposed by a number of people including: Tits [**118**], Kapranov-Smirnov [**86**], Soulé [**115**], Deitmar [**56**], Kurokawa et.al. [**93**], Connes-Consani [**55**], Manin-Marcolli [**117**], Lorscheid [**117**], Haran [**70**], etc. The non-vanishing curvature in our theory also prevents our arithmetic differential geometry from directly fitting into Borger's λ-ring framework [**13**] for $\mathbb{F}_1$; indeed, roughly speaking, λ-ring structure leads to zero curvature. For each individual prime, though, our theory is consistent with Borger's philosophy of $\mathbb{F}_1$; cf. [**13, 102**] and the Introduction to [**35**]. In spite of the differences between these theories and ours we expect interesting analogies and interactions.

This being said what is, after all, the position of our theory among more established mathematical theories? The answer we would like to suggest is that our curvature of $\mathbb{Z}$, encoded in the Lie algebra $\mathfrak{hol}_{\mathbb{Q}}$, could be viewed as an infinitesimal counterpart of the absolute Galois group $\Gamma_{\mathbb{Q}} := Gal(\mathbb{Q}^a/\mathbb{Q})$ of $\mathbb{Q}$: our Lie algebra $\mathfrak{hol}_{\mathbb{Q}}$ should be to the absolute Galois group $\Gamma_{\mathbb{Q}}$ what the identity component Hol^0 of the holonomy group Hol is to the monodromy group Hol/Hol^0 in classical differential geometry. As such our $\mathfrak{hol}_{\mathbb{Q}}$ could be viewed as an object of study in its own right, as fundamental, perhaps, as the absolute Galois group $\Gamma_{\mathbb{Q}}$ itself. A "unification" of $\mathfrak{hol}_{\mathbb{Q}}$ and $\Gamma_{\mathbb{Q}}$ may be expected in the same way in which Hol^0 and Hol/Hol^0 are "unified" by Hol. Such a unification of $\mathfrak{hol}_{\mathbb{Q}}$ with $\Gamma_{\mathbb{Q}}$ might involve interesting Galois representations in the same way in which the unification of $\Gamma_{\mathbb{Q}}$ with the geometric fundamental group via the arithmetic fundamental group in Grothendieck's theory yields interesting Galois representations [**77, 57**].

Organization of the book. The book starts with an Introduction in which we give an outline of our theory and we briefly compare our theory with other theories.

In Chapter 1 we briefly review the basic algebra and superalgebra concepts that we are going to use throughout the book. Most of this material is standard so the reader may choose to skip this chapter and consult it later as needed.

Chapter 2 will be devoted to revisiting classical differential geometry from an algebraic standpoint. We will be using the language of differential algebra [**91, 112**], i.e., the language of rings with derivations. We call the attention upon the fact that some of the classical differential geometric concepts will be presented in a somewhat non-conventional way in order to facilitate and motivate the transition to the arithmetic setting. This chapter plays no role in our book other than that of a motivational and referential framework; so, again, the reader may choose to skip this chapter and then go back to it as needed or just in order to compare the arithmetic theory with (the algebraic version of) the classical one.

Chapter 3 is where the exposition of our theory properly begins; here we present the basic notions of arithmetic differential geometry and, in particular, we introduce our arithmetic analogues of connection and curvature. The theory will be presented in the framework of arbitrary (group) schemes.

Chapter 4 specializes the theory in Chapter 3 to the case of the group scheme GL_n; here we prove, in particular, our main existence results for Ehresmann, Chern, Levi-Cività, and Lax connections respectively.

Chapters 5, 6, 7, 8 are devoted to the in-depth analysis of these connections; in particular we prove here the existence of the analytic continuation between primes necessary to define curvature for these connections and we give our vanishing/non-vanishing results for these curvatures. In Chapter 5, we also take first steps towards a corresponding (arithmetic differential) Galois theory.

The last Chapter 9 lists some of the natural problems, both technical and conceptual, that one faces in the further development of the theory.

Chapters 1, 2, 3, 4 should be read in a sequence (with Chapters 1 and 2 possibly skipped and consulted later as needed); Chapters 5, 6, 7, 8 depend on Chapters 1, 2, 3, 4 but are essentially independent of each other. Chapter 9 can be read right after the Introduction.

Cross references are organized in a series of sequences as follows. Sections are numbered in one sequence and will be referred to as "section $x.y$." Definitions, Theorems, Propositions, Lemmas, Remarks, and Examples are numbered in a separate sequence and are referred to as "Theorem $x.y$, Example $x.y$," etc. Finally equations are numbered in yet another separate sequence and are referred to simply as "$x.y$." For all three sequences x denotes the number of the chapter.

Readership and prerequisites. The present book addresses graduate students and researchers interested in algebra, number theory, differential geometry, and the analogies between these fields. The only prerequisites are some familiarity with commutative algebra (cf., e.g., the Atiyah-MacDonald book [**4**] or, for more specialized material, Matsumura's book [**103**]) and with foundational scheme theoretic algebraic geometry (e.g., the first two chapters of Hartshorne's book [**72**]). The text also contains a series of remarks that assume familiarity with basic concepts of classical differential geometry (as presented in [**89**], for instance); but these remarks are not essential for the understanding of the book and can actually be skipped.

Acknowledgments

Over the years the author has greatly profited from inspiring discussions and/or collaborations with a number of people including: M. Barrett, D. Bertrand, J. Borger, J.-B. Bost, C. P. Boyer, P. Cartier, P. J. Cassidy, P. Deligne, T. Dupuy, I. Fesenko, E. Hrushovski, Y. Ihara, M. Kim, E. R. Kolchin, S. Lang, Yu. I. Manin, H. Moscovici, A. Pillay, B. Poonen, F. Pop, E. Previato, D. C. Rössler, A. Saha, T. Scanlon, S. R. Simanca, J. Tate, D. Thakur, D. Ulmer, D. Vassilev, F. Voloch. Also, while writing this book, the author was partially supported by the Simons Foundation through grant 311773. Finally the author is grateful to the Max Planck Institut für Mathematik (Bonn), the Institut des Hautes Études Scientifiques (Bures sur Yvette), the Australian National University (Canberra), and the Institute of Mathematics of the Romanian Academy (Bucharest), for hospitality and support during the writing of various parts of this book.

Alexandru Buium

Introduction

By *classical differential geometry* we will understand, in this book, the theory of differentiable manifolds equipped with various "geometric structures" such as metrics, connections, curvature, etc. Our standard reference for this theory is [**89**]. However we shall only be interested in the algebraic aspects of this theory so we will allow ourselves below to reintroduce some of its concepts using the language of differential algebra [**91, 112**] (i.e., the language of rings equipped with derivations) and we will still refer to this "algebraized" version of differential geometry as *classical differential geometry*. On the other hand our purpose, in this book, is to develop an arithmetic analogue of classical differential geometry, which we will refer to as *arithmetic differential geometry*; in this latter theory derivation operators acting on functions will be replaced by Fermat quotient type operators acting on numbers, in the spirit of [**23, 35**]. In this Introduction we present an outline of our theory and we provide a quick comparison of our theory with other theories.

0.1. Outline of the theory

0.1.1. Classical versus arithmetic differentiation. In the classical theory of smooth (i.e., differentiable) manifolds one considers, for any m-dimensional smooth manifold M, the ring of smooth real valued functions $C^{\infty}(M,\mathbb{R})$. For the purposes of this Introduction it is enough to think of M as being the Euclidean space $M=\mathbb{R}^m$. In this book the arithmetic analogue of the manifold $\mathbb{R}^m$ will be the scheme *Spec* $\mathbb{Z}$. In order to stress our analogy between functions and numbers it is convenient to further fix a subring

$$A\subset C^{\infty}(\mathbb{R}^m,\mathbb{R}) \tag{0.1}$$

that is stable under partial differentiation and to frame the classical differential geometric definitions in terms of this ring. The arithmetic analogue of A in 0.1 will then be the ring of integers $\mathbb{Z}$ or, more generally, rings of fractions A of rings of integers in an abelian extension of $\mathbb{Q}$. (The theory can be easily extended to the case when $\mathbb{Q}$ is replaced by an arbitrary number field; we will not consider this more general situation in the book.) A prototypical example of such a ring is the ring

$$A=\mathbb{Z}[1/M,\zeta_N], \tag{0.2}$$

where M is an even integer, N is a positive integer, and ζ_N is a primitive N-th root of unity. Inverting M allows, as usual, to discard a set of "bad primes" (the divisors of M); adjoining ζ_N amounts to adjoining "new constants" to $\mathbb{Z}$.

Most of the times the ring A in 0.1 can be thought of as equal to $C^\infty(\mathbb{R}^m, \mathbb{R})$. But it will be sometimes useful to think of the ring A in 0.1 as consisting of analytic functions; indeed we will sometimes view analytic/algebraic functions as corresponding, in our theory, to global arithmetic objects in which case C^∞ objects will correspond, in our theory, to adelic objects. We will adopt these various viewpoints according to the various situations at hand.

The analogue of the set of coordinate functions

$$\mathcal{U} = \{\xi_1, \xi_2, ..., \xi_m\} \subset C^\infty(\mathbb{R}^m, \mathbb{R}) \tag{0.3}$$

on $\mathbb{R}^m$ will be a (possibly infinite) set of primes,

$$\mathcal{V} = \{p_1, p_2, p_3, ...\} \subset \mathbb{Z} \tag{0.4}$$

not dividing MN. We denote by $m = |\mathcal{V}| \leq \infty$ the cardinality of $\mathcal{V}$. (In case $\mathbb{Q}$ is replaced by an arbitrary number field the set $\mathcal{V}$ needs to be replaced by a set of finite places of that field.)

Next one considers the partial derivative operators on $A \subset C^\infty(\mathbb{R}^m, \mathbb{R})$,

$$\delta_i := \delta_i^A : A \to A, \quad \delta_i f := \frac{\partial f}{\partial \xi_i}, \quad i \in \{1, ..., m\}. \tag{0.5}$$

Following [**23**] we propose to take, as an analogue of the operators 0.5, the operators δ_p on $A = \mathbb{Z}[1/M, \zeta_N]$ defined by

$$\delta_p := \delta_p^A : A \to A, \quad \delta_p(a) = \frac{\phi_p(a) - a^p}{p}, \quad p \in \mathcal{V}, \tag{0.6}$$

where $\phi_p := \phi_p^A : A \to A$ is the unique ring automorphism sending ζ_N into ζ_N^p.

More generally recall that a *derivation* on a ring B is an additive map $B \to B$ that satisfies the Leibniz rule. This concept has, as an arithmetic analogue, the concept of *p-derivation* defined as follows; cf. [**23, 35, 83**].

DEFINITION 0.1. Assume B is a ring and assume, for simplicity, that p is a non-zero divisor in B; then a *p-derivation* on B is a set theoretic map

$$\delta_p := \delta_p^B : B \to B \tag{0.7}$$

with the property that the map

$$\phi_p := \phi_p^B : B \to B \tag{0.8}$$

defined by

$$\phi_p(b) := b^p + p\delta_p b \tag{0.9}$$

is a ring homomorphism.

We will always denote by ϕ_p the ring homomorphism 0.8 attached to a p-derivation δ_p as in 0.7 via the formula 0.9 and we shall refer to ϕ_p as the *Frobenius lift* attached to δ_p; note that ϕ_p induces the p-power Frobenius on B/pB.

0.1.2. Classical differential geometry revisited. Classical differential geometric concepts are often introduced by considering frame bundles $P \to M$ of rank n vector bundles over m-dimensional manifolds M. For our purposes we take here $M = \mathbb{R}^m$. Such a P is a principal homogeneous space for the group $GL_n(\mathbb{R})$; and if the vector bundle is trivial (which, for simplicity, we assume in what follows) then P is identified with $M \times GL_n(\mathbb{R})$. Note that the rank n of the vector bundle and the dimension m of M in this picture are unrelated.

We want to review the classical concept of connection in P; we shall do it in a somewhat non-standard way so that the transition to arithmetic becomes more transparent. Indeed consider an $n \times n$ matrix $x = (x_{ij})$ of indeterminates and consider the ring of polynomials in these n^2 indeterminates over some $A \subset C^\infty(\mathbb{R}^m, \mathbb{R})$ as in 0.1, with the determinant inverted,

$$B = A[x, \det(x)^{-1}]. \tag{0.10}$$

Note that B is naturally a subring of the ring $C^\infty(P, \mathbb{R})$. Denote by $\text{End}(B)$ the Lie ring of $\mathbb{Z}$-module endomorphisms of B. Also consider $G = GL_n = Spec\ B$ as a scheme over A.

DEFINITION 0.2. A *connection* on P (or on G, or on B) is a tuple $\delta = (\delta_i)$ of derivations

$$\delta_i := \delta_i^B : B \to B, \quad i \in \{1, ..., m\}, \tag{0.11}$$

extending the derivations 0.5. The *curvature* of δ is the matrix (φ_{ij}) of commutators $\varphi_{ij} \in \text{End}(B)$,

$$\varphi_{ij} := [\delta_i, \delta_j] : B \to B, \quad i, j \in \{1, ..., m\}. \tag{0.12}$$

The *holonomy ring* of δ is the $\mathbb{Z}$-span $\mathfrak{hol}$ in $\text{End}(B)$ of the commutators

$$[\delta_{i_1}, [\delta_{i_2}, [...., [\delta_{i_{s-1}}, \delta_{i_s}]...]]], \tag{0.13}$$

where $s \geq 2$; it is a Lie subring of $\text{End}(B)$.

In particular one can consider the *trivial connection* $\delta_0 = (\delta_{0i})$ defined by

$$\delta_{0i} x = 0. \tag{0.14}$$

Here $\delta_{0i} x = (\delta_{0i} x_{kl})$ is the matrix with entries $\delta_{0i} x_{kl}$; this, and similar, notation will be constantly used in the sequel. A connection is called *flat* if its curvature vanishes: $\varphi_{ij} = 0$ for all $i, j = 1, ..., m$. For instance δ_0 is flat. For a flat δ the holonomy ring $\mathfrak{hol}$ vanishes: $\mathfrak{hol} = 0$.

More generally let us define a *connection* on an arbitrary A-algebra B as a tuple of derivations 0.11 extending the derivations 0.5; one can then define the *curvature* by the same formula 0.12.

Note that, in classical differential geometry, a prominent role is played by connections on vector bundles; the framework of vector bundles is, however, too "linear" to have a useful arithmetic analogue: our arithmetic theory will be essentially a "non-linear" theory in which vector bundles (more generally modules) need to be replaced by principal bundles (more generally by rings).

There are various types of connections that we shall be interested in and for which we will seek arithmetic analogues; they will be referred to as:

(1) Ehresmann connections,
(2) Chern connections,
(3) Levi-Cività connections,
(4) Fedosov connections,
(5) Lax connections,
(6) Hamiltonian connections,
(7) Cartan connections,
(8) Riccati connections,
(9) Weierstrass connections,
(10) Painlevé connections.

In some cases our terminology above is not entirely standard. For instance what we call *Fedosov connection* is usually called *symplectic connection* [**96, 62, 66**]. We changed the name by which we refer to these connections in order to avoid confusion with the symplectic paradigm underlying the Hamiltonian case; our choice of name is based on the fact that symplectic manifolds equipped with a symplectic connection are usually called *Fedosov manifolds* [**66**]. Also there are a number of objects in classical differential geometry that are known under the name of *Cartan connection*; what we call *Cartan connection* is an object corresponding to what is usually called *Cartan distribution* [**1**], p. 133, and "lives" on jet bundles; our Cartan connection is unrelated, for instance, to the Maurer-Cartan connection. The Maurer-Cartan connection, on the other hand, will play a role in our book through the consideration of the *logarithmic derivative* map from an algebraic group to its Lie algebra. Finally what we call *connections* in cases (5), (6), (8), (9), (10) are classically referred to as *equations* or *flows*; e.g. what we call *Lax connection* is classically called *Lax equation*. Indeed connections classically refer to *linear* equations whereas the equations involved in (5)-(10) are *non-linear*. So what we call *connection* will not always refer to linear equations; this is in line with what will happen later in the arithmetic case where connections will be "intrinsically non-linear."

In what follows we discuss the connections (1) through (7) in some detail. For these connections we assume A as in 0.1; for the connections (1) through (5) we assume, in addition, that B is as in 0.10; for connections (6) and (7) B will be defined when we get to discuss these connections. Connections (8), (9), (10) will not be discussed in this Introduction but will appear in the body of the book; these are connections on curves (for (8), (9)) or surfaces (for (10)) appearing in relation to the classical theory of differential equations with no movable singularities. The Painlevé connections (10) fit into the Hamiltonian paradigm (6). Both (9) and (10) lead to elliptic curves.

DEFINITION 0.3. A connection (δ_i) on B in 0.10 is an *Ehresmann connection* if it satisfies one of the following two equivalent conditions:

(a) There exist $n \times n$ matrices A_i with coefficients in A such that

$$\delta_i x = A_i x \tag{0.15}$$

(b) The following diagrams are commutative:

$$\begin{array}{ccc} B & \stackrel{\mu}{\longrightarrow} & B \otimes_A B \\ \delta_i \downarrow & & \downarrow \delta_i \otimes 1 + 1 \otimes \delta_{0i} \\ B & \stackrel{\mu}{\longrightarrow} & B \otimes B \end{array} \tag{0.16}$$

Here μ is the comultiplication. Condition (a) can be referred to as saying that (δ_i) is *right linear*. Condition (b) can be referred to as saying that δ is *right invariant*. (As we shall see the arithmetic analogues of conditions (a) and (b) will cease to be equivalent.) If (δ_i) is an Ehresmann connection the curvature satisfies $\varphi_{ij}(x) = F_{ij}x$ where F_{ij} is the matrix given by the classical formula

$$F_{ij} := \delta_i A_j - \delta_j A_i - [A_i, A_j]. \tag{0.17}$$

Also there is a Galois theory attached to flat Ehresmann connections, the Picard-Vessiot theory. Indeed, for a flat Ehresmann connection $\delta = (\delta_i)$ consider the *logarithmic derivative map*, $l\delta : GL_n(A) \to \mathfrak{gl}_n(A)^m$, with coordinates

$$l\delta_i(u) = \delta_i u \cdot u^{-1}, \tag{0.18}$$

where $\mathfrak{gl}_n$ is the Lie algebra of GL_n. The fibers of the map $l\delta$ are solution sets of systems of linear equations

$$\delta_i u = A_i \cdot u. \tag{0.19}$$

And if one replaces A by a ring of complex analytic functions then Galois groups can be classically attached to such systems; these groups are algebraic subgroups of $GL_n(\mathbb{C})$ measuring the algebraic relations among the solutions to the corresponding systems.

We next discuss Chern connections. We will introduce two types of Chern connections: *real Chern connections* and *complex Chern connections*. Complex Chern connections will be modeled on the classical hermitian connections on hermitian vector bundles [**69, 88**] which play a central role in complex geometry; real Chern connections are a real analogue of the complex ones and do not seem to play an important role in differential geometry (although see Duistermaat's paper [**61**] for a special case of these connections). Both real and complex Chern connections will have arithmetic analogues; the arithmetic analogue of real Chern connections will be more subtle (and hence more interesting) than the arithmetic analogue of the complex Chern connection so we will mostly concentrate on the former than on the latter.

To introduce these concepts let us consider again the rings A as in 0.1 and B as in 0.10, and let $q \in GL_n(A)$ be an $n \times n$ invertible matrix with coefficients in A which is either symmetric or anti-symmetric,

$$q^t = \pm q; \tag{0.20}$$

so the t superscript means here *transposition*. Of course, a symmetric q as above is viewed as defining a (semi-Riemannian) metric on the trivial vector bundle

$$M \times \mathbb{R}^n \to M, \tag{0.21}$$

on $M = \mathbb{R}^m$ while an anti-symmetric q is viewed as defining a 2-form on 0.21; here metrics/2-forms on vector bundles mean smoothly varying non-degenerate symmetric/anti-symmetric bilinear forms on the fibers of the bundle. With B be as in 0.10 let $G = GL_n = Spec\ B$ be viewed as a scheme over A and consider the maps of schemes over A,

$$\mathcal{H}_q : G \to G, \quad \mathcal{B}_q : G \times G \to G, \tag{0.22}$$

defined on points by

$$\mathcal{H}_q(x) = x^t q x, \quad \mathcal{B}_q(x, y) = x^t q y. \tag{0.23}$$

We continue to denote by the same letters the corresponding maps of rings $B \to B$ and $B \to B \otimes_A B$. Consider the *trivial* connection $\delta_0 = (\delta_{0i})$ on G defined by $\delta_{0i} x = 0$. Then one has the following easy results:

THEOREM 0.4. *There is a unique connection $\delta = (\delta_i)$ on G such that the following diagrams are commutative:*

$$\begin{array}{ccc} B & \stackrel{\delta_i}{\longleftarrow} & B \\ \mathcal{H}_q \uparrow & & \uparrow \mathcal{H}_q \\ B & \stackrel{\delta_{0i}}{\longleftarrow} & B \end{array} \qquad \begin{array}{ccc} B & \stackrel{\delta_{0i}\otimes 1 + 1 \otimes \delta_i}{\longleftarrow} & B \otimes_A B \\ \delta_i \otimes 1 + 1 \otimes \delta_{0i} \uparrow & & \uparrow \mathcal{B}_q \\ B \otimes_A B & \stackrel{\mathcal{B}_q}{\longleftarrow} & B \end{array} \tag{0.24}$$

THEOREM 0.5. *There is a unique connection $\delta = (\delta_i)$ on G such that the following diagrams are commutative:*

$$
\begin{array}{ccc}
B & \stackrel{\delta_{0i}\otimes 1 + 1\otimes\delta_i}{\longleftarrow} & B\otimes_A B \\
\mathcal{H}_q \uparrow & & \uparrow \mathcal{B}_q \\
B & \stackrel{\delta_{0i}}{\longleftarrow} & B.
\end{array}
\tag{0.25}
$$

DEFINITION 0.6. A connection δ is $\mathcal{H}_q$*-horizontal* (respectively $\mathcal{B}_q$*-symmetric*) with respect to δ_0 if the left (respectively right) diagrams in 0.24 are commutative for all i. The unique connection that is $\mathcal{H}_q$*-horizontal* and $\mathcal{B}_q$*-symmetric* with respect to δ_0 (cf. Theorem 0.4) is called the *real Chern connection* attached to q.

DEFINITION 0.7. A connection δ is $(\mathcal{H}_q, \mathcal{B}_q)$*-horizontal* with respect to δ_0 if the diagrams 0.25 are commutative for all i. The unique connection that is $(\mathcal{H}_q, \mathcal{B}_q)$-*horizontal* with respect to δ_0 (cf. Theorem 0.5) is called the *complex Chern connection* attached to q.

Real and complex Chern connections turn out to be automatically Ehresmann connections. The arithmetic analogues of the real and complex Chern connections will not be "Ehresmann" in general.

The definitions just given may look non-standard. To see the analogy between our (real and complex) Chern connections and the classical hermitian Chern connections in [**69, 88**] we need to look at the explicit formulas for these connections; so we need to introduce more notation as follows.

For any connection (δ_i) on B in 0.10 set

$$
\begin{array}{ccccc}
A_i & = & A_i(x) & := & \delta_i x \cdot x^{-1} \in \mathfrak{gl}_n(B), \\
\Gamma^k_{ij} & = & -A_{ikj} & = & (k,j) \text{ entry of } (-A_i).
\end{array}
\tag{0.26}
$$

The quantities Γ^k_{ij} will be referred to as *Christoffel symbols of the 2nd kind* of the connection. Also, for q as in 0.20, set

$$\Gamma_{ijk} := \Gamma^l_{ij} q_{lk}, \tag{0.27}$$

which we refer to as the *Christoffel symbols of the 1st kind.* Here we use Einstein summation.

It is trivial to check that δ is $\mathcal{H}_q$-horizontal if and only if the followng condition is satisfied:

$$\delta_i q_{jk} = \Gamma_{ijk} \pm \Gamma_{ikj}. \tag{0.28}$$

Classically if 0.28 holds one says that q is *parallel* with respect to δ.

Similarly it is trivial to check that δ is $\mathcal{B}_q$-symmetric if and only if the following condition is satisfied:

$$\Gamma_{ijk} = \pm \Gamma_{ikj}. \tag{0.29}$$

This is, as we shall see, a condition different from the classical condition of *symmetry* for a connection on the tangent bundle.

In 0.28 and 0.29 the upper (respectively lower) sign correspond to the upper (respectively lower) sign in 0.20.

Using 0.28 and 0.29 it is trivial to see that the real Chern connection attached to q exists and is unique, being given by

$$\Gamma_{ijk} = \frac{1}{2}\delta_i q_{jk}. \tag{0.30}$$

This proves Theorem 0.4. Note that, as promised earlier, the Chern connection is an Ehresmann connection, i.e. A_i belong to $\mathfrak{gl}_n(A)$ rather than $\mathfrak{gl}_n(B)$.

By the way the real Chern connection has the following compatibility with the special linear group SL_n: if $\delta_i(\det(q)) = 0$ then the real Chern connection $\delta_i : B \to B$ attached to q sends the ideal of SL_n into itself and hence induces a "connection on SL_n." (This compatibility with SL_n will fail to hold in the arithmetic case.)

Similarly it is easy to see that δ is $(\mathcal{H}_q, \mathcal{B}_q)$-horizontal with respect to δ_0 if and only if

$$\Gamma_{ijk} = \delta_i q_{jk}. \tag{0.31}$$

Hence the complex Chern connection exists and is unique, being given by 0.31, which proves Theorem 0.5. Comparing 0.30 and 0.31 we see that the Christoffel symbols of the real and complex Chern connections coincide up to a $\frac{1}{2}$ factor and in this sense they are analogous to each other; on the other hand formula 0.31 is analogous to the formula for the classical hermitian Chern connection in [**69, 88**].

We next discuss Levi-Cività connections. Assume A as in 0.1, B as in 0.10, and assume that $n = m$. (Note we also implicitly assume here that a bijection is given between the set indexing the derivations δ_i and the set indexing the rows and columns of the matrix $x = (x_{ij})$; such a bijection plays the role of what is classically called a *soldering*.)

DEFINITION 0.8. The connection (δ_i) is *symmetric* or *torsion free* if

$$\Gamma_{ijk} = \Gamma_{jik}. \tag{0.32}$$

Note the difference between the condition 0.32 defining *symmetry* and the condition 0.29 defining $\mathcal{B}_q$*-symmetry*: the two types of symmetry involve different pairs of indices. To avoid any confusion we will use the term *torsion free* rather than *symmetric* in what follows. The fundamental theorem of Riemannian geometry is, in this setting, the following (completely elementary) statement:

THEOREM 0.9. *Let $q \in GL_n(A)$, $q^t = q$. Then there is a unique connection δ that is $\mathcal{H}_q$-horizontal (i.e., satisfies equation* 0.28 *with the $+$ sign) and is torsion free (i.e., it satisfies equation* 0.32*); it is given by*

$$\Gamma_{kij} = \frac{1}{2}\left(\delta_k q_{ij} + \delta_i q_{jk} - \delta_j q_{ki}\right) \tag{0.33}$$

DEFINITION 0.10. The connection in Theorem 0.9 is called the *Levi-Cività* connection attached to q.

Note that, in particular, the Levi-Cività connection is an Ehresmann connection, i.e. A_i belong to $\mathfrak{gl}_n(A)$ rather than $\mathfrak{gl}_n(B)$. (The latter will fail to hold in the arithmetic case.) The Levi-Cività connection is, in a precise sense to be discussed later, "dual" to the real Chern connection and is generally different from the real Chern connection; it coincides with the real Chern connection if and only if

$$\delta_i q_{jk} = \delta_j q_{ik} \tag{0.34}$$

in which case q is called *Hessian* (the "real" analogue of *Kähler*).

We next discuss Fedosov connections partially following [**96, 62, 66, 119**] where the term *symplectic* rather than *Fedosov* is being used. To explain these we start, again, with A as in 0.1 and B as in 0.10.

DEFINITION 0.11. Let $q \in GL_n(A)$, $q^t = -q$, so n is even. A *Fedosov connection* relative to q is a connection δ that is $\mathcal{H}_q$-horizontal (i.e., satisfies equation 0.28 with the $-$ sign) and is torsion free (i.e. it satisfies 0.32).

DEFINITION 0.12. $q \in GL_n(A)$, $q^t = -q$, is *symplectic* if it satisfies

$$\delta_i q_{jk} + \delta_j q_{ki} + \delta_k q_{ij} = 0. \tag{0.35}$$

One trivially checks that a Fedosov connection relative to q exists if and only if q is symplectic.

Unlike in [**62, 66, 119**] our Fedosov connections here are not necessarily Ehresmann. And, for a given symplectic matrix q, Fedosov connections that are Ehresmann exist but are not unique; one Fedosov connection which is an Ehresmann connection is given by

$$\Gamma_{ijk} = \frac{1}{3}\left(\delta_i q_{jk} + \delta_j q_{ik}\right). \tag{0.36}$$

We next discuss Lax connections. Let A be as in 0.1 and B as in 0.10.

DEFINITION 0.13. A connection (δ_i) on B is called a *Lax connection* if it satisfies

$$\delta_i x = [A_i(x), x] := A_i(x)x - xA_i(x) \tag{0.37}$$

for some $n \times n$ matrix $A_i(x)$ with coefficients in B.

Note that, unlike Chern and Levi-Cività connections, Lax connections are *not* a subclass of the Ehresmann connections. For a Lax connection, the following diagrams are commutative:

$$\begin{array}{ccc} B & \xleftarrow{\delta_i} & B \\ \mathcal{P}\uparrow & & \uparrow\mathcal{P} \\ A[z] & \xleftarrow{\delta_{0i}} & A[z] \end{array} \tag{0.38}$$

where $A[z] = A[z_1, ..., z_n]$ is a ring of polynomials in the variables z_j, δ_{0i} are the unique derivations extending the corresponding derivations on A with $\delta_{0i}z_j = 0$, and $\mathcal{P}$ is the A-algebra homomorphism with $\mathcal{P}(z_j) = \mathcal{P}_j(x)$,

$$\det(s \cdot 1 - x) = \sum_{j=0}^{n} (-1)^j \mathcal{P}_j(x) s^{n-j}. \tag{0.39}$$

The commutativity of 0.38 expresses the fact that the Lax connections describe "isospectral flows" on GL_n.

By the way the "real" theory summarized above has a "complex" analogue (and hence a "(1, 1)"-analogue) for which we refer to the body of the book. Suffices to say here that, for the complex theory, one may start with $M = \mathbb{C}^m$ and with a subring

$$A \subset C^\infty(M, \mathbb{C}) \tag{0.40}$$

of the ring of smooth complex valued functions on M which is stable under the derivations

$$\delta_i := \frac{\partial}{\partial z_i}, \quad \delta_{\overline{i}} := \frac{\partial}{\partial \overline{z}_i}, \quad i = 1, ..., m, \tag{0.41}$$

where $z_1, ..., z_m$ are the complex coordinates on M. A connection on $G = Spec\ B$, $B = A[x, \det(x)^{-1}]$, is then an m-tuple of derivations $\delta_i : B \to B$ extending the derivations $\delta_i : A \to A$. Consider the unique derivations $\delta_{\overline{i}} : B \to B$, extending the derivations $\delta_{\bar{i}} : A \to A$, such that $\delta_{\bar{i}} x = 0$. Then one defines the $(1,1)$-*curvature* of $\delta = (\delta_i)$ as the matrix $(\varphi_{i\bar{j}})$ with entries the A-derivations

$$\varphi_{i\bar{j}} := [\delta_i, \delta_{\bar{j}}] : B \to B. \tag{0.42}$$

The theory proceeds from here.

So far we discussed connections on the ring B in 0.10. In what follows we informally discuss Hamiltonian and Cartan connections; for a precise discussion we refer to the main text. These are connections on rings B other than 0.10.

To discuss Hamiltonian connections consider the ring

$$B = A[x], \tag{0.43}$$

with A as in 0.1 and $x = \{x_1, ..., x_d\}$ a d-tuple of variables. Elements of the exterior products of the module of Kähler differentials $\Omega_{B/A}$ will be referred to as (vertical) *forms*. Also consider the derivations $\partial^j = \frac{\partial}{\partial x_j}$ on B.

DEFINITION 0.14. A 2-*form*

$$\omega := \omega^{ij} \cdot dx_i \wedge dx_j, \tag{0.44}$$

$\omega^{ij} \in B$, $\omega_{ij} + \omega_{ji} = 0$, is *symplectic* if $\det(\omega_{ij}) \in B^\times$ and

$$\partial^k \omega^{ij} + \partial^j \omega^{ki} + \partial^i \omega^{jk} = 0. \tag{0.45}$$

Note the contrast with the concept of *symplectic matrices* q: q are symplectic "in the horizontal directions"

$$\delta_j = \frac{\partial}{\partial \xi_j},$$

cf. 0.35, while 2-forms ω in 0.44 are symplectic in the "vertical directions"

$$\partial^j = \frac{\partial}{\partial x_j}.$$

For any 2-form 0.44 one defines the *Lie derivative* with respect to ξ_k as

$$\delta_k \omega := (\delta_k \omega^{ij}) \cdot dx_i \wedge dx_j + \omega^{ij} \cdot d(\delta_k x_i) \wedge dx_j + \omega^{ij} \cdot dx_i \wedge d(\delta_k x_j). \tag{0.46}$$

DEFINITION 0.15. A *Hamiltonian* connection with respect to a symplectic 2-form ω is a connection on a B as in 0.43 whose Lie derivatives vanish:

$$\delta_k \omega = 0.$$

Hamiltonian connections with respect to symplectic forms naturally appear, by the way, in the background of some of the basic differential equations of mathematical physics, in particular in the background of the Painlevé VI equations. This kind of Hamiltonian connections turn out to have an arithmetic analogue [**42**] which will be discussed in the body of our book but not in this Introduction.

On the other hand one can consider *Hamiltonian connections* on a B as in 0.43 with respect to a *Poisson structure*. The Poisson story will be reviewed in the main body of the book. The two concepts of Hamiltonian connections (relative to symplectic forms and relative to Poisson structures) are related in case

$$\delta_k \omega^{ij} = 0.$$

An example of Poisson structure is provided by *Lie-Poisson structures* on Lie algebras in which case the corresponding Hamiltonian connections typically lead to Lax connections. A classical example of equations arising from a Lie-Poisson structure are the *Euler equations* for the rigid body which will be reviewed in the body of the book; the Euler equations are related to Lax equations in at least two different ways (via 3×3 and 2×2 matrices respectively). Euler equations will have an arithmetic analogue, although this arithmetic analogue will not be a priori related to our arithmetic analogues of Lax equations.

Finally we mention *Cartan connections.* In the same way we considered principal bundles and attached to them algebras B as in 0.10 one can consider "infinite jet bundles" (which we do not define here) and attach to them polynomial algebras,

$$B = A[x_j^{(\alpha)};\ \alpha \in \mathbb{Z}^m_{\geq 0}, j = 1, ..., d], \tag{0.47}$$

where A is as in 0.1 and $x_j^{(\alpha)}$ are indeterminates; these algebras come equipped with a natural flat connection

$$\delta_i := \frac{\partial}{\partial \xi_i} + \sum_j \sum_\alpha x_j^{(\alpha+e_i)} \frac{\partial}{\partial x_j^{(\alpha)}}, \tag{0.48}$$

where e_i is the canonical basis of $\mathbb{Z}^m$.

DEFINITION 0.16. The flat connection (δ_i) on the ring B in 0.47 defined by 0.48 is called the *Cartan connection.*

Other names for (δ_i) above are: the *total derivative* or the *Cartan distribution.* The Cartan connections have an arithmetic analogue which was thoroughly studied in [**35, 34**], will be reviewed in the main text of the present book in case $m = 1$, and plays a central role in the whole theory.

0.1.3. Arithmetic differential geometry. Roughly speaking arithmetic differential geometry is obtained from classical differential geometry by replacing classical differentiation (derivations) with arithmetic differentiation (p-derivations). This is often a convoluted process that we now explain.

We start with the analogue of connections in the real case. The first step is to consider the ring

$$B = A[x, \det(x)^{-1}]$$

defined as in 0.10, with $x = (x_{ij})$ an $n \times n$ matrix of indeterminates, but where A is given now by

$$A = \mathbb{Z}[1/M, \zeta_N]$$

as in 0.2. We again consider the group scheme over A,

$$G = GL_n := Spec\ B.$$

A first attempt to define arithmetic analogues of connections would be to consider families of p-derivations

$$\delta_p : B \to B, \quad p \in \mathcal{V},$$

extending the p-derivations 0.6; one would then proceed by considering their commutators on B (or, if necessary, expressions derived from these commutators). But the point is that the examples of "arithmetic analogues of connections" we will encounter in practice will almost never lead to p-derivations $B \to B$! What we shall be led to is, rather, an adelic concept we next introduce. (Our guiding "principle" here is that, as mentioned before, C^∞ geometric objects should correspond to adelic objects in arithmetic while analytic/algebraic geometric objects correspond to global objects in arithmetic.)

To introduce our adelic concept let us consider, for each $p \in \mathcal{V}$, the p-adic completion of B:

$$B^{\widehat{p}} := \lim_{\leftarrow} B/p^n B. \tag{0.49}$$

Then we make the following:

DEFINITION 0.17. An *adelic connection* on $G = GL_n$ is a family $\delta = (\delta_p)$ of p-derivations

$$\delta_p := \delta_p^B : B^{\widehat{p}} \to B^{\widehat{p}}, \quad p \in \mathcal{V}, \tag{0.50}$$

extending the p-derivations in 0.6.

If $\phi_p : B^{\widehat{p}} \to B^{\widehat{p}}$ are the Frobenius lifts attached to δ_p and $G^{\widehat{p}} = Spf\ B^{\widehat{p}}$ is the p-adic completion of $G = GL_n = Spec\ B$ then we still denote by $\phi_p : G^{\widehat{p}} \to G^{\widehat{p}}$ the induced morphisms of p-adic formal schemes.

Next we will explore analogues of the various types of connections encountered in classical differential geometry: Ehresmann, Chern, Levi-Cività, Fedosov, and Lax. (The stories of Hamiltonian connections and Cartan connections will be discussed in the main body of the book and will be skipped here.)

In what follows we need an analogue of the trivial connection in 0.14. It will be given by the adelic connection (δ_{0p}) defined by

$$\delta_{0p} x = 0. \tag{0.51}$$

The attached Frobenius lifts will be denoted by (ϕ_{0p}); they satisfy

$$\phi_{0p}(x) = x^{(p)} \tag{0.52}$$

where $x^{(p)}$ is the matrix (x_{ij}^p). We call $\delta_0 = (\delta_{0p})$ the *trivial* adelic connection.

To introduce arithmetic analogues of Ehresmann connections one starts by noting that, for $n \geq 2$, there are *no* adelic connections $\delta = (\delta_p)$ whose attached Frobenius lifts (ϕ_p) make the following diagrams commute:

$$\begin{array}{ccc} G^{\widehat{p}} \times G^{\widehat{p}} & \stackrel{\mu}{\longrightarrow} & G^{\widehat{p}} \\ \phi_p \times \phi_{0p} \downarrow & & \downarrow \phi_p \\ G^{\widehat{p}} \times G^{\widehat{p}} & \stackrel{\mu}{\longrightarrow} & G^{\widehat{p}} \end{array} \tag{0.53}$$

Since 0.53 is an analogue of 0.16 one can view this as saying that there are no adelic connections that are analogues of right invariant connections. This is an elementary observation; we will in fact prove a less elementary result:

THEOREM 0.18. *For $n \geq 2$ and $p \nmid n$ there are no adelic connections (δ_p) and (δ_{1p}) whose attached Frobenius lifts (ϕ_p) and (ϕ_{1p}) make the following diagrams commute:*

$$\begin{array}{ccc} G^{\widehat{p}} \times G^{\widehat{p}} & \stackrel{\mu}{\longrightarrow} & G^{\widehat{p}} \\ \phi_p \times \phi_{1p} \downarrow & & \downarrow \phi_p \\ G^{\widehat{p}} \times G^{\widehat{p}} & \stackrel{\mu}{\longrightarrow} & G^{\widehat{p}} \end{array} \tag{0.54}$$

There is a useful property, weaker than the commutativity of 0.53, namely an invariance property with respect to the action on G by right translation of the group

$$N := (\text{normalizer of the diagonal maximal torus } T \text{ of } G). \tag{0.55}$$

Indeed we will say that an adelic connection (δ_p) with associated Frobenius lifts (ϕ_p) is *right invariant with respect to* N if the following diagrams are commutative:

$$\begin{array}{ccc} G^{\widehat{p}} \times N^{\widehat{p}} & \stackrel{\mu}{\longrightarrow} & G^{\widehat{p}} \\ \phi_p \times \phi_{0p} \downarrow & & \downarrow \phi_p \\ G^{\widehat{p}} \times N^{\widehat{p}} & \stackrel{\mu}{\longrightarrow} & G^{\widehat{p}} \end{array} \tag{0.56}$$

This latter property has its own merits but is too weak to function appropriately as a defining property of Ehresmann connections in arithmetic. Instead, we will consider an appropriate analogue of "linearity," 0.15. What we will do will be to replace the Lie algebra $\mathfrak{gl}_n$ by an arithmetic analogue of it, $\mathfrak{gl}_{n,\delta_p}$, and then we will introduce an arithmetic analogue of the logarithmic derivative. This new framework will naturally lead us to the following:

DEFINITION 0.19. An adelic connection (δ_p) is an *Ehresmann connection* if

$$\delta_p x = \alpha_p \cdot x^{(p)}, \tag{0.57}$$

where α_p are matrices with coefficients in A.

By the way, clearly, Ehresmann connections are right invariant with respect to N. We will attach Galois groups to such Ehresmann connections and develop the basics of their theory. A natural expectation is that these Galois groups belong to the group $N(A)^{\delta}$ of all matrices in $N(A)$ whose entries are roots of unity or 0. This expectation is not always realized but we will prove that something close to it is realized for (α_p) "sufficiently general." The above expectation is justified by the fact that, according to the general philosophy of the field with one element $\mathbb{F}_1$, the union of the $N(A)^{\delta}$'s, as A varies, plays the role of "$GL_n(\mathbb{F}_1^a)$," where $\mathbb{F}_1^a$ is the "algebraic closure of $\mathbb{F}_1$."

Next we explain our arithmetic analogue of Chern connections.

Let $q \in GL_n(A)$ with $q^t = \pm q$. Attached to q we have, again, maps

$$\mathcal{H}_q : G \to G, \quad \mathcal{B}_q : G \times G \to G; \quad \mathcal{H}_q(x) = x^t q x, \quad \mathcal{B}_q(x,y) = x^t q y.$$

We continue to denote by $\mathcal{H}_q, \mathcal{B}_q$ the maps induced on the p-adic completions $G^{\widehat{p}}$ and $G^{\widehat{p}} \times G^{\widehat{p}}$. Consider again the trivial adelic connection $\delta_0 = (\delta_{0p})$ on G (so $\delta_{0p} x = 0$) and denote by (ϕ_{0p}) the attached Frobenius lifts (so $\phi_{0p}(x) = x^{(p)}$). Then we will prove the following:

THEOREM 0.20. *For any $q \in GL_n(A)$ with $q^t = \pm q$ there exists a unique adelic connection $\delta = (\delta_p)$ whose attached Frobenius lifts (ϕ_p) make the following diagrams commute:*

$$\begin{array}{ccc} G^{\widehat{p}} & \stackrel{\phi_p}{\longrightarrow} & G^{\widehat{p}} \\ \mathcal{H}_q \downarrow & & \downarrow \mathcal{H}_q \\ G^{\widehat{p}} & \stackrel{\phi_{0p}}{\longrightarrow} & G^{\widehat{p}} \end{array} \qquad \begin{array}{ccc} G^{\widehat{p}} & \stackrel{\phi_{0p} \times \phi_p}{\longrightarrow} & G^{\widehat{p}} \times G^{\widehat{p}} \\ \phi_p \times \phi_{0p} \downarrow & & \downarrow \mathcal{B}_q \\ G^{\widehat{p}} \times G^{\widehat{p}} & \stackrel{\mathcal{B}_q}{\longrightarrow} & G^{\widehat{p}} \end{array} \tag{0.58}$$

DEFINITION 0.21. An adelic connection $\delta = (\delta_p)$ is *$\mathcal{H}_q$-horizontal* (respectively *$\mathcal{B}_q$-symmetric*) with respect to $\delta_0 = (\delta_{0p})$ if the left (respectively right) diagrams in 0.58 are commutative. The unique connection which is *$\mathcal{H}_q$-horizontal* and *$\mathcal{B}_q$-symmetric* with respect to δ_0 (cf. Theorem 0.20) is called the *real Chern connection* attached to q.

The real Chern connection introduced above is an arithmetic analogue of the real Chern connection in classical differential geometry. Unlike in the case of classical differential geometry our adelic Chern connections will *not* be special cases of Ehresmann connections (although they *will* be right invariant with respect to N).

On the other hand we will have the following easy:

THEOREM 0.22. *For any $q \in GL_n(A)$ with $q^t = \pm q$ there is a unique adelic connection (δ_p) on GL_n whose attached Frobenius lifts (ϕ_p) make the following diagrams commute:*

$$\begin{array}{ccc} G^{\widehat{p}} & \stackrel{\phi_{0p}\times\phi_p}{\longrightarrow} & G^{\widehat{p}}\times G^{\widehat{p}} \\ \mathcal{H}_q\downarrow & & \downarrow\mathcal{B}_q \\ G^{\widehat{p}} & \stackrel{\phi_{0p}}{\longrightarrow} & G^{\widehat{p}} \end{array} \tag{0.59}$$

DEFINITION 0.23. An adelic connection $\delta = (\delta_p)$ is $(\mathcal{H}_q, \mathcal{B}_q)$*-horizontal* with respect to $\delta_0 = (\delta_{0p})$ if the diagrams 0.59 are commutative. The unique connection which is $(\mathcal{H}_q, \mathcal{B}_q)$*-horizontal* with respect to δ_0 (cf. Theorem 0.22) is called the *complex Chern connection* attached to q.

The concept of real Chern connection introduced above is, from an arithmetic point of view, more subtle than the concept of complex Chern connection. This can be seen already in the case of GL_1. Indeed let

$$q \in GL_1(A) = A^\times, \quad A = \mathbb{Z}[1/M].$$

In this case it turns out that the "Christoffel symbols" defining the real Chern connection are related to the Legendre symbol; explicitly the Frobenius lift $\phi_p : G^{\widehat{p}} \to G^{\widehat{p}}$ corresponding to the real Chern connection attached to q is induced by the homomorphism

$$\phi_p : \mathbb{Z}_p[x, x^{-1}]^{\widehat{p}} \to \mathbb{Z}_p[x, x^{-1}]^{\widehat{p}} \tag{0.60}$$

satisfying

$$\phi_p(x) = q^{(p-1)/2}\left(\frac{q}{p}\right)x^p, \tag{0.61}$$

where $\left(\frac{q}{p}\right)$ is the Legendre symbol of $q \in A^\times \subset \mathbb{Z}_{(p)}$. On the other hand the Frobenius lift $\phi_p : G^{\widehat{p}} \to G^{\widehat{p}}$ corresponding to the complex Chern connection attached to q is simply defined by a homomorphism 0.60 satisfying

$$\phi_p(x) = q^{p-1}x^p. \tag{0.62}$$

In view of the above we will mainly concentrate, in this book, on the study of the real Chern connection while leaving the simpler case of the complex Chern connection to the reader. Note by the way that the quantity $\phi(x)/x^p$ in 0.62 is the square of the quantity $\phi(x)/x^p$ in 0.61; this has an analogue for GL_n with n arbitrary and is analogous to the fact that the Christoffel symbols of the classical complex Chern connection are twice the Christoffel symbols of the classical real Chern connection; cf. 0.30 and 0.31.

We shall also introduce, in our book, adelic connections that are analogues of Levi-Cività connections. They are already relevant in case $\mathcal{V}$ consists of one prime only. So assume, in the theorem below, that $\mathcal{V} = \{p\}$ consists of one prime p.

THEOREM 0.24. *For any $q \in GL_n(A)$ with $q^t = q$ there is a unique n-tuple $(\delta_{1p}, ..., \delta_{np})$ of adelic connections on $G = GL_n$ with attached Frobenius lifts*

$$(\phi_{1p}, ..., \phi_{np}),$$

such that for all $i = 1, ..., n$ the following diagrams are commutative

$$\begin{array}{ccc} G^{\widehat{p}} & \xrightarrow{\phi_{ip}} & G^{\widehat{p}} \\ \mathcal{H}_q \downarrow & & \downarrow \mathcal{H}_q \\ G^{\widehat{p}} & \xrightarrow{\phi_{0p}} & G^{\widehat{p}} \end{array} \tag{0.63}$$

and such that, for all $i, j = 1, ..., n$, we have

$$\delta_{ip} x_{kj} = \delta_{jp} x_{ki}. \tag{0.64}$$

DEFINITION 0.25. The tuple $(\delta_{1p}, ..., \delta_{np})$ in Theorem 0.24 is called the *Levi-Cività connection* attached to q,

The condition 0.63 says of course that for each i, (δ_{ip}) is $\mathcal{H}_q$-horizontal with respect to (δ_{0p}) and hence is analogous to the condition of parallelism 0.28 (with the $+$ sign). The condition 0.64 is analogous to the condition of torsion freeness 0.32. This justifies our Definition 0.25 of the Levi-Cività connection. But note that, unlike in the case of classical differential geometry, our adelic Levi-Cività connections are *not* special cases of Ehresmann connections.

The adelic Levi-Cività connection and the adelic real Chern connection attached to q will be related by certain congruences mod p that are reminiscent of the relation between the two connections in classical differential geometry.

The one prime paradigm of the Levi-Cività connection above can be viewed as corresponding to the classical Levi-Cività connections attached to *(translational) cohomogeneity one* metrics, by which we understand here metrics that, in appropriate coordinates, are invariant under a codimension one group of translations; equivalently metrics which in appropriate coordinates have the form $g = \sum g_{ij} d\xi_i d\xi_j$ with

$$\delta_k g_{ij} = \delta_l g_{ij} \tag{0.65}$$

for all i, j, k, l. As a rule we will omit, for simplicity, the adjective *translational* from now on.

Continuing to assume $\mathcal{V} = \{p\}$ one can also attempt to develop an arithmetic analogue of Fedosov connections as follows. Consider an anti-symmetric $q \in GL_n(A)$, $q^t = -q$. Let us say that an n-tuple of $(\delta_{1p}, ..., \delta_{np})$ of adelic connections on $G = GL_n$ is a *Fedosov connection* relative to q if the attached Frobenius lifts $(\phi_{1p}, ..., \phi_{np})$ make the diagrams 0.63 commutative and, in addition, the equalities 0.64 hold. We will prove that for $n = 2$ and any anti-symmetric q Fedosov connections relative to q exist. However, in contrast with the Levi-Cività story, we will prove that for $n \geq 4$ there is no Fedosov connection relative to the split q, for instance.

Finally there are adelic connections that are analogous to Lax connections. In fact there are two such analogues which we call *isospectral* and *isocharacteristic* Lax connections. They offer two rather different arithmetic analogues of isospectral flows in the space of matrices. Indeed the isocharacteristic property essentially says that a certain characteristic polynomial has "δ-constant" coefficients whereas the isospectrality property essentially says that the characteristic polynomial has "δ-constant" roots. (Here *δ-constant* means "killed by all δ_p" which amounts to "being a root of unity or 0".) In usual calculus the two properties are equivalent but in our arithmetic calculus these two properties are quite different. By the way isospectral and isocharacteristic Lax connections will not be defined on the whole of G but rather on certain Zariski open sets G^* and G^{**} of G respectively. Let us give some details of this in what follows. Let $G^* \subset G$ be the open set of *regular* matrices, consider the diagonal maximal torus T in G, and let $T^* = G^* \cap T$; *regular* means here "with distinct diagonal entries." We will prove:

THEOREM 0.26. *There exists a unique adelic connection $\delta = (\delta_p)$ on G^* with attached Frobenius lifts (ϕ_p) such that the following diagrams commute:*

$$\begin{array}{ccc} (T^*)^{\widehat{p}} \times G^{\widehat{p}} & \overset{\phi_{0p} \times \phi_{0p}}{\longrightarrow} & (T^*)^{\widehat{p}} \times G^{\widehat{p}} \\ \mathcal{C} \downarrow & & \downarrow \mathcal{C} \\ (G^*)^{\widehat{p}} & \overset{\phi_p}{\longrightarrow} & (G^*)^{\widehat{p}}, \end{array} \tag{0.66}$$

where $\mathcal{C}(t, x) := x^{-1}tx$.

DEFINITION 0.27. The adelic connection δ in Theorem 0.26 is called the *canonical isospectral Lax connection.*

More generally let (δ_p) be the canonical isospectral Lax connection, with attached Frobenius lifts (ϕ_p), $\phi_p(x) = \Phi_p$, i.e., $\Phi_p = (\Phi_{pij}(x))$ where $\Phi_{pij}(x) = \phi_p(x_{ij})$. Let $\alpha_p(x)$ be $n \times n$ matrices with entries in $\mathcal{O}(G^*)^{\widehat{p}}$. Then, setting $\epsilon_p(x) = 1 + p\alpha_p(x)$, one can consider the *isospectral Lax connection* attached to (α_p), defined by the family of Frobenius lifts $(\phi_p^{(\epsilon)})$,

$$\phi_p^{(\epsilon)}(x) = \Phi_p^{(\epsilon)}(x) := \epsilon_p(x) \cdot \Phi_p(x) \cdot \epsilon_p(x)^{-1}.$$

The latter has the following property that justifies the term *isospectral.* Let

$$u_p = a_p^{-1} b_p a_p, \quad a_p \in GL_n(A^{\widehat{p}}), \quad b_p = diag(b_{p1}, ..., b_{pn}) \in T^*(A^{\widehat{p}})$$

be such that $\phi_p^{(\epsilon)}(u) = \Phi_p^{(\epsilon)}(u)$. Then

$$\delta_p b_{pi} = 0, \quad i = 1, ..., n.$$

So the eigenvalues b_{pi} of u_p are δ_p-constant.

On the other we will prove:

THEOREM 0.28. *There is an adelic connection $\delta = (\delta_p)$ on an open set $G^{**} \subset G$, $(G^{**})^{\widehat{p}} \neq \emptyset$, with attached Frobenius lifts (ϕ_p), such that the following diagrams commute:*

$$\begin{array}{ccc} (G^{**})^{\widehat{p}} & \overset{\phi_p}{\longrightarrow} & (G^{**})^{\widehat{p}} \\ \mathcal{P} \downarrow & & \downarrow \mathcal{P} \\ (\mathbb{A}^n)^{\widehat{p}} & \overset{\phi_{0p}}{\longrightarrow} & (\mathbb{A}^n)^{\widehat{p}}, \end{array} \tag{0.67}$$

where $\mathbb{A}^n = Spec\ A[z]$.

The diagram 0.67 should be viewed as analogous to diagram 0.38. The adelic connection δ is, of course, not unique.

DEFINITION 0.29. An adelic connection δ as in Theorem 0.28 is called an *isocharacteristic Lax connection.*

Among isocharacteristic Lax connections there will be a *canonical* one.

For any isocharacteristic Lax connection (δ_p), if $\phi_p(x) = \Phi_p(x)$ and $u_p \in G^{**}(A^{\widehat{p}})$ satisfies $\phi_p(u) = \Phi_p(u)$ we have that

$$\delta_p(\mathcal{P}_j(u)) = 0, \quad j = 1, ..., n.$$

So the coefficients of the characteristic polynomial of u_p are δ_p-constant.

Note that the canonical Frobenius lifts ϕ_p in the two diagrams 0.66 and 0.67 will *not* coincide on the intersection $G^* \cap G^{**}$; so the isospectral and isocharacteristic stories are really different.

Next we would like to explore the curvature of adelic connections.

Consider first the case of Ehresmann connections, 0.57. Since $\alpha_p \in A$ for all p our Frobenius lifts $\phi_p : B^{\widehat{p}} \to B^{\widehat{p}}$ induce Frobenius lifts $\phi_p : A[x] \to A[x]$ and hence one can consider the "divided" commutators

$$\varphi_{pp'} := \frac{1}{pp'}[\phi_p, \phi_{p'}] : A[x] \to A[x], \quad p, p' \in \mathcal{V}. \tag{0.68}$$

The family $(\varphi_{pp'})$ will be referred to as the *curvature* of the Ehresmann connection (δ_p).

The situation for general adelic connections (including the cases of real Chern and Lax connections) will be quite different. Indeed, in defining curvature we face the following dilemma: our p-derivations δ_p in 0.50 do not act on the same ring, so there is no a priori way of considering their commutators and, hence, it does not seem possible to define, in this way, the notion of curvature. It will turn out, however, that some of our adelic connections will satisfy a remarkable property which we call *globality along the identity* (more generally along various subvarieties); this property will allow us to define curvature via commutators.

DEFINITION 0.30. Consider the matrix $T = x - 1$, where 1 is the identity matrix. An adelic connection $\delta = (\delta_p)$ on GL_n, with attached family of Frobenius lifts (ϕ_p), is *global along* 1 if, for all p, $\phi_p : B^{\widehat{p}} \to B^{\widehat{p}}$ sends the ideal of 1 into itself and, moreover, the induced homomorphism $\phi_p : A^{\widehat{p}}[[T]] \to A^{\widehat{p}}[[T]]$ sends the ring $A[[T]]$ into itself. If the above holds then the *curvature* of (δ_p) is defined as the family of "divided" commutators $(\varphi_{pp'})$,

$$\varphi_{pp'} := \frac{1}{pp'}[\phi_p, \phi_{p'}] : A[[T]] \to A[[T]], \tag{0.69}$$

where $p, p' \in \mathcal{V}$. Let $\text{End}(A[[T]])$ denote the Lie ring of $\mathbb{Z}$-module endomorphisms of $A[[T]]$. Then define the *holonomy ring* $\mathfrak{hol}$ of δ as the $\mathbb{Z}$-linear span in $\text{End}(A[[T]])$ of all the *Lie monomials*

$$[\phi_{p_1}, [\phi_{p_2}, ..., [\phi_{p_{s-1}}, \phi_{p_s}]...]] : A[[T]] \to A[[T]]$$

where $s \geq 2$, $p_i \in \mathcal{V}$. Similarly define the *holonomy $\mathbb{Q}$-algebra* $\mathfrak{hol}_{\mathbb{Q}}$ of δ as the $\mathbb{Q}$-linear span of $\mathfrak{hol}$ in $\text{End}(A[[T]]) \otimes \mathbb{Q}$. Finally define the *completed holonomy ring*,

$$\widehat{\mathfrak{hol}} = \lim_{\leftarrow} \mathfrak{hol}_n,$$

where $\mathfrak{hol}_n$ is the image of the map

$$\mathfrak{hol} \to \text{End}(A[[T]]/(T)^n). \tag{0.70}$$

The various maps referred to above can be traced on the following diagram:

$$\begin{array}{ccccccccc} B^{\widehat{p}} & \subset & A^{\widehat{p}}[[T]] & \supset & A[[T]] & \subset & A^{\widehat{p'}}[[T]] & \supset & B^{\widehat{p'}} \\ \phi_p \downarrow & & \phi_p \downarrow & & \phi_p \downarrow\downarrow \phi_{p'} & & \downarrow \phi_{p'} & & \downarrow \phi_{p'} \\ B^{\widehat{p}} & \subset & A^{\widehat{p}}[[T]] & \supset & A[[T]] & \subset & A^{\widehat{p'}}[[T]] & \supset & B^{\widehat{p'}} \end{array}$$

The idea of comparing p-adic phenomena for different p's by "moving along the identity section" was introduced in [**34**] where it was referred to as *analytic continuation between primes.* Analytic continuation is taken here in the sense of Zariski [**124**], Preface, pp. xii-xiii, who was the first to use completions of varieties along subvarieties as a substitute for classical analytic continuation over the complex numbers. The technique of analytic continuation is also used, in the form of *formal patching,* in inverse Galois theory [**71**]. But note that our use of formal patching is rather different from the one used in inverse Galois theory. Indeed, in our context, we are patching data defined on "tubular neighborhoods"

$$Spf\ B^{\widehat{p}} \quad \text{and} \quad Spf\ A[[T]]$$

of two *closed* subsets

$$Spec\ B/pB \quad \text{and} \quad Spec\ B/(T)$$

of the scheme $Spec\ B$, where the data are required to coincide on the "tubular neighborhood"

$$Spf\ A^{\widehat{p}}[[T]]$$

of the closed subset

$$Spec\ B/(p,T) = (Spec\ B/pB) \cap (Spec\ B/(T))$$

of $Spec\ B$. This is in contrast with the use of patching in Galois theory [**71**] where one patches data defined on two *open* sets covering a formal scheme.

Of course, the trivial adelic connection $\delta_0 = (\delta_{0p})$, $\delta_{0p}x = 0$, is global along 1 so it induces ring endomorphisms $\phi_{0p} : A[[T]] \to A[[T]]$,

$$\phi_{0p}(T) = (1+T)^{(p)} - 1.$$

We may morally view δ_0 as an analogue of a flat connection in real geometry (where $A \subset C^\infty(\mathbb{R}^m, \mathbb{R})$). Alternatively we may view δ_0 as an arithmetic analogue of the derivations $\delta_{\overline{i}} = \partial/\partial \overline{z}_i$ on $A[x, \det(x)^{-1}]$ which kill x, where $A \subset C^\infty(\mathbb{C}^m, \mathbb{C})$. Following this second analogy we may consider an arbitrary adelic connection $\delta = (\delta_p)$ global along 1, with attached Frobenius lifts (ϕ_p), and introduce the following:

DEFINITION 0.31. The $(1,1)$-*curvature* of δ is the matrix of "divided commutators" $(\varphi_{p\overline{p}'})$

$$\varphi_{p\overline{p}'} := \frac{1}{pp'}[\phi_p, \phi_{0p'}] : A[[T]] \to A[[T]], \quad p \neq p', \tag{0.71}$$

$$\varphi_{p\overline{p}} := \frac{1}{p}[\phi_p, \phi_{0p}] : A[[T]] \to A[[T]]. \tag{0.72}$$

Let us discuss next the curvature of real Chern connections. (The case of complex Chern connections is easier and will be skipped here.) We will prove:

THEOREM 0.32. *Let $q \in GL_n(A)$ with $q^t = \pm q$. If all the entries of q are roots of unity or 0 then the real Chern connection δ attached to q is global along 1; in particular δ has a well defined curvature and $(1,1)$-curvature.*

So we may address the question of computing the curvature and $(1,1)$-curvature of real Chern connections for various q's whose entries are 0 or roots of unity. A special case of such q's is given by the following:

DEFINITION 0.33. A matrix $q \in GL_n(A)$ is *split* if it is one of the following:

$$\left(\begin{array}{cc} 0 & 1_r \\ -1_r & 0 \end{array}\right), \left(\begin{array}{cc} 0 & 1_r \\ 1_r & 0 \end{array}\right), \left(\begin{array}{ccc} 1 & 0 & 0 \\ 0 & 0 & 1_r \\ 0 & 1_r & 0 \end{array}\right), \tag{0.73}$$

where 1_r is the $r \times r$ identity matrix and $n = 2r, 2r, 2r+1$ respectively.

We will prove:

THEOREM 0.34. *Let q be split and let $(\varphi_{pp'})$ and $(\varphi_{p\overline{p}'})$ be the curvature and the $(1,1)$-curvature of the real Chern connection on G attached to q. Then the following hold:*

(1) *Assume $n \geq 4$. Then for all $p \neq p'$ we have $\varphi_{pp'} \neq 0$.*
(2) *Assume $n = 2r \geq 2$. Then for all p, p' we have $\varphi_{pp'}(T) \equiv 0 \mod (T)^3$.*
(3) *Assume $n = 2$ and $q^t = -q$. Then for all p, p' we have $\varphi_{pp'} = 0$.*
(4) *Assume $n \geq 2$. Then for all p, p' we have $\varphi_{p\overline{p}'} \neq 0$.*
(5) *Assume $n = 1$. Then for all p, p' we have $\varphi_{pp'} = \varphi_{p\overline{p}'} = 0$.*

Assertion (1) morally says that *Spec* $\mathbb{Z}$ is "curved," while assertion (2) morally says that *Spec* $\mathbb{Z}$ is only "mildly curved." Assertions (1) and (2) will imply, in particular, assertion (1) in the following:

THEOREM 0.35. *Assume q split and $n \geq 4$ is even. Then, for the real Chern connection attached to q, the following hold:*

(1) $\widehat{\mathfrak{hol}}$ *is non-zero and pronilpotent.*
(2) $\mathfrak{hol}_{\mathbb{Q}}$ *is not spanned over $\mathbb{Q}$ by the components of the curvature.*

Assertion (1) is in stark contrast with the fact, to be explained in the body of the book, that *holonomy Lie algebras* arising from Galois theory are never nilpotent unless they vanish. Assertion (2) should be viewed as a statement suggesting that the flavor of our arithmetic situation is rather different from that of classical *locally symmetric* spaces; indeed, for the latter, the Lie algebra of holonomy *is* spanned by the components of the curvature.

Note that the above theorem says nothing about the vanishing of the curvature $\varphi_{pp'}$ in case $n = 2, 3$ and $q^t = q$; our method of proof does not seem to apply to these cases.

By the way if $q \in GL_{2r}$ is symmetric and split then the real Chern connection $\delta_p : B^{\widehat{p}} \to B^{\widehat{p}}$ attached to q *does not* send the ideal of SL_n into itself; this is in contrast with the situation encountered in classical differential geometry. To remedy this situation we will construct connections on GL_n that *do* send the ideal of SL_n into itself; there are many such connections and the "simplest" one will be called the *special linear connection.*

Other curvatures can be introduced and vanishing/non-vanishing results for them will be proved; they will be referred to as 3-*curvature, first Chern form, first Chern* $(1,1)$-*form, mean curvature, scalar curvature,* etc. We will also introduce *real Chern connections* attached to hermitian metrics and results will be proved for their curvature. A similar theory can be developed for complex Chern connections.

Similar results will also be proved for the curvature of Lax connections. As we shall see the open sets G^* and G^{**} where isospectral and isocharacteristic Lax connections are defined (cf. 0.66 and 0.67) do not contain the identity of the group $G = GL_n$ hence curvature cannot be defined by analytic continuation along the identity; however these open sets will contain certain torsion points of the diagonal maximal torus of G and we will use analytic continuation along such torsion points to define curvature and $(1,1)$-curvature. We will then prove the non-vanishing of the $(1,1)$-curvature of isocharacteristic Lax connection for $n = 2$. The (canonical) isospectral Lax connection has, immediately from its definition, a vanishing curvature.

Note that the concept of curvature discussed above was based on what we called *analytic continuation between primes*; this was the key to making Frobenius lifts corresponding to different primes act on a same ring and note that it only works as stated for adelic connections that are global along 1. This restricts the applicability of our method to "metrics" q with components roots of unity or 0. One can generalize our method to include q's with more general entries by replacing the condition of being global along 1 with the condition of "being global along certain tori;" this will be discussed in the body of the book. However, there is a different approach towards making Frobenius lifts comparable; this approach is based on *algebraizing Frobenius lifts via correspondences* and works for adelic connections that are not necessarily global along 1 (or along a torus). The price to pay for allowing this generality is that endomorphisms (of $A[[T]]$) are replaced by correspondences (on GL_n). Let us explain this alternative road to curvature in what follows.

We will prove the following:

THEOREM 0.36. *Let $\delta = (\delta_p)$ be the real Chern connection on $G = GL_n$ attached to a matrix $q \in GL_n(A)$ with $q^t = \pm q$. Then there exists a family of diagrams of A-schemes*

$$\begin{array}{ccccc} & & Y_p & & \\ & \stackrel{\pi_p}{\swarrow} & & \stackrel{\varphi_p}{\searrow} & \\ G & & & & G \end{array} \tag{0.74}$$

where π_p is affine and étale, inducing, via p-adic completion, a family of diagrams

$$\begin{array}{ccccc} & & Y_p^{\widehat{p}} & & \\ & \stackrel{\pi_p^{\widehat{p}}}{\swarrow} & & \stackrel{\varphi_p^{\widehat{p}}}{\searrow} & \\ G^{\widehat{p}} & & \stackrel{\phi_p}{\longrightarrow} & & G^{\widehat{p}} \end{array} \tag{0.75}$$

where $\pi_p^{\widehat{p}}$ are isomorphisms.

The triples

$$\Gamma_p = (Y_p, \pi_p, \varphi_p) \tag{0.76}$$

can be viewed as *correspondences* on G which provide "algebraizations" of our Frobenius lifts ϕ_p.

The family (Γ_p) is not uniquely attached to (δ_p) but does have some "uniqueness features" to be explained in the body of the book.

On the other hand if E is the field of rational functions on the scheme G and F_p is the finite E-algebra

$$F_p := \mathcal{O}(Y_p \otimes E)$$

then each correspondence Γ_p induces, via tensorization by E, a diagram

$$\begin{array}{ccccc} & & Spec\ F_p & & \\ & \overset{\pi_p}{\swarrow} & & \overset{\varphi_p}{\searrow} & \\ Spec\ E & & & & Spec\ E \end{array} \tag{0.77}$$

The family of diagrams 0.77 will be referred to as a *correspondence structure* for (δ_p). Note that each Γ_p induces an additive map $\Gamma_p^* : E \to E$,

$$\Gamma_p^*(z) := \mathrm{Tr}_{\pi_p}(\varphi_p^*(z)), \quad z \in E, \tag{0.78}$$

where

$$\mathrm{Tr}_{\pi_p} : F_p \to E$$

is the trace of the finite algebra map

$$\pi_p^* : E \to F_p$$

and

$$\varphi_p^* : E \to F_p$$

is induced by φ_p.

DEFINITION 0.37. The *curvature* of (Γ_p) is the matrix $(\varphi_{pp'}^*)$ with entries the additive homomorphisms

$$\varphi_{pp'}^* := \frac{1}{pp'}[\Gamma_p^*, \Gamma_{p'}^*] : E \to E, \quad p, p' \in \mathcal{V}. \tag{0.79}$$

Note that, in this way, we have defined a concept of curvature for real Chern connections attached to arbitrary q's (that do not necessarily have entries zeroes or roots of unity). There is a $(1,1)$-version of the above as follows. Indeed the trivial adelic connection $\delta_0 = (\delta_{0p})$ has a canonical correspondence structure (Γ_{0p}) induced by

$$\Gamma_{0p} = (G, \pi_{0p}, \varphi_{0p}),$$

where π_{0p} is the identity, and $\varphi_{0p}(x) = x^{(p)}$.

DEFINITION 0.38. The $(1,1)$-*curvature* of (Γ_p) is the family $(\varphi_{p\overline{p}'}^*)$ where $\varphi_{p\overline{p}'}^*$ is the additive endomorphism

$$\varphi_{p\overline{p}'}^* := \frac{1}{pp'}[\Gamma_{0p'}^*, \Gamma_p^*] : E \to E \ \text{ for } p \neq p', \tag{0.80}$$

$$\varphi_{p\overline{p}}^* := \frac{1}{p}[\Gamma_{0p}^*, \Gamma_p^*] : E \to E. \tag{0.81}$$

Then we will prove the following:

THEOREM 0.39. *Let $q \in GL_2(A)$ be split. Then, for the real Chern connection attached to q, the following hold:*

(1) *Assume $q^t = -q$. Then for all p, p' we have $\varphi_{pp'}^* = 0$ and $\varphi_{p\overline{p}'}^* \neq 0$.*
(2) *Assume $q^t = q$. Then for all p, p' we have $\varphi_{p\overline{p}'}^* \neq 0$.*

Once again our results say nothing about curvature in case $n = 2$ and $q^t = q$; our method of proof does not seem to apply to this case.

Later in the body of the book the curvature and $(1,1)$-curvature of (Γ_p) introduced above will be called *upper $*$-curvatures* and we will also introduce the concepts of *lower $*$-curvature* and *lower $*$-$(1,1)$-curvature* coming from actions on "cycles."

Finally note that one can define curvature for the Levi-Cività connection.

DEFINITION 0.40. Let $(\delta_{1p}, ..., \delta_{np})$ be the Levi-Cività connection attached to a symmetric $q \in GL_n(A)$. The *curvature* is defined as the family (φ_p^{ij}), indexed by $i, j = 1, ..., n$ given by the divided commutators

$$\varphi_p^{ij} := \frac{1}{p}[\phi_{ip}, \phi_{jp}] : \mathcal{O}(G^{\widehat{p}}) \to \mathcal{O}(G^{\widehat{p}}). \tag{0.82}$$

This is a "vertical" curvature (indexed by the index set of the columns and rows of x) rather than a "horizontal" curvature, in the style of the previously introduced curvatures (which are indexed by primes). We will prove non-vanishing results for these curvatures. For instance we have:

THEOREM 0.41. *Let i, j, k, l be fixed indices between 1 and n. Then, for the Levi-Cività connection, the following hold:*

(1) *Assume $\delta_p q_{jk} + \delta_p q_{il} \not\equiv \delta_p q_{ik} + \delta_p q_{jl}$ mod p. Then $\varphi_p^{ij}, \varphi_p^{kl} \not\equiv 0$ mod p.*
(2) *Assume $n = 2r$ and q is split. Then $\varphi_p^{ij} \not\equiv 0$ mod p for $i \neq j$.*

We will actually prove a result more precise than assertion (1) above. What we will have is a congruence mod p relating φ_p^{ij} to the expression

$$\delta_p q_{jk} + \delta_p q_{il} - \delta_p q_{ik} - \delta_p q_{jl}.$$

The latter can be viewed as an "infinitesimal" analogue of the expression

$$q_{jk} q_{il} - q_{ik} q_{jl}$$

which gives, up to a constant, the classical covariant Riemann tensor of a metric with constant sectional curvature. So, in some sense, all metrics, in the arithmetic setting, behave "mod p" like metrics with constant sectional curvature.

In case the entries of q are roots of unity or 0 we will prove that, for each i, the adelic connection (δ_{ip}) appearing in the Levi-Cività connection is global along 1 so we will be able to define a *mixed curvature* $(\varphi_{pp'}^{ij})$ indexed by $i, j = 1, ..., n$ and $p, p' \in \mathcal{V}$ given by the divided commutators

$$\varphi_{pp'}^{ij} := \frac{1}{pp'}[\phi_{ip}, \phi_{jp'}] : A[[T]] \to A[[T]], \quad p \neq p', \tag{0.83}$$

$$\varphi_{pp}^{ij} := \frac{1}{p}[\phi_{ip}, \phi_{jp}] : A[[T]] \to A[[T]]. \tag{0.84}$$

For a fixed p and the Fedosov connection $(\delta_{1p}, \delta_{2p})$ relative to any anti-symmetric $q \in GL_2(A)$ the formula 0.82 defines, again, a *curvature*; we will prove that this curvature does not vanish in general even if q is split.

0.2. Comparison with other theories

0.2.1. Analogies. From a number theoretic viewpoint there are at least two perspectives on the integers:

(A) Integers can be viewed as numbers.
(B) Integers can be viewed as analogous to functions.

The two perspectives sometimes coexist in various specific contexts in spite of the obvious tensions between them.

This being said, Perspective A is mainly the basis for subjects such as: the *geometry of numbers* [**49**]; *diophantine approximation* [**50**]; *p-adic analysis* [**90**]; *Fourier analysis* on adeles [**94**]; *p-adic mathematical physics* [**120**]; etc.

On the other hand perspective B interprets $\mathbb{Z}$ as a ring of "functions" on a hypothetical "space;" cf. [**102**]. Here are some possibilities that have been suggested:

(B1) Integers are analogous to functions on a Riemann surface.
(B2) Integers are analogous to functions on a 3-dimensional manifold.
(B3) Integers are analogous to functions on an infinite dimensional manifold.

Accordingly the hypothetical "space" referred to above should be an analogue of a Riemann surface, a 3-manifold, or an infinite dimensional manifold, respectively.

Analogy B1 is classical, it has a complex geometric flavor, and is the basis of work stretching from Dedekind to Arakelov and beyond. According to this viewpoint the primes p in $\mathbb{Z}$ are analogous to complex functions on a Riemann surface that vanish at one point only and hence they define points on the surface. A similar picture holds for integers in number fields. Note that Riemann surfaces are objects of complex dimension 1 (or real dimension 2). Genera of number fields are classically defined and finite, as in the case of Riemann surfaces. There is a related viewpoint according to which $\mathbb{Z}$ is analogous to a ring of functions on a Riemann surface of infinite genus; cf., e.g., [**55**].

Analogy B2, in its topological form, originates in suggestions of Mazur, Manin, Kapranov, and others. According to this viewpoint the primes p in $\mathbb{Z}$ correspond to embedded circles (knots) in a 3-dimensional manifold; so primes can be viewed as analogues of complex functions on the manifold that vanish along these knots. The Legendre symbol is then an analogue of linking numbers. Cf. Morishita's monograph [**104**] and the references therein. A different form of analogy B2, with a dynamical flavor, was proposed by Deninger [**59**].

By contrast, the present book and previous work by the author adopt the analogy B3 and have a differential geometric flavor. The primes, in this setting, are analogous to global coordinate functions on an infinite dimensional manifold. Other viewpoints that fit into analogy B3 have also been proposed; cf. the discussion below on the field with one element.

There is a possibility that our theory has connections with viewpoints B1 and B2, as shown, for instance, by the presence of the Legendre symbol in our real Chern connections. Indeed the underlying Galois theory of reciprocity is an analogue of the monodromy in either the 2 or 3-dimensional picture; in the same way our arithmetic curvature theory could be an analogue of the identity component of a natural "holonomy" in one of these pictures.

There is also a possibility that our arithmetic differential geometry could play the role of framework for an *arithmetic physics* in the sense of Manin [**100**] which would then have perspective B in its background (as opposed to [**120**] which has perspective A in its background).

0.2.2. Field with one element. According to the viewpoint B described above integers can be viewed as functions on a hypothetical "space." As noted, at least in its variant B1, this viewpoint is entirely classical. More recently this viewpoint was adopted in the search for a geometry over the *field with one element*, $\mathbb{F}_1$; cf. the pioneering suggestions of Tits [**118**]. See also the articles in [**117**], and the bibliography therein, for a survey of some approaches to the subject.

A remarkable idea appearing in the context of $\mathbb{F}_1$ is the consideration by Kurokawa et. al., [**93**] and Haran [**70**] of the *absolute derivation operators*,

$$\partial_p : \mathbb{Z} \to \mathbb{Z}, \quad \partial_p a := v_p(a)\frac{a}{p},$$

where $v_p(a)$ is the p-adic valuation of a. This is in line with paradigm B3. However the flavor of the operators ∂_p is rather different from that of the Fermat quotient operators δ_p and it seems unlikely that Haran's theory and ours are directly related. Even more remote from our theory are the approaches to $\mathbb{F}_1$ in papers of Kapranov-Smirnov [**86**], Soulé [**115**], Deitmar [**56, 117**], Manin-Marcolli [**117**], Connes-Consani [**55**], etc., which do not directly provide an analogue of derivation for the integers.

Borger's philosophy of $\mathbb{F}_1$, cf. [**11**], is, in some sense, perpendicular to the above mentioned approaches to $\mathbb{F}_1$. On the other hand, in the "case of one prime" Borger's approach is consistent with ours: roughly speaking, in the case of one prime, Borger's theory [**11, 13**] can be viewed as an algebraization of our analytic theory in [**23, 35**]. In the case of more (all) primes Borger's $\mathbb{F}_1$ theory can also be viewed as a viewpoint consistent with B3 above: indeed Borger's remarkable suggestion is to take λ-*structures* (in the sense of Grothendieck) as *descent data* from $\mathbb{Z}$ to $\mathbb{F}_1$. Recall that a λ-*structure* on a scheme X flat over $\mathbb{Z}$ is the same as a commuting family (ϕ_p) of Frobenius lifts $\phi_p : X \to X$. So our theory would fit into "λ-geometry" as long as:

(1) the Frobenius lifts are defined on the schemes X themselves (rather than on the various p-adic completions $X^{\hat{p}}$) and
(2) the Frobenius lifts commute.

However conditions (1) and (2) are almost never satisfied in our theory: the failure of condition (2) is precisely the origin of our curvature, while finding substitutes for condition (1) requires taking various convoluted paths (such as analytic continuation between primes or algebraization by correspondences). So in practice our approach places us, most of the times, outside the paradigm of λ-geometry.

0.2.3. Ihara's differential. Next we would like to point out what we think is an important difference between our viewpoint here and the viewpoint proposed by Ihara in [**78**]. Our approach, in its simplest form, proposes to see the operator

$$\delta = \delta_p : \mathbb{Z} \to \mathbb{Z}, \ a \mapsto \delta a = \frac{a - a^p}{p},$$

where p is a fixed prime, as an analogue of a derivation with respect to p. In [**78**] Ihara proposed to see the map

$$d : \mathbb{Z} \to \prod_p \mathbb{F}_p, \ a \mapsto \left(\frac{a - a^p}{p} \ mod \ p \right) \tag{0.85}$$

as an analogue of differentiation for integers and he proposed a series of conjectures concerning the "zeroes" of the differential of an integer. These conjectures are still completely open; they are in the spirit of paradigm B1 listed above, in the sense that counting zeroes of 1-forms is a Riemann surface concept. But what we see as the main difference between Ihara's viewpoint and ours is that *we do not consider the reduction mod p of the Fermat quotients but the Fermat quotients themselves.* This allows the possibility of considering compositions between our δ_p's which leads to the possibility of considering arithmetic analogues of differential equations, curvature, etc.

0.2.4. Fontaine-Colmez calculus. The Fontaine-Colmez theory of p-adic periods [**64**] also speaks of a *differential calculus with numbers.* Their calculus is perpendicular to ours in the following precise sense. For a fixed prime p our calculus based on the Fermat quotient operator δ_p should be viewed as a a differential calculus in the "unramified direction," i.e., the "direction" given by the extension

$$\mathbb{Q} \subset \bigcup_{p \nmid N} \mathbb{Q}(\zeta_N),$$

whereas the Fontaine-Colmez calculus should be viewed as a differential calculus in the "totally ramified direction," i.e., roughly, in the "direction" given by the extension

$$\mathbb{Q} \subset \bigcup_n \mathbb{Q}(\zeta_{p^n}).$$

The Fontaine-Colmez theory is based on the *usual* Kähler differentials of totally ramified extensions and hence, unlike ours, it is about *usual* derivations. It is not unlikely, however, that, for a fixed p, a theory unifying the unramified and the totally ramified cases, involving two "perpendicular" directions, could be developed leading to arithmetic partial differential equations in two variables. Signs of possibility of such a theory can be found in our papers [**32, 33**]. This two variable setting would fit into paradigm B1 listed above but would be rather different from the classical forms (Dedekind style) of paradigm B1 discussed above.

0.2.5. Grothendieck's p-curvature. It is worth pointing out that the study of our *curvature* here resembles the study of the *p-curvature* appearing in the arithmetic theory of differential equations that has been developed around the Grothendieck conjecture (cf., e.g., [**87**]). Both curvatures measure the lack of commutation of certain operators and both theories rely on technical matrix computations. However we should also point out that the nature of the above mentioned operators in the two theories is quite different. Indeed our curvature here involves the "p-differentiation" of numbers with respect to primes p (in other words it is about d/dp and fits into paradigm B) whereas the theory in [**87**], and related papers, is about usual differentiation d/dt of power series in a variable t and essentially fits into paradigm A. In spite of these differences the two types of curvatures could interact; a model for such an interaction between d/dp and d/dt is in the papers [**32, 33**]. A similar remark can be made about the difference between our approach and that in [**17**] which, again, is about usual Kähler differentials, hence about usual derivations.

0.2.6. Discrete geometry as Euclidean geometry. Finally we would like to point out that the theory in this book is a priori unrelated to topics such as *the geometry of numbers* [**49**] on the one hand and *discrete differential geometry* [**10**] on the other. Indeed in both these geometries what is being studied are discrete configurations of points in the Euclidean space $\mathbb{R}^m$; in the geometry of numbers the configurations of points typically represent rings of algebraic numbers (making the subject fit into paradigm A) while in discrete differential geometry the configurations of points approximate smooth submanifolds of the Euclidean space (making the subject a priori unrelated to number theory). This framework is, therefore, that of the classical geometry of Euclidean space, based on $\mathbb{R}$-coordinates, and *not* that of an analogue of this geometry, based on "prime coordinates." It may very well happen, however, that (one or both of) the above topics are a natural home for some (yet to be discovered) Archimedian counterpart of our (finite) adelic theory.

CHAPTER 1

Algebraic background

In this chapter we review some basic concepts of algebra and superalgebra that are going to be used throughout the book. Most of our terminology and notation is standard so this chapter could in principle be skipped and later consulted as needed.

1.1. Algebra

1.1.1. Rings. By a *ring* we will always understand an associative commutative unital ring; the identity is denoted by 1.

The symbols

$$\mathbb{Z},\ \ \mathbb{Q},\ \ \mathbb{R},\ \ \mathbb{C}$$

denote, as usual, the rings of integer, rational, real, and complex numbers respectively.

The letters $p, p', p'', \ldots$ will denote primes in $\mathbb{Z}$.

We denote by ζ_N a primitive N-th root of unity in $\mathbb{C}$.

For any field F we denote by F^a an algebraic closure of F.

So for a prime integer p we denote by

$$\mathbb{F}_p \quad \text{and} \quad \mathbb{F}_p^a$$

the field with p elements and an algebraic closure of it.

For a ring A and $n \in \mathbb{Z}$ we say that A is *n-torsion free* if n is a non-zero divisor in A, i.e., the following holds: for all $a \in A$ if $na = 0$ then $a = 0$. We say A is *torsion free* if it is torsion free as a $\mathbb{Z}$-module i.e., if it is n-torsion free for all $n \in \mathbb{Z}\backslash\{0\}$.

For a ring A and an ideal I in A generated by a set S we denote by

$$A^{\widehat{S}} = A^{\widehat{I}} = \lim_{\leftarrow} A/I^n A$$

the I-adic completion of A. (This notation is not standard but will be quite convenient for our purposes.) We say A is *I-adically complete* if the natural homomorphism $A \to A^{\widehat{I}}$ is an isomorphism. A local ring with maximal ideal M is called *complete* if it is M-adically complete.

In particular

$$A^{\widehat{p}} = \lim_{\leftarrow} A/p^n A$$

denotes the p-adic completion of A. As usual we denote by

$$\mathbb{Z}_p = \mathbb{Z}^{\widehat{p}}$$

the ring of p-adic integers.

For the notion of completion see [**103**], pp. 55-63. Suffices to recall here a few general properties which will be repeatedly used in the book. For instance for any Noetherian ring A and any ideal I the ring $A^{\widehat{I}}$ is flat over A; cf. [**103**], p.60. Moreover Krull's intersection theorem ([**103**], p. 60) states that for any Noetherian integral domain and any proper ideal I we have

$$\bigcap_{n=1}^{\infty} I^n = (0).$$

Also for any Noetherian ring A and any ideals $J \subset I$ in A if A is I-adically complete then A is J-adically complete; cf. [**103**], p. 63. Finally if a ring A is I-adically complete for some ideal I in A then, for x a variable, the power series ring $A[[x]]$ is (I, x)-adically complete; cf. [**103**], p. 63.

Recall also (cf. [**103**], p. 123) that a sequence of elements $a_1, ..., a_n$ in a ring A is called a *regular sequence* if it generates a proper ideal of A and, for all $i = 1, ..., n$,

$$a_i \quad \text{is a non-zero divisor in} \quad A/(a_1, ..., a_{i-1}).$$

Recall that a *discrete valuation ring* is a local principal ideal domain.

Let A and B be complete discrete valuation rings with maximal ideals generated by p. Let Hom denote homomorphisms in the category of rings. Then recall that the natural map

$$\text{Hom}(A, B) \to \text{Hom}(A/pA, B/pB) \tag{1.1}$$

is surjective; cf. [**103**], p. 224. If, in addition, the residue fields A/pA and B/pB are perfect then the above map is an isomorphism; cf. [**113**], p. 38. In particular, if A is a complete discrete valuation ring with maximal ideal generated by p and residue field A/pA then the $p-$power Frobenius endomorphism of A/pA lifts to an endomorphism ϕ^A of A; if, in addition, A/pA is perfect then ϕ^A is unique.

If $\mathbb{F}$ is a field of characteristic p we denote by

$$W(\mathbb{F})$$

the unique (up to isomorphism) complete discrete valuation ring of characteristic zero with maximal ideal generated by p and residue field $\mathbb{F}$. So if $\mathbb{F}$ is perfect then $W(\mathbb{F})$ possesses a unique ring automorphism $\phi^{W(\mathbb{F})}$ whose reduction mod p is the p-power Frobenius on $\mathbb{F}$. (For $\mathbb{F}$ perfect $W(\mathbb{F})$ is isomorphic to the *Witt ring* on $\mathbb{F}$ [**113**]; for our purposes we will not need to review Witt rings here.) In particular we have

$$\mathbb{Z}_p = W(\mathbb{F}_p).$$

On the other hand, in our arithmetic theory (starting with Chapter 3), we will constantly denote by R the ring

$$R = W(\mathbb{F}_p^a).$$

For the above concepts we refer to [**103**], section 29.

A *number field* will mean a finite field extension K of $\mathbb{Q}$ contained in $\mathbb{C}$, usually denoted by

$$K/\mathbb{Q}.$$

We denote by

$$\mathcal{O}_K$$

the ring of integers of a number field K.

In particular for the number field $K = \mathbb{Q}(\zeta_N)$ one has $\mathcal{O}_K = \mathbb{Z}[\zeta_N]$.

We will also encounter examples of natural rings appearing in analytic contexts: the rings of real and complex valued functions on a smooth real manifold M will be denoted by

$$C^{\infty}(M, \mathbb{R}) \quad \text{and} \quad C^{\infty}(M, \mathbb{C})$$

respectively. For an open set $D \subset \mathbb{C}^m$ we denote by

$$\text{Mer}(D)$$

the field of meromorphic functions on D.

1.1.2. Algebras. By a $\mathbb{Z}$-*algebra* we mean a $\mathbb{Z}$-module equipped with a binary $\mathbb{Z}$-bilinear operation.

By an *associative ring* we mean an associative $\mathbb{Z}$-algebra (that is not necessarily commutative or with 1); we denote multiplication by a dot or simply by juxtaposition.

By a *Lie ring* we will mean a $\mathbb{Z}$-algebra whose multiplication is anti-symmetric and satisfies the Jacobi identity; the bracket will be denoted, as usual, by $[\ ,\ ]$.

As already mentioned commutative associative rings with 1 (unital) will be referred to simply as *rings*. Modules over rings are always assumed to be unital (multiplication by 1 is the identity).

If A is a ring by an A-*algebra* we mean an A-module equipped with an A-bilinear operation; similar definition for associative and Lie A-algebras. To give an associative commutative unital A-algebra B is the same as to give a ring homomorphism $A \to B$. In the context of commutative algebra or algebraic geometry, and if no confusion is feared, associative commutative unital A-algebras will simply be referred to as A-*algebras*.

For a Lie ring L and two $\mathbb{Z}$-submodules M_1, M_2 of L we denote, as usual, by $[M_1, M_2]$ the $\mathbb{Z}$-submodule of L spanned by all the elements of the form $[m_1, m_2]$ with $m_1 \in M_1$, $m_2 \in M_2$. We can then consider the *central series*:

$$L = \mathfrak{c}^0(L) \supset \mathfrak{c}^1(L) \supset \mathfrak{c}^2(L) \supset \ldots,$$

defined by

$$\mathfrak{c}^s(L) := [L, \mathfrak{c}^{s-1}(L)], \quad s \geq 1,$$

and the *derived series*

$$L = \mathfrak{d}^0(L) \supset \mathfrak{d}^1(L) \supset \mathfrak{d}^2(L) \supset \ldots,$$

defined by

$$\mathfrak{d}^s(L) := [\mathfrak{c}^{s-1}(L), \mathfrak{c}^{s-1}(L)], \quad s \geq 1.$$

A Lie ring L is called *nilpotent* if one of the members of its central series is zero. A Lie ring is *solvable* if one of the members of its derived series is zero.

For any monoid Γ we denote by

$$\mathbb{Z}[\Gamma]$$

the monoid ring of Γ; we view it as an associative ring and also as a Lie ring with respect to the induced bracket.

If Γ is a group we refer to $\mathbb{Z}[\Gamma]$ as the *group ring* of Γ.

For any $\mathbb{Z}$-module A set theoretic maps $A \to A$ (which are not necessarily $\mathbb{Z}$-module homomorphisms) will be referred to as *operators* on A. For any two operators

$$f, g : A \to A$$

the *commutator* is defined, as usual, as the operator

$$[f, g] := f \circ g - g \circ f : A \to A.$$

For any $\mathbb{Z}$-module A we denote by

$$\mathrm{End}(A) \quad \text{or} \quad \mathrm{End}_{\mathrm{gr}}(A) \quad \text{or} \quad \mathrm{End}_{\mathbb{Z}}(A)$$

the set of $\mathbb{Z}$-module endomorphisms of A. We view $\mathrm{End}(A)$ as an associative ring and also as a Lie ring with respect to the induced bracket.

For any associative ring A we denote by $A^\times$ the group of invertible elements. If A is a field we denote by

$$\mathrm{Aut}_{\mathrm{field}}(A) \quad \text{and} \quad \mathrm{End}_{\mathrm{field}}(A)$$

the group of field automorphisms of A and the monoid of field endomorphisms of A respectively.

Recall Artin's *independence of characters* [**95**], p. 283: any set of distinct monoid homomorphisms from a monoid into the group of invertible elements of a field is linearly independent over that field. This implies, for instance, that if A is a field of characteristic zero then the associative ring homomorphism

$$\mathbb{Z}[\mathrm{End}_{\mathrm{field}}(A)] \to \mathrm{End}_{\mathrm{gr}}(A) \tag{1.2}$$

is injective.

We denote by $\mathfrak{gl}_n(A)$ the associative ring of $n \times n$ matrices with entries in a ring A. We denote by

$$GL_n(A) := \mathfrak{gl}_n(A)^\times$$

the group of invertible elements of $\mathfrak{gl}_n(A)$ and we denote by 1_n, or simply by 1, the identity matrix in $GL_n(A)$. We also view $\mathfrak{gl}_n(A)$ as a Lie ring. If $M = (M_{ij}) \in \mathfrak{gl}_n(A)$ we usually denote by

$$(M) \subset A$$

the ideal of A generated by the entries M_{ij} of M. For a set theoretic map

$$f : A \to B$$

between two rings we continue to denote by

$$f : \mathfrak{gl}_n(A) \to \mathfrak{gl}_n(B)$$

the induced map: if $a = (a_{ij})$ is a matrix in $\mathfrak{gl}_n(A)$ we let $f(a)$ be the matrix

$$(f(a)_{ij})$$

with entries

$$f(a)_{ij} = f(a_{ij}).$$

For $a = (a_{ij})$ a matrix with entries in A and p a prime integer we denote by $a^{(p)} = (a_{ij}^{(p)})$ the matrix with entries

$$a_{ij}^{(p)} = a_{ij}^p$$

and we denote by $a^t = (a_{ij}^t)$ the transpose of a,

$$a_{ij}^t := a_{ji}.$$

We sometimes (although not always) use Einstein summation convention: in sums repeated indices must be summed over. Usually this convention is applied for pairs of indices appearing one as an upper index and one as a lower index; since we do not always want to keep track of the distinction between upper and lower indices we will sometimes apply the Einstein summation convention for indices that are both lower or both upper indices. E.g.,

$$a_{ijk}^{lm}b_{ijm}^{n} := \sum_{ijm} a_{ijk}^{lm}b_{ijm}^{n}.$$

1.1.3. Derivations. Let A be a ring, B an associative commutative unital A-algebra and E a B-module. Recall the following:

DEFINITION 1.1. A *derivation* from B to E is a map $\delta : B \to E$ which is additive and satisfies the *Leibniz rule*:

$$\begin{aligned} \delta(a+b) &= \delta a + \delta b \\ \delta(ab) &= a \cdot \delta b + b \cdot \delta a. \end{aligned} \tag{1.3}$$

If $B \to E$ is a ring homomorphism we say δ is a derivation of $B \to E$; if $E = B$ and $B \to E$ is the identity we say δ is a derivation on B. A derivation is an A-*derivation* if it is A-linear.

We denote by $\mathrm{Der}_A(B) = \mathrm{Der}_A(B, B)$ the B-module of all A-derivations on B; it is a Lie A-algebra with respect to the usual commutator.

We denote by

$$\Omega_{B/A} = I/I^2, \quad I = \mathrm{Ker}(\mathrm{mult} : B \otimes_A B \to B),$$

the B-module of *Kähler differentials* of B over A and we let $d : B \to \Omega_{B/A}$ be the universal derivation. We write $\Omega_B = \Omega_{B/\mathbb{Z}}$. We have

$$\mathrm{Hom}_B(\Omega_{B/A}, B) \simeq \mathrm{Der}_A(B).$$

For the theory of Kähler differentials we refer to [**103**], Chapter 9.

We will later repeatedly use the following *Jacobi formula* valid for any derivation δ on a ring B and any invertible matrix $u \in GL_n(B)$ with entries in B:

$$\delta(\det(u)) \cdot \det(u)^{-1} = \mathrm{tr}\ (\delta u \cdot u^{-1}). \tag{1.4}$$

We will also repeatedly use the following fact and notation. Let

$$\begin{array}{ccc} A & \to & B_1 \\ \downarrow & & \downarrow \\ B_2 & \to & C \end{array}$$

be a diagram of rings and

$$\delta_1 : B_1 \to C, \quad \delta_2 : B_2 \to C$$

be derivations whose restriction to A coincide. Then there is a unique derivation denoted by

$$\delta_1 \otimes 1 + 1 \otimes \delta_2 : B_1 \otimes_A B_2 \to C \tag{1.5}$$

whose action on simple tensors is

$$b_1 \otimes b_2 \mapsto (\delta_1 b_1) \cdot b_2 + b_1 \cdot (\delta_2 b_2), \quad b_1 \in B_1, \quad b_2 \in B_2.$$

Similarly, for a Lie A-algebra L, we have:

DEFINITION 1.2. A *derivation* on L is a map $\delta : L \to L$ which is additive and satisfies the *Leibniz rule*:

$$\begin{aligned} \delta(a+b) &= \delta a + \delta b \\ \delta[a,b] &= [\delta a, b] + [a, \delta b]. \end{aligned} \tag{1.6}$$

A derivation is an A-*derivation* if it is A-linear.

We denote by $\mathrm{Der}_A(L)$ the A-module of all A-derivations on L; it is a Lie A-algebra with respect to the usual commutator.

1.1.4. Extensions. Let

$$1 \to \Gamma_1 \to \Gamma_2 \to \Gamma_3 \to 1$$

be an exact sequence of groups. Then there is a naturally induced group homomorphism

$$\Gamma_3 \to \mathrm{Out}(\Gamma_2) := \frac{\mathrm{Aut}(\Gamma_1)}{\mathrm{Int}(\Gamma_1)}, \tag{1.7}$$

where $\mathrm{Aut}(\Gamma_1)$ is the group of group automorphisms of Γ_1 and $\mathrm{Int}(\Gamma_1)$ is the normal subgroup of inner automorphisms of Γ_1.

The above construction arises, for instance, in the case when Γ_1 is a geometric fundamental group, Γ_2 is an arithmetic fundamental group, and Γ_3 is the Galois group of a number field extension; then 1.7 yields interesting Galois representations as in [**78**].

An analogue of the above is as follows. Let

$$0 \to L_1 \to L_2 \to L_3 \to 0$$

be an exact sequence of Lie A-algebras. Then there is an induced morphism of Lie A-algebras

$$L_3 \to \mathrm{Out}_A(L_1) := \frac{\mathrm{Der}_A(L_1)}{\mathrm{Inn}_A(L_1)}, \tag{1.8}$$

where $\mathrm{Der}_A(L_1)$ is the Lie A-algebra of A-derivations of L_1 and $\mathrm{Inn}_A(L_1)$ is the ideal (and sub Lie A-algebra) of inner derivations (i.e., derivations of the form $x \mapsto [x, y]$ for some $y \in L_1$).

1.2. Algebraic geometry

1.2.1. Schemes and formal schemes. For the language of schemes and formal schemes we refer to [**72**], Chapter 2. Schemes will always be assumed separated. Formal schemes will always be assumed separated and Noetherian. For a scheme (or formal scheme) X we denote by $\mathcal{O}_X$ the structure sheaf and we often write $\mathcal{O}$ instead of $\mathcal{O}_X$.

For a ring A we denote by

$$Spec\ A$$

the associated affine scheme; if A is complete with respect to an adic topology we denote by

$$Spf\ A$$

the corresponding affine formal scheme.

For any ring homomorphism $f : A \to B$ we continue to denote by

$$f : Spec\ B \to Spec\ A$$

the induced morphism of schemes. Conversely, for any morphism of affine schemes

$$f : X \to Y$$

we denote by

$$f^* : \mathcal{O}(Y) \to \mathcal{O}(X)$$

the homomorphism between the corresponding rings; however we often write f instead of f^*. Similar conventions will be adopted for formal schemes.

For any scheme X over a ring A and any A-algebra B we denote by

$$X(B) = \mathrm{Hom}_{Spec\ A}(Spec\ B, X)$$

the set of B-points of X (over A). We adopt a similar notation for formal schemes.

A *group scheme* over a ring A is a scheme G over A equipped with morphisms over A, $G \times G \to G$, $G \to G$, $Spec\ A \to G$, satisfying the usual axioms. A *principal homogeneous space* for G is a scheme P over A equipped with a morphism $P \times G \to P$ over A, $(x, g) \mapsto xg$, satisfying the usual action axioms, such that the induced morphism $P \times G \to P \times P$, $(x, g) \mapsto (xg, x)$ is an isomorphism.

For any Noetherian scheme X over a Noetherian ring A and any ideal I in A we denote by $X^{\widehat{I}}$ the formal scheme over $A^{\widehat{I}}$ obtained from X via I-adic completion. Also we denote by $X^{\widehat{Y}}$ the completion of X along a closed subscheme Y.

We recall a form of Grothendieck's existence theorem in formal geometry. Assume X and Y are projective schemes over an I-adically complete Noetherian ring A. Then Grothendieck's existence theorem states, in particular, that any morphism $X^{\widehat{I}} \to Y^{\widehat{I}}$ of formal schemes over A comes from a morphism $X \to Y$ of schemes over A.

1.2.2. Smoothness. We will need, so we recall here, some basic definitions and properties of smoothness. We need a slightly more general framework than that of [**72**] for which we refer to [**16**], Chapter 2. Here are the basic concepts.

Let A be a ring and B a finitely generated A–algebra. (As mentioned before, in the context of algebraic geometry, algebras are always assumed associative, commutative, and unital.) One says B is *smooth* (resp. *étale*) over A if for any A–algebra C and any ideal $I \subset C$ with $I^2 = 0$ the canonical map

$$\mathrm{Hom}_A(B, C) \to \mathrm{Hom}_A(B, C/I)$$

is surjective (resp. bijective). A scheme X over A is called *smooth* (resp. *étale*) if it can be covered with open affine schemes $Spec\ B$ with B finitely generated and smooth (resp. étale) over A. If B is smooth over A then $\Omega_{B/A}$ is locally free. A morphism $f : X \to Y$ of schemes is called *smooth* (resp. *étale*) if Y can be covered by affine open schemes $U = Spec\ A$ such that $f^{-1}(U)$ are smooth (resp. *étale*) over A. A morphism $X \to Y$ of finite type of Noetherian schemes is smooth if and only if it is flat and all geometric fibers are regular; cf. [**103**], [**72**] for these concepts. If $Spec\ B \to Spec\ A$ is of finite type and smooth (resp. *étale*) then B is smooth (resp. *étale*) over A. If a scheme X over A is smooth then each point in X has an affine open neighborhood $U = Spec\ B$ such that B is étale over an algebra of polynomials $A[x_1, ..., x_n]$ contained in B; $x_1, ..., x_n$ are then called *étale coordinates* on U.

1.3. Superalgebra

In this section we quickly review some standard superalgebra formalism. This formalism can be used, in classical differential geometry, to encode the deRham calculus; this point of view is crucial, for instance, in non-commutative geometry [**53**]. Note that in arithmetic differential geometry there will be no deRham calculus but Lie superalgebra formalism will still be available; so in the arithmetic context the Lie superalgebra formalism will play, in particular, the role of deRham calculus.

1.3.1. Graded Lie superalgebra. Let $\mathcal{U}$ be a finite or countable set. (In classical differential geometry $\mathcal{U}$ is usually assumed finite. But the case when $\mathcal{U}$ is infinite will naturally play a role later, in the arithmetic case.) Fix an ordering

$$\mathcal{U} = \{\xi_1, \xi_2, ...\}. \tag{1.9}$$

The definitions below will not depend on this ordering.

DEFINITION 1.3. The set $\mathcal{U}$ will be referred to as the set of *basic variables.*

Intuitively the basic variables play the role of "coordinates on a basis manifold."

Denote by Λ^1_m the free $\mathbb{Z}$-module with basis the family of symbols

$$\{d\xi_1, d\xi_2, ...d\xi_m\}. \tag{1.10}$$

The Λ^1_m's form a projective system with projections $\Lambda^1_{m+1} \to \Lambda^1_m$ sending

$$d\xi_1 \mapsto d\xi_1, ..., d\xi_m \mapsto d\xi_m, \ d\xi_{m+1} \mapsto 0.$$

Let $\Lambda^k_m = \wedge^k \Lambda^1_m$ be the exterior product $\mathbb{Z}$-modules of Λ^1_m; so $\Lambda^0_m = \mathbb{Z}$.

DEFINITION 1.4. For any $\mathbb{Z}$-module E we let

$$\Lambda^k(E) = \lim_{\overleftarrow{m}} (E \otimes_{\mathbb{Z}} \Lambda^k_m), \tag{1.11}$$

$$\Lambda(E) := \bigoplus_{k=0}^{\infty} \Lambda^k(E), \tag{1.12}$$

$$\Lambda^{even}(E) := \bigoplus_{k=0}^{\infty} \Lambda^{2k}(E). \tag{1.13}$$

For $I \subset \mathbb{N}$ a set of natural numbers, with

$$|I| = k, \quad I = \{i_1, ..., i_k\}, \quad i_1 < ... < i_k,$$

we write

$$d\xi_I := d\xi_{i_1} \wedge ... \wedge d\xi_{i_k}.$$

So the elements of $\Lambda^k(E)$ can be written uniquely as formal (infinite if $\mathcal{U}$ is infinite) sums

$$\eta = \sum_{|I|=k} \eta_I \cdot d\xi_I, \tag{1.14}$$

with $\eta_I \in E$. The elements η will be referred to as *k-forms* with coefficients in E.

REMARK 1.5. $\Lambda^{even}(\mathbb{Z})$ has a (commutative !) ring structure while $\Lambda^{even}(E)$ has a natural structure of $\Lambda^{even}(\mathbb{Z})$-module. If $E = C$ with C a commutative ring with unit then $\Lambda^{even}(C)$ has a natural structure of associative commutative unital $\Lambda^{even}(\mathbb{Z})$-algebra.

Assume now $E = L$, where L is a Lie ring. For simplicity we will assume L is 6-torsion free. Note that $\Lambda^{even}(L)$ has a unique structure of Lie $\Lambda^{even}(\mathbb{Z})$-algebra with the property that $L = \Lambda^0(L)$ is a Lie $\mathbb{Z}$-subalgebra of $\Lambda^{even}(L)$. Moreover there is a unique $\mathbb{Z}$-bilinear map

$$[\ ,\ ] : \Lambda(L) \times \Lambda(L) \to \Lambda(L) \tag{1.15}$$

such that for all $\eta, \eta' \in L$ we have

$$[\eta \cdot d\xi_I, \eta' \cdot d\xi_{I'}] = [\eta, \eta'] \cdot d\xi_I \wedge d\xi_{I'}.$$

DEFINITION 1.6. A *graded Lie superalgebra* (over $\mathbb{Z}$) is a $\mathbb{Z}$-module $\mathfrak{L}$ equipped with a direct sum decomposition

$$\mathfrak{L} = \bigoplus_{k\in\mathbb{Z}} \mathfrak{L}^k \tag{1.16}$$

and with a $\mathbb{Z}$-bilinear map

$$[\ ,\ ] : \mathfrak{L} \times \mathfrak{L} \to \mathfrak{L} \tag{1.17}$$

such that for any $\alpha_i \in \mathfrak{L}^{k_i}$, $i = 1, 2, 3$, the following hold:

(1) $[\alpha_1, \alpha_2] \in \mathfrak{L}^{k_1+k_2}$;
(2) $[\alpha_1, \alpha_2] = -(-1)^{k_1k_2}[\alpha_2, \alpha_1]$;
(3) $\sum_{(123)}(-1)^{k_1k_3}[\alpha_1, [\alpha_2, \alpha_3]] = 0$.

Here the index (123) means "circular permutations."

Then it is trivial to check that $\mathfrak{L} := \Lambda(L)$ equipped with the bracket $[\ ,\ ]$ in 1.15 is a *graded Lie superalgebra*. Moreover, for any $\alpha \in \Lambda^1(L)$, we have:

(i) $[\alpha, \alpha] = 2\beta$ for a unique $\beta \in \Lambda^2(L)$;
(ii) $[\alpha, \beta] = 0$ (Bianchi identity);
(iii) $[\alpha, [\alpha, \eta]] = [\beta, \eta]$, for all $\eta \in \Lambda(L)$.

DEFINITION 1.7. In the notation above β is the *curvature* of α.

1.3.2. Higher curvature and holonomy. Let L be a Lie ring and $X \subset L$ a subset. A *Lie monomial* on X of degree $s \geq 2$ is an element of L of the form

$$[x_1, [x_2, ...[x_{s-1}, x_s]...]]$$

with $x_i \in X$. A *Lie monomial* on X of degree 1 is simply an element of X. If y and z are two Lie monomials on X of degrees $s \geq 1$ and $t \geq 1$ respectively then $[y, z]$ is a $\mathbb{Z}$-linear combination of Lie monomials on X of degree $s + t$.

DEFINITION 1.8. The smallest Lie subring $\mathfrak{lie}(X)$ of L containing X is called the *Lie hull* of X; it is the $\mathbb{Z}$-span of all Lie monomials on X of degree ≥ 1. The commutator

$$\mathfrak{hol}(X) = [\mathfrak{lie}(X), \mathfrak{lie}(X)]$$

is called the *holonomy ring* of X; it is the $\mathbb{Z}$-span of Lie monomials on X of degree ≥ 2.

More generally one can consider the *central series* of $\mathfrak{lie}(X)$:

$$\mathfrak{lie}(X) = \mathfrak{c}^0(\mathfrak{lie}(X)) \supset \mathfrak{c}^1(\mathfrak{lie}(X)) \supset \mathfrak{c}^2(\mathfrak{lie}(X)) \supset ...,$$

$$\mathfrak{c}^s(\mathfrak{lie}(X)) := [\mathfrak{lie}(X), \mathfrak{c}^{s-1}(\mathfrak{lie}(X))], \quad s \geq 1;$$

clearly $\mathfrak{c}^{s-1}(\mathfrak{lie}(X))$ is spanned over $\mathbb{Z}$ by the Lie monomials on X of degree $\geq s$.

Now if $\alpha = \sum \alpha_i d\xi_i \in \Lambda^1(L)$ one can consider the set

$$X_\alpha = \{\alpha_i;\ i \geq 1\} \subset L$$

and make the following:

DEFINITION 1.9. For any $s \geq 2$, the s-*curvature* of α is the family $(\varphi_{i_1i_2...i_s})$ of Lie monomials on X_α,

$$\varphi_{i_1i_2...i_s} = [\alpha_{i_1}, [\alpha_{i_2}, ...[\alpha_{i_{s-1}}, \alpha_{i_s}]...]] \in L.$$

Of course, the 2-curvature (φ_{ij}) and the curvature β of α are related by

$$\beta = \sum_{i<j} \varphi_{ij} \cdot d\xi_i \wedge d\xi_j. \tag{1.18}$$

By the above discussion any commutator

$$[\varphi_{i_1 i_2 \dots i_s}, \varphi_{j_1 j_2 \dots j_t}]$$

is a $\mathbb{Z}$-linear combination of the components of the $s+t$-curvature

$$(\varphi_{k_1 k_2 \dots k_{s+t}}).$$

DEFINITION 1.10. The Lie rings

$$\mathfrak{lie}(\alpha) := \mathfrak{lie}(X_\alpha) \quad \text{and} \quad \mathfrak{hol}(\alpha) := \mathfrak{hol}(X_\alpha) = [\mathfrak{lie}(\alpha), \mathfrak{lie}(\alpha)]$$

are called the *Lie hull* and *holonomy ring* of α, respectively.

The use of the term *holonomy* is meant to evoke the analogy with classical holonomy, via the results of Ambrose-Singer and Nijenhuis; cf. [**89**], Chapter III, section 9. Consider also the following:

DEFINITION 1.11. α has a *horizontal* curvature if one of the following equivalent conditions holds:

(1) $\mathfrak{hol}$ is spanned as a $\mathbb{Z}$-module by the components of the 2-curvature.
(2) all the components of the 3-curvature are $\mathbb{Z}$-linear combinations of the components of the 2-curvature.

As it is the above definition is too naive for the applications we have in mind; but two variants of this definition (one in classical differential geometry and another in arithmetic differential geometry) will play a role later. The variant of this definition that plays a role in classical differential geometry will be related to the concept of *locally symmetric space* and hence to connections with *parallel* curvature.

1.3.3. Chern forms. Assume now that we are given a ring C and a finitely generated free C-module M; we then let $L = \mathrm{End}(M) := \mathrm{End}_{\mathbb{Z}}(M)$ be the Lie ring of $\mathbb{Z}$-module endomorphisms with the usual commutator.

To any form $\alpha \in \Lambda^1(L)$ and any C-basis ω of M we will attach *Chern forms* $c_k[\alpha, \omega] \in \Lambda^{2k}(C)$ as follows. View $\omega = (\omega_1, \dots, \omega_d)^t$ as a column vector with entries $\omega_i \in M$. We have an additive group homomorphism

$$L = \mathrm{End}_{\mathbb{Z}}(M) \to \mathfrak{gl}_d(C), \quad \alpha \mapsto (\alpha)_\omega = (\alpha), \tag{1.19}$$

where $\alpha(\omega) = (\alpha) \cdot \omega$. Consider now the induced additive group homomorphism

$$\Lambda(L) \to \Lambda(\mathfrak{gl}_d(C)) \simeq \mathfrak{gl}_d(\Lambda(C)). \tag{1.20}$$

Then consider the curvature $\beta := \frac{1}{2}[\alpha, \alpha] \in \Lambda^2(L)$ of α, and its image

$$(\beta) \in \mathfrak{gl}_d(\Lambda^2(C)) \subset \mathfrak{gl}_d(\Lambda^{even}(C)). \tag{1.21}$$

Since $\Lambda^{even}(C)$ has a natural structure of (commutative) ring one can consider the characteristic polynomial

$$\det(t \cdot 1_d - (\beta)) = \sum_{k=0}^{d} (-1)^k c_k t^{d-k}, \quad c_k = c_k[\alpha, \omega] \in \Lambda^{2k}(C). \tag{1.22}$$

DEFINITION 1.12. $c_k[\alpha, \omega]$ is called k-th *Chern form* attached to α and ω.

If ω and ω' are two C-bases of M that generate the same $\mathbb{Z}$-submodule of M then $c_k[\alpha, \omega] = c_k[\alpha, \omega']$.

Note that if $C = \mathbb{Z}$ the group homomorphism 1.19 is a Lie ring isomomorphism. However, if C is not $\mathbb{Z}$, the group homomorphism 1.19 is not generally a Lie ring homomorphism and is not generally bijective. Nevertheless we will be interested later in elements of L that act in special ways on M, e.g., as "connections" 1.24 or, alternatively, "semi-linearly" 1.29; on such elements of L the additive map 1.19, although still not generally a Lie ring homomorphism, will exhibit various "compatibilites" with respect to brackets; also the images of such elements via 1.19 will "remember enough information" about the elements themselves.

1.3.4. "Complex" case. The following refinement of the above situation may be considered. Assume there is a map $u \to \bar{u}$ on $\mathcal{U}$ which is *involutive* (i.e. $\bar{\bar{u}} = u$ for all u) and without fixed points (i.e. $\bar{u} \neq u$ for all u) and select a subset $\mathcal{Z} \subset \mathcal{U}$ such that

$$\mathcal{U} = \mathcal{Z} \cup \overline{\mathcal{Z}}. \tag{1.23}$$

Fix an ordering

$$\mathcal{Z} = \{z_1, z_2, \ldots\};$$

this ordering is not assumed compatible with the ordering on $\mathcal{U}$ and the concepts below will be independent of this ordering of $\mathcal{Z}$. Then an element $\eta \in \Lambda^k(E)$ is called of type (i, j) if it can be written as

$$\eta = \sum_{|I|+|J|=k} \eta_{IJ} \cdot dz_I \wedge d\bar{z}_J$$

with $\eta_{IJ} = 0$ whenever $(|I|, |J|) \neq (i, j)$. We denote by $\Lambda^{i,j}(E)$ the subgroup of $\Lambda^k(E)$ of elements of type (i, j); we have that

$$\Lambda^k(E) = \bigoplus_{i+j=k} \Lambda^{i,j}(E),$$

and for any $\eta \in \Lambda^k(E)$ we have a *type decomposition*

$$\eta = \sum \eta^{(i,j)}, \quad \eta^{(i,j)} \in \Lambda^{i,j}(E).$$

The bracket $[\ ,\ ]$ on $\Lambda(E)$ is compatible in the obvious way with this type decomposition.

1.3.5. Case of derivations. In classical differential geometry one usually specializes the discussion in subsection 1.3.3 to the following situation. One considers a ring C equipped with derivations

$$\delta_i = \delta_i^C : C \to C, \quad 1 \leq i \leq m,$$

and one considers a finitely generated free C-module M equipped with $\mathbb{Z}$-module homomorphisms $\delta_i^M : M \to M$ satisfying the following *connection-like* property:

$$\delta_i^M(cm) = \delta_i(c)m + c\delta_i^M(m), \quad c \in C, \quad m \in M. \tag{1.24}$$

View $\delta_i^M \in L := \mathrm{End}(M)$. We have an embedding $C \to L$ sending any $c \in C$ into the multiplication by c which will still be denoted by c. Then we have

$$[\delta_i^M, c] = \delta_i c \tag{1.25}$$

in L. Define

$$\delta := \sum_i \delta_i^M \cdot d\xi_i \in \Lambda(L) \tag{1.26}$$

and consider the embedding $\Lambda(C) \to \Lambda(L)$ induced by the embedding $C \to L$. One has a commutative diagram

$$\begin{array}{ccc} \Lambda(C) & \stackrel{d^\delta}{\longrightarrow} & \Lambda(C) \\ \downarrow & & \downarrow \\ \Lambda(L) & \stackrel{[\delta,\ ]}{\longrightarrow} & \Lambda(L) \end{array} \tag{1.27}$$

where $d^\delta = d$ is given by the familiar "exterior differentiation" formula,

$$d(\sum_I c_I d\xi_I) = \sum_I (\sum_i \delta_i c_I \cdot d\xi_i) \wedge d\xi_I. \tag{1.28}$$

Of course, $d \circ d = 0$. In this way one is able to "embed" the usual deRham formalism into a formalism involving rings of operators. By the way replacing derivations on rings of functions by inner derivations of rings of operators is a standard method of quantization; cf., for instance, [**53**], Introduction.

1.3.6. Case of ring endomorphisms. In our arithmetic differential geometry we will be led to consider the formalism of subsection 1.3.3 in the following setting. Consider a ring C equipped with ring endomorphisms $\phi_i : C \to C$, and a finitely generated free C-module M equipped with $\mathbb{Z}$-module homomorphisms $\phi_i^M : M \to M$ that act *semi-linearly* in the sense that

$$\phi_i^M(cm) = \phi_i(c) \cdot \phi_i^M(m), \quad c \in C, \quad m \in M. \tag{1.29}$$

Again we view $\phi_i^M \in L = \text{End}(M)$ and we have an embedding $C \to L$ sending any $c \in C$ into the multiplication by c which will still be denoted by c. Then we have

$$[\phi_i^M, c] = (\phi_i(c) - c) \cdot \phi_i^M \tag{1.30}$$

in L. Typically we will have that $\xi_i = p_i$ for primes $p_i \in \mathbb{Z}$, ϕ_i will reduce mod p_i to the p_i-power Frobenius on C/p_iC, and there will be a fixed integer ν such that

$$\phi_i^M(M) \subset p_i^\nu M$$

for all i; also C will be p_i-torsion free. We will then be led to consider the following analogue of 1.26:

$$\frac{\phi^M}{\mathbf{p}^\nu} := \sum_i \frac{\phi_i^M}{p_i^\nu} \cdot dp_i \in \Lambda(L) \tag{1.31}$$

However there will be no analogue of exterior differentiation and hence no analogue of the diagram 1.27. So in some sense the situation in our arithmetic theory will be "quantized" from the beginning.

Intuitively M equipped with the structure above will be an object over $\mathbb{Z}$ which "looks like a crystal of slope ν" for each p_i. If, in addition, p_i runs through all the primes and all ϕ_i^M commute, M becomes a λ-module in the sense of [**74**]; however, the whole point in our curvature formalism will be, of course, that the ϕ_i^M are *not* commuting in general.

CHAPTER 2

Classical differential geometry revisited

The aim of this chapter is to revisit some basic concepts of classical differential geometry [**89**] from an algebraic standpoint. In principle this chapter may be skipped and later consulted as needed. Although the present chapter is not logically necessary for understanding the following chapters, the material here contains our key motivation/blueprint for the arithmetic theory to be developed in our book and, possibly, in subsequent research.

We will work over arbitrary rings equipped with derivations in the style of differential algebra [**91**]. But note that our framework in this chapter *does not always reduce* to the framework of classical differential geometry in [**89**] even if one takes our ground ring to be ring of smooth functions on a manifold. Rather, our framework here is intermediate between that of [**89**] and the arithmetic framework to be introduced in Chapter 3. Nevertheless we will still refer to our framework here as *classical differential geometry*. By the way, since we place ourselves in a differential algebra setting, our framework here will have the flavor of local (rather than global) differential geometry.

We will start by introducing connections and curvature, as well as some basic concepts related to classical groups and their symmetric spaces. We will then proceed to a discussion of some special types of classical connections: Ehresmann, Chern, Levi-Cività, Fedosov, Lax, and Hamiltonian, whose arithmetic analogues will be our main concern in this book. We will end with a quick discussion of Cartan connections (i.e., connections on infinite jet spaces) and a discussion of differential groups. In particular we will review Cassidy's fundamental theorem on differential subgroups of simple groups [**52**] and Manin's existence of differential characters of elliptic curves [**99**]. The latter are directly related to the Painlevé VI equations which have a Hamiltonian structure; cf. [**101**].

We would like to stress the fact that the present chapter focuses on revisiting the main *concepts*, and *not* the main *theorems*, of classical differential geometry. Most of the theorems we mention (and sometimes prove) are derived via straightforward computations using the definitions. This is all we need in this book; for our main concern in later chapters will be to discover the "right" arithmetic analogues of the differential geometric concepts rather than the arithmetic analogues of the classical differential geometric theorems and proofs. The theorems we will encounter in the arithmetic case, and their proofs, will turn out to have their own, predominantly arithmetic, flavor.

2.1. Connections in principal bundles and curvature

In this section we review the classical picture of connections and curvature, including holonomy and first Chern forms.

2.1.1. Connections in principal bundles.

We start by recalling:

DEFINITION 2.1. [**91**] A *δ-ring* is a ring A equipped with commuting derivations,

$$(2.1) \qquad \delta_i^A : A \to A, \quad i = 1, ..., m.$$

A *δ-subring* of a δ-ring A is a subring A' of A such that $\delta_i^A(A') \subset A'$. Any such A' has a naturally induced structure of δ-ring given by $\delta_i^{A'} = (\delta_i^A)_{|A'}$. A *$\delta$-ring homomorphism* is a ring homomorphism $u : A \to R$ between two δ-rings such that $u \circ \delta_i^A = \delta_i^R \circ u$.

We will typically write δ_i in place of δ_i^A if A is understood from context.

EXAMPLE 2.2. The basic example we have in mind is as follows. Let $M \subset \mathbb{R}^m$ be an open set and let $C^\infty(M, \mathbb{R})$ be the ring of smooth real valued functions on M. Let $\xi_1, ..., \xi_m$ be the real coordinates on M. Then one may view $C^\infty(M, \mathbb{R})$ as a δ-ring with derivations $\delta_i = \frac{\partial}{\partial \xi_i}$ and one may consider any δ-subring $A \subset C^\infty(M, \mathbb{R})$.

EXAMPLE 2.3. Another basic example is as follows. Let $M \subset \mathbb{C}^m$ be an open set and let $C^\infty(M, \mathbb{C})$ be the ring of smooth complex valued functions on M. Let $z_1, ..., z_m$ be the complex coordinates on M. Then one may view $C^\infty(M, \mathbb{C})$ as a δ-ring with derivations $\delta_i = \frac{\partial}{\partial z_i}$ and one may consider any δ-subring $A \subset C^\infty(M, \mathbb{C})$.

EXAMPLE 2.4. The derivations in 2.1 need not be "independent". Here is a natural example of this situation. One can consider $M = \mathbb{R}^m$, one can view $C^\infty(M, \mathbb{R})$ as a δ-ring with derivations $\delta_i = \frac{\partial}{\partial \xi_i}$, and one can consider the δ-subring

$$A \subset C^\infty(M, \mathbb{R})$$

consisting of all functions of the form

$$f(\xi_1, ..., \xi_m) = F(\xi_1 + ... + \xi_m)$$

where $F \in C^\infty(\mathbb{R}, \mathbb{R})$. Then A consists exactly of those functions $f \in C^\infty(M, \mathbb{R})$ for which

$$\delta_i f = \delta_j f, \quad i, j = 1, ..., m.$$

In particular the derivations $\delta_i^A : A \to A$ coincide. This paradigm is the natural framework for *cohomogeneity one* metrics to be discussed later; cf. Example 2.53. In spite of the fact that the elements of the ring A "depend on one parameter only" (rather than on m parameters) it turns out, as we shall see, that metrics with coefficients in this ring retain certain m-dimensional features that make them interesting, e.g., from the viewpoint of curvature.

EXAMPLE 2.5. The basic example in *differential algebra* [**112, 91, 92, 51, 52**] is that in which A is a field of characteristic zero equipped with m commuting derivations. Explicit fields of this sort, appearing in complex analysis, are δ-subfields

$$A \subset \text{Mer}(D)$$

in the field $\text{Mer}(D)$ of meromorphic functions on a domain $D \subset \mathbb{C}^m$, where, if $z_1, ..., z_n$ are the complex coordinates, then $\text{Mer}(D)$ is viewed as δ-field with the derivations $\delta_i = \frac{\partial}{\partial z_i}$.

Returning to the setting of an abstract δ-ring A as in 2.1 we have:

DEFINITION 2.6. Let B be an (associative commutative unital) A-algebra. A *connection* on B (equivalently, on the scheme $X := Spec\ B$) is an m-tuple $\delta^B = (\delta_i^B)$ where for each i,

$$\delta_i^B : B \to B \tag{2.2}$$

is a derivation extending the corresponding derivation δ_i^A on A. The connection δ^B is *flat* if the derivations δ_i^B commute on B.

Again, we will tend to write δ, δ_i instead of δ^B, δ_i^B if B is understood; sometimes we will write δ_i^X instead of δ_i^B if $X = Spec\ B$.

So to give a flat connection on B is the same as giving a δ-ring structure on B such that $A \to B$ is a δ-ring homomorphism.

REMARK 2.7. In the notation of Definition 2.6 we may consider the Lie ring $\text{End}(B)$ of $\mathbb{Z}$-module endomorphisms of B and we may consider the smallest Lie subring $\mathfrak{lie}(\delta^B)$ of $\text{End}(B)$ containing all the δ_i^B's. Similarly we may consider the smallest Lie subring $\mathfrak{lie}(\delta^A)$ of $\text{End}(A)$ containing all δ_i^A's; then $\mathfrak{lie}(\delta^A)$ is the $\mathbb{Z}$-span of the δ_i^A's and is a Lie ring with zero bracket. We have at our disposal the restriction map (or *anchor map*)

$$\chi : \mathfrak{lie}(\delta^B) \to \mathfrak{lie}(\delta^A). \tag{2.3}$$

The map χ is a map of Lie rings. If, in addition $\delta_1^A, ..., \delta_m^A$ are $\mathbb{Z}$-linearly independent in $\text{End}(A)$ then χ has a section in the category of $\mathbb{Z}$-modules given by $\delta_i^A \mapsto \delta_i^B$ and the surjective map χ in 2.3 has kernel the commutator

$$\mathfrak{hol}(\delta^B) = [\mathfrak{lie}(\delta^B), \mathfrak{lie}(\delta^B)]. \tag{2.4}$$

We are mainly interested in the following special case. Set

$$B := A[x, \det(x)^{-1}] \tag{2.5}$$

and denote by

$$G = GL_n = Spec\ B \tag{2.6}$$

the *general linear group scheme* over A, where $x = (x_{kl})$ is an $n \times n$ matrix of indeterminates. Here the size n of our matrix x and the size m of the set of derivations in δ are unrelated.

So $G(A) = GL_n(A)$ is the group of all invertible $n \times n$ matrices with entries in A and $\mathcal{O}(G) = A[x, \det(x)^{-1}]$. Consider the scheme

$$\mathfrak{g} = \mathfrak{gl}_n = Spec\ A[x] \tag{2.7}$$

so $\mathfrak{g}(A) = \mathfrak{gl}_n(A)$ is the set of all $n \times n$ matrices with entries in A; we view this set both as associative algebra and as Lie algebra over A. By the above a *connection* on G is an m-tuple $\delta = (\delta_i)$, where

$$\delta_i : \mathcal{O}(G) \to \mathcal{O}(G) \tag{2.8}$$

are derivations, extending the corresponding derivations on A. The set of connections on G is in a natural bijection with the set of m-tuples (Δ_i) where Δ_i is an $n \times n$ matrix with entries in $A[x, \det(x)^{-1}]$; the bijection is given by $\delta_i x = \Delta_i$.

DEFINITION 2.8. A connection $\delta = (\delta_i)$ on G is called *right linear* if for all i,

$$\delta_i x := \Delta_i = A_i x$$

for some $A_i \in \mathfrak{g}(A)$.

DEFINITION 2.9. The *trivial connection* $\delta_0 = (\delta_{0i})$ on G is the connection defined by $\delta_{0i}x = 0$.

DEFINITION 2.10. A connection $\delta = (\delta_i)$ is called *right invariant* if the following diagrams are commutative for all i:

$$\begin{array}{ccc} \mathcal{O}(G) & \xrightarrow{\mu} & \mathcal{O}(G) \otimes \mathcal{O}(G) \\ \delta_i \downarrow & & \downarrow \delta_i \otimes 1 + 1 \otimes \delta_{0i} \\ \mathcal{O}(G) & \xrightarrow{\mu} & \mathcal{O}(G) \otimes \mathcal{O}(G) \end{array} \tag{2.9}$$

Here μ is induced by multiplication.
It is trivial to check that:

PROPOSITION 2.11. *A connection is right linear if and only if it is right invariant.*

DEFINITION 2.12. A right linear (equivalently right invariant) connection will be called an *Ehresmann connection*. We denote by $\Lambda^1(\mathfrak{g})$ the set of all Ehresmann connections on G.

REMARK 2.13. The above definitions correspond to the case of connections in trivial principal bundles in classical differential geometry. The case of not necessarily trivial principal bundles corresponds to the following situation (which we will not pursue in this book). Consider a fixed projective A-module V of rank n, consider the scheme $\mathfrak{p}$ representing the functor attaching to any A-algebra A' the set of n-tuples

$$\{(v_1, ..., v_n); v_i \in V \otimes_A A'\}, \tag{2.10}$$

and consider the open subscheme $P \subset \mathfrak{p}$ representing the functor that attaches to any A' the subset of 2.10 of all tuples which are bases of the A'-module $V \otimes_A A'$. One calls P the *frame bundle* of V. Clearly P has a structure

$$P \times_A G \to P$$

of principal bundle over *Spec* A. If in addition V is a free A-module then the inclusion

$$P \subset \mathfrak{p}$$

is (non-canonically) isomorphic to

$$G \subset \mathfrak{g}.$$

REMARK 2.14. The term *linear connection* is usually (cf. [**89**]) but not always (cf. [**6**], p. 22) reserved to the case of connections on the tangent bundle (or frame bundle of the tangent bundle); this corresponds to the case $n = m$. Here we follow [**6**] and talk about *(right) linear connections* in case n and m are unrelated.

REMARK 2.15. One has a similar concept of *left invariant* connection; then a connection is left invariant if and only if it is *left linear* in the sense that

$$\Delta_i = xA_i, \quad A_i \in \mathfrak{g}(A).$$

Left linear/invariant connections will play no role in what follows.

REMARK 2.16. Note the following property of Ehresmann connections; this property is trivially satisfied here but later, in the arithmetic context, an analogue of this property will be satisfied, in various situations, in a less trivial way. The property of Ehresmann connections referred to is that the matrices $\delta_i x$ are obtained from the matrices A_i and x via "size independent" *matrix* operations; in this case the only operation needed is matrix multiplication. This is in the spirit of *free non-commutative function theory* [**85**] and is stronger than saying that the entries of $\delta_i x$ are obtained from the entries of A_i and x via *scalar* operations on these entries.

2.1.2. Horizontal sections, prime integrals. Let $\delta = (\delta_i)$ be a connection on $G = GL_n$,

$$\delta_i x = \Delta_i, \quad \Delta_i = \Delta_i(x) \in \mathfrak{g}(B), \quad B = A[x, \det(x)^{-1}].$$

Let $A \to R$ be a homomorphism of δ-rings; i.e. R be an A-algebra equipped with commuting derivations $\delta_i = \delta_i^R : R \to R$ compatible with the corresponding derivations 2.1 on A.

DEFINITION 2.17. A matrix $u \in GL_n(R)$ is called *horizontal* (or a *horizontal section*) for the connection δ if, for all i,

$$\delta_i u = \Delta_i(u). \tag{2.11}$$

We also say that, for u an *unknown*, 2.11 is the *differential equation* attached to the connection (δ_i); if u satisfies 2.11 we say u is a *solution* to this differential equation.

The solutions to 2.11 in $GL_n(R)$ are in bijection with the δ-ring homomorphisms $B \to R$ over A; the bijection attaches to any solution u of 2.11 the homomorphism $x \mapsto u$.

DEFINITION 2.18. A *prime integral* for a connection δ on G is an element $H \in \mathcal{O}(G)$ such that $\delta_i H = 0$ for all i.

The following is trivial to check:

PROPOSITION 2.19. *Assume δ is flat. A function $H \in \mathcal{O}(G) = A[x, \det(x)^{-1}]$ is a prime integral for δ if and only if for any R and any solution $u \in GL_n(R)$ of 2.11 we have*

$$\delta_i(H(u)) = 0, \tag{2.12}$$

for all i.

In other words, for flat connections (in particular if the number of derivations is $m = 1$) a function H is a prime integral if and only if it is "constant" along all solutions of 2.11.

If δ is an Ehresmann connection, $\delta_i x = A_i x$, then the equations 2.11 read

$$\delta_i u = A_i u, \tag{2.13}$$

which is what is classically called a system of *linear differential equations*. The analogue of linear equations in arithmetic will not be obvious.

2.1.3. Dual connection. Let $\delta = (\delta_i)$ be a connection on $G = GL_n$, and set

$$A_i = A_i(x) = \Delta_i \cdot x^{-1} = \delta_i x \cdot x^{-1}.$$

So A_i have entries in $\mathcal{O}(G)$ and

$$\delta_i x = A_i x.$$

(Note that δ is Ehresmann if and only if A_i have entries in A.) Given a connection δ as above one can define a connection $\delta^* = (\delta_i^*)$ by the formula

$$\delta_i^* x = A_i^* x, \quad A_i^* := -A_i^t. \tag{2.14}$$

We let

$$A_i = (A_{ijk}), \quad A_{ijk} = (j,k) \text{ entry of } A_i. \tag{2.15}$$

We will then usually write

$$\Gamma_{ij}^k := -A_{ikj} = (j,k) \text{ entry of } A_i^*. \tag{2.16}$$

DEFINITION 2.20. The connection δ^* is called the *dual* of δ. The elements Γ_{ij}^k are called the *Christoffel symbols of the* 2nd *kind* of δ.

We will later also consider *Christoffel symbols of the* 1st *kind.*

REMARK 2.21. Clearly δ^* is Ehresmann if and only if δ is Ehresmann.

REMARK 2.22. Assume in what follows that δ is Ehresmann. Let

$$M \subset B = A[x, \det(x)^{-1}]$$

be the A-submodule generated by the entries of the matrix x,

$$M = \{\ell_\alpha := \mathrm{tr}(\alpha x) \in B;\ \alpha \in \mathfrak{gl}_n(A)\},$$

and let

$$\langle\ ,\ \rangle : M \times M \to A \tag{2.17}$$

be the pairing defined by

$$\langle \ell_\alpha, \ell_\beta \rangle := \mathrm{tr}(\alpha\beta); \tag{2.18}$$

hence,

$$\langle x_{ij}, x_{kl} \rangle = \delta_{ik}\delta_{jl}, \tag{2.19}$$

where δ_{ij} is the Kronecker symbol. Then the dual connection $\delta^* = (\delta_i^*)$ is the unique Ehresmann connection on G, such that

$$\delta_i \langle \ell_\alpha, \ell_\beta \rangle = \langle \delta_i \ell_\alpha, \ell_\beta \rangle + \langle \ell_\alpha, \delta_i^* \ell_\beta \rangle. \tag{2.20}$$

2.1.4. Curvature and 3-curvature.

DEFINITION 2.23. For a connection $\delta = (\delta_i)_{1 \le i \le m}$ on $G = GL_n$ the *curvature* of δ is the $m \times m$ matrix (φ_{ij}) where φ_{ij} are the A-derivations

$$\varphi_{ij} := [\delta_i, \delta_j] := \delta_i\delta_j - \delta_j\delta_i : \mathcal{O}(G) \to \mathcal{O}(G). \tag{2.21}$$

So δ is flat if and only if the curvature vanishes: $\varphi_{ij} = 0$ for all i, j.

Set $\Phi_{ij} := \varphi_{ij}(x)$; so Φ_{ij} are $n \times n$ matrices with entries in $A[x, \det(x)^{-1}]$.

If δ is Ehresmann, with $\Delta_i = \delta_i x = A_i x$, $A_i \in \mathfrak{g}(A)$, then

$$\Phi_{ij} = F_{ij}x, \quad F_{ij} := \delta_i A_j - \delta_j A_i - [A_i, A_j] \in \mathfrak{g}(A), \tag{2.22}$$

where $[A_i, A_j] = A_i A_j - A_j A_i$ is the usual commutator of matrices. The matrix

$$F = (F_{ij}) = (F_{ij}[\delta]) = F[\delta]$$

is also referred to, in this case, as the *curvature* of the Ehresmann connection $\delta = (\delta_i)$; clearly

$$F_{ij} = -F_{ji}.$$

So δ is flat if and only if its curvature vanishes, $F = 0$. We let

$$F_{ij} = (F_{ijkl}), \tag{2.23}$$

i.e., F_{ijkl} is the (k, l) entry of F_{ij}.

If (F^*_{ij}) is the curvature of the dual connection δ^* of δ then we have

$$F^*_{ij} = -(F_{ij})^t. \tag{2.24}$$

DEFINITION 2.24. If δ is an Ehresmann connection we define the *Riemann tensor* of δ as the collection (R^k_{lij}) of elements of A,

$$\begin{aligned} R^k_{lij} := -F_{ijkl} &= (k,l) \text{ entry of } (-F_{ij}) \\ &= (l,k) \text{ entry of } F^*_{ij}. \end{aligned}$$

Classically the term *Riemann tensor* is reserved to the case when $n = m$ (and sometimes only for the Levi-Cività connection); here we will use this term in the most general case when n and m are arbitrary.

In terms of the Christoffel symbols the Riemann tensor is given by

$$R^k_{lij} = \delta_i \Gamma^k_{jl} - \delta_j \Gamma^k_{il} + \Gamma^s_{jl}\Gamma^k_{is} - \Gamma^s_{il}\Gamma^k_{js}. \tag{2.25}$$

In particular

$$R^k_{lij} = -R^k_{lji}. \tag{2.26}$$

Note the following trivially checked link between curvature and "integrability."

PROPOSITION 2.25. *Assume $\delta = (\delta_i)$ is an Ehresmann connection $\delta_i x = A_i x$ with curvature (F_{ij}). The following are equivalent:*

(1) *There is an injective homomorphism of δ-rings $A \to R$ and a horizontal section $u \in GL_n(R)$ for δ, i.e., a solution $u \in GL_n(R)$ of the differential equation*

$$\delta_i u = A_i u.$$

(2) *δ is flat, i.e.,*

$$F_{ij} = 0.$$

DEFINITION 2.26. For a connection $\delta = (\delta_i)$ the 3-*curvature* of δ is the collection (φ_{ijk}) of A-derivations

$$\varphi_{ijk} := [\delta_i, [\delta_j, \delta_k]] = [\delta_i, \varphi_{jk}] : \mathcal{O}(G) \to \mathcal{O}(G). \tag{2.27}$$

REMARK 2.27.

(1) The 3-curvature φ_{ijk} is anti-symmetric in jk, i.e.

$$\varphi_{ijk}+\varphi_{ikj}=0$$

and satisfies the *Bianchi identity*:

$$\sum_{(ijk)}\varphi_{ijk}=0$$

where (ijk) under $\sum$ means summation over the cyclic permutations of i,j,k.

(2) If δ is Ehresmann, with $\delta_i x = A_i x$, we have

$$\varphi_{ijk}(x)=(\delta_i F_{jk}-[A_i,F_{jk}])x \tag{2.28}$$

and the Bianchi identity reads

$$\sum_{(ijk)}(\delta_i F_{jk}-[A_i,F_{jk}])=0,\quad i,j,k=1,...,m \tag{2.29}$$

in $\mathfrak{g}(A)$.

(3) If δ is Ehresmann, with with $\delta_i x = A_i x$, and $g=(g_{ij})\in GL_n(A)$ is a symmetric matrix with inverse (g^{ij}) one says that the *Yang-Mills equation* are satisfied if (with Einstein summation convention) we have the following equalities in $\text{End}(\mathcal{O}(G))$:

$$g^{ij}\varphi_{ijk}=0,\quad k=1,...,m; \tag{2.30}$$

equivalently, if we have the following equalities in $\mathfrak{g}(A)$:

$$Y_k:=g^{ij}(\delta_i F_{jk}-[A_i,F_{jk}])=0,\quad k=1,...,m. \tag{2.31}$$

Cf. [**54**].

DEFINITION 2.28. The family (Y_k) in 2.31 is the *Yang-Mills form* of δ.

The 3-curvature, as well as the Yang-Mills form, will have an arithmetic analogue.

Note that if, in the discussion above, the A_is belong to the set of A-points $\mathfrak{h}(A)$ of the Lie algebra $\mathfrak{h}$ of a closed subgroup H of G defined by polynomials with coefficients annihilated by the δ_is then we also have

$$F_{ij},Y_k\in\mathfrak{h}(A). \tag{2.32}$$

2.1.5. s-curvature and holonomy. 3-curvature can be generalized as follows.

DEFINITION 2.29. For a connection $\delta=(\delta_i)$ and $s\geq 2$ the *s-curvature* of δ is the collection $(\varphi_{i_1 i_2 \ldots i_s})$ of A-derivations

$$\varphi_{i_1 i_2\ldots i_s}:=[\delta_{i_1},[\delta_{i_2},...[\delta_{i_{s-1}},\delta_{i_s}]...]]:\mathcal{O}(G)\to\mathcal{O}(G). \tag{2.33}$$

The $\mathbb{Z}$-span $\mathfrak{hol}=\mathfrak{hol}(\delta)$ in $L:=\text{End}(\mathcal{O}(G))$ of all the components of all s-curvatures for all $s\geq 2$ is called the *holonomy ring* of δ. Moreover the A-linear span $\mathfrak{hol}_A$ of $\mathfrak{hol}$ in L is called the *holonomy A-algebra* of δ.

REMARK 2.30. So 2-curvature is what we earlier called *curvature*. Also recall that by general properties of Lie rings the commutator

$$[\varphi_{i_1 i_2 \dots i_s}, \varphi_{j_1 j_2 \dots j_t}]$$

of any two components of the s-curvature and t-curvature respectively is a $\mathbb{Z}$-linear combination of the components

$$\varphi_{k_1 k_2 \dots k_{s+t}}$$

of the $s+t$-curvature. So $\mathfrak{hol}$ is indeed a Lie subring of L. Note also that $\mathfrak{hol}_A$ is indeed a Lie A-subalgebra of L and, in fact, even a Lie A-subalgebra of the Lie A-algebra $Der_A(B)$ of A-derivations on $B = \mathcal{O}(G)$.

REMARK 2.31. Assume δ is Ehresmann on $G = GL_n$ and $B = \mathcal{O}(G) = A[x \det(x)^{-1}]$. Note that $\mathfrak{hol}$ is then a Lie A-subalgebra of the Lie A-algebra

$$Der_A(B)^0 := \{D \in Der_A(B) \ ; \ D(x) \cdot x^{-1} \in \mathfrak{g}(A)\}.$$

The map

$$Der_A(B)^0 \to \mathfrak{g}(A), \quad D \mapsto D(x) \cdot x^{-1}$$

is an injective Lie A-algebra homomorphism so it defines a Lie A-algebra embedding of $\mathfrak{hol}_A$ into $\mathfrak{g}(A)$; we can view this embedding as an inclusion so we have:

$$\mathfrak{hol} \subset \mathfrak{hol}_A \subset \mathfrak{g}(A), \tag{2.34}$$

which is the classical way of looking at the holonomy algebra.

The significance of 3-curvature comes, in particular, from the following:

LEMMA 2.32. *In the notation of Remark 2.31 the following are equivalent:*

(1) $\mathfrak{hol}_A$ *is the A-linear span of the components of the curvature* (φ_{ij}).
(2) *All the components of the 3-curvature* (φ_{ijk}) *are A-linear combinations of the components of the curvature* (φ_{ij}).

PROOF. All we need is to prove, by induction on $n \geq 3$, that if assertion (2) holds, then the components $\varphi_{i_1 \dots i_n}$ of the n-curvature are A-linear combinations of the components φ_{ij} of the curvature. So assume

$$\varphi_{i_1 \dots i_n} = \sum_{kl} a_{kl} \varphi_{kl}, \qquad a_{kl} \in A.$$

We get

$$\begin{aligned} \varphi_{i i_1 \dots i_n} &= [\delta_i, \varphi_{i_1 \dots i_n}] \\ &= [\delta_i, \textstyle\sum_{kl} a_{kl} \varphi_{kl}] \\ &= \textstyle\sum_{kl} (\delta_i a_{kl}) \varphi_{kl} + \sum_{kl} a_{kl} \varphi_{ikl}. \end{aligned}$$

But, by assertion (2), each φ_{ikl} is an A-linear combination of φ_{kl}'s and we are done. □

DEFINITION 2.33. An Ehresmann connection δ on GL_n is said to have an *A-horizontal* curvature if it satisfies the equivalent conditions in Lemma 2.32.

As we shall see Ehremann connections with A-horizontal curvature will turn out to be related to locally symmetric connections (in particular to connections with *parallel* curvature).

2.1.6. First Chern form. Let $\mathrm{tr} : \mathfrak{g}(A) \to A$ be the trace. Assume furthermore that $n \in A^{\times}$.

DEFINITION 2.34. For an Ehresmann connection $\delta = (\delta_i)$ with curvature (F_{ij}), the *first Chern form* of δ is the matrix

$$\rho = (\rho_{ij}) = (\rho_{ij}[\delta]) = \rho[\delta]$$

where

$$\rho_{ij} = \mathrm{tr}(F_{ij}) = \delta_i(\mathrm{tr}(A_j)) - \delta_j(\mathrm{tr}(A_i)) \in A. \tag{2.35}$$

Morally (ρ_{ij}) "defines a deRham exact form," in particular it "defines a deRham closed form," in the sense that

$$\sum_{(ijk)} \delta_k \rho_{ij} = 0. \tag{2.36}$$

Note that the first Chern form is given, in terms of the Riemann tensor, by

$$\rho_{ij} = R^k_{kji}. \tag{2.37}$$

DEFINITION 2.35. The *traceless curvature* of δ is the collection (W_{ij}) with entries the matrices

$$W_{ij} := F_{ij} - \frac{1}{n} \cdot \rho_{ij} \cdot 1_n. \tag{2.38}$$

Of course, $\mathrm{tr}(W_{ij}) = 0$.

The formalism of curvature and first Chern form fits into our superalgebra formalism as follows. We will place ourselves in the context of subsection 1.3.5. So consider the set of basic variables

$$\mathcal{U} = \{\xi_1, \xi_2, ..., \xi_m\}$$

and for any $\mathbb{Z}$-module E we consider the $\mathbb{Z}$-module of *n-forms*,

$$\Lambda^n(E) = \{ \sum_{i_1 < ... < i_n} \alpha_{i_1 ... i_n} \cdot d\xi_{i_1} \wedge ... \wedge d\xi_{i_n};\ \alpha_{i_1 ... i_n} \in E\}.$$

If E is a Lie ring then $\Lambda(E) = \bigoplus \Lambda^n(E)$ is a graded Lie superalgebra. Cf. 1.12. In particular we may consider the $\mathbb{Z}$-modules of forms

$$\Lambda^n(A), \quad \Lambda^n(\mathrm{End}(\mathcal{O}(G))), \quad \Lambda^n(\mathfrak{g}(A))$$

and the graded Lie superalgebras

$$\Lambda(\mathrm{End}(\mathcal{O}(G))), \quad \Lambda(\mathfrak{g}(A))$$

attached to the $\mathbb{Z}$-module A and to the Lie rings $\mathrm{End}(\mathcal{O}(G))$ and $\mathfrak{g}(A)$, respectively. Recall, on the other hand, that we denoted by $\Lambda^1(\mathfrak{g})$ the set of all Ehresmann connections on G. (In $\Lambda^1(\mathfrak{g})$, the letter $\mathfrak{g}$ stands for a scheme, rather than for a $\mathbb{Z}$-module or a Lie ring, so there should be no confusion between our set of Ehresmann connections $\Lambda^1(\mathfrak{g})$ and the $\mathbb{Z}$-module of 1-forms $\Lambda^1(\mathfrak{g}(A))$.)

Then any Ehresmann connection $\delta = (\delta_i) \in \Lambda^1(\mathfrak{g})$ can be identified with its *connection form*,

$$\delta = \sum_i \delta_i \cdot d\xi_i \in \Lambda^1(\mathrm{End}(\mathcal{O}(G))). \tag{2.39}$$

Similarly one can identify the families

$$\varphi = (\varphi_{ij}), \quad F = (F_{ij}), \quad \rho = (\rho_{ij}), \quad \tau := (\mathrm{tr}(A_i))$$

with forms

$$\varphi = \sum_{i<j} \varphi_{ij} \cdot d\xi_i \wedge d\xi_j \in \Lambda^2(\mathrm{End}(\mathcal{O}(G))), \tag{2.40}$$

$$F = \sum_{i<j} F_{ij} \cdot d\xi_i \wedge d\xi_j \in \Lambda^2(\mathfrak{g}(A)), \tag{2.41}$$

$$\rho = \sum_{i<j} \rho_{ij} \cdot d\xi_i \wedge d\xi_j \in \Lambda^2(A). \tag{2.42}$$

$$\tau = \sum_{i} \mathrm{tr}(A_i) \cdot d\xi_i \in \Lambda^1(A). \tag{2.43}$$

2.36 we have that

$$\varphi = \frac{1}{2}[\delta, \delta], \quad [\delta, \varphi] = 0, \quad d\tau = \rho = \mathrm{tr}(F), \quad d\rho = 0.$$

Also, under the natural embedding $\mathfrak{g}(A) \subset \mathrm{End}(\mathcal{O}(G))$ (sending any matrix γ into the unique A-derivation on $\mathcal{O}(G)$ that sends x into γx) we have $\varphi = F$. Finally note that the two definitions of $\mathfrak{hol}(\delta)$ in Definitions 1.10 and 2.29 agree under the above identifications.

There is an alternative way to look at the first Chern form which will be the one we shall imitate in the arithmetic case. Indeed, for any $n \times n$ matrix $M = (m_{ij})$, let M^{col} be the column $n^2 \times 1$ matrix

$$M^{col} := (m_{11}, m_{21}, ..., m_{n-1,n}, m_{nn})^t. \tag{2.44}$$

Assume now that δ is an Ehresmann connection with curvature (F_{ij}) and consider, as before, the $n \times n$ matrix

$$\Phi_{ij} = F_{ij}x.$$

Then we can consider the $n^2 \times n^2$ Jacobian matrix

$$\frac{\partial \Phi_{ij}}{\partial x} := \left(\left(\frac{\partial}{\partial x} \right)^{col} \cdot (\Phi_{ij}^{col})^t \right)^t$$

and consider the trace

$$\mathrm{tr}\left(\frac{\partial \Phi_{ij}}{\partial x} \right) \in \mathcal{O}(G). \tag{2.45}$$

Note that one has a block representation

$$\frac{\partial \Phi_{ij}}{\partial x} = \mathcal{F}_{ij} := diag(F_{ij}, ..., F_{ij}), \tag{2.46}$$

hence, for the first Chern form 2.35, we have

$$\rho_{ij} = \frac{1}{n} \mathrm{tr}\left(\frac{\partial \Phi_{ij}}{\partial x} \right). \tag{2.47}$$

As we shall see the elements 2.45 will have natural arithmetic analogues.

Note that there is a different way of taking traces in the above construction; indeed, setting, as before, $F_{ij} = (F_{ijkl})$, we have:

DEFINITION 2.36. The *curvature potential* of $\delta = (\delta_i)$ is the family (ψ_{ij}) where $\psi_{ij} \in A$ are defined (using Einstein summation convention for pairs of lower indices) by:

$$\psi_{ij} := \mathrm{tr}(\Phi_{ij}) = \mathrm{tr}(F_{ij}x) = F_{ijkl}x_{lk}. \tag{2.48}$$

The family (ψ_{ij}) completely determines the curvature (F_{ij}) by the formula:

$$F_{ij}^t = \frac{\partial \psi_{ij}}{\partial x},$$

where, as usual, upper t means "transpose." This curvature potential plays no role in the present classical situation but will have an arithmetic analogue.

Finally note the following invariant way of seeing the Jacobian matrix

$$\frac{\partial \Phi_{ij}}{\partial x};$$

this will also have an arithmetic analogue. Denote by

$$\Omega := \Omega_{\mathcal{O}(G)/A}$$

the module of Kähler differentials and let $\delta = (\delta_i)$ be an Ehresmann connection with curvature (F_{ij}). Then δ_i induce unique additive operators ("Lie derivatives") δ_i^{Ω} on Ω, satisfying

$$\delta_i^{\Omega}(fdg) = (\delta_i f)dg + fd(\delta_i g)$$

for all $f, g \in \mathcal{O}(G)$. Consider the endomorphism

$$\varphi_{ij}^{\Omega} := \delta_i^{\Omega}\delta_j^{\Omega} - \delta_j^{\Omega}\delta_i^{\Omega} : \Omega \to \Omega.$$

We then have the following formula:

$$\varphi_{ij}^{\Omega}(dx) = F_{ij}dx. \tag{2.49}$$

In other words 2.46 is the matrix of the endomorphism φ_{ij}^{Ω} of Ω with respect to the $\mathcal{O}(G)$-basis (consisting of the entries of) dx. Note that dx is *not* a basis of invariant 1-forms! If we consider the matrix of left invariant forms $x^{-1} \cdot dx$ we get

$$\varphi_{ij}^{\Omega}(x^{-1} \cdot dx) = 0.$$

Also if we consider the matrix of right invariant forms $dx \cdot x^{-1}$ we get

$$\varphi_{ij}^{\Omega}(dx \cdot x^{-1}) = F_{ij} \cdot dx \cdot x^{-1} - dx \cdot x^{-1} \cdot F_{ij} = [F_{ij}, dx \cdot x^{-1}].$$

In particular the traceless curvature W_{ij} is completely determined by

$$\varphi_{ij}^{\Omega}(dx \cdot x^{-1}).$$

One can ask what is a natural interpretation of dx in 2.49; the answer is that dx is the pull-back of the matrix of forms dx on $\mathfrak{g} = Spec\ A[x]$ which can be given a natural interpretation in terms of the free A-module V in 2.10.

In our superalgebra formalism (cf. subsection 1.3.5) some of the above discussion can be rephrased as follows. Consider the forms

$$\delta^{\Omega} = \sum_i \delta_i^{\Omega} \cdot d\xi_i \in \Lambda^1(\mathrm{End}(\Omega)), \tag{2.50}$$

$$\varphi^{\Omega} = \sum_{i<j} \varphi_{ij}^{\Omega} \cdot d\xi_i \wedge d\xi_j \in \Lambda^2(\mathrm{End}(\Omega)). \tag{2.51}$$

Then:

$$[\delta^{\Omega}, \delta^{\Omega}] = 2 \cdot \varphi^{\Omega}, \quad \rho = \frac{1}{n} \cdot c_1[\delta^{\Omega}, dx], \tag{2.52}$$

where we recall that c_1 denotes the first Chern form of δ^{Ω} with respect to dx (cf. 1.22) and $\rho = \rho[\delta]$ is defined via the first Chern form (ρ_{ij}) of δ (cf. 2.42). One can further consider the holonomy ring $\mathfrak{hol}^{\Omega} := \mathfrak{hol}(\delta^{\Omega})$ inside $\mathrm{End}(\Omega)$.

2.1.7. $(1,1)$-curvature and first Chern $(1,1)$-form. Assume the ring A is equipped with two structures of δ-ring, i.e., with two families

$$\delta^A = \delta = (\delta_i)_{1 \le i \le m}, \quad \bar{\delta}^A = \bar{\delta} = (\bar{\delta}_i)_{1 \le i \le m}. \tag{2.53}$$

Assume in addition that

$$\{\delta_1, ..., \delta_m, \bar{\delta}_1, ..., \bar{\delta}_m\}$$

are pairwise commuting on A; we then say that $\bar{\delta}$ *commutes* with δ on A. Also assume that we have two Ehresmann connections on G,

$$\delta^G = \delta = (\delta_i)_{1 \le i \le m}, \quad \bar{\delta}^G = \bar{\delta} = (\bar{\delta}_i)_{1 \le i \le m} \tag{2.54}$$

with respect to the two structures of δ-ring on A respectively. Set $\bar{i} = i + m$ for $i = 1, ..., m$ and write $\delta_{\bar{i}} := \bar{\delta}_i$.

EXAMPLE 2.37. The typical example we have in mind is when $A \subset C^{\infty}(M, \mathbb{C})$, $M \subset \mathbb{C}^m$ open, A stable under

$$\delta_i = \frac{\partial}{\partial z_i}, \quad \delta_{\bar{i}} = \frac{\partial}{\partial \bar{z}_i}.$$

Here z_i are the complex coordinates on M and in practice we often assume that, on $B = A[x, \det(x)^{-1}]$, we have $\delta_{\bar{i}} x = 0$.

Going back to our abstract situation in 2.54, we have at our disposal the curvature (F_{ij}) and first Chern form (ρ_{ij}) of δ and also the curvature $(F_{\bar{i}\bar{j}})$ and the first Chern form $(\rho_{\bar{i}\bar{j}})$ of $\bar{\delta}$. On the other hand consider the commutators

$$\varphi_{i\bar{j}} := [\delta_i, \delta_{\bar{j}}] : \mathcal{O}(G) \to \mathcal{O}(G)$$

and set $\Phi_{i\bar{j}} := \varphi_{i\bar{j}}(x)$. If $\delta_i x = A_i x$, $\delta_{\bar{i}} x = A_{\bar{i}} x$, then

$$\Phi_{i\bar{j}} = F_{i\bar{j}} x, \quad F_{i\bar{j}} := \delta_i A_{\bar{j}} - \delta_{\bar{j}} A_i - [A_i, A_{\bar{j}}].$$

Set

$$\rho_{i\bar{j}} = \mathrm{tr}(F_{i\bar{j}}). \tag{2.55}$$

DEFINITION 2.38. We call

$$(F_{i\bar{j}}) = (F_{i\bar{j}}[\delta])$$

the $(1,1)$-*curvature* of δ with respect to $\bar{\delta}$; also we call

$$(\rho_{i\bar{j}}) = (\rho_{i\bar{j}}[\delta])$$

the *first Chern* $(1,1)$-*form* of δ with respect to $\bar{\delta}$.

We can interpret these objects as follows. The family

$$(\delta,\bar{\delta}) = (\delta_\alpha)_{1\leq\alpha\leq 2m}$$

is an Ehresmann connection with curvature

$$(F_{\alpha\beta})_{1\leq\alpha,\beta\leq 2m},$$

where

$$F_{\bar{j}i} := -F_{i\bar{j}}.$$

We may refer to (F_{ij}), $(F_{i\bar{j}})$, $(F_{\bar{i}\bar{j}})$ as the $(2,0)$, $(1,1)$, $(0,2)$ *components* of the curvature of $(\delta,\bar{\delta})$, respectively. We have a similar picture for the first Chern forms. Assuming in addition $\bar{\delta}x = 0$ (i.e. $A_{\bar{j}} = 0$ for all j) we have

$$\rho_{i\bar{j}} = -\delta_{\bar{j}}(\mathrm{tr}(A_i)). \tag{2.56}$$

In particular we morally have that $(\rho_{i\bar{j}})$ defines a "Dolbeault exact," hence a "Dolbeault closed," form, in the sense that

$$\delta_{\bar{j}}\rho_{i\bar{k}} = \delta_{\bar{k}}\rho_{i\bar{j}}. \tag{2.57}$$

The above can be interpreted in terms of superalgebra as follows. Assume $\mathcal{U} = \mathcal{Z}\cup\overline{\mathcal{Z}}$ as in 1.23 and let

$$\delta := \sum_i \delta_i \cdot dz_i \in \Lambda^{1,0}(\mathrm{End}(\mathcal{O}(G))),\quad \bar{\delta} := \sum_i \bar{\delta}_i \cdot d\bar{z}_i \in \Lambda^{0,1}(\mathrm{End}(\mathcal{O}(G))). \tag{2.58}$$

Then we have the following equality in $\Lambda^2(\mathrm{End}(\mathcal{O}(G)))$:

$$\begin{aligned}\tfrac{1}{2}[\delta+\bar{\delta},\delta+\bar{\delta}] &= \textstyle\sum_{i<j}\varphi_{ij}\cdot dz_i\wedge dz_j\\ &\quad +\textstyle\sum_{ij}\varphi_{i\bar{j}}\cdot dz_i\wedge d\bar{z}_j\\ &\quad +\textstyle\sum_{i<j}\varphi_{\bar{i}\bar{j}}\cdot d\bar{z}_i\wedge d\bar{z}_j.\end{aligned} \tag{2.59}$$

Also, for

$$\begin{aligned}\tau &:= \textstyle\sum_i \mathrm{tr}(A_i)\cdot dz_i &&\in \Lambda^{1,0}(A)\subset\Lambda^{1,1}(\mathrm{End}(\mathcal{O}(G))),\\ \rho^c &:= \textstyle\sum_{ij}\rho_{i\bar{j}}\cdot dz_i\wedge d\bar{z}_j &&\in \Lambda^{1,1}(\mathrm{End}(\mathcal{O}(G))),\end{aligned} \tag{2.60}$$

we have the following equalities in $\Lambda^{1,1}(\mathrm{End}(\mathcal{O}(G)))$:

$$\rho^c = [\bar{\delta},\tau] = \frac{1}{n}\cdot\{c_1[\delta^\Omega+\bar{\delta}^\Omega, dx]\}^{(1,1)}. \tag{2.61}$$

2.1.8. Gauge action. First for any $g\in G(A) = GL_n(A)$ we let

$$L_g : \mathcal{O}(G)\to\mathcal{O}(G)$$

be the "left translation by g" i.e. unique A-algebra homomorphism such that $L_g(x) = gx$. Then note that there is an action $\star$ of $G(A)$ on the set $\Lambda^1(\mathfrak{g})$ of Ehresmann connections on G defined by

$$g\star(\delta_1,\ldots,\delta_m) = (L_g^{-1}\circ\delta_1\circ L_g,\ldots,L_g^{-1}\circ\delta_m\circ L_g). \tag{2.62}$$

On the other hand let

$$\Lambda^1(\mathfrak{g})\to\mathfrak{g}(A)^m \tag{2.63}$$

be the bijection

$$(\delta_1,\ldots,\delta_m)\mapsto(\delta_1 x\cdot x^{-1},\ldots,\delta_m x\cdot x^{-1}).$$

DEFINITION 2.39. The $G(A)$-action $\star$ on $\mathfrak{g}(A)^m$ induced by 2.62 and 2.63 is called the *gauge action.*

The gauge action is, of course, given by

$$g \star (A_1, ..., A_m) = (g \star_1 A_1, ..., g \star_m A_m)$$

where

$$g \star_i A_i = (\delta_i g)g^{-1} + gA_i g^{-1}. \tag{2.64}$$

Also we have the following covariance property for the curvature:

$$F_{ij}[g \star \delta] = g \cdot F_{ij}[\delta] \cdot g^{-1}. \tag{2.65}$$

In particular we have the following gauge invariance of the first Chern form:

$$\rho[g \star \delta] = \rho[\delta].$$

The above considerations will have an arithmetic analogue.

2.1.9. Translation action. There is a natural action of $\mathfrak{g}(A)^m$ on $\Lambda^1(\mathfrak{g})$ induced, via the bijection 2.63, by the action of $\mathfrak{g}(A)^m$ on itself by (additive) translation. If $\mathfrak{A} = (A_i) \in \mathfrak{g}(A)^m$ and $\delta \in \Lambda^1(\mathfrak{g})$ is such that $\delta_i x = B_i x$ then the action

$$(\mathfrak{A}, \delta) \mapsto \delta^{\mathfrak{A}} \tag{2.66}$$

is defined by $\delta^{\mathfrak{A}} = (\delta_i^{A_i})$, where

$$\delta_i^{A_i} x = (A_i + B_i)x.$$

2.2. Lie algebra and classical groups

Let G be a group scheme over a ring A. We do not assume A is equipped with derivations. For simplicity we assume G is a closed subgroup scheme of GL_n; we then say G is *linear.* In particular G can be GL_n itself.

DEFINITION 2.40. The *Lie algebra* of G is the scheme $L(G)$ over A representing the functor that attaches to any A-algebra C the (abelian) group

$$\mathrm{Ker}(G(C[\epsilon]) \to G(C)). \tag{2.67}$$

Here $C[\epsilon] = C \oplus C\epsilon$, $\epsilon^2 = 0$ is the ring of dual numbers on C and $C[\epsilon] \to C$ is defined by $\epsilon \mapsto 0$. So if

$$G = Spec\ A[x, \det(x)^{-1}]/(f_1, ..., f_m)$$

and $1 = 1_n$ is the identity $n \times n$ matrix then

$$L(G) = Spec\ A[x']/(d_1 f_1, ..., d_1 f_m), \tag{2.68}$$

where x' is an $n \times n$ matrix of indeterminates (x'_{ij}) and

$$d_1 f_j := \sum_{kl} \frac{\partial f_j}{\partial x_{kl}}(1) x'_{kl}. \tag{2.69}$$

In particular the Lie algebra of GL_n is

$$L(GL_n) = \mathfrak{gl}_n = Spec\ A[x']. \tag{2.70}$$

Of course we have a canonical open embedding

$$GL_n \subset \mathfrak{gl}_n, \quad x' \mapsto x.$$

Such an embedding does not exist, of course, for more general groups.

Assume now 2 is invertible in A and let the superscript t for matrices mean "transpose."

DEFINITION 2.41. A matrix $q \in GL_n(A)$ that is symmetric, i.e., $q^t = q$, is called a *metric*. A matrix $q \in GL_n(A)$ that is anti-symmetric, i.e., $q^t = -q$, is called a (non-degenerate) *2-form*.

DEFINITION 2.42. Two metrics $q_1, q_2 \in GL_n(A)$ are called *conformally equivalent* if there exists an invertible element $\lambda \in A^\times$ such that

$$q_2 = \lambda \cdot q_1.$$

DEFINITION 2.43. Let $q \in GL_n(A)$, $q^t = \pm q$. The *special orthogonal group of* q is the identity component, $SO(q)$, of the closed subgroup scheme of GL_n over A defined by the ideal generated by the entries of the matrix $x^t q x - q$.

One checks $SO(q)$ is a smooth group scheme over A. Then the Lie algebra of $SO(q)$ is given by $L(SO(q)) =: \mathfrak{so}(q)$,

$$\mathfrak{so}(q)(C) = \{v \in \mathfrak{gl}_n(C);\ v^t q + qv = 0\}. \tag{2.71}$$

Similarly if SL_n is, as usual, defined in GL_n by $\det(x) - 1$ its Lie algebra is $L(SL_n) = \mathfrak{sl}_n$,

$$\mathfrak{sl}_n(C) = \{v \in \mathfrak{gl}_n(C);\ \mathrm{tr}(v) = 0\}. \tag{2.72}$$

DEFINITION 2.44. A *split* matrix is a matrix $q \in GL_n(A)$ of one of the following forms

$$\left(\begin{array}{cc} 0 & 1_r \\ -1_r & 0 \end{array}\right), \quad \left(\begin{array}{cc} 0 & 1_r \\ 1_r & 0 \end{array}\right), \quad \left(\begin{array}{ccc} 1 & 0 & 0 \\ 0 & 0 & 1_r \\ 0 & 1_r & 0 \end{array}\right), \tag{2.73}$$

where $n = 2r, 2r, 2r+1$ respectively, and 1_r is the $r \times r$ identity matrix.

If q is split we denote $SO(q)$ by

$$Sp_{2r}, \quad SO_{2r}, \quad SO_{2r+1},$$

respectively; these groups are referred to as the *split classical groups*.

A similar discussion can be made for unitary groups as follows. Let $q_0 \in GL_{2r}$ be the split anti-symmetric matrix, i.e.,

$$q_0 = \left(\begin{array}{cc} 0 & 1_r \\ -1_r & 0 \end{array}\right).$$

(In what follows q_0 will intuitively play the role of a complex structure.) Denote by

$$GL_r^c := C(q_0) \tag{2.74}$$

the centralizer of q_0 in GL_{2r}; it is a closed (smooth) subgroup scheme of GL_{2r} and its points are exactly the points of GL_{2r} of the form

$$z := \left(\begin{array}{cc} a & b \\ -b & a \end{array}\right). \tag{2.75}$$

Assume there is a square root $\sqrt{-1} \in A$ of -1 (which we fix throughout our discussion) and assume 2 is invertible in A. We then have a natural isomorphism over A,

$$GL_r^c \to GL_r \times GL_r, \quad z \mapsto (z^c, z^{tct}), \tag{2.76}$$

where z is as in 2.75 and

$$z^c := a + \sqrt{-1} \cdot b, \qquad z^{tct} := ((z^t)^c)^t = a - \sqrt{-1} \cdot b. \tag{2.77}$$

Recall that, since

$$\begin{pmatrix} a - \sqrt{-1} \cdot b & 0 \\ 0 & a + \sqrt{-1} \cdot b \end{pmatrix} \begin{pmatrix} 1 & \sqrt{-1} \\ 1 & -\sqrt{-1} \end{pmatrix} = \begin{pmatrix} 1 & \sqrt{-1} \\ 1 & -\sqrt{-1} \end{pmatrix} \begin{pmatrix} a & b \\ -b & a \end{pmatrix},$$

we have

$$\det \begin{pmatrix} a & b \\ -b & a \end{pmatrix} = \det(a + \sqrt{-1} \cdot b) \cdot \det(a - \sqrt{-1} \cdot b). \tag{2.78}$$

Note also that GL_r^c is stable under transposition. We denote by

$$x \mapsto x^\dagger$$

the restriction to GL_r^c of the transposition $GL_{2r} \to GL_{2r}$,

$$x \to x^t$$

and consider the automorphism of $GL_r \times GL_r$ given by

$$(u, v)^\dagger = (v^t, u^t).$$

Then the isomorphism 2.76 commutes with $\dagger$.

DEFINITION 2.45. A matrix $q \in GL_{2r}(A)$ is called a *hermitian metric* if $q^t = q$ and q commutes with the split anti-symmetric matrix q_0. A matrix $q \in GL_{2r}(A)$ is called a (non-degenerate) *hermitian form* if $q^t = -q$ and q commutes with the split anti-symmetric matrix q_0.

So $q \in GL_{2r}(A)$ is a *hermitian metric* if and only if $q \in GL_r^c(A)$ and $q^\dagger = q$, in other words, if and only if

$$q = \begin{pmatrix} q_1 & q_2 \\ -q_2 & q_1 \end{pmatrix}, \quad q_1^t = q_1, \quad q_2^t = -q_2. \tag{2.79}$$

If q in 2.79 is a hermitian metric then the matrix

$$\check{q} = -q q_0 = \begin{pmatrix} q_2 & -q_1 \\ q_1 & q_2 \end{pmatrix} \tag{2.80}$$

is a hermitian form called the hermitian form *associated* to the hermitian metric q.

For a hermitian metric q, and under the assumption that A contains $\sqrt{-1}$, we have

$$\det(q_1 - \sqrt{-1} \cdot q_2) = \det((q_1 - \sqrt{-1} \cdot q_2)^t) = \det(q_1 + \sqrt{-1} \cdot q_2)$$

hence, by 2.78, setting as usual $q^c = q_1 + \sqrt{-1} \cdot q_2$, we have

$$\det(q) = \det(q^c)^2. \tag{2.81}$$

DEFINITION 2.46. The *complexified unitary group* of q, $U^c(q)$, is the closed subgroup scheme of GL_r^c defined by the equations $x^\dagger q x = q$.

Note that, for $q = 1$, the identity matrix, the induced map

$$U_r^c := U^c(q) \to GL_r, \quad z \mapsto z^c = a + \sqrt{-1} \cdot b, \tag{2.82}$$

with z as in 2.75, is an isomorphism; its inverse is given by $g \mapsto z$ with z as in 2.75 and

$$a = \frac{1}{2}(g + (g^t)^{-1}), \quad b = \frac{1}{2\sqrt{-1}}(g - (g^t)^{-1}).$$

In particular U_r^c is connected.

Our terminology is justified by the following considerations.

DEFINITION 2.47. Let A be a ring equipped with a square root of -1, $\sqrt{-1} \in A^\times$. A *complex conjugation* on A is a ring automorphism $z \mapsto \overline{z}$ which is *involutive* (i.e., $\overline{\overline{z}} = z$) and maps $\sqrt{-1}$ into $-\sqrt{-1}$. Set

$$A^{real} = \{z \in A; \overline{z} = z\},$$

the subring of *real elements* of A. For a matrix $g \in \mathfrak{gl}_r(A)$ one sets

$$g^* := \overline{g}^t.$$

We say $h \in \mathfrak{gl}_n(A)$ is a *hermitian matrix* if

$$h^* = h.$$

We say $u \in GL_r(A)$ is *unitary* if

$$uu^* = 1.$$

Clearly any element $z \in A$ can be written uniquely as

$$z = a + \sqrt{-1} \cdot b$$

with $a, b \in A^{real}$; and similarly for matrices.

EXAMPLE 2.48. The natural example we have in mind is that of any subring $A \subset C^\infty(M, \mathbb{C})$ that is closed under complex conjugation, where $M \subset \mathbb{C}^m$ is open; in this case

$$A^{real} = A \cap C^\infty(M, \mathbb{R}).$$

Returning to a general ring A equipped with a complex conjugation we have:

DEFINITION 2.49. The *unitary* group over A is the group

$$U_r(A) := \{u \in GL_r(A);\ uu^* = 1\}. \tag{2.83}$$

Then the following is trivial to check:

PROPOSITION 2.50. *The map $GL_r^c(A) \to GL_r(A)$, $z \mapsto z^c$, induces isomorphisms*

$$GL_r^c(A^{real}) \to GL_r(A), \tag{2.84}$$

$$U_r^c(A) \cap GL_r^c(A^{real}) \to U_r(A). \tag{2.85}$$

Moreover, let $q \in GL_{2r}(A)$ be such that $\overline{q} = q$; then q is a hermitian metric if and only if $h := q^c$ is a hermitian matrix, i.e., $h^* = h$.

We end this section with a few more comments on metrics and 2-forms.

First we consider a class of metrics whose arithmetic analogues will play a role later.

DEFINITION 2.51. A *cohomogeneity one metric* is a metric $q = (q_{ij}) \in GL_n(A)$ such that

$$\delta_i q_{kl} = \delta_j q_{kl}, \quad i, j = 1, ..., m, \quad k, l = 1, ..., n.$$

The following Lemma is trivial to check.

LEMMA 2.52. *q is a cohomogeneity one metric if and only if there exists a δ-subring $C \subset A$ such that $q \in GL_n(C)$ and such that for all i and j the induced derivations $\delta_i^C : C \to C$ and $\delta_j^C : C \to C$ coincide.*

Next we explain how the definitions given in this section, involving metrics and 2-forms, are abstractions of familiar concepts in classical differential geometry.

EXAMPLE 2.53. Let $M \subset \mathbb{R}^n$ be an open set with coordinates $\xi_1, ..., \xi_n$ and let $A \subset C^\infty(M, \mathbb{R})$ be a subring closed under all $\frac{\partial}{\partial \xi_i}$.

To give a symmetric $q = (q_{ij}(\xi)) \in GL_n(A)$ is equivalent to giving what is classically called a (semi-Riemannian) *bundle metric* on the trivial bundle $M \times \mathbb{R}^n \to M$, i.e., a smoothly varying family of non-degenerate symmetric bilinear forms on the fibers of this bundle; if our bundle is the tangent bundle the bundle metric is simply a metric on M,

$$g := \sum_{ij} q_{ij}(\xi) d\xi_i d\xi_j. \tag{2.86}$$

To give an anti-symmetric $q = (q_{ij}(\xi)) \in GL_n(A)$ is equivalent to giving what is classically called a *non-degenerate 2-form* on the trivial vector bundle $M \times \mathbb{R}^n \to M$, i.e., a smoothly varying family of non-degenerate anti-symmetric bilinear forms on the fibers of this bundle; if our bundle is the tangent bundle our 2-form on the bundle is simply a 2-form on M,

$$\omega := \frac{1}{2} \sum_{ij} q_{ij}(\xi) d\xi_i \wedge d\xi_j. \tag{2.87}$$

Finally, assuming $n = 2r$ and $\mathbb{R}^{2r}$ is identified with $\mathbb{C}^r$ via complex coordinates $z_i = \xi_i + \sqrt{-1}\xi_{i+r}$, $i = 1, ..., r$, to give a hermitian metric $q = (q_{ij}(\xi)) \in GL_{2r}(A)$ is equivalent to giving what is classically called a *hermitian metric* on the trivial vector bundle $M \times \mathbb{C}^n \to M$; if our bundle is the tangent bundle one is reduced to giving a hermitian metric on M,

$$\sum_{i,j=1}^{r} h_{i\bar{j}}(\xi) dz_i d\bar{z}_j = \sum_{i,j=1}^{2r} q_{ij}(\xi) d\xi_i d\xi_j, \tag{2.88}$$

where $h_{i\bar{j}} := q_{ij}^c := q_{ij} + \sqrt{-1} q_{i,j+r}$ for $i, j = 1, ..., r$. If $\check{q} = (\check{q}_{ij})$ is the hermitian form associated to the hermitian metric q as in 2.80 then

$$-\sqrt{-1} \sum_{i,j=1}^{r} h_{i\bar{j}}(\xi) dz_i \wedge d\bar{z}_j = \sum_{i,j=1}^{2r} \check{q}_{ij}(\xi) d\xi_i \wedge d\xi_j \tag{2.89}$$

is what is classically called the *fundamental 2-form* associated to the hermitian metric 2.88 [**89**].

Assume, in what follows, that our trivial bundle is the tangent bundle, in particular $n = m$; this corresponds to what classically is referred to as a *soldering*. Then conformal equivalence of matrices corresponds to the usual notion of conformal equivalence of metrics on M. Let us look at the manifold counterpart of *cohomogeneity one metrics*.

For a metric g on $\mathbb{R}^n$ the following conditions are equivalent:

(1) In appropriate coordinates the metric is invariant under an $n-1$ dimensional group of translations.
(2) In appropriate coordinates $y_1, y_2, ..., y_n$, the metric g the has the form

$$g := \sum_{i,j=1}^{n} g_{ij}^*(y_1) dy_i dy_j, \quad g_{ij}^* \in C^\infty(\mathbb{R}, \mathbb{R}). \tag{2.90}$$

(3) In appropriate coordinates $\xi_1, ..., \xi_n$ the metric g has the form

$$g = \sum_{i,j=1}^{n} g_{ij}^{**}(\xi) d\xi_i d\xi_j, \quad \frac{\partial g_{ij}^{**}}{\partial \xi_1} = ... = \frac{\partial g_{ij}^{**}}{\partial \xi_n}. \tag{2.91}$$

The link between the ξ and y coordinates above is given by

$$y_1 = \xi_1 + ... + \xi_n, \quad y_2 = \xi_2, ..., y_n = \xi_n.$$

A g satisfying one of (1), (2), (3) above will be called, for our purposes here, a *(translational) cohomogeneity one* metric; we will drop the adjective *translational* in what follows. (This is an abelian version of the usual global geometric concept of *cohomogeneity one* in which $\mathbb{R}^n$ is replaced by an arbitrary n dimensional manifold and the group of translations is replaced by an arbitrary Lie group acting with $n-1$ dimensional orbits.)

There are many remarkable examples of cohomogeneity one metrics (in our sense above) appearing in geometry; e.g.,

(1) the round spheres S^n for $n = 2, 3$, [**110**], pp. 13-15,
(2) the standard hyperbolic space [**110**], p. 74,

$$\mathbb{H}^n = \{(y_1, ..., y_n); y_n > 0\},$$

for $n \geq 2$, with metric

$$y_n^{-2}(dy_1^2 + ... + dy_n^2),$$

(3) the 3-dimensional Heisenberg group

$$\mathfrak{H}^3 = \{\begin{pmatrix} 1 & y_3 & y_2 \\ 0 & 1 & y_1 \\ 0 & 0 & 1 \end{pmatrix}\}$$

with left invariant metric

$$dy_3^2 + dy_1^2 + (dy_2 - y_1 dy_3)^2,$$

(4) the half plane

$$\{(y_1, y_2); \; y_2 > 1\}$$

with the Schwarzschild metric [**108**], p. 152,

$$-hdy_1^2 + h^{-1}dy_2^2, \quad h = 1 - \frac{1}{y_2}.$$

The framework above (in ξ coordinates) will have, as an analogue in our arithmetic differential geometry, the case of *one* prime. The *one* prime will correspond to the *one* derivation on the ring C of all functions of the form $F(\xi_1 + ... + \xi_n)$ that is obtained by restricting to C the derivations $\delta_i = \frac{\partial}{\partial \xi_i}$.

2.3. Involutions and symmetric spaces

In our next discussion we let A be an arbitrary ring (not necessarily equipped with derivations) and we let G be a smooth connected linear group scheme over A (i.e., a smooth closed subgroup scheme of GL_n over A).

DEFINITION 2.54. By an *automorphism* of G we understand an automorphism $\tau : G \to G$ of G viewed as a group scheme over A; so

$$(gh)^{\tau} = g^{\tau} h^{\tau}$$

on points $g, h \in G(C)$, C any A-algebra. We say τ is an *involution* if

$$g^{\tau\tau} = g$$

on points. (Note that we allow the identity to be an involution.)

Let τ be an automorphism (not necessarily an involution). One can attach to τ a closed subgroup scheme G^+ defined as the fixed closed subscheme of the automorphism τ; on points,

$$G^+(C) = \{g \in G(C); g^{\tau} = g\}. \tag{2.92}$$

DEFINITION 2.55. The identity component $(G^+)^{\circ}$ is called the *group (scheme) defined by* τ.

Similarly one defines the closed subscheme G^- by

$$G^-(C) = \{g \in G(C); g^{\tau} = g^{-1}\}. \tag{2.93}$$

note that G^- is not a subgroup scheme of G in general; but it is stable under $g \mapsto g^{\nu}$ for all $\nu \in \mathbb{Z}$. For any $g \in G(C)$ write

$$g^{-\tau} := (g^{-1})^{\tau} = (g^{\tau})^{-1}.$$

DEFINITION 2.56. The *quadratic map* attached to τ is the morphism of schemes

$$\mathcal{H} : G \to G, \quad \mathcal{H}(x) = x^{-\tau} x \tag{2.94}$$

Here we denoted by x an arbitrary C-point of G; we also previously denoted by x the matrix of indeterminates defining GL_n but no confusion should arise from this since C-points arise as specializations in C of the matrix of indeterminates x. More generally, for any $g \in G(A)$, we may consider the left translation by g,

$$L_g : G \to G, \quad L_g(x) = gx,$$

and the composition $\mathcal{H}_g := L_g \circ \mathcal{H}$ so

$$\mathcal{H}_g : G \to G, \quad \mathcal{H}_g(x) = g x^{-\tau} x. \tag{2.95}$$

In particular, for $g = 1$, we have $\mathcal{H}_1 = \mathcal{H}$. Note that

$$G^+ = \mathcal{H}^{-1}(1).$$

DEFINITION 2.57. The *bilinear map* attached to τ is the morphism of schemes

$$\mathcal{B} : G \times G \to G, \quad \mathcal{B}(x, y) = x^{-\tau} y. \tag{2.96}$$

More generally for $g \in G(A)$ we may define a morphism

$$\mathcal{B}_g : G \times G \to G, \quad \mathcal{B}(x, y) = g x^{-\tau} y. \tag{2.97}$$

Assume next that τ is an involution. Then $\mathcal{H}$ factors through a map

$$\mathcal{H} : G \to G^-.$$

DEFINITION 2.58. The pair (G, G^+) is called the *symmetric pair* attached to τ. If A is an algebraically closed field of characteristic zero the orbit space $G^+ \backslash G$ is called the *symmetric space* attached to τ. Cf. [**73**], p.209.

Let us examine the geometry of these objects more closely. To simplify assume, in this discussion, that A is an algebraically closed field of characteristic $\neq 2$. (In classical differential geometry we only care about characteristic zero; however, in arithmetic differential geometry, we will care about the picture in positive characteristic as well so for the sake of subsequent developments we do not restrict here to characteristic zero.) For convenience we will identify our varieties over A with their sets of A-points. Then, in our notation above, for any point $a \in G$ we have an equality

$$\mathcal{H}^{-1}(\mathcal{H}(a)) = G^+ \cdot a.$$

Moreover the morphism

$$\mathcal{H} : G \to G$$

factors through the embedding $G^- \subset G$ hence induces a morphism

$$\mathcal{H} : G \to G^-.$$

In particular $\mathcal{H}$ induces an injection of sets

$$G^+ \backslash G \to G^-. \tag{2.98}$$

Furthermore $(G^-)^\tau \subset G^-$, and we have a commutative diagram

$$\begin{array}{ccc} G & \xrightarrow{\tau} & G \\ \mathcal{H}\downarrow & & \downarrow\mathcal{H} \\ G^- & \xrightarrow{\tau} & G^- \end{array} \tag{2.99}$$

Let us denote by $L(Y)$ the tangent space at the identity 1 of a closed subscheme Y of G that passes through 1, i.e., the scheme representing the functor sending any A-algebra C into the preimage of 1 under the map

$$Y(C[\epsilon]) \to Y(C).$$

Then $L(G^+)$ identifies with the $+1$ eigenspace $L(G)^+$ of $L(\tau)$ in $L(G)$ while $L(G^-)$ identifies with the -1 eigenspace $L(G)^-$ of $L(\tau)$ in $L(G)$. Furthermore (since the characteristic is not 2) we have a direct sum decomposition,

$$L(G) = L(G)^+ \oplus L(G)^-. \tag{2.100}$$

DEFINITION 2.59. The decomposition 2.100 is called the *Cartan decomposition.*

On 2.100 $L(\tau)$ acts as $(b^+, b^-) \mapsto (b^+, -b^-)$; moreover the map

$$L(\mathcal{H}) : L(G) \to L(G^-)$$

sends $(b^+, b^-) \mapsto 2b^-$ so it is surjective with kernel $L(G^+)$. Assume in what follows that G is connected and G^+ is smooth and denote by $(G^-)^\circ$ the connected component of G^- containing 1.

PROPOSITION 2.60. *The map $\mathcal{H} : G \to (G^-)^\circ$ is dominant and its image is contained in the smooth locus $(G^-)^\circ_s$ of $(G^-)^\circ$.So one has an induced dominant morphism*

$$\mathcal{H} : G \to (G^-)^\circ_s. \tag{2.101}$$

PROOF. The set theoretic image of $\mathcal{H} : G \to (G^-)^\circ$ can be identified with the quotient $G^+\backslash G$; on the other hand we have

$$\begin{aligned}\dim(G^+\backslash G) &= \dim(G) - \dim(G^+)\\ &= \dim(L(G)) - \dim(L(G)^+), \quad \text{by smoothness of } G^+\\ &= \dim(L(G)^-)\\ &\geq \dim_1(G^-)\\ &\geq \dim(G^+\backslash G)\end{aligned}$$

which forces the inequalities above to be equalities. (In the above $\dim_1$ is the dimension of the local ring at 1.) This implies that the local ring at 1 of the scheme G^- is regular so $1 \in (G^-)^\circ_s$. So $(G^-)^\circ$ is irreducible and of the same dimension as $G^+\backslash G$, hence the latter is dense in $(G^-)^\circ$. Since G acts on the right on $(G^-)^\circ$ via $(a, b) \mapsto b^{-\tau}ab$ we get that the whole of the set $G^+\backslash G$ is contained in $(G^-)^\circ_s$ which ends the proof. □

PROPOSITION 2.61. *Let $q \in GL_n(A)$, $n \geq 2$, and assume that A is an algebraically closed field of characteristic zero. Then the following hold.*

(1) *$x \mapsto q^{-1}(x^t)^{-1}q$ is an involution of GL_n if and only if $q^t = \pm q$.*
(2) *$x \mapsto q^{-1}xq$ is an involution of GL_n if and only if $q = cu$ where $c \in GL_n(A)$ is in the center of GL_n and $u \in GL_n(A)$, $u^2 = 1$.*
(3) *$x \mapsto \det(x)q^{-1}(x^t)^{-1}q$ is an involution of GL_n if and only if $n = 2$ and $q^t = \pm q$.*
(4) *$x \mapsto \det(x)^{-1}q^{-1}xq$ is an involution of GL_n if and only if $n = 2$ and $q = cu$, where $c \in GL_2(A)$ is in the center of GL_2 and $u \in GL_2(A)$, $u^2 = 1$.*
(5) *Any involution of GL_n is of one of the forms (1), (2), (3), (4) above.*

PROOF. Assertions (1), (2), (3), (4) are trivially checked. Assertion (5) follows easily from the well known fact [**60**], pp. 90-96, that the automorphism group of SL_n over a algebraically closed field is generated by the inner automorphisms together with $x \mapsto (x^t)^{-1}$. □

DEFINITION 2.62. Involutions of the form (1) above are referred to as *outer* involutions. Involutions of the form (2) are referred to as *inner* involutions.

More generally, of course, automorphisms of the form $x \mapsto q^{-1}xq$ are referred to as *inner* automorphisms. Note that if $q \in GL_n(A)$, $q^t = \pm q$, and $G = GL_n$ then the group G^+ defined by the outer involution $x \mapsto q^{-1}(x^t)^{-1}q$ of type (1) identifies, by definition, with the special orthogonal group $SO(q)$. Inner automorphisms are, on the other hand, closely connected to conjugacy classes. Each of these two types of homogeneous spaces will be considered from the viewpoint of our theory in what follows. The involutions of the form (3) and (4) will not be considered in this book but can be treated along the same lines as the other types (1) and (2).

REMARK 2.63. If A morally plays the role of a ring of smooth functions on a manifold M, for instance on $M = \mathbb{R}^m$, then our group schemes G over A play the role of products

$$M \times G(\mathbb{R}),$$

where $G(\mathbb{R})$ is a Lie group. Then our symmetric spaces $G^+\backslash G$ over A morally play the role of products

$$M \times (G^+(\mathbb{R})^0\backslash G(\mathbb{R})) \tag{2.102}$$

where $G^+(\mathbb{R})^0$ is the identity component of the fixed group $G^+(\mathbb{R})$ of an involution on $G(\mathbb{R})$. For the purpose of what is going to be explained later it is useful at this point to note that there are at least two distinct situations that involve symmetric spaces in classical differential geometry; we can refer to them as the "vertical" and the "horizontal" pictures as follows:

(1) The "vertical" picture is the one we shall be mainly interested in this book. This situation corresponds to looking at connections on the trivial principal bundle

$$P := M \times G(\mathbb{R}) \to M$$

that are compatible with a given involution of P over M. The products 2.102 implicitly appear when one expresses these compatibilities via the quadratic map $\mathcal{H}$ attached to the involution; cf. Definition 2.69 below. The above situation can be globalized and here is an important example of such a globalization. Let $G(\mathbb{R}) = GL_n(\mathbb{R})$ and $x^\tau = (x^t)^{-1}$ so the identity component $G^+(\mathbb{R})^0$ is the group $SO_n'(\mathbb{R})$ of all matrices $a \in SL_n(\mathbb{R})$ with $aa^t = 1$. Let $P \to M$ be the frame bundle of (the tangent bundle of) an arbitrary orientable manifold M of dimension n. Then metrics on M can be viewed as sections of

$$SO_n'(\mathbb{R})\backslash P \to M; \tag{2.103}$$

for a given metric the section sends a point of M into the class of an oriented orthonormal frame at that point. Note that the fibers

$$F \simeq SO_n'(\mathbb{R})\backslash GL_n(\mathbb{R}) \tag{2.104}$$

of the fibration 2.103 are symmetric spaces in the sense of Riemannian geometry [**89**]. By the way metrics on M, viewed in coordinates as symmetric matrices with entries smooth functions on M, define (locally on M) involutions of the frame bundle P over M. Although the involution defined by a metric g on M is only defined locally the corresponding quadratic and bilinear maps induce global maps

$$P \to GL_n(\mathbb{R}), \quad P \times_M P \to GL_n(\mathbb{R}),$$

given by the formulas

$$(v_1, ..., v_n) \mapsto (g_P(v_i, v_j)), \quad (v_1, ..., v_n, w_1, ..., w_n) \mapsto (g_P(v_i, w_j)),$$

where $(v_1, ..., v_n)$ and $(w_1, ..., w_n)$ are frames at an arbitrary point $P \in M$.

(2) The "horizontal" picture corresponds to the case when one is looking at metrics on (the tangent bundle of)

$$M := G^+(\mathbb{R})^0\backslash G(\mathbb{R}) \tag{2.105}$$

that make this space *symmetric* in the sense of Riemannian geometry [**89**]. This is historically the origin of symmetric spaces à la Cartan and this is the main focus of classical treatments such as the one in [**89**]; this aspect of symmetric spaces will only appear indirectly in our book (via the notion of connections whose curvature is parallel to the metric; cf Definition 2.126 below).

The pictures (1) and (2) above coexist in the following situation. Let M be as in 2.105 (with $G(\mathbb{R})$ and the involution arbitrary) and let $P \to M$ be the frame bundle of M. Then note that both the base 2.105 and the fibers 2.104 of the fibration 2.103 are symmetric spaces in the sense of [**89**]; the base M of 2.103 fits into the horizontal picture (2) while the fibers F of 2.103 fit into the vertical picture (1).

2.4. Logarithmic derivative and differential Galois groups

For simplicity we assume, in this section, that A is a field of characteristic 0 equipped with one derivation $\delta = \delta^A$.

DEFINITION 2.64. The *constant field* of A is the subfield of A defined by:

$$A^\delta = \{a \in A;\ \delta a = 0\}. \tag{2.106}$$

DEFINITION 2.65. For any A-algebra R equipped with a derivation $\delta = \delta^R$ extending δ^A the map of sets

$$l\delta : GL_n(R) \to \mathfrak{gl}_n(R), \quad l\delta(u) = \delta u \cdot u^{-1}, \tag{2.107}$$

is called the *logarithmic derivative* attached to δ.

It is easy to check that, for any closed subgroup scheme G of GL_n defined by equations with coefficients in A^δ, the map 2.107 sends $G(R)$ into $\mathfrak{g}(R)$ where $\mathfrak{g} := L(G)$ is the Lie algebra of G; hence we have an induced map

$$l\delta^G : G(R) \to \mathfrak{g}(R), \tag{2.108}$$

called the *logarithmic derivative* attached to δ on G. We have the following *cocycle identity* for $u, v \in GL_n(R)$:

$$l\delta(uv) = l\delta u + u \cdot l\delta v \cdot u^{-1}. \tag{2.109}$$

Note that for any $\alpha \in \mathfrak{gl}(A)$ the set

$$l\delta^{-1}(\alpha) = \{u \in GL_n(R); \delta u = \alpha \cdot u\} \tag{2.110}$$

is the set of all $u \in GL_n(R)$ that are horizontal for the Ehresmann connection δ on GL_n defined by $\delta x = \alpha \cdot x$. Also note that, by 2.109, we have

$$l\delta u = l\delta(uv) \quad \text{iff} \quad v \in GL_n(R^\delta). \tag{2.111}$$

Assume, in what follows, that the field of constants A^δ is algebraically closed and fix $\alpha \in GL_n(A)$. One has the following:

THEOREM 2.66. [**91**]

(1) *There exists a field extension R of A equipped with a derivation $\delta = \delta^R$ extending $\delta = \delta^A$ and there exists $u \in GL_n(R)$ such that $R^\delta = A^\delta$, R is generated as a field over A by the entries of u (write $R = A(u)$), and $\delta u = \alpha \cdot u$.*

(2) *Let R and u be as in assertion (1) and let $G_{u/A}$ be the group of all field automorphisms of R over A that commute with δ^R. Then the map*

$$G_{u/A} \to GL_n(R), \qquad \sigma \mapsto u^{-1} \cdot \sigma(u) \tag{2.112}$$

is injective, takes values in $GL_n(A^\delta)$, and its image is an algebraic subgroup of $GL_n(A^\delta)$.

(3) *The isomorphism class of the algebraic group $G_{u/A}$ over A^δ depends only on A and α (but not on R and u).*

REMARK 2.67. In assertion (2) above we identify A^δ-varieties with their sets of A^δ-points. The injectivity of the map 2.112 follows from the condition $R = A(u)$. The fact that this map takes values in $GL_n(A^\delta)$ follows from 2.111.

DEFINITION 2.68. The group $G_{u/A}$ in Theorem 2.66 is called the *differential Galois group* of the equation $\delta u = \alpha \cdot u$ (or of the Ehresmann connection δ on GL_n defined by $\delta x = \alpha x$.)

There is a well developed Galois theory involving these groups; cf. [**91, 114**].

2.5. Chern connections: the symmetric/anti-symmetric case

In this section A is a ring equipped with m derivations $\delta_1, ..., \delta_m$ and n is not necessarily equal to m. Let $q \in G(A) = GL_n(A)$ be a metric or a 2-form; hence

$$q^t = \pm q, \tag{2.113}$$

respectively. We would like to attach connections to q, called the *real* and *complex Chern connection.* Later we will look at the case when q is a hermitian metric.

Consider the maps

$$\mathcal{H}_q : G \to G, \quad \mathcal{B}_q : G \times G \to G \tag{2.114}$$

defined by

$$\mathcal{H}_q(x) = x^t q x, \quad \mathcal{B}_q(x, y) = x^t q y. \tag{2.115}$$

These are the maps in 2.95 and 2.97 corresponding to the involution

$$x^\tau := q^{-1}(x^t)^{-1} q$$

on GL_n. Now as usual, $\mathcal{O}(G) = A[x, \det(x)^{-1}]$ and let us identify $\mathcal{O}(G) \otimes \mathcal{O}(G)$ with

$$A[x, y, \det(x)^{-1}, \det(y)^{-1}], \text{ by identifying } x = x \otimes 1, \ \ y = 1 \otimes x.$$

Then we still denote by

$$\mathcal{H}_q : \mathcal{O}(G) \to \mathcal{O}(G), \quad \mathcal{B}_q : \mathcal{O}(G) \to \mathcal{O}(G) \otimes \mathcal{O}(G),$$

the induced maps given by

$$\mathcal{H}_q(x) = x^t q x, \quad \mathcal{B}_q(x) = x^t q y. \tag{2.116}$$

(Of course x and y in 2.115 denote points of schemes whereas x and y in 2.116 denote functions on schemes; we will often adopt this abuse of notation.) Consider the *trivial* connection $\delta_0 = (\delta_{0i})$ on G defined by $\delta_{0i} x = 0$ and consider also an arbitrary (not necessarily Ehresmann) connection (δ_i) on G.

DEFINITION 2.69. One says that δ is $\mathcal{H}_q$*-horizontal* with respect to δ_0 if for all i the following diagram is commutative:

$$\begin{array}{ccc} \mathcal{O}(G) & \stackrel{\delta_i}{\longleftarrow} & \mathcal{O}(G) \\ \mathcal{H}_q \uparrow & & \uparrow \mathcal{H}_q \\ \mathcal{O}(G) & \stackrel{\delta_{0i}}{\longleftarrow} & \mathcal{O}(G) \end{array} \tag{2.117}$$

DEFINITION 2.70. One says that δ is $\mathcal{B}_q$-*symmetric* with respect to δ_0 if for all i the following diagram is commutative:

$$\begin{array}{ccc} \mathcal{O}(G) & \stackrel{\delta_{0i}\otimes 1+1\otimes\delta_i}{\longleftarrow} & \mathcal{O}(G)\otimes_A\mathcal{O}(G) \\ \delta_i\otimes 1+1\otimes\delta_{0i}\uparrow & & \uparrow\mathcal{B}_q \\ \mathcal{O}(G)\otimes_A\mathcal{O}(G) & \stackrel{\mathcal{B}_q}{\longleftarrow} & \mathcal{O}(G) \end{array} \tag{2.118}$$

DEFINITION 2.71. One says that δ is $(\mathcal{H}_q,\mathcal{B}_q)$-*horizontal* with respect to δ_0 if for all i the following diagram is commutative:

$$\begin{array}{ccc} \mathcal{O}(G) & \stackrel{\delta_{0i}\otimes 1+1\otimes\delta_i}{\longleftarrow} & \mathcal{O}(G)\otimes_A\mathcal{O}(G) \\ \mathcal{H}_q\uparrow & & \uparrow\mathcal{B}_q \\ \mathcal{O}(G) & \stackrel{\delta_{0i}}{\longleftarrow} & \mathcal{O}(G) \end{array} \tag{2.119}$$

LEMMA 2.72. *If $\delta_i x = A_i x$, $A_i = A_i(x)\in\mathfrak{gl}_n(\mathcal{O}(G))$, then δ is $\mathcal{H}_q$-horizontal with respect to δ_0 if and only if for all i,*

$$\delta_i q + A_i^t q + qA_i = 0. \tag{2.120}$$

By the way, if the conditions of Lemma 2.72 are satisfied and $A_i\in\mathfrak{gl}_n(A)$ then the curvature (F_{ij}) is trivially seen to satisfy:

$$F_{ij}^t q + qF_{ij} = 0,$$

i.e., F_{ij} belong to the Lie algebra $\mathfrak{so}(q)(A)$.

PROOF. Indeed we have $\mathcal{H}_q\delta_{i0}x = 0$ while

$$\begin{aligned} \delta_i\mathcal{H}_q x &= \delta_i(x^t qx) \\ &= (\delta_i x)^t qx + x^t(\delta_i q)x + x^t q(\delta_i x) \\ &= x^t(A_i^t q + \delta_i q + qA_i)x. \end{aligned}$$

So δ is $\mathcal{H}_q$-*horizontal* with respect to δ_0 if and only if

$$\mathcal{H}_q\delta_{i0}x = \delta_i\mathcal{H}_q x$$

hence if and only if 2.120 holds. □

Similarly we have:

LEMMA 2.73. *δ is $\mathcal{B}_q$-symmetric with respect to δ_0 if and only if for all i,*

$$A_i^t q = qA_i. \tag{2.121}$$

Finally we have:

LEMMA 2.74. *δ is $(\mathcal{H}_q,\mathcal{B}_q)$-horizontal with respect to δ_0 if and only if for all i,*

$$A_i = -q^{-1}\cdot\delta_i q. \tag{2.122}$$

PROOF. Recall that

$$\mathcal{O}(G)\otimes_A\mathcal{O}(G) = A[x,y,\det(x)^{-1},\det(y)^{-1}],$$

where

$$x = x\otimes 1,\quad y = 1\otimes x.$$

Then $\mathcal{H}_q(\delta_{0i}x) = 0$. On the other hand

$$\begin{aligned}(\delta_{0i} \otimes 1 + 1 \otimes \delta_i)(\mathcal{B}_q(x)) &= (\delta_{0i} \otimes 1 + 1 \otimes \delta_i)(x^t q y)\\ &= x^t(\delta_i q)x + x^t q A_i(x) x\end{aligned}$$

So δ is $(\mathcal{H}_q, \mathcal{B}_q)$-horizontal with respect to δ_0 if and only if

$$\delta_i q + q A_i(x) = 0$$

which is equivalent to 2.122. □

Assume in what follows that 2 is invertible in A.
The following is then trivial to check:

PROPOSITION 2.75. *For any given q with $q^t = \pm q$ there is a unique connection δ on G that is $\mathcal{H}_q$-horizontal and $\mathcal{B}_q$-symmetric with respect to δ_0; it is given by*

$$A_i = -\frac{1}{2} q^{-1} \delta_i q. \tag{2.123}$$

On the other hand, by Lemma 2.74 we have:

PROPOSITION 2.76. *For any given q with $q^t = \pm q$ there is a unique connection δ on G that is $(\mathcal{H}_q, \mathcal{B}_q)$-horizontal with respect to δ_0; it is given by*

$$A_i = -q^{-1} \delta_i q. \tag{2.124}$$

DEFINITION 2.77. The connection δ in Proposition 2.75 is called the *real Chern connection* attached to q. The connection δ in Proposition 2.76 is called the *complex Chern connection* attached to q

REMARK 2.78. Note that the real and complex Chern connections are, in particular, Ehresmann connections; i.e. A_i belong to $\mathfrak{gl}_n(A)$ rather than to $\mathfrak{gl}_n(\mathcal{O}(G))$.

REMARK 2.79. The real Chern connection is a real analogue of Chern connections in hermitian vector bundles, cf. [**69**], p. 73., and also a generalization of the connection introduced by Duistermaat in [**61**]. The real Chern connection has an arithmetic analogue [**45**] which will play a key role in our book. On the other hand the complex Chern connection is a direct analogue of the Chern connection in hermitian vector bundles, it has an arithmetic analogue, but its arithmetic analogue will be less subtle than the arithmetic analogue of the real Chern connection.

REMARK 2.80. Note that the commutativity of 2.117 is trivially seen to be equivalent to the condition that

$$\delta_i(\mathcal{H}_q(x)) = 0 \tag{2.125}$$

for all i; the latter condition is equivalent to the condition that the entries of $\mathcal{H}_q(x) = x^t q x$ be prime integrals for δ.

REMARK 2.81. Note the following property of Chern connections; this property is trivially satisfied here but later, in the arithmetic context, an analogue of this property will be satisfied in a less trivial way. The property of Chern connections referred to is that the matrices $\delta_i x$ are obtained from the matrices q, $\delta_i q$, and x via "size independent" *matrix* operations, namely:

$$\times, \ -\frac{1}{2}\times, \ (\ )^{-1}.$$

This is in the spirit of free non-commutative function theory [**85**] and is stronger than saying that the entries of $\delta_i x$ are obtained from the entries of $q, \delta_i q$ and x via *scalar* operations on these entries.

REMARK 2.82. For the real Chern connection δ attached to q the curvature and first Chern form are easily seen to be given, for $1 \leq i, j \leq m$, by

$$(2.126)\qquad \begin{array}{rcl} F_{ij} & = & \frac{1}{4}\{q^{-1}(\delta_i q)q^{-1}(\delta_j q) - q^{-1}(\delta_j q)q^{-1}(\delta_i q)\} \\ \rho_{ij} & = & 0. \end{array}$$

Note that F_{ij} depends only of the first (rather than the first and second) order derivatives of q; cf. the discussion in [**61**]. Interestingly the arithmetic analogue of the real Chern connection, to be introduced later, will have a nonvanishing $(2,0)$ component of the first Chern form for $n \geq 4$.

On the other hand assume $\overline{\delta}$ is a connection commuting with δ on A such that $\overline{\delta}x = 0$; then the $(1,1)$-curvature and first Chern $(1,1)$-form of the real Chern connection δ with respect to $\overline{\delta}$ are given by

$$(2.127)\qquad \begin{array}{rcl} F_{i\bar{j}} & = & \frac{1}{2}\delta_{\bar{j}}(q^{-1}\delta_i q). \\ \rho_{i\bar{j}} & = & \frac{1}{2}\text{tr}(\delta_{\bar{j}}(q^{-1}\delta_i q)) = \frac{1}{2}\delta_{\bar{j}}(\det(q)^{-1}\delta_i(\det(q))) \\ & =: & \frac{1}{2}\delta_{\bar{j}}\delta_i \log\det(q) \\ & = & \frac{1}{2}\delta_i\delta_{\bar{j}} \log\det(q). \end{array}$$

Note that the $(1,1)$-curvature and Chern $(1,1)$-form depend on the second derivatives of q; the same will happen in the arithmetic case.

Similar computations can be made for the complex Chern connection.

Let us note the following compatibility of the real Chern connection with the determinant; the arithmetic analogue of the Proposition below will turn out to be false.

PROPOSITION 2.83. *Let $q^t = \pm q \in GL_n(A)$ and let $\delta = (\delta_i)$ be the real Chern connection attached to q. Assume $\delta_i(\det(q)) = 0$. Then*

$$(2.128)\qquad \delta_i(\det(x)) = 0.$$

In particular $\delta_i : \mathcal{O}(GL_n) \to \mathcal{O}(GL_n)$ send the ideal of SL_n into itself and hence induce a connection on SL_n.

PROOF. Set, as usual $\delta_i x = A_i x$. By Jacobi's formula 1.4 and Proposition 2.75 we have

$$\begin{array}{rcl} (\det(x))^{-1}\delta_i(\det(x)) & = & \text{tr}(\delta_i x \cdot x^{-1}) \\ & = & \text{tr}(A_i) \\ & = & -\frac{1}{2}\text{tr}(q^{-1}\delta_i q) \\ & = & -\frac{1}{2}\delta_i(\det(q)) \cdot \det(q)^{-1} = 0. \end{array}$$

$\square$

For our next discussion it is convenient to introduce more (classical) notation. Recall from Definition 2.20 that for any connection (δ_i) on $G = GL_n$, setting

$$A_i = A_i(x) := \delta_i x \cdot x^{-1} \in \mathfrak{gl}_n(\mathcal{O}(G)),$$

we defined the Christoffel symbols of the second kind by the formula:

$$\begin{aligned} \Gamma^k_{ij} := -A_{ikj} &= (k,j) \text{ entry of } (-A_i) \\ &= (j,k) \text{ entry of } A_i^*, \end{aligned}$$

where $A_i^* = -A_i^t$ defines the dual connection $\delta_i^* x = A_i^* x$.

DEFINITION 2.84. Let $\delta = (\delta_i)$ be a connection on $G = GL_n$ and let $q \in GL_n(A)$, $q^t = \pm q$. Then define

$$\Gamma_{ijk} := \Gamma^l_{ij} q_{lk} \text{ (\textit{Christoffel symbols of the 1st kind})}.$$

REMARK 2.85. With our notation condition 2.120 can be rewritten as

$$\delta_i q_{jk} = \Gamma_{ijk} \pm \Gamma_{ikj}; \tag{2.129}$$

and the condition 2.121 can be rewritten as

$$\Gamma_{ijk} = \pm \Gamma_{ikj}. \tag{2.130}$$

In 2.129 and 2.130 the upper (respectively lower) signs correspond to the upper (respectively lower) sign in 2.113. Finally the condition 2.122 can be rewritten as

$$\Gamma_{ijk} = \delta_i q_{jk}. \tag{2.131}$$

Our formulas above agree with the formulas appearing in some of the standard literature, e.g. in [**89**] and [**81**] pp. 11-13; however not all standard literature adopts these index conventions.

With the above notation Proposition 2.75 can be rephrased as follows:

PROPOSITION 2.86. *For any q with $q^t = \pm q$ there is a unique connection δ satisfying the equations* 2.129 *and* 2.130 *(which we called the real Chern connection attached to q); it is given by*

$$\Gamma_{ijk} = \frac{1}{2} \delta_i q_{jk} \tag{2.132}$$

and it is, in particular, an Ehresmann connection.

Similarly we have:

PROPOSITION 2.87. *For any q with $q^t = \pm q$ the complex Chern connection attached to q is given by*

$$\Gamma_{ijk} = \delta_i q_{jk} \tag{2.133}$$

and it is, in particular, an Ehresmann connection.

DEFINITION 2.88. We let

$$q^{kl} := (k,l) \text{ entry of } q^{-1} \text{ (\textit{inverse of the metric})}.$$

REMARK 2.89. In terms of the Christoffel symbols the Riemann tensor of the real Chern connection attached to q is given by:

$$R^k_{lji} = \Gamma^s_{il}\Gamma^k_{js} - \Gamma^s_{jl}\Gamma^k_{is} = \frac{1}{4} q^{km} q^{st} \{(\delta_i q_{ms})(\delta_j q_{tl}) - (\delta_j q_{ms})(\delta_i q_{tl})\}. \tag{2.134}$$

Let us discuss, in what follows, the concepts of mean and scalar $(1,1)$-curvature.

Start with two Ehresmann connections δ and $\overline{\delta}$ on $G = GL_n$, let $(F_{i\overline{j}})$ be the $(1,1)$-curvature, and let $R^k_{li\overline{j}}$ be the (k,l) entry of the matrix $F_{i\overline{j}}$. In particular,

$$\rho_{i\overline{j}} = R^k_{ki\overline{j}}. \tag{2.135}$$

Furthermore, assume A has a complex conjugation, fix a hermitian matrix $h = (h_{i\overline{j}}) \in GL_m(A)$, $h^* = h$, and denote by $h^{i\overline{j}}$ the (i,j) entry of h^{-1}. Following [**88**], p. 99, one can introduce the following:

DEFINITION 2.90. The *mean* $(1,1)$-*curvature* of $(\delta, \overline{\delta})$ with respect to h is the $m \times m$ matrix $K = (K^k_l)$ with A-coefficients,

$$K^k_l := h^{i\overline{j}} R^k_{li\overline{j}}. \tag{2.136}$$

One says that the *weak* $(1,1)$-*Einstein condition* holds if there exists $a \in A$ with the property that

$$K^k_l = a \cdot \delta^k_l, \tag{2.137}$$

where δ^k_l is the Kronecker symbol. The *scalar* $(1,1)$-*curvature* is

$$\sigma := h^{i\overline{j}} \rho_{i\overline{j}}. \tag{2.138}$$

REMARK 2.91. Note that the $(2,0)$ counterpart of the above concepts is uninteresting. Indeed if one is given a symmetric matrix $g = (g_{ij}) \in GL_m(A)$ and a connection $\delta = (\delta_i)$ we have that, for each k, l the sum

$$g^{ij} R^k_{lij}$$

vanishes because R^k_{lij} is anti-symmetric in i, j.

REMARK 2.92. As a final remark about the discussion above we recall that Chern forms of C^∞ connections in complex bundles on manifolds give rise to Chern classes of the bundles in question, generating *characteristic rings* inside the cohomology ring of the manifold. In the case of smooth projective manifolds X and algebraic vector bundles (locally free sheaves) E the picture is as follows. If P is the principal GL_n-bundle attached to E and T is the tangent bundle of X then there is a natural extension (the *Atiyah extension*):

$$0 \to \text{End}_{\mathcal{O}}(E) \to Q \to T \to 0 \tag{2.139}$$

where Q is the vector bundle on X whose sections are the invariant vector fields on P and $\text{End}_{\mathcal{O}}(E)$ is the sheaf of $\mathcal{O}$-module endomorphisms of E. The algebraic connections on X are in bijection with the splittings of 2.139. And the obstruction to the splitting of 2.139, viewed as a class in

$$Ext^1(T, \text{End}_{\mathcal{O}}(E)) = H^1(X, \text{End}_{\mathcal{O}}(E) \otimes \Omega),$$

generates (via the operation of invariant polynomials plus Hodge decomposition) the characteristic ring of E inside the cohomology ring

$$H^*(X, \mathbb{C});$$

cf. [**3**]. It would be interesting to find an arithmetic analogue, in the spirit of the present book, of this circle of ideas. (What we have in mind is, therefore, quite different from, and more radical than, the approach in [**17**] which involves "usual" derivations in the geometric direction rather than our operators $\delta_p = d/dp$ in the arithmetic direction.) Our arithmetic analogues of connections will be adelic, hence "rather non-algebraic," so they may be viewed as analogues of C^∞ connections encoding cohomology-type information about $Spec\ \mathbb{Z}$.

2.6. Chern connections: the hermitian case

We introduce here Chern connections attached to hermitian metrics; they are analogues of Chern connections in hermitian vector bundles [**69, 88**]. We continue to assume we have pairwise commuting derivations $\delta_1, ..., \delta_m$ on A and we continue to assume n and m are not necessarily equal. We consider the case $n = 2r$ is even. Also we assume 2 is invertible in A and A contains a square root $\sqrt{-1}$ of -1 which we fix in our discussion. Recall the closed subgroup GL_r^c of GL_{2r} whose points are the matrices 2.75. Let q be a hermitian metric in $GL_{2r}(A)$, cf. 2.79; recall that by this we mean

$$q = \begin{pmatrix} q_1 & q_2 \\ -q_2 & q_1 \end{pmatrix} \in GL_r^c(A) \subset GL_{2r}(A) \tag{2.140}$$

satisfies $q^t = q$, equivalently

$$q_1^t = q_1, \quad q_2^t = -q_2. \tag{2.141}$$

Moreover consider the matrix

$$q^c := q_1 + \sqrt{-1} \cdot q_2 \in GL_r(A). \tag{2.142}$$

Let $x = \begin{pmatrix} a & b \\ c & d \end{pmatrix}$ be a $2r \times 2r$ matrix of variables, hence

$$\mathcal{O}(GL_{2r}) = A[x, \det(x)^{-1}], \tag{2.143}$$

and let v, w be matrices of $r \times r$ variables. Consider now the natural isomorphism

$$GL_r^c \to GL_r \times GL_r \tag{2.144}$$

given by the A-algebra map from

$$\mathcal{O}(GL_r \times GL_r) = A[v, \det(v)^{-1}] \otimes_A A[w, \det(w)^{-1}] \tag{2.145}$$

to

$$\mathcal{O}(GL_r^c) = A[x, \det(x)^{-1}]/(a - d, b + c) = A[a, b, \det \begin{pmatrix} a & b \\ -b & a \end{pmatrix}^{-1}] \tag{2.146}$$

sending v and w into the classes of

$$x^c := a + \sqrt{-1} \cdot b, \quad \text{and} \quad x^{tct} := a - \sqrt{-1} \cdot b,$$

respectively; cf. 2.76.

Let $(\delta_i^{GL_{2r}})$ be the real Chern connection attached to a hermitian metric $q \in GL_{2r}(A)$. By 2.123,

$$\delta_i^{GL_{2r}} x = \alpha_i x \tag{2.147}$$

where

$$\alpha_i := -\frac{1}{2} q^{-1} \delta_i q. \tag{2.148}$$

Note, by the way, that if $\check{q} := -q q_0 = -q_0 q$ is the hermitian form attached to q (where q_0 is the split antisymmetric matrix) then

$$-\frac{1}{2} \check{q}^{-1} \delta_i \check{q} = -\frac{1}{2} (-q^{-1} q_0^{-1}) \delta_i (-q_0 q) = -\frac{1}{2} q^{-1} \delta_i q \tag{2.149}$$

so the real Chern connection on GL_{2r} attached to q coincides with the real Chern connection attached to $\check{q}$.

On the other hand note that

$$\alpha_i^c = -\frac{1}{2}(q^c)^{-1}(\delta_i q)^c = -\frac{1}{2}(q^c)^{-1}\delta_i(q^c). \tag{2.150}$$

Denote by $\delta_i^{GL_r}$ the derivation on $A[v, \det(v)^{-1}]$ which is δ_i on A and satisfies

$$\delta_i^{GL_r} v = \alpha_i^c v. \tag{2.151}$$

DEFINITION 2.93. The *real Chern connection* on GL_r attached to q^c is the connection $(\delta_i^{GL_r})$ satisfying 2.151.

The real Chern connection attached to q^c was introduced above analytically, by formula 2.151. However there is an intrinsic meaning to this connection which we now describe by giving a geometric link between the real Chern connection $(\delta_i^{GL_r})$ attached to q^c and the real Chern connection $(\delta_i^{GL_{2r}})$ attached to q. Indeed the derivation $\delta_i^{GL_{2r}}$ on the ring 2.143 induces a derivation, which we denote by $\delta_i^{GL_r^c}$, on the ring 2.146 (which we could refer to as the *real Chern connection* on GL_r^c attached to q). Let $c : GL_r^c \to GL_r$ be the composition of the isomorphism 2.144 with the first projection and continue to denote by $c : \mathcal{O}(GL_r) \to \mathcal{O}(GL_r^c)$ the induced algebra map. Then we have:

PROPOSITION 2.94. *The following diagrams are commutative:*

$$\begin{array}{ccc} \mathcal{O}(GL_r^c) & \stackrel{\delta_i^{GL_r^c}}{\longleftarrow} & \mathcal{O}(GL_r^c) \\ c\uparrow & & \uparrow c \\ \mathcal{O}(GL_r) & \stackrel{\delta_i^{GL_r}}{\longleftarrow} & \mathcal{O}(GL_r) \end{array} \tag{2.152}$$

Note that since $\delta_i^{GL_{2r}}$ has an intrinsic meaning, so does $\delta_i^{GL_r^c}$, and hence so does $\delta_i^{GL_r}$.

PROOF OF PROPOSITION 2.94. View the algebra map c as an inclusion. We need to show that the derivation $\delta_i^{GL_r^c}$ sends $A[v, \det(v)^{-1}]$ into itself and its restriction to $A[v, \det(v)^{-1}]$ equals $\delta_i^{GL_r}$. Set

$$\alpha_i = \begin{pmatrix} \alpha_{i1} & \alpha_{i2} \\ -\alpha_{i2} & \alpha_{i1} \end{pmatrix}.$$

Then, by 2.147, we have

$$\begin{aligned} \delta_i^{GL_{2r}} v &= \delta_i^{GL_{2r}}(a + \sqrt{-1} \cdot b) \\ &= \alpha_{i1}a - \alpha_{i2}b + \sqrt{-1}(\alpha_{i1}b + \alpha_{i2}a) \\ &= \alpha_i^c \cdot v, \end{aligned}$$

which proves our Proposition. □

REMARK 2.95. Assume in addition that A contains a square root $\sqrt{-1}$ of -1 and has a complex conjugation. Let $\delta = (\delta_i^{GL_r})$ be the real Chern connection on GL_r attached to q^c, where q is a hermitian metric with $\bar{q} = q$. Moreover assume $\bar{\delta} = (\delta_{\bar{i}})$ is a connection commuting with δ on A such that $\bar{\delta}x = 0$; then, exactly as in 2.127 the $(1,1)$-curvature and first Chern $(1,1)$-form of δ with respect to $\bar{\delta}$ are given by

$$
\begin{aligned}
F_{i\bar{j}} &= \tfrac{1}{2}\delta_{\bar{j}}((q^c)^{-1}\delta_i q^c). \\
\rho_{i\bar{j}} &= \tfrac{1}{2}\delta_i\delta_{\bar{j}} \log\det(q^c).
\end{aligned} \tag{2.153}
$$

The right hand side of the second equation 2.153 is (up to a factor) formally the same as the expression of the Ricci form of a hermitian vector bundle in [**89**], Vol. II, p. 184. Note that the $\rho_{i\bar{j}}$ in 2.153 is directly related to $\rho_{i\bar{j}}$ in 2.127 because by 2.81 we have

$$\frac{1}{2}\delta_i\delta_{\bar{j}} \log\det(q) = \delta_i\delta_{\bar{j}} \log\det(q^c).$$

A similar discussion can be made for the complex Chern connection.

Indeed let $(\delta_i^{GL_{2r}})$ be the complex Chern connection attached to a hermitian metric $q \in GL_{2r}(A)$. So

$$\delta_i^{GL_{2r}} x = \beta_i x \tag{2.154}$$

where

$$\beta_i := -q^{-1}\delta_i q. \tag{2.155}$$

Again, if $\check{q} := -qq_0 = -q_0 q$ is the hermitian form attached to q (where q_0 is the split antisymmetric matrix) then the complex Chern connection on GL_{2r} attached to q coincides with the complex Chern connection attached to $\check{q}$. Also, again,

$$\beta_i^c = -(q^c)^{-1}(\delta_i q)^c = -(q^c)^{-1}\delta_i(q^c). \tag{2.156}$$

Denote by $\delta_i^{GL_r}$ the derivation on $A[v, \det(v)^{-1}]$ which is δ_i on A and satisfies

$$\delta_i^{GL_r} v = \beta_i^c v. \tag{2.157}$$

DEFINITION 2.96. The *complex Chern connection* on GL_r attached to q^c is the connection $(\delta_i^{GL_r})$ satisfying 2.157.

Then Proposition 2.94 still holds for the complex Chern connections.

In the discussion of Chern connections n and m were unrelated; morally this corresponds to the fact that we are considering a "gauge theoretic" situation with group GL_n and "basis" of "dimension" m. On the other hand, in case $n = m$, one can consider extra-symmetries (as is customary in the study of connections on the frame bundle of the tangent bundle of a manifold), leading, for instance, to Levi-Cività connections or Fedosov connections. We quickly review this case below.

2.7. Levi-Cività connection and Fedosov connection

Assume, in this section, that A is a ring in which 2 is invertible and assume A is equipped with m derivations $\delta = (\delta_i)$. As usual, set $G = GL_n$ and assume that $n = m$. (More invariantly we assume that a bijection is given between the set labeling the derivations in δ and the set labeling the rows/columns in the matrix x; fixing such a bijection can be viewed as an analogue of Cartan's *soldering* [**116**], p. 351.)

PROPOSITION 2.97. *Let $\delta = (\delta_i)$ be a connection on G. Let Γ_{ij}^k be its Christoffel symbols of the second kind, Γ_{ijk} its Christoffel symbols of the first kind with respect to some $q \in GL_n(A)$ with $q^t = \pm q$, let $\delta^* = (\delta_i^*)$ be the dual connection, and let $x^{-1} = (x^{kj})$ be the inverse of $x = (x_{kj})$. Then the following conditions are equivalent:*

$$\Gamma_{ijk} = \Gamma_{jik}; \tag{2.158}$$

$$\Gamma_{ij}^k = \Gamma_{ji}^k; \tag{2.159}$$

$$\delta_i^* x_{jk} = \delta_j^* x_{ik}; \tag{2.160}$$

$$\delta_i x^{kj} = \delta_j x^{ki}. \tag{2.161}$$

DEFINITION 2.98. A connection on G is *torsion free* if the conditions in Proposition 2.97 are satisfied.

Note that in the definition above δ is not assumed to be Ehresmann. Classically one only makes this definition for Ehresmann connections and one refers to torsion free connections also as *symmetric* connections; we will not use here the term *symmetric* for these connections in order to avoid confusion with our concept of $\mathcal{B}_q$*-symmetry*.

PROPOSITION 2.99. *If δ is an Ehresmann torsion free connection then the Riemann tensor satisfies*

$$R_{ijk}^m + R_{jki}^m + R_{kij}^m = 0 \quad \textit{(Bianchi)}.$$

PROOF. The "standard argument" applies: write

$$\delta_i^* \delta_j^* x_{kl} - \delta_j^* \delta_i^* x_{kl} = R_{kij}^m x_{ml},$$

permute i, j, k cyclically, take the sum, and use 2.160. □

REMARK 2.100. Note the difference between the symmetry conditions 2.130 and 2.158: the first condition is a symmetry in the last 2 indices (labeling the rows and columns of matrices in GL_n), whereas the second condition is a symmetry in the first index (labeling derivations) and the second index (labeling the rows of matrices in GL_n). By the way, for an Ehresmann connection, the system consisting of 2.129, 2.130, 2.158, (with q given and with Γ_{ijk} viewed as unknowns in A) has a solution if and only if q satisfies the following equations:

$$\delta_i q_{jk} = \delta_j q_{ik}. \tag{2.162}$$

DEFINITION 2.101.

(1) A matrix $u \in GL_n(A)$ is called *Jacobian* if

$$\delta_i u_{jk} = \delta_j u_{ik}. \tag{2.163}$$

(2) A matrix $q \in GL_n(A)$ is called *Hessian* if it is symmetric (i.e., $q^t = q$) and Jacobian (i.e., 0.34 holds).

DEFINITION 2.102.

(1) A tuple $(\eta_1, ..., \eta_n) \in A^n$ is called a *Jacobian potential* for $u \in GL_n(A)$ if

$$u_{jk} = \delta_j \eta_k. \tag{2.164}$$

(2) An element $\psi \in A$ is called a *Hessian potential* for $q \in GL_n(A)$ if

$$q_{ij} = \delta_i \delta_j \psi. \tag{2.165}$$

Clearly, we have:

PROPOSITION 2.103.

(1) *If u admits a Jacobian potential then u is Jacobian.*
(2) *If q admits a Hessian potential then q is Hessian.*

DEFINITION 2.104. The *deRham* complex of the δ-ring A is the complex of $\mathbb{Z}$-modules

$$A \xrightarrow{d} \Lambda^1(A) \xrightarrow{d} \Lambda^2(A) \xrightarrow{d} ...,$$

$$\begin{array}{rcl} da & = & \sum(\delta_i a) d\xi_i, \\ d(\sum a_i d\xi_i) & = & \sum da_i \wedge d\xi_i, \\ d(\sum a_{ij} d\xi_i \wedge d\xi_j) & = & \sum(da_{ij}) \wedge d\xi_i \wedge d\xi_j, ... \end{array}$$

where $a, a_i, a_{ij}, ... \in A$. We say that the *$i$-Poincaré lemma holds* in A if the *deRham* complex is exact in $\Lambda^i(A)$.

REMARK 2.105. $(\eta_1, ..., \eta_n)$ is a Jacobian potential if and only if

$$u_{jk} d\xi_j = d\eta_k \quad \text{(Einstein summation)} \tag{2.166}$$

in the deRham group $\Lambda^1(A)$. In matrix notation

$$d\xi \cdot u = d\eta.$$

We trivially have:

PROPOSITION 2.106.

(1) *If u is Jacobian and the 1-Poincaré lemma holds in A then u admits a Jacobian potential.*
(2) *If q is Hessian and the 1-Poincaré lemma holds in A then q admits a Hessian potential.*

Clearly we have:

PROPOSITION 2.107. *Assume $u \in GL_n(A)$ is a horizontal section for δ^* (hence, in particular, δ^*, and hence δ, is flat). Assume furthermore that δ is torsion free. Then u is Jacobian. So if in addition the 1-Poincaré lemma holds in A then u admits a Jacobian potential.*

REMARK 2.108. The significance of Jacobian potentials is as follows. Assume, in Proposition 2.107 that

$$\delta_i x = A_i x, \quad A_i = (A_{ijk}), \quad \delta_i u = -A_i^t u, \quad u_{jk} d\xi_j = d\eta_k, \quad \eta_k \in A,$$

and define "connection-like" maps

$$\nabla_i : \Lambda^1(A) \to \Lambda^1(A), \quad \nabla : \Lambda^1(A) \to \Lambda^1(A) \otimes \Lambda^1(A), \tag{2.167}$$

by the formulas

$$\nabla_i(a_j d\xi_j) = (\delta_i a_j) d\xi_j + a_j A_{ijk} d\xi_k, \quad \nabla\alpha = (\nabla_i \alpha) \otimes d\xi_i. \tag{2.168}$$

Then

$$\begin{aligned} \nabla_i(d\eta_k) &= \nabla_i(u_{jk} d\xi_j) \\ &= (\delta_i u_{jk}) d\xi_j + u_{jk} A_{ijl} d\xi_l \\ &= -A_{ilj} u_{lk} d\xi_j + u_{jk} A_{ijl} d\xi_l = 0, \end{aligned}$$

hence

$$\nabla(d\eta_k) = 0. \tag{2.169}$$

This is an algebraic analogue of the classical fact that if a flat torsion free connection is given on the tangent bundle of a manifold then locally there are coordinates with respect to which the Cristoffel symbols vanish; cf. [**82**], p. 143.

The following is trivial to check and is referred to as the Fundamental Theorem of Riemannian Geometry:

THEOREM 2.109. *If $q^t = q$, there is a unique connection δ that is $\mathcal{H}_q$-horizontal with respect to δ_0 (equivalently, satisfies the equations 2.129 with the + sign) and torsion free (i.e., satisfies 2.158). It is given by*

$$\Gamma_{kij} = \frac{1}{2}\left(\delta_k q_{ij} + \delta_i q_{jk} - \delta_j q_{ki}\right). \tag{2.170}$$

DEFINITION 2.110. The connection δ in Theorem 2.109 is called the *Levi-Cività* connection attached to q.

REMARK 2.111. Note that the Levi-Cività connection is, in particular, an Ehresmann connection. Also the formulae 0.33 imply that $\Gamma_{ijk} = 0$ for all i, j, k if and only if $\delta_i q_{jk} = 0$ for all i, j, k.

On the other hand, by Remark 2.100, we have:

PROPOSITION 2.112. *The Levi-Cività connection and the real Chern connection attached to a given $q = q^t$ coincide if and only if q is Hessian.*

REMARK 2.113. For any ring A in which 2 is invertible denote by A^{n^3} the collection of all families (Γ_{ijk}) indexed by $i, j, k = 1, ..., n$ with $\Gamma_{ijk} \in A$; we can refer to these families as 3-*tensors* with components in A. Then we can consider the spaces of 3-tensors

$$A^{n^3}_{12} = \{(\Gamma_{ijk}) \in A^{n^3} \mid \Gamma_{ijk} = \Gamma_{jik}\}, \tag{2.171}$$

$$A^{n^3}_{23} = \{(\Gamma_{ijk}) \in A^{n^3} \mid \Gamma_{ijk} = \Gamma_{ikj}\}, \tag{2.172}$$

$$A^{n^3}_{13} = \{(\Gamma_{ijk}) \in A^{n^3} \mid \Gamma_{ijk} = \Gamma_{kji}\}. \tag{2.173}$$

The intersection

$$A^{n^3}_s := A^{n^3}_{12} \cap A^{n^3}_{23} \cap A^{n^3}_{13} \tag{2.174}$$

is the space of *symmetric* 3-tensors. (Of course any of the 3 sets in the right hand side of 2.174 can be removed.) It is trivial to check that the map

$$A_{12}^{n^3} \to A_{23}^{n^3}, \tag{2.175}$$

given by

$$(\Gamma_{ijk}) \mapsto \left(\frac{1}{2}(\Gamma_{ijk} + \Gamma_{ikj})\right) \tag{2.176}$$

is an A-module isomorphism; its inverse is the map

$$A_{23}^{n^3} \to A_{12}^{n^3}, \tag{2.177}$$

given by

$$(\Gamma_{ijk}) \mapsto (\Gamma_{ijk} + \Gamma_{jik} - \Gamma_{kij}). \tag{2.178}$$

Two elements of $A_{12}^{n^3}$ and $A_{23}^{n^3}$ respectively that correspond to each other via the above isomorphism will be called *dual* to each other. So if we denote by

$$(\Gamma'_{ijk}) \in A_{23}^{n^3}, \quad (\Gamma''_{ijk}) \in A_{12}^{n^3}$$

the Christoffel symbols of the 1st kind of the real Chern and Levi-Cività connections attached to some q with $q^t = q$ respectively then (Γ'_{ijk}) and (Γ''_{ijk}) are dual to each other.

REMARK 2.114. Assume, again, the notation of Remark 2.113. For any

$$(\Gamma_{ijk}) \in A^{n^3}$$

let Γ_i be the $n \times n$ matrix with entries $(\Gamma_i)_{jk} := \Gamma_{ijk}$. Clearly the map

$$A^{n^3} \to A^{n^3}, \quad ((\Gamma_i)_{jk}) \mapsto ((\Gamma_i^t)_{jk}) \tag{2.179}$$

induces a bijection

$$A_{12}^{n^3} \to A_{13}^{n^3}. \tag{2.180}$$

Let E be any $n \times n$ matrix with entries in A. Then the map

$$A^{n^3} \to A^{n^3}, \quad ((\Gamma_i)_{jk}) \mapsto ((\Gamma_i E)_{jk}) \tag{2.181}$$

sends $A_{12}^{n^3}$ into itself. Similarly the map

$$A^{n^3} \to A^{n^3}, \quad ((\Gamma_i)_{jk}) \mapsto ((E\Gamma_i)_{jk}) \tag{2.182}$$

sends $A_{13}^{n^3}$ into itself.

DEFINITION 2.115. In the case of the Levi-Cività connection set:

$$\begin{aligned} R_{ij} &:= R^k_{ikj} \quad (\textit{Ricci tensor}) \\ R &:= q^{ij}R_{ij} \quad (\textit{scalar curvature}) \end{aligned} \tag{2.183}$$

We say q is *Einstein* if there exists $a \in A$, $\delta_i a = 0$, such that

$$R_{ij} = a \cdot q_{ij}. \tag{2.184}$$

We say that q has *constant sectional curvature* if there exists $\kappa \in A$, $\delta_i \kappa = 0$, such that

$$q_{im} R^m_{jkl} = \kappa(q_{ik}q_{jl} - q_{jk}q_{il}).$$

Clearly if q has constant sectional curvature then q is Einstein.

Note also that if one defines

$$R_{ijkl} := q_{im}R^m_{jkl} \quad \textit{(covariant Riemann tensor)}$$

then one has the following classical symmetries:

$$R_{ijkl} = -R_{ijlk}, \quad R_{ijkl} = -R_{jikl}, \quad R_{lijk} + R_{ljki} + R_{lkij} = 0, \quad R_{ijkl} = R_{klij}.$$

Finally we introduce Fedosov connections; cf. [**62, 66, 119**]. The natural hypothesis here is that 6 is invertible in A which we assume from now on.

DEFINITION 2.116. If $q^t = -q$, a connection δ is called a *Fedosov connection* (relative to q) if it is $\mathcal{H}_q$-horizontal with respect to δ_0 (i.e., it satisfies equations 2.129 with the $-$ sign) and torsion free (i.e., satisfies 2.158).

DEFINITION 2.117. A matrix $q \in GL_n(A)$ with $q^t = -q$ is called *symplectic* if

$$\delta_i q_{jk} + \delta_j q_{ki} + \delta_k q_{ij} = 0. \tag{2.185}$$

Condition 2.185 is of course equivalent to the condition

$$d(q_{ij}d\xi_i \wedge d\xi_j) = 0 \tag{2.186}$$

in the group $\Lambda^3(A)$ of the deRham complex of A.

The following are trivial to check:

PROPOSITION 2.118. *If $q^t = -q$ and there exists a Fedosov connection relative to q then q is symplectic.*

PROPOSITION 2.119. *If $q^t = -q$ is symplectic then there exists an Ehresmann connection which is Fedosov relative to q. One such connection is given by:*

$$\Gamma_{ijk} = \frac{1}{3}(\delta_i q_{jk} + \delta_j q_{ik}). \tag{2.187}$$

Unlike the Levi-Cività connection the Fedosov connections for a given q are not necessarily Ehresmann; and those that are Ehresmann are not unique. Indeed:

PROPOSITION 2.120. *If $q^t = -q$ is symplectic then the set of Ehresmann connections that are Fedosov relative to q is a principal homogenous space for the additive group of symmetric 3-tensors $A^{n^3}_s$; cf. 2.174. Also the set of (not necessarily Ehresmann) connections that are Fedosov relative to q is a principal homogenous space for the corresponding additive group of symmetric 3-tensors $B^{n^3}_s$, $B = A[x, \det(x)^{-1}]$.*

In particular the Christoffel symbols of the first kind of the Ehresmann connections that are Fedosov relative to q are obtained by adding to 2.187 an arbitrary element of the group 2.174.

Finally we have:

PROPOSITION 2.121. *If $q^t = -q$, conditions 2.129 (with the $-$ sign), 2.130 (with the $-$ sign), and 2.158 imply*

$$\delta_k q_{ij} = 0, \quad \Gamma_{ijk} = 0, \quad \delta = \delta_0. \tag{2.188}$$

In other words if the real Chern connection attached to q is Fedosov then δ must be trivial.

PROOF. The first equality in 2.188 follows by combining 0.34, 2.185; the second follows from 2.132; the third follows from the second. $\square$

We end by recording the Kähler condition. We will not pursue the Kähler paradigm further, though, because no arithmetic analogue of it seems to be available at this point.

DEFINITION 2.122. Assume $n = m = 2r$. Let $q \in GL_{2r}(A)$ be a hermitian metric, i.e. $q^t = q$ and q commutes with the split antisymmetric matrix q_0. Let $\check{q} = -qq_0 = (\check{q}_{ij})$ be the associated hermitian 2-form. Then q is a *Kähler metric* if $\check{q}$ is symplectic, i.e., if

$$\delta_i \check{q}_{jk} + \delta_j \check{q}_{ki} + \delta_k \check{q}_{ij} = 0. \tag{2.189}$$

Assume in addition that A contains $\sqrt{-1}$ and has a complex conjugation. Assume moreover that $\delta_i(A^{real}) \subset A^{real}$, $i = 1, ..., 2r$, and consider the following derivations on A:

$$\delta_i^c = \frac{1}{2}(\delta_i - \sqrt{-1}\delta_{i+r}), \quad \delta_{\bar{i}}^c = \frac{1}{2}(\delta_i + \sqrt{-1}\delta_{i+r}), \quad i = 1, ..., r. \tag{2.190}$$

A formal computation yields:

PROPOSITION 2.123. *Let $q \in GL_{2r}(A)$ be a hermitian metric with $\bar{q} = q$ and let $h_{i\bar{j}} = q_{i\bar{j}}^c$ for $i, j = 1, ..., r$. If q is Kähler then we have*

$$\delta_k^c h_{i\bar{j}} = \delta_i^c h_{k\bar{j}}, \quad \delta_{\bar{k}}^c h_{j\bar{i}} = \delta_{\bar{i}}^c h_{j\bar{k}}.$$

The latter makes the Kähler condition on q an analogue of the Hessian condition and implies, under an obvious Poincaré lemma-type condition, the existence of a "Kähler potential," i.e., a $\psi \in A$ such that $h_{i\bar{j}} = \delta_i^c \delta_{\bar{j}}^c \psi$. Then the Levi-Cività connection of q, and hence its curvature, are "nicely" expressible in terms of the one function ψ.

2.8. Locally symmetric connections

We continue to assume $n = m$. Recall that this morally corresponds to the case when our principal bundle is the frame bundle (of the tangent bundle). First we recall the definition of covariant derivative of a tensor.

DEFINITION 2.124. Let $\delta = (\delta_i)$ be an Ehresmann connection with Christoffel symbols of the second kind (Γ_{ij}^k). A family of elements of A,

$$(K_{li...j}^{uv...w})$$

will be called a *tensor* of type (a, b) if there are a indices $u, v, ..., w$ and b indices $l, i, ..., j$. The *covariant derivative* of such a tensor is the tensor of type $(a, b+1)$,

$$(\nabla_s K_{li...j}^{uv...w})$$

defined by

$$\begin{aligned} \nabla_s K_{li...j}^{uv...w} \quad := \quad & \delta_s K_{li...j}^{uv...w} \\ & -K_{mi...j}^{uv...w}\Gamma_{sl}^m - K_{lm...j}^{uv...w}\Gamma_{si}^m - ... \\ & +K_{li...j}^{mv...w}\Gamma_{sm}^u + K_{li...j}^{um...w}\Gamma_{sm}^v + ... \end{aligned}$$

For the next Lemma let δ be an Ehresmann connection,

$$\delta_i x = A_i x, \quad A_i = (A_{ikj}) = (-\Gamma_{ij}^k)$$

and consider the curvature

$$\varphi_{ij}(x) = F_{ij}x, \quad F_{ij} = (F_{ijkl}) = (-R^k_{lij}),$$

and the 3-curvature

$$\varphi_{ijk}(x) = (\delta_i F_{jk} - [A_i, F_{jk}])x.$$

A trivial computation yields:

LEMMA 2.125. *The following are equivalent:*

(1) $\nabla_s R^k_{lij} = 0$ *for all* s, l, i, j, k.
(2) $\delta_s F_{ij} - [A_s, F_{ij}] = \sum_m A_{smj} F_{mi} - \sum_m A_{smi} F_{mj}$ *for all* s, i, j.
(3) $\varphi_{sij} = \sum_m A_{smj} \varphi_{mi} - \sum_m A_{smi} \varphi_{mj}$ *for all* s, i, j.

DEFINITION 2.126. ([**89**], p. 303). An Ehresmann connection is said to have *parallel* curvature if the equivalent conditions in Lemma 2.125 are satisfied. An Ehresmann connection is called *locally symmetric* if it is torsion free and has parallel curvature.

Trivially, if a metric q has constant sectional curvature then the Levi-Cività connection attached to q is locally symmetric.

COROLLARY 2.127. *If an Ehresmann connection has parallel curvature (in the sense of Definition* 2.126*) then it has A-horizontal curvature (in the sense of Definition* 2.33*).*

REMARK 2.128. The concept of a connection having parallel curvature is only defined in case $n = m$. On the other hand the concept of a connection having A-horizontal curvature is defined in case n and m are not necessarily related.

REMARK 2.129. Our tensors in Definition 2.124 can be of course identified with elements in corresponding tensor products of copies of $\Lambda^1(A)$ and its A-dual. Then the covariant derivative in Definition 2.124 defines "connection-like" maps between the corresponding tensor products (as in 2.167, 2.168) in the style of classical tensor calculus.

2.9. Ehresmann connections attached to inner involutions

Chern and Levi-Cività connections are Ehresmann connections attached to certain outer involutions on GL_n. Here is a remark on the analogue of Chern connections for inner involutions. We continue to assume n and m are not necessarily equal. Consider an inner involution on $G = GL_n$ defined by

$$x^\tau = q^{-1}xq \tag{2.191}$$

with $q \in GL_n(A)$, $q^2 = 1$. Then the quadratic map $\mathcal{H}$ attached to τ (cf. 2.94) is given by

$$\mathcal{H}(x) = q^{-1}x^{-1}qx \tag{2.192}$$

and the fixed group G^+ of τ is the centralizer $C(q)$ of q,

$$G^+ = C(q) := \{x \in G; xq = qx\}. \tag{2.193}$$

The Lie algebra of $C(q)$ is, of course,

$$\mathfrak{c}(q) = \{\alpha \in \mathfrak{gl}_n; \alpha q = q\alpha\}. \tag{2.194}$$

Now if $\delta = \delta^G = (\delta_i)$ is the Ehresmann connection on $G = GL_n$ defined by

$$\delta_i x = \alpha_i \cdot x, \qquad \alpha_i \in \mathfrak{gl}_n(A),$$

and if $\delta_i q = 0$ for all i we have

$$\begin{aligned} \delta_i(\mathcal{H}(x)) &= -q^{-1}x^{-1}(\alpha_i x)x^{-1}qx + q^{-1}x^{-1}q(\alpha_i x) \\ &= q^{-1}x^{-1}[q, \alpha_i]x. \end{aligned} \tag{2.195}$$

Hence

$$\delta_i(\mathcal{H}(x)) = 0 \quad \text{iff} \quad \alpha_i \in \mathfrak{c}(q). \tag{2.196}$$

Recall that the condition

$$\delta_i(\mathcal{H}(x)) = 0, \quad i = 1, ..., m, \tag{2.197}$$

is equivalent to the condition that the entries of $\mathcal{H}$ be prime integrals for δ and also equivalent to the condition that the following diagrams be commutative:

$$\begin{array}{ccc} \mathcal{O}(G) & \stackrel{\delta_i}{\longleftarrow} & \mathcal{O}(G) \\ \mathcal{H}\uparrow & & \uparrow\mathcal{H} \\ \mathcal{O}(G) & \stackrel{\delta_{0i}}{\longleftarrow} & \mathcal{O}(G) \end{array} \tag{2.198}$$

where $\delta_{0i}x = 0$. Clearly 2.197 is analogous to 2.125. Now 2.125 will have an arithmetic analogue; but 2.197 will not have a direct analogue in the arithmetic setting.

2.10. Connections in vector bundles

We include, in what follows, a short digression on the "vector bundle" version of the above "principal bundle" formalism. For simplicity we assume A is a field of characteristic 0 equipped with one derivation δ; so $m = 1$. Start with an n-dimensional vector space V over A.

DEFINITION 2.130. A *connection* on V is an additive group homomorphism $\nabla : V \to V$ such that

$$\nabla(av) = (\delta a)v + a\nabla v$$

for all $v \in V$, $a \in A$.

Assume we are given a non-degenerate A-bilinear map

$$B : V \times V \to V$$

which is either symmetric or anti-symmetric (which we view as an analogue of a metric or a 2-form respectively).

DEFINITION 2.131. ∇ is *B-horizontal* if

$$\delta(B(u,v)) = B(\nabla u, v) + B(u, \nabla v) \tag{2.199}$$

for all $u, v \in V$.

If this is the case we view ∇ as an analogue of a connection that is compatible with a metric or a 2-form, respectively.

DEFINITION 2.132. Given an A-linear endomorphism Λ of V one says that Λ is *B-symmetric* if

$$B(\Lambda u, v) = B(u, \Lambda v) \tag{2.200}$$

for all $u, v \in V$.

PROPOSITION 2.133. *For any given non-degenerate bilinear map B and any connection ∇_0 there is exactly one connection ∇ such that ∇ is B-horizontal and $\nabla - \nabla_0$ is B-symmetric.*

PROOF. The uniqueness is clear. The existence is proved exactly in the same way one proves the existence of Chern connections in differential geometry. Indeed, let V^* be the A-linear dual of V, equipped with the *dual connection*

$$\nabla_0^* : V^* \to V^*$$

defined as the unique connection with the property that

$$\delta\langle u^*, v\rangle = \langle \nabla_0^* u^*, v\rangle + \langle u^*, \nabla_0 v\rangle,$$

for $u^* \in V^*$ and $v \in V$; here $\langle\ ,\ \rangle : V^* \times V \to A$ is the duality pairing. Let $B^* : V \to V^*$ be the linear map defined by

$$\langle B^*(u), v\rangle := B(u, v).$$

Then define the connection $\nabla : V \to V$ by

$$\nabla := \frac{1}{2}\nabla_0 + \frac{1}{2}\cdot (B^*)^{-1} \circ \nabla_0^* \circ B^*. \tag{2.201}$$

It is easy to see that ∇ is B-horizontal and $\nabla - \nabla_0$ is B-symmetric. □

The concepts of B-horizontality and B-symmetry relate to our concepts of $\mathcal{H}_q$-horizontality and $\mathcal{B}_q$-symmetry as follows. Start with a matrix $\alpha \in \mathfrak{gl}_n(A)$ and consider the A-linear space $V = A^n$ whose elements we view as column vectors. Consider the connection

$$\nabla : V \to V, \quad \nabla v = \delta v - \alpha v,$$

for $v \in V$, where if $v = (v_i)$ has components v_i then $\delta v := (\delta v_i)$. Also consider the connections δ and δ_0 on $G = GL_n = Spec\ A[x, \det(x)^{-1}]$ defined by $\delta x = \alpha x$ and $\delta_0 x = 0$ respectively. Now for $q \in GL_n$ with $q^t = \pm q$ we consider the non-degenerate bilinear (symmetric, respectively anti-symmetric) map

$$B : V \times V \to A, \quad B(u, v) := u^t q v.$$

It is trivial then to see that, with the notation above, ∇ is B-horizontal if and only if δ is $\mathcal{H}_q$-horizontal with respect to δ_0. On the other hand we may consider the connection

$$\nabla_0 : V \to V, \quad \nabla_0 v = \delta v;$$

this is the unique connection that kills the standard basis of A^n. Then $\nabla - \nabla_0$ is B-symmetric if and only if δ is $\mathcal{B}_q$-symmetric with respect to δ_0.

Let us also record the hermitian paradigm. Assume the field A has a complex conjugation. Let V is an n-dimensional vector space over A and let δ and $\overline{\delta}$ be two commuting derivations on A such that $\overline{\delta}(\overline{a}) = \overline{\delta a}$ for all $a \in A$. Also let $H : V \times V \to A$ be a hermitian form with respect to the involution on A, i.e. a map which is A-linear in the first argument and

$$H(v, u) = \overline{H(u, v)}, \quad u, v \in V.$$

DEFINITION 2.134. A *δ-connection* on V is an additive operator $\nabla_\delta : V \to V$ such that

$$\nabla_\delta(av) = \delta a \cdot v + a \cdot \nabla_\delta v$$

for $v \in V$ and $a \in A$. A *$\overline{\delta}$-connection* on V is an additive operator $\nabla_{\overline{\delta}} : V \to V$ such that

$$\nabla_{\overline{\delta}}(av) = \overline{\delta} a \cdot v + a \cdot \nabla_{\overline{\delta}} v.$$

DEFINITION 2.135. A pair $(\nabla_\delta, \nabla_{\overline{\delta}})$ consisting of a δ-connection and a $\overline{\delta}$-connection on V is said to be *compatible* with H if one of the following two equivalent conditions is satisfied:

$$\delta H(v,w) = H(\nabla_\delta v, w) + H(v, \nabla_{\overline{\delta}} w), \quad v, w \in V$$

$$\overline{\delta} H(v,w) = H(\nabla_{\overline{\delta}} v, w) + H(v, \nabla_\delta w), \quad v, w \in V.$$

PROPOSITION 2.136. *Assume we are given a hermitian form H and a $\overline{\delta}$-connection $\nabla_{\overline{\delta}}$ on V such that the kernel of $\nabla_{\overline{\delta}}$ spans the A-vector space V. Then there exists a unique δ-connection ∇_δ on V such that the pair $(\nabla_\delta, \nabla_{\overline{\delta}})$ is compatible with H.*

PROOF. Take an A-basis (e_i) of V killed by $\nabla_{\overline{\delta}}$ and consider the hermitian matrix $h = (h_{ij})$, where $h_{ij} = H(e_i, e_j)$; then define ∇_δ by

$$\nabla_\delta e_i = \sum_j \gamma_{ij} e_j, \quad \gamma = (\gamma_{ij}), \quad \gamma = \delta h \cdot h^{-1}. \tag{2.202}$$

□

Rewriting 2.202 as

$$-\gamma^t = -(h^t)^{-1} \delta h^t \tag{2.203}$$

notice the analogy between 2.203 on the one hand and 2.156 and 2.157 on the other: the matrices $h^t, -\gamma^t$ are analogous to the matrices q^c, β^c; and the matrix $-\gamma^t$ corresponds to the connection dual to the one defined by the matrix γ.

2.11. Lax connections

Assume here A is an arbitrary ring equipped with m commuting derivations. We assume n and m are arbitrary.

DEFINITION 2.137. Assume $\delta = (\delta_i)$ is a connection on $G = GL_n$, $\delta_i x = \Delta_i$. The *Lax connection* $\delta^{Lax} = (\delta_i^{Lax})$ on G attached to δ is the connection with components the derivations $\delta_i^{Lax} : \mathcal{O}(G) \to \mathcal{O}(G)$ defined by the formula

$$\delta_i^{Lax} x := \Delta_i - x \cdot \Delta_i \cdot x^{-1} = [\alpha_i(x), x], \quad \alpha_i(x) := \Delta_i(x) \cdot x^{-1}. \tag{2.204}$$

A connection on G is called a *Lax connection* if it is of the form δ^{Lax} for some connection δ on G.

REMARK 2.138. In particular if δ is Ehresmann, $\delta_i x = A_i x$, with curvature (F_{ij}) and first Chern form $\rho = (\rho_{ij})$, we have:

$$\delta_i^{Lax} x := A_i x - x A_i = [A_i, x], \tag{2.205}$$

hence

$$[\delta_i^{Lax}, \delta_j^{Lax}] x = [F_{ij}, x].$$

So the traceless curvature (W_{ij}) of δ and the curvature of δ^{Lax} can be recovered from each other. In particular δ is flat if and only if δ^{Lax} is flat and $\rho = 0$. (In the arithmetic case both Ehresmann connections and Lax connections will have analogues although these analogues will not be directly related; therefore the curvatures of these two analogues will offer two alternative paths towards curvature in arithmetic.)

REMARK 2.139. Note that even if δ on GL_n is Ehresmann, δ^{Lax} on GL_n is neither right nor left invariant in general. On the other hand one can attach to any Ehresmann connection δ on GL_n an Ehresmann connection δ^{ad} on GL_{n^2} (its *transform under the adjoint representation*) that is closely related to δ^{Lax}; this is done as follows. For any $n \times n$ matrix B with entries in some ring denote by B^{ad} the $n^2 \times n^2$ matrix such that for any $n \times n$ matrix C we have

$$B^{ad}C^{col} = (BC - CB)^{col};$$

cf. 2.44 for the "col" superscript. Also, for any $n^2 \times 1$ matrix v denote by $v^{\times n^2}$ the $n^2 \times n^2$ matrix with columns $v, ..., v$. Now given an Ehresmann connection $\delta = (\delta_i)$ on GL_n, $\delta_i x = A_i x$, we may define the Ehresmann connection $\delta^{ad} = (\delta_i^{ad})$ on GL_{n^2} by setting

$$\delta_i^{ad} y = A_i^{ad} y$$

where y is an $n^2 \times n^2$ matrix of indeterminates which are coordinates on GL_{n^2}. It is easy to see that if (F_{ij}) is the curvature of (δ_i) then (F_{ij}^{ad}) is the curvature of (δ_i^{ad}). Moreover δ^{Lax} and δ^{ad} are related as follows:

$$((\delta_i^{Lax} x)^{col})^{\times n^2} = \delta_i^{ad}((x^{col})^{\times n^2}).$$

REMARK 2.140. There are infinite dimensional analogues of the Lax formalism; cf., e.g., the algebraic theory of the KP system [**105**], section 4. Note that the finite dimensional theory of GL_n involves the associative ring $\mathfrak{g}(A) = \text{End}_A(V)$ of A-module endomorphisms of a free A-module V of rank n; by contrast the infinite dimensional theory involves (subrings of) the ring $\text{End}_A(V)$ of A-module endomorphisms of an A-module V of infinite rank. Typically one takes $V = A((t))$, the ring of Laurent power series in an intedeterminate t, and one considers subrings of $\text{End}_A(A((t)))$ consisting of (pseudo) differential operators

$$\sum a_{ij} t^i (d/dt)^j$$

with A-coefficients $a_{ij} \in A$ acting on $A((t))$.

Going back to the finite dimensional case considered above, we would like to discuss the relation between curvature, integrability, and isospectrality. First we introduce the following notation.

DEFINITION 2.141. Let s be a variable. The *characteristic polynomial* of the matrix x is:

$$\det(s \cdot 1 - x) = \sum_{j=0}^{n} (-1)^j \mathcal{P}_j(x) s^{n-j}. \tag{2.206}$$

In particular,

$$\mathcal{P}_1(x) = \text{tr}(x), ..., \mathcal{P}_n(x) = \det(x). \tag{2.207}$$

PROPOSITION 2.142. *Let (δ_i) be a (not necessarily Ehresmann) connection. Set $A[z] = A[z_1, ..., z_n]$, a polynomial ring in n variables, let $\delta_{0i} : A[z] \to A[z]$ be the unique derivations extending the corresponding derivations on A with $\delta_{0i} z_j = 0$, and let $\mathcal{P} : A[z] \to \mathcal{O}(G)$ be the A-algebra homomorphism with $\mathcal{P}(z_j) = \mathcal{P}_j(x)$. Then the following diagram is commutative:*

$$\begin{array}{ccc} \mathcal{O}(G) & \stackrel{\delta_i^{Lax}}{\longleftarrow} & \mathcal{O}(G) \\ \mathcal{P}\uparrow & & \uparrow\mathcal{P} \\ A[z] & \stackrel{\delta_{0i}}{\longleftarrow} & A[z] \end{array} \tag{2.208}$$

In other words $\delta_i(\mathcal{P}_j(x)) = 0$, i.e., $\mathcal{P}_j(x)$ are prime integrals for δ^{Lax}.

PROOF. Consider the ring

$$\mathcal{O}(G)[s, \det(s \cdot 1 - x)^{-1}]$$

and the unique derivations δ_i^{Lax} on this ring that extend $\delta_i^{Lax} : \mathcal{O}(G) \to \mathcal{O}(G)$ and send $s \mapsto 0$. We want to show that

$$\delta_i^{Lax}(\det(s \cdot 1 - x)) = 0.$$

By Jacobi's formula 1.4 we have

$$\delta_i^{Lax}(\det(s \cdot 1 - x)) = \text{tr}(\delta_i^{Lax}(s \cdot 1 - x) \cdot (s \cdot 1 - x)^{-1}) \det(s \cdot 1 - x);$$

so it is enough to show that the above trace vanishes. Now, if $\delta_i x = \alpha_i x$, $\alpha_i = \alpha_i(x)$, we have $\delta_i^{Lax} x = \alpha_i x - x\alpha_i$; hence we have:

$$\begin{aligned} \text{tr}(\delta_i^{Lax}(s \cdot 1 - x) \cdot (s \cdot 1 - x)^{-1}) &= \text{tr}((x\alpha_i - \alpha_i x) \cdot (s \cdot 1 - x)^{-1}) \\ &= \text{tr}(x\alpha_i(s \cdot 1 - x)^{-1}) - \text{tr}(\alpha_i x(s \cdot 1 - x)^{-1}) \\ &= \text{tr}(\alpha_i(s \cdot 1 - x)^{-1}x) - \text{tr}(\alpha_i x(s \cdot 1 - x)^{-1}) \\ &= 0, \end{aligned}$$

because x and $(s \cdot 1 - x)^{-1}$ commute, and the Proposition is proved. □

In the discussion below let $G = GL_n$ and $\mathfrak{g} = \mathfrak{gl}_n$.

DEFINITION 2.143. Let (δ_i) be Ehresmann, with $\delta_i x = A_i x$, and let $S \in \mathfrak{g}(A)$. We say S is *isospectral* (with respect to the δ_i's) if

$$\delta_i(\mathcal{P}_j(S)) = 0$$

for all i and j.

COROLLARY 2.144. *Let (δ_i) be Ehresmann, with $\delta_i x = A_i x$, let $S \in \mathfrak{g}(A)$, and let $U \in G(A)$. Consider the following conditions:*

(1) *$S = US_0U^{-1}$, $S_0 \in \mathfrak{g}(A)$, $\delta_i S_0 = 0$ for all i;*
(2) *$\delta_i U = A_i U$ for all i;*
(3) *$\delta_i S = [A_i, S]$ for all i;*
(4) *S is isospectral;*
(5) *(δ_i) is flat.*

Then we have the following implications

$$\begin{array}{rcl} 1+2 & \Rightarrow & 3 \\ 1 & \Rightarrow & 4 \\ 3 & \Rightarrow & 4 \\ 2 & \Rightarrow & 5 \end{array}$$

PROOF. $3 \Rightarrow 4$ follows from Proposition 2.142. $2 \Rightarrow 5$ is Proposition 2.25. The rest of the implications are trivial. □

REMARK 2.145. Let us note that the differential equation

$$\delta_i u = [\alpha(u), u] \tag{2.209}$$

corresponding to the Lax connection $\delta_i^{Lax} x = [\alpha_i(x), x]$ is equivalent to

$$l\delta_i u = \alpha(u) - u \cdot \alpha_i(u) \cdot u^{-1}, \tag{2.210}$$

where $l\delta_i u = \delta_i u \cdot u^{-1}$ is the logarithmic derivative. In particular, for any solution $u \in GL_n(R)$ of 2.209 or 2.210 we have that u is isospectral:

$$\delta_i(\mathcal{P}_j(u)) = 0. \tag{2.211}$$

2.12. Hamiltonian connections

Lax connections arise naturally in the context of Hamiltonian equations for certain Poisson structures. We briefly review the latter in what follows.

DEFINITION 2.146. Given a ring A and an (associative commutative unital) A-algebra B, a *Poisson structure* (or a *Poisson bracket*) on B is a Lie A-algebra structure

$$\{\ ,\ \} : B \times B \to B$$

on B such that for any $f \in B$ the map

$$B \to B, \quad g \to \{f, g\}$$

is an A-derivation.

Let now $x_1, ..., x_N$ be variables over A and assume we are in one of the following two cases:

(1) $B = A[x_1, ..., x_N]$ (the *polynomial case*);
(2) $B = A[[x_1, ..., x_N]]$ (the *series case*).

We have

$$Der_A(B) = \bigoplus_{i=1}^{n} B \cdot \partial^i, \quad \partial^i := \frac{\partial}{\partial x_i}.$$

On the other hand set

$$\Omega = \bigoplus_{i=1}^{N} B \cdot dx_i, \quad \Omega^i := \wedge^i \Omega.$$

Of course, in the polynomial case, Ω is the usual module of Kähler differentials $\Omega_{B/A}$; in the series case Ω is the projective limit of the modules of Kähler differentials $\Omega_{B_\nu/A}$, where $B_\nu = B/(x)^\nu$. We refer to the elements of Ω^i as (vertical) *i-forms*; compare these with the "horizontal" i-forms in $\Lambda^i(A)$; cf. Definition 2.104.

DEFINITION 2.147. The *deRham complex* of B/A is the complex

$$B \xrightarrow{d} \Omega^1 \xrightarrow{d} \Omega^2 \xrightarrow{d} \ldots$$

where

$$\begin{aligned} db &= \sum(\partial^i b)dx_i, \\ d(\sum b_i dx_i) &= \sum db_i \wedge dx_i, \\ d(\sum b_{ij} dx_i \wedge dx_j) &= \sum(db_{ij}) \wedge dx_i \wedge dx_j, \ldots \end{aligned}$$

where $b, b_i, b_{ij}, \ldots \in B$.

Let B be given a Poisson structure $\{\ ,\ \}$ and set

$$\omega_{ij} = \{x_i, x_j\} \in B.$$

Then for $f, g \in B$ we have (with Einstein summation),

$$\{f, g\} = \omega_{ij}(\partial^i f)(\partial^j g). \tag{2.212}$$

Conversely formula 2.212 defines a Poisson bracket if and only if ω_{ij} satisfy:

$$\omega_{ij} = -\omega_{ji}, \tag{2.213}$$

$$\omega_{il}\partial^l \omega_{jk} + \omega_{kl}\partial^l \omega_{ij} + \omega_{jl}\partial^l \omega_{ki} = 0 \tag{2.214}$$

for all i, j, k.

DEFINITION 2.148. A 2-form

$$\omega = \omega^{ij} dx_i \wedge dx_j$$

in Ω^2, where $\omega^{ij} = -\omega^{ji}$, is *symplectic* if

$$\det(\omega^{ij}) \in B^\times \tag{2.215}$$

and

$$\partial^k \omega^{ij} + \partial^j \omega^{ki} + \partial^i \omega^{jk} = 0 \tag{2.216}$$

for all i, j, k.

Condition 2.216 is, of course, equivalent to the condition that

$$d\omega = 0 \tag{2.217}$$

in the group Ω^3 of the deRham complex of B/A.

The following follows by a direct (classical) computation:

PROPOSITION 2.149. *Assume $(\omega_{ij}) \in GL_N(B)$, $\omega_{ij} = -\omega_{ji}$, and let (ω^{ij}) be the inverse of (ω_{ij}). Then Equation 2.214 is equivalent to Equation 2.216. In other words (ω_{ij}) defines a Poisson structure (via 2.212) if and only if the 2-form $\omega^{ij} dx_i \wedge dx_j$ is symplectic.*

REMARK 2.150. Note that the closedness condition 2.216 defining a symplectic form ω is entirely different from the condition 2.185 defining a symplectic matrix q; indeed the condition 2.185 is a "closedness in the horizontal directions,"

$$\delta_i = \partial/\partial\xi_i,$$

whereas condition 2.216 is a "closedness in the vertical directions,"

$$\partial^j = \partial/\partial x_j.$$

Going back to the case when (ω_{ij}) is not necessarily invertible, assume A is equipped with a derivation δ. (Intuitively δ controls "time dependence." One can trivially generalize the discussion below to the case when A is equipped with several commuting derivations.) In the spirit of previous terminology we make the following:

DEFINITION 2.151. A derivation δ on B extending the derivation δ on A is called a *connection* on B (or on *Spec* B).

Set $A_0 = A^\delta = \{a \in A; \delta a = 0\}$ and let

$$B_0 = A_0[x_1, ..., x_N]$$

in the polynomial case and

$$B_0 = A_0[[x_1, ..., x_N]]$$

in the series case; intuitively the elements of B_0 are the "time independent" elements of B.

DEFINITION 2.152. A connection δ on B is *Hamiltonian* with respect to the Poisson structure $\{\ ,\ \}$ if there exists $H \in B$ (referred to as a *Hamiltonian*) such that, for all i,

$$\delta x_i = \{x_i, H\}.$$

If this is the case then we also have $\delta f = \{f, H\}$ for all $f \in B_0$.

REMARK 2.153. For a connection the property of being *Hamiltonian* with respect to $\{\ ,\ \}$ depends on $\{\ ,\ \}$, δ, and the subalgebra B_0 of B but not on the choice of the "coordinates" x_i.

On the other hand recall that, for any connection δ on B there is a unique additive map $\delta : \Omega \to \Omega$ (the *Lie derivative*) satisfying

$$\delta(fdg) = \delta f \cdot dg + fd(\delta g) \tag{2.218}$$

for $f, g \in B$. This map δ induces an A-derivation (still called δ) on the exterior algebra of Ω.

DEFINITION 2.154. A connection δ on B is *Hamiltonian* with respect to a symplectic form ω on B if $\delta\omega = 0$.

Again this property depends only on B, δ, and ω, and not on the choice of the coordinates x_i.

For a connection on B the two concepts of being Hamiltonian (one with respect to a Poisson structure, the other with respect to a symplectic form) are related in the "time independent" case. Indeed assume we are given a Poisson structure $\{\ ,\ \}$ on B, let $\omega_{ij} = \{x_i, x_j\}$, $\det(\omega_{ij}) \in B^\times$, let (ω^{ij}) be the inverse of (ω_{ij}), and let $\omega = \omega^{ij} dx_i \wedge dx_j$ be the corresponding symplectic form. Then the following hold:

PROPOSITION 2.155. *Assume δ is Hamiltonian with respect to $\{\ ,\ \}$. Then δ is Hamiltonian with respect to ω if and only if $\omega_{ij} \in B_0$.*

PROOF. A direct computation. □

PROPOSITION 2.156. *Assume δ is Hamiltonian with respect to ω. Assume furthermore that $\omega_{ij} \in A_0$ in the polynomial case and $\omega_{ij} \in B_0$ in the series case. Assume also that A_0 contains the field of rationals in the polynomial case and equals the real field in the series case. Then δ is Hamiltonian with respect to $\{\ ,\ \}$.*

PROOF. The case $\omega_{ij} \in A_0$ is a direct computation. The case $\omega_{ij} \in B_0$ follows, for instance, by reduction to the case $\omega_{ij} \in A_0$ via Darboux's theorem on the local equivalence of symplectic C^∞ forms [**68**] plus Borel's theorem on the existence of C^∞ functions on $\mathbb{R}^N$ with any prescribed Taylor series at the origin. $\square$

Standard examples of Poisson brackets arise from *Lie-Poisson* structures as follows. Let L be a (free finite rank) Lie A-algebra with basis $e_1, ..., e_N$ and bracket $[\ ,\ ]$, with

$$[e_i, e_j] = \sum_m c_{ijm} e_m, \quad c_{ijm} \in A.$$

Now let $B = A[x_1, ..., x_N]$.

DEFINITION 2.157. The *Lie-Poisson* structure on B (induced by L) is the Poisson structure on B defined by

$$\omega_{ij} = \{x_i, x_j\} := \sum_m c_{ijm} x_m.$$

In its turn the Lie-Poisson structures are closely related to the Lax formalism and hence to isospectrality. We shall explain this in examples below.

EXAMPLE 2.158. A "generic" example of the relation between the Lie-Poisson structures and the Lax formalism is the following. Let us take the Lie algebra $L = \mathfrak{gl}_n$ with basis $\{e_{ij}\}$ where e_{ij} is the matrix with a 1 in the (i,j) entry and 0 in the other entries. One has

$$[e_{kl}, e_{rs}] = \delta_{rl} \cdot e_{ks} - \delta_{ks} \cdot e_{rl},$$

where δ_{rl}, etc., are the Kronecker symbols. Let $B = A[x]$, where $x = (x_{ij})$ is a square matrix of indeterminates. The Lie-Poisson structure on B is given by

$$\{x_{kl}, x_{rs}\} = \delta_{rl} \cdot x_{ks} - \delta_{ks} \cdot x_{rl} = [e_{lk}, x]_{rs},$$

the (r,s) entry of $[e_{lk}, x]$, in other words

$$\{x_{kl}, x\} = [e_{kl}, x].$$

So the A-derivation $\{x_{kl},\ \} : A[x] \to A[x]$ vanishes on the coefficients $\mathcal{P}_j(x) \in A[x]$ of the characteristic polynomial $\det(s \cdot 1 - x)$ (cf. our discussion on Lax connections with the role of A there being played here by $A[x]$). In particular if, to simplify, A is an algebraically closed field of characteristic zero then the Poisson bracket $\{\ ,\ \}$ on $A[x]$ induces Poisson brackets on every regular conjugacy class; a conjugacy class is called here *regular* if its eigenvalues are distinct. (This is of course an easy instance of the Kostant-Kirilov Poisson structure on coadjoint orbits [**68**].)

Going back to an arbitrary A equipped with a derivation δ and letting $x = (x_{ij})$ be a square matrix of indeterminates one has:

PROPOSITION 2.159. *Any connection on $A[x]$ that is Hamiltonian with respect to the Lie-Poisson bracket induces on GL_n a Lax connection.*

PROOF. Consider a Hamiltonian connection δ on $A[x]$ with respect to the Lie-Poisson bracket $\{\ ,\ \}$, hence $\delta x = \{\mathcal{H}, x\}$ for some $\mathcal{H} \in A[x]$. Set

$$\frac{\partial \mathcal{H}}{\partial x} = \left(\frac{\partial \mathcal{H}}{\partial x_{kl}}\right)$$

Then, with Einstein summation notation, we have

$$\begin{aligned} \delta x &= \tfrac{\partial \mathcal{H}}{\partial x_{kl}} \cdot \{x_{kl}, x\} \\ &= \tfrac{\partial \mathcal{H}}{\partial x_{kl}} \cdot [e_{kl}, x] \\ &= \left[\tfrac{\partial \mathcal{H}}{\partial x_{kl}} \cdot e_{kl}, x\right] \\ &= \left[\tfrac{\partial \mathcal{H}}{\partial x}, x\right], \end{aligned}$$

So the connection δ is a Lax connection. □

EXAMPLE 2.160. Take $L = \mathfrak{so}_3$ with basis e_1, e_2, e_3 with

$$\begin{aligned} \{e_1, e_2\} &= e_3 \\ \{e_2, e_3\} &= e_1 \\ \{e_3, e_1\} &= e_2. \end{aligned}$$

Let us take any $H \in B = A[x_1, x_2, x_3]$ and let us consider the connection δ on B, Hamiltonian with respect to the Lie-Poisson structure $\{\ ,\ \}$, with Hamiltonian H; it is defined by $\delta x_i = \{H, x_i\}$. We would like to analyze this connection more closely. One can embed *Spec* B into *Spec* $A[x]$, $x = (x_{ij})$ a 3×3 matrix of indeterminates, by sending x into

$$\widehat{x} := \begin{pmatrix} 0 & x_3 & x_2 \\ -x_3 & 0 & x_1 \\ -x_2 & -x_1 & 0 \end{pmatrix}. \tag{2.219}$$

One trivially checks that the derivation δ on B is induced (via the embedding above) by the derivation δ on *Spec* $A[x]$ defined by

$$\delta x = \left[\frac{\partial \mathcal{H}}{\partial x}, x\right] \tag{2.220}$$

where

$$\mathcal{H} := H(x_{23}, x_{13}, x_{12}) + H(-x_{32}, -x_{31}, -x_{21}), \qquad \frac{\partial \mathcal{H}}{\partial x} := \left(\frac{\partial \mathcal{H}}{\partial x_{kl}}\right).$$

This connection 2.220 is (by the proof of Proposition 2.159) the Hamiltonian connection on $\mathfrak{gl}_3$ attached to the Lie-Poisson structure on $\mathfrak{gl}_3$ and Hamiltonian $\mathcal{H}$; it induces, of course, a Lax connection on GL_3. Note that, in particular, any solution to these equations is isospectral.

Assume now, for simplicity, that A is an algebraically closed field. Let

$$H := H_1 := \frac{1}{2}\sum a_i x_i^2,$$

with $a_i \in A_0 = \{a \in A;\ \delta a = 0\}$. Then the derivation δ on B is given by

$$\begin{aligned} \delta x_1 &= (a_2 - a_3)x_2x_3, \\ \delta x_2 &= (a_3 - a_1)x_3x_1, \\ \delta x_3 &= (a_1 - a_2)x_1x_2, \end{aligned} \tag{2.221}$$

which are the *Euler equations* for the rigid body. By the way the derivation δ on B induces a derivation δ on $B_0 = A_0[x_1, x_2, x_3]$. If, in addition, one defines

$$H_2 := \sum x_i^2,$$

then one has

$$\delta H_1 = \delta H_2 = 0, \tag{2.222}$$

i.e., H_1, H_2 are *prime integrals* for δ. Moreover we have

$$\{H_1, H_2\} = 0.$$

This prime integral H_2 is, by the way, a coefficient of the characteristic polynomial of 2.219,

$$\det(z - \widehat{x}) = z^3 + (x_1^2 + x_2^2 + x_3^2)z.$$

Assume, in what follows that $c := (c_1, c_2) \in A_0^2$ are such that all 2×2 minors of the matrix

$$\begin{pmatrix} a_1 & a_2 & a_3 & c_1 \\ 1 & 1 & 1 & c_2 \end{pmatrix} \tag{2.223}$$

are non-zero. Then the scheme

$$E_c := Spec\ (B_0/(H_1 - c_1, H_2 - c_2))$$

is an affine smooth elliptic curve in the affine 3-space over A_0,

$$\mathbb{A}^3 = Spec\ B_0,$$

with the property that its projective closure $\overline{E}_c$ in the projective space

$$\mathbb{P}^3 = Proj\ A[t_0, t_1, t_2, t_3], \quad x_i = t_i/t_0,$$

is a smooth projective elliptic curve. The derivation δ_c on

$$B_0/(H_1 - c_1, H_2 - c_2)$$

induced by the derivation δ on B_0 extends to a vector field on (i.e., a derivation of the structure sheaf of) $\overline{E}_c$. Indeed, there is a global 1-form on $\overline{E}_c$ which restricted to E_c is given by

$$\omega_c = \frac{dx_1}{(a_2 - a_3)x_2x_3} = \frac{dx_2}{(a_3 - a_1)x_3x_1} = \frac{dx_3}{(a_1 - a_2)x_1x_2}, \tag{2.224}$$

cf. [**18**], pp. 136-139, and the pairing between δ_c and ω_c is

$$\langle \delta_c, \omega_c \rangle = 1. \tag{2.225}$$

The conjunction of the conditions 2.222 and 2.225 is an instance of *algebraic complete integrability*; cf. [**2**]. By the way it is trivial to see that 2.225 implies

$$\delta_c \omega_c = 0 \tag{2.226}$$

where δ_c acts on 1-forms on E_c as in 2.218. We will later give an arithmetic analogue of the above construction, more precisely an analogue of 2.222 and 2.226.

On the other hand the Euler equations 2.221 are equivalent to a *Lax equation with parameter* involving 2×2 matrices; cf. [**75**], p. 5. The latter is very different from the 3×3 matrix equation in 2.220, it will not be reviewed here, and is the prototype for a rich theory underlying algebraic complete integrability.

REMARK 2.161. Let us pursue the formalism of Hamiltonian connections relative to a symplectic form in the case of series in 2 variables. Let A be a field of characteristic zero equipped with a derivation $\delta : A \to A$ (e.g., $A = \mathbb{C}((t))$, $\delta = d/dt$) and let B be a ring isomorphic to a power series ring in 2 variables with coefficients in A. Choose formal coordinates $x, y \in B$, i.e., elements in B such that $B = A[[x, y]]$. Then set, as before,

$$\Omega = B \cdot dx + B \cdot dy, \qquad \Omega^2 = B \cdot dx \wedge dy.$$

These definitions are independent of the choice of the formal coordinates x, y. Of course, a form

$$\omega = u \cdot dx \wedge dy \in \Omega^2$$

is symplectic if and only if $u(0, 0) \neq 0$.

Assume now one is given a connection on B, i.e., a derivation $\delta : B \to B$ extending δ on A and assume δ on B sends the maximal ideal of B into itself. Then the following are easy to check:

(1) Assume we are given a symplectic $\omega \in \Omega^2$. Then there exist formal coordinates $x, y \in B$ such that

$$\omega = dy \wedge dx.$$

Furthermore if δ is Hamiltonian with respect to ω, i.e., if $\delta\omega = 0$, then and there exists an element $\mathcal{H} \in B$ such that

$$\delta y = -\frac{\partial \mathcal{H}}{\partial x}, \qquad \delta x = \frac{\partial \mathcal{H}}{\partial y}.$$

(2) Assume there exist formal coordinates $x, y \in B$, and a series $\mathcal{H} \in B$ (interpreted as a "Hamiltonian") such that

$$\delta y = -\frac{\partial \mathcal{H}}{\partial x}, \qquad \delta x = \frac{\partial \mathcal{H}}{\partial y}.$$

Set

$$\omega = dy \wedge dx, \qquad \nu = y dx, \qquad \epsilon = \delta\nu,$$

and

$$\mathcal{L} = y\frac{\partial \mathcal{H}}{\partial y} - \mathcal{H} = y\delta x - \mathcal{H}.$$

(So $\mathcal{L}$ is interpreted as the "Lagrangian.") Then

$$d\nu = \omega, \qquad \delta\omega = 0, \qquad d\mathcal{L} = \epsilon.$$

(3) Assume the notation in (2) and assume the series

$$\frac{\partial^2 \mathcal{H}}{\partial y^2}$$

is invertible in $A[[x, y]]$. Then there exists a unique A-derivation, denoted by $\frac{\partial}{\partial \delta x}$, on $A[[x, y]]$ sending

$$x \mapsto 0, \qquad \delta x \mapsto 1.$$

Moreover we have

$$\frac{\partial \mathcal{L}}{\partial \delta x} = y, \qquad \frac{\partial \mathcal{L}}{\partial x} = -\frac{\partial \mathcal{H}}{\partial x},$$

and hence

$$\delta\left(\frac{\partial \mathcal{L}}{\partial \delta x}\right)=\frac{\partial \mathcal{L}}{\partial x}, \quad \delta x \cdot \frac{\partial \mathcal{L}}{\partial \delta x}-\mathcal{L}=\mathcal{H}.$$

These equations can be interpreted as the "Euler-Lagrange equation" and the identification of the "Hamiltonian" with the "energy," respectively.

The above (and their higher dimensional versions) provide, of course, the basic paradigm for Lagrangian and Hamiltonian mechanics. Some basic equations of mathematical physics fit into this paradigm; among them are the Painlevé VI equations [**101**]; cf. Remark 2.176 below. An arithmetic analogue of the Hamiltonian formalism related to the Painlevé VI equations was introduced in [**42**] and will be reviewed later; it is based on the arithmetic analogue in [**23**] of the Manin map for abelian varieties over function fields [**99**].

REMARK 2.162. It would be interesting to find arithmetic analogues of connections that are Hamiltonian with respect to Poisson structures in the polynomial case. In view of Proposition 2.159 one way to get them could be via our arithmetic analogues of Lax connections. Another way is by generalizing our arithmetic analogue of the Euler equations to be explained in the body of the book.

REMARK 2.163. The polynomial case of the theory reviewed in this section can be easily generalized to the case when B is a smooth A-algebra.

2.13. Cartan connection

The geometrization of the theory of differential equations pioneered by Lie and Cartan is based on seeing differential equations on a manifold M as certain sequences of submanifolds of the various jet spaces of M. Let us recall this point of view here using the language of differential algebra [**91, 92, 51, 20**]. Let A be a ring equipped with commuting derivations

$$\delta_i : A \to A, \quad i = 1, ..., m.$$

We denote by

$$A^{\delta} := A_0 = \{a \in A;\ \delta_i a = 0,\ i = 1, ..., m\} \tag{2.227}$$

the *constant ring* of A. Let x be an N-tuple of variables $x_1, ..., x_N$ and define

$$A\{x\} := A[x_j^{(\alpha)};\ \alpha \in \mathbb{Z}^m_{\geq 0}, j = 1, ..., N], \tag{2.228}$$

where $x_j^{(\alpha)}$ are indeterminates, $x_j^{(0)} = x_j$; $A\{x\}$ comes equipped with a "universal" flat connection $\delta = (\delta_i)$ defined by the derivations $\delta_i : A\{x\} \to A\{x\}$,

$$\delta_i := \delta_{0i} + \sum_j \sum_\alpha x_j^{(\alpha+e_i)} \frac{\partial}{\partial x_j^{(\alpha)}}, \tag{2.229}$$

where e_i is the canonical basis of $\mathbb{Z}^m$ and δ_{0i} is the unique derivation on $A\{x\}$ extending δ_i on A, with $\delta_{0i}x = 0$ for all i. The δ_i's are called the *total derivative operators* and correspond to what is called in classical differential geometry the *Cartan distribution*, cf. [**1**], p. 133. Assume for simplicity in what follows that A is Noetherian. Consider an affine scheme of finite type

$$X = Spec\ \frac{A[x]}{(f)}, \tag{2.230}$$

where x is an N-tuple of variables $x_1, ..., x_N$ as above and f is an s-tuple of polynomials $f_1, ..., f_s \in A[x]$. For

$$\alpha = (\alpha_1, ..., \alpha_m) \in \mathbb{Z}^m_{\geq 0}$$

write

$$|\alpha| = \alpha_1 + ... + \alpha_m, \qquad \delta^\alpha f = \delta_1^{\alpha_1} ... \delta_m^{\alpha_m} f.$$

DEFINITION 2.164. The n-th *jet space* (or *jet bundle*) of X is the scheme

$$J^n(X) := Spec\ \frac{A[x^{(\alpha)};\ |\alpha| \leq n]}{(\delta^\alpha f;\ |\alpha| \leq n)}.$$

If X is a scheme of finite type over A which is not necessarily affine one takes a finite affine cover

$$X = \bigcup X_i$$

and one defines

$$J^n(X) = \bigcup J^n(X_i)$$

with $J^n(X_i)$ glued naturally. The definition does not depend on the cover and is, indeed, functorial in X.

For the theory and some Diophantine applications of $J^n(X)$ we refer to [**19**], [**20**]. One has natural morphisms of A-schemes,

$$(2.231) \qquad ... \to J^n(X) \to J^{n-1}(X) \to ... \to J^1(X) \to J^0(X) = X.$$

The projective system above will be denoted by $J^*(X)$. Set $\mathcal{O}^n(X) = \mathcal{O}(J^n(X))$ and

$$\mathcal{O}^\infty(X) := \lim_{\to} \mathcal{O}^n(X), \qquad J^\infty(X) = \lim_{\leftarrow} J^n(X).$$

The commuting derivations $\delta_i : A\{x\} \to A\{x\}$ induce commuting derivations

$$(2.232) \qquad \delta_i : \mathcal{O}^\infty(X) \to \mathcal{O}^\infty(X).$$

These derivations can be viewed as an infinite dimensional analogue of a connection in classical differential geometry.

DEFINITION 2.165. The flat connection $\delta = (\delta_i)$ on $\mathcal{O}^\infty(X)$ is called the *Cartan connection* on $\mathcal{O}^\infty(X)$ or on $J^\infty(X)$.

We have the following functoriality property: for any morphism $X \to Y$ of affine schemes of finite type there is a unique A-algebra morphism

$$\mathcal{O}^\infty(Y) \to \mathcal{O}^\infty(X)$$

commuting with all δ_i and compatible with the morphism $\mathcal{O}(Y) \to \mathcal{O}(X)$.

In what follows, in order to simplify our discussion, let us assume that A (and hence also A_0) is an algebraically closed field of characteristic zero. Also assume $m = 1$ so A is equipped with one derivation $\delta = \delta^A$. In this case, for X as in 2.230,

$$(2.233) \qquad J^n(X) = Spec\ \frac{A[x, x', ..., x^{(n)}]}{(f, \delta f, ..., \delta^n f)}$$

where $x', x'', ..., x^{(n)}$ are N-tuples of variables and

$$\delta : A\{x\} \to A\{x\} := A[x, x', x'', ..., x^{(n)}, ...]$$

is given by

$$\delta := \delta_0 + \sum_j \sum_i x_j^{(i+1)} \frac{\partial}{\partial x_j^{(i)}}, \tag{2.234}$$

where $\delta_0 a = \delta^A a$ for $a \in A$ and $\delta_0 x = 0$.

If X is smooth over A, the morphisms 2.231 are smooth and surjective; indeed they are Zariski locally trivial fiber bundles with fibers affine spaces. In particular we have natural injections

$$\mathcal{O}^n(X) \to \mathcal{O}^\infty(X).$$

Also, in this case, clearly,

$$\delta(\mathcal{O}^n(X)) \subset \mathcal{O}^{n+1}(X)$$

for all n; and for a morphism $X \to Y$ with X and Y smooth the induced $\mathcal{O}^\infty(Y) \to \mathcal{O}^\infty(X)$ is compatible with the filtration by $\mathcal{O}^n$ i.e., sends $\mathcal{O}^n(Y)$ into $\mathcal{O}^n(X)$.

If, in addition, $X = G$ is a linear algebraic group [**76**] (i.e. an affine group scheme of finite type, automatically smooth) then $J^n(G)$ are linear algebraic groups and the morphisms 2.231 are surjective algebraic group homomorphisms.

For $m = 1$ the spaces $J^n(X)$ and the Cartan connection δ have arithmetic analogues which were extensively studied in [**35**], will be reviewed later in the present book, and play a key role in our arithmetic theory.

2.14. Weierstrass and Riccati connections

We continue to assume A is an algebraically closed field of characteristic zero and $m = 1$ so A is equipped with one derivation $\delta = \delta^A$ and the constant field $A_0 = A^\delta = \{a \in A; \delta a = 0\}$ is also algebraically closed.

DEFINITION 2.166. A *connection* on a scheme of finite type X over A is a derivation δ^X on the sheaf $\mathcal{O}_X$ extending the derivation δ^A on A.

REMARK 2.167. If we denote by $\mathcal{F}(X)$ the set of connections on X then we have a natural bijection

$$\mathcal{F}(X) \simeq \{\text{sections of } J^1(X) \to X\}.$$

Recall a basic isotriviality result:

THEOREM 2.168. [**18**] *Let X be a projective variety over A. Then X possesses a connection if and only if X descends to the constant field A_0 (by which we mean that there exists a projective variety X_0 over A_0 such that $X \simeq X_0 \otimes_{A_0} A$).*

EXAMPLE 2.169. Let X be a smooth projective curve over A of genus g which descends to A_0.

(1) If $g \geq 2$ then X has a unique connection.
(2) If $g = 1$ then the connections on X are in bijection with A: if X is the smooth projective model over A of the plane curve

$$y^2 = x^3 + c_2 x + c_3, \quad c_2, c_3 \in A_0,$$

every connection on X is given by

$$\delta x = ay, \quad a \in A.$$

One can refer to these connections as *Weierstrass connections*. Indeed, for any such connection,

$$(\delta x)^2 = a^2(x^3 + c_2 x + c_3),$$

which defines a classical *Weierstrass equation.* Cf. 2.238 below for more on this.

(3) If $g = 0$ then the connections on X are in bijection with A^3: if

$$X = Spec\ A[x] \cup Spec\ A[x^{-1}]$$

every connection on X is given by

$$\delta x = ax^2 + bx + c, \quad a, b, c \in A.$$

We can refer to such connections as *Riccati connections* since they define the classical *Riccati equations.*

2.15. Differential groups: Cassidy and Painlevé

We continue to assume here that A is an algebraically closed field of characteristic zero and $m = 1$ so A is equipped with one derivation $\delta = \delta^A$ and has an algebraically closed constant field $A_0 = A^\delta = \{a \in A;\ \delta a = 0\}$.

Let G be an algebraic group over A and consider the natural homomorphisms

$$\mu_n^* : \mathcal{O}^n(G) \to \mathcal{O}^n(G \times G),\ s_n^* : \mathcal{O}^n(G) \to \mathcal{O}^n(G),\ \epsilon_n^* : \mathcal{O}^n(G) \to R$$

induced by multiplication, inverse and unit,

$$\mu : G \times G \to G,\ s : G \to G,\ \epsilon : Spec\ A \to G.$$

We also have homomorphisms

$$pr_{1n}^*, pr_{2n}^* : \mathcal{O}^n(G) \to \mathcal{O}^n(G \times G)$$

induced by the projections $pr_1, pr_2 : G \times G \to G$.

The following is a reformulation of a basic definition in [**51**], [**92**].

DEFINITION 2.170. A *δ-subgroup* of a linear algebraic group G is a sequence $S^* = (S^n)$ of schemes $S^n := Spec\ \mathcal{O}(G)/J_n$, where $J_n \subset \mathcal{O}^n(G)$, $n \geq 0$, are ideals satisfying the following conditions:

(1) $J_{n+1} \cap \mathcal{O}^n(G) = J_n$,
(2) $\mu_n^* J_n \subset (pr_{1n}^* J_n, pr_{2n}^* J_n)$,
(3) $s_n^* J_n = J_n$,
(4) $\epsilon_n^* J_n = 0$.
We say that a δ-subgroup S^* of G is *saturated* if, in addition, we have:
(5) $\delta J_n \subset J_{n+1}$.

Intuitively, such a sequence should be viewed as a system of "global algebraic differential equations" (with respect to our operator δ) that is "compatible with the group law". Clearly if S^* is a δ-subgroup of G then S^n are algebraic subgroups of $J^n(G)$ and form a projective system with surjective transition homomorphisms,

$$\ldots \to S^n \to S^{n-1} \to \ldots \to S^0.$$

DEFINITION 2.171. A δ-subgroup S^* of G is *Zariski dense* in G if $S^0 = G$.

DEFINITION 2.172. A δ-subgroup S^* of G is *split* if there exists an algebraic group isomorphism $\sigma : G \to G_0 \otimes_{A_0} A$, with $G_0 = Spec\ A_0[x]/(f)$ a linear algebraic group over A_0, such that, for each n, J_n is the image, via the isomorphism

$$\mathcal{O}^n(G_0 \otimes A) \to \mathcal{O}^n(G)$$

induced by σ, of the ideal generated by $x', \ldots, x^{(n)}$.

If S^* is split then it is saturated.

Also recall from [**76**], pp. 89, 168, the following standard:

DEFINITION 2.173. A linear algebraic group is called *simple* if it is connected, non-commutative, and has no non-trivial closed connected normal subgroups other than itself.

The following result, due to P. Cassidy, is a deep contribution to the program, initiated through work of S. Lie, E. Cartan, and J.F.Ritt, of understanding the "infinite dimensional simple groups". Cassidy proved in [**52**] a more general result in the framework of fields with several commuting derivations. Her arguments depend on a careful analysis of the interaction between differential algebra and the Cartan matrices of simple groups. A proof of the theorem below, that does not involve Cartan matrices was given in [**20**].

THEOREM 2.174. [**52**] *Let G be a simple linear algebraic group over A and let S^* be a saturated δ-subgroup of G that is Zariski dense in G. Then either S^* is split or $S^* = J^*(G)$.*

An arithmetic analogue of this theorem holds and will be proved later in the book.

We end with a discussion of a situation that leads to *differential subgroups of (non-affine) algebraic groups*; we will not need to define the latter in what follows but the reader interested in a general definition is referred to [**91, 20**]. The construction below goes back to work of Manin [**99**] (with roots in classical work of Fuchs).

Let $\mathbb{G}_a$ be, as usual, the line $\mathbb{A}^1 := Spec\ A[z]$, with the additive group structure. We have the following:

THEOREM 2.175. [**99**] *Let E be an elliptic curve over A whose j-invariant is non-constant: $j(A) \not\subset A_0$. Then the following hold:*

(1) $\mathcal{O}(J^1(E)) = A$.
(2) *There exists $\psi \in \mathcal{O}(J^2(E))$, $\psi \neq 0$, such that*

$$\mu_2^*\psi = pr_{12}^*\psi + pr_{22}^*\psi$$

in $\mathcal{O}^2(E \times E)$. Equivalently, the induced map

$$\psi : J^2(E) \to \mathbb{G}_a$$

is a group homomorphism. This homomorphism is unique up to multiplication by an element of A.

The map $\psi : J^2(E) \to \mathbb{G}_a$ is sometimes referred to as the *Manin map*.

REMARK 2.176. Manin, following Fuchs, showed in [**101**] that the classical Painlevé VI equation is "defined" by an element of the form

$$\psi - r \in \mathcal{O}(J^2(E \backslash E[2])), \tag{2.235}$$

where $r \in \mathcal{O}(E\backslash E[2])$ is an appropriate function; here $E[2] \subset E$ denotes, as usual, the kernel of the multiplication by 2, $[2] : E \to E$. Specifically

$$r = \sum_{i=1}^{r} c_i L_{P_i}^* y$$

where $c_i \in A_0$ are constants, y is the coordinate function in the plane affine equation,

$$y^2 = x^3 + ax + b,$$

P_i are the four 2-torsion points of E, $L_{P_i} : E \to E$ are the translations by P_i, and $L_{P_i}^*$ are the induced ring automorphisms of $\mathcal{O}(E \backslash E[2])$.

Manin also showed in [**101**] how the Painlevé VI equation is "Hamiltonian with respect to a natural symplectic form" in a sense similar to the one we adopted in Definition 2.154. We will only mention here the classical complex analytic picture behind the Painlevé VI equation.

Indeed the classical family of Painlevé VI equations depends on four arbitrary complex constants $(\alpha, \beta, \gamma, \delta)$ and is written as an ordinary differential equation satisfied by a function X of one variable t as follows:

$$\frac{d^2X}{dt^2} = \frac{1}{2}\left(\frac{1}{X} + \frac{1}{X-1} + \frac{1}{X-t}\right)\left(\frac{dX}{dt}\right)^2 - \left(\frac{1}{t} + \frac{1}{t-1} + \frac{1}{X-t}\right)\frac{dX}{dt} +$$
$$+ \frac{X(X-1)(X-t)}{t^2(t-1)^2}\left[\alpha + \beta\frac{t}{X^2} + \gamma\frac{t-1}{(X-1)^2} + \delta\frac{t(t-1)}{(X-t)^2}\right].$$

As Fuchs remarked in 1907, this equation can be rewritten as the differenttial equation for a (local) holomorphic section $P := (X(t), Y(t))$ of the generic elliptic curve

$$E = E(t) : \; Y^2 = X(X-1)(X-t)$$

in the form:

$$t(1-t)\left[t(1-t)\frac{d^2}{dt^2} + (1-2t)\frac{d}{dt} - \frac{1}{4}\right]\int_{\infty}^{(X,Y)} \frac{dx}{\sqrt{x(x-1)(x-t)}} = \tag{2.236}$$
$$= \alpha Y + \beta\frac{tY}{X^2} + \gamma\frac{(t-1)Y}{(X-1)^2} + (\delta - \frac{1}{2})\frac{t(t-1)Y}{(X-t)^2}.$$

If we pass to the analytic picture replacing the algebraic family of curves $E(t)$ by the analytic one $E_\tau := \mathbb{C}/(\mathbb{Z} + \mathbb{Z}\tau) \mapsto \tau \in \mathbb{H}$, $\mathbb{H}$ the upper half plane, and if we denote by z a fixed coordinate on $\mathbb{C}$, then 2.236 can be equivalently written in the form

$$\frac{d^2 z}{d\tau^2} = \frac{1}{(2\pi i)^2}\sum_{j=0}^{3} \alpha_j \wp_z(z + \frac{T_j}{2}, \tau)$$

where

$$(\alpha_0, \ldots, \alpha_3) := (\alpha, -\beta, \gamma, \frac{1}{2} - \delta), \qquad (T_0, \ldots, T_3) = (0, 1, \tau, 1+\tau),$$

$$\wp(z, \tau) := \frac{1}{z^2} + \sum_{(m,n)\neq(0,0)} \left(\frac{1}{(z + m\tau + n)^2} - \frac{1}{(m\tau + n)^2}\right),$$

and

$$\wp_z := \frac{\partial \wp}{\partial z}.$$

By the way recall that $\wp$ satisfies the *Weiestrass equation:*

$$\wp_z(z,\tau)^2 = 4(\wp(z,\tau) - e_1(\tau))(\wp(z,\tau) - e_2(\tau))(\wp(z,\tau) - e_3(\tau)), \tag{2.237}$$

where

$$e_i(\tau) = \wp(\frac{T_i}{2}, \tau),$$

so $e_1+e_2+e_3=0$. More generally, for any complex analytic function $a(z)$ defined in a neighborhood of some $z_0\in\mathbb{C}$ which is not a pole of $\wp$, and for any fixed τ, the function

$$u(z):=\wp\left(\frac{1}{2}\int_{z_0}^{z}a(\xi)d\xi\right)$$

is a solution to the *Weierstrass equation*

$$u'(z)^2=a^2(z)(u(z)-e_1(\tau))(u(z)-e_2(\tau))(u(z)-e_3(\tau)). \tag{2.238}$$

On the other hand, the classical PVI equation 2.236 can be represented as a Hamiltonian flow, with time-dependent Hamiltonian:

$$\frac{dz}{d\tau}=\frac{\partial\mathcal{H}}{\partial y},\ \frac{dy}{d\tau}=-\frac{\partial\mathcal{H}}{\partial z},$$

where

$$\mathcal{H}:=\mathcal{H}(\alpha_0,\dots,\alpha_3):=\frac{y^2}{2}-\frac{1}{(2\pi i)^2}\sum_{j=0}^{3}\alpha_j\wp(z+\frac{T_j}{2},\tau).$$

The link between the analytic picture here and the differential algebraic picture in 2.235 is given by taking A to be the algebraic closure of the field $\mathbb{C}(t)$ of rational functions; then the left hand side of 2.236 here corresponds to ψ in Remark 2.176 while the right hand side of 2.236 corresponds to the function r in 2.235.

The arithmetic analogue of this theory will be provided later in the book, following [**42**].

CHAPTER 3

Arithmetic differential geometry: generalities

This chapter is where the exposition of the theory in this book properly begins. In this chapter we present some of the basic concepts of what we call *arithmetic differential geometry*. They will be introduced by analogy with the *classical differential geometry* that was revisited in the previous chapter from the viewpoint of differential (super)algebra. We restrict ourselves in the present chapter to the "abstract case" of general rings and schemes. In the next chapter we will specialize to the case of the scheme GL_n which is the main object of the present book.

3.1. Global connections and their curvature

We introduce here an arithmetic analogue of connection. This analogue will be referred to as *global connection*. For global connections it will be easy to define *curvature*. However the concept of global connection is too naive to be useful by itself: it is just a stepping stone towards a more subtle concept, namely that of *adelic connection* which will be introduced later.

3.1.1. p-derivations. [**83, 23**] Let B be a ring and $p \in \mathbb{Z}$ a prime. Recall:

DEFINITION 3.1. A *Frobenius lift* on B is a ring homomorphism $\phi_p : B \to B$ whose reduction mod p is the p-power Frobenius $B/pB \to B/pB$.

A *Frobenius lift* on a scheme X is a morphism of schemes $\phi_p : X \to X$ whose reduction mod p is the absolute Frobenius $X \otimes \mathbb{F}_p \to X \otimes \mathbb{F}_p$ (i.e., the identity on the topological space $X \otimes \mathbb{F}_p$ and the p-power Frobenius on the structure sheaf of $X \otimes \mathbb{F}_p$).

Clearly, to give a a Frobenius lift on B is the same as to give a Frobenius lift on *Spec* B.

For the next definition consider the polynomial

$$C_p(X,Y) := \frac{X^p + Y^p - (X+Y)^p}{p} \in \mathbb{Z}[X,Y].$$

DEFINITION 3.2. An operator $\delta_p : B \to B$ is a *p-derivation* if $\delta_p 1 = 0$ and, for all $a, b \in B$,

(1) $\delta_p(a+b) = \delta_p a + \delta_p b + C_p(a,b)$,
(2) $\delta_p(ab) = a^p \delta_p b + b^p \delta_p a + p(\delta_p a)(\delta_p b)$.

If δ_p is a p-derivation, then the operator $\phi_p : B \to B$,

$$\phi_p(a) := a^p + p\delta_p a,$$

is a Frobenius lift. Conversely, if $\phi_p : B \to B$ is a Frobenius lift and B is p-torsion free (i.e. p is a non-zero divisor in B) then one can define a p-derivation $\delta_p : B \to B$

by

$$\delta_p a := \frac{\phi(a) - a^p}{p};$$

we say that δ_p and ϕ_p are *attached* to each other. If δ_p is a p-derivation on B and I is an ideal in B such that $\delta_p(I) \subset I$ then $\delta_p(I^\nu) \subset I^\nu$ for all ν.

If B is an A-algebra (always assumed, in this context, to be associative, commutative and unital) we have an obvious notion of p-derivation $\delta_p : A \to B$: it is a map of sets that sends 1 into 0 and satisfies the rules in Definition 3.2 for all $a, b \in A$.

A ring equipped with a p-derivation is referred to as a *δ_p-ring*. By a *morphism of δ_p-rings* we mean a morphism of rings which commutes with δ_p. Let B be a δ_p-ring and let Ω_B be the B-module of Kähler differentials. Then there is a unique additive map

$$\frac{\phi_p}{p} : \Omega_B \to \Omega_B$$

satisfying

$$\frac{\phi_p}{p}(b \cdot dc) = \phi_p(b) \cdot (c^{p-1} \cdot dc + d(\delta_p c))$$

for all $b, c \in B$; cf. **[11, 34]**. The map $\frac{\phi_p}{p}$ is a characteristic zero version of the inverse of the Cartier operator and is, of course, directly related to crystalline theory; this map will be viewed as a "linearization" of our p-derivation δ_p. The maps $\frac{\phi_p}{p}$ are functorial in the δ_p-ring B i.e., for a morphism of δ_p-rings, $A \to B$, the maps $\frac{\phi_p}{p}$ commute with the natural map $\Omega_A \to \Omega_B$; moreover, in this situation, we have an induced additive map

$$\frac{\phi_p}{p} : \Omega_{B/A} \to \Omega_{B/A}.$$

If, in addition, $I \subset B$ is an ideal in a Noetherian ring B set $B_\nu = B/I^\nu$ and

$$\Omega^{for}_{B/A} := \underset{\leftarrow}{\lim}\, \Omega_{B_\nu/A} = \underset{\leftarrow}{\lim}\, (\Omega_{B/A} \otimes_B B_\nu).$$

For any p-derivation δ_p on B such that $\delta_p I \subset I$ we get an induced operator

$$\frac{\phi_p}{p} : \Omega^{for}_{B/A} \to \Omega^{for}_{B/A}.$$

3.1.2. Global connections. Let $\mathcal{V}$ be a set of primes in $\mathbb{Z}$ which we fix once and for all. We denote by $m := |\mathcal{V}| \leq \infty$ the cardinality of $\mathcal{V}$. The elements of $\mathcal{V}$ will be denoted by $p, p', p'', \ldots$. We also write

$$\mathcal{V} = \{p_1, p_2, p_3, \ldots\}.$$

Taking the set $\mathcal{V}$ above as the set $\mathcal{U}$ of basic variables in our superalgebra formalism of Chapter 1 (cf. Definition 1.3) and letting E be any $\mathbb{Z}$-module, we may consider the spaces $\Lambda^k(E)$ in 1.11 whose elements are formal (infinite if $\mathcal{V}$ is infinite) sums

$$\sum_{i_1 < \ldots < i_k} \eta_{i_1 \ldots i_k} \cdot dp_{i_1} \wedge \ldots \wedge dp_{i_k}, \quad \eta_{i_1 \ldots i_k} \in E.$$

DEFINITION 3.3. A *global connection* on a ring B (or on the scheme *Spec* B) is a family $\delta = (\delta_p)$ indexed by $\mathcal{V}$ where, for each $p \in \mathcal{V}$, $\delta_p : B \to B$ is a p-derivation.

Since it will be always clear whether our context is classical differential algebra (derivations) or the arithmetic setting (p-derivations) no confusion should arise from the above definition (compared to the corresponding definition in the setting of differential algebra).

DEFINITION 3.4. Given a global connection $\delta = (\delta_p)$ on a ring B we say that an element $b \in B$ is *δ-constant* if $\delta_p b = 0$ for all $p \in \mathcal{V}$.

Given a global connection δ as above we will often denote by $\phi = (\phi_p)$ the family of attached Frobenius lifts $\phi_p : B \to B$. In particular we may consider (cf. subsection 1.3.6) the 1-form

$$\phi := \sum_p \phi_p \cdot dp \in \Lambda^1(\mathrm{End}(B)), \tag{3.1}$$

where, as usual, End stands for the $\mathbb{Z}$-module endomorphisms.

If we want to emphasize that the global connection is on B (or on $X = Spec\ B$) we write $\delta = \delta^B = \delta^X$, $\delta_p = \delta_p^B = \delta_p^X$ and $\phi = \phi^B = \phi^X$, $\phi_p = \phi_p^B = \phi_p^X$.

The concept of *global connection* introduced here is, as its name indicates, a "global" one which, intuitively, should be viewed as an "analytic/algebraic" concept; it is far too restrictive for the theory we have in mind and later we will be led to introduce a less global notion, that of *adelic connection* which, intuitively, should be viewed as an analogue of a C^∞ concept.

3.1.3. Curvatures, holonomy, first Chern form.

Given two distinct primes $p, p' \in \mathbb{Z}$ we consider the polynomial $C_{pp'}$ in the ring $\mathbb{Z}[X_0, X_1, X_2]$ defined by

$$C_{pp'}(X_0, X_1, X_2) := \frac{1}{pp'}\left((X_0^{p'} + p'X_2)^p - (X_0^p + pX_1)^{p'} + pX_1^{p'} - p'X_2^p\right) \tag{3.2}$$

DEFINITION 3.5. Let $\delta = (\delta_p)$ be a global connection on a ring B, indexed by $p \in \mathcal{V}$. The *curvature* of the global connection $\delta = (\delta_p)$ is the family $\varphi = \varphi[\delta] = (\varphi_{pp'})$, indexed by $p, p' \in \mathcal{V}$, where $\varphi_{pp'} : B \to B$ are the maps given by $\varphi_{pp} = 0$ and

$$\varphi_{pp'}(a) := [\delta_p, \delta_{p'}](a) - C_{pp'}(a, \delta_p a, \delta_{p'} a), \quad p \neq p'. \tag{3.3}$$

The global connection δ will be called *flat* if $\varphi_{pp'} = 0$ for all p, p'.

Note that

$$pp'\varphi_{pp'} = [\phi_p, \phi_{p'}]. \tag{3.4}$$

We claim that for all $a, b \in B$

(1) $\varphi_{pp'}(a + b) = \varphi_{pp'}(a) + \varphi_{pp'}(b)$ (additivity),
(2) $\varphi_{pp'}(a) + \varphi_{p'p}(a) = 0$ (antisymmetry)
(3) $\varphi_{pp'}(ab) = \phi_p\phi_{p'}(b) \cdot \varphi_{pp'}(a) + \phi_{p'}\phi_p(a) \cdot \varphi_{pp'}(b)$ ("Leibniz")

The above are clear if B is pp'-torsion free due to 3.4. In general one can write B as a quotient of a ring of polynomials with $\mathbb{Z}$-coefficients (which is pp'-torsion free) modulo some ideal, lift δ_p to a p-derivation on the polynomial ring (and similarly for p') and reduce to the torsion free case. Note that if B is pp'-torsion free then $\varphi_{pp'} = 0$ if and only if ϕ_p and $\phi_{p'}$ commute. Note also that (3) is, of course, not a genuine Leibniz rule so $\varphi_{pp'}$ is, of course, not a derivation; on the other hand (3) implies that $\varphi_{pp'}$ sends 1 into 0.

In the superalgebra language we have the following equality in $\Lambda^2(\mathrm{End}(B))$:

$$\frac{1}{2}[\phi,\phi] = \sum_{p<p'} pp'\varphi_{pp'} \cdot dp \wedge dp'. \tag{3.5}$$

DEFINITION 3.6. Assume B is an A-algebra and B and A are equipped with global connections. We say that the global connections on A and B are *compatible* if the structure morphism $A \to B$ is a morphism of δ_p-rings for all $p \in \mathcal{V}$.

In the situation of Definition 3.6 we may consider the commutators

$$\varphi_{pp'}^{\Omega} = \left[\frac{\phi_p}{p}, \frac{\phi_{p'}}{p'}\right] : \Omega_{B/A} \to \Omega_{B/A}. \tag{3.6}$$

Finally if we assume that $I \subset B$ is an ideal sent into itself by δ_p and $\delta_{p'}$ then, for all $a \in I$,

$$\varphi_{pp'}(a) \equiv [\delta_p, \delta_{p'}](a) \quad mod \quad I^{\min\{p,p'\}}. \tag{3.7}$$

Also by the above discussion, $\varphi_{pp'}$ sends any power I^ν into itself hence there are induced additive operators,

$$\varphi_{pp'}^{(\nu)} : I/I^{\nu+1} \to I/I^{\nu+1}. \tag{3.8}$$

Clearly if $T_1, ..., T_d$ is a system of generators for I then $\varphi_{pp'}^{(\nu)} = 0$ if and only $\varphi_{pp'}(T_i) \in I^{\nu+1}$ for all i. Also, in the situation of Definition 3.6, if all δ_p send I into itself, we have induced additive maps

$$\varphi_{pp'}^{\Omega} = \left[\frac{\phi_p}{p}, \frac{\phi_{p'}}{p'}\right] : \Omega_{B/A}^{for} \to \Omega_{B/A}^{for}. \tag{3.9}$$

In the notation above we may also consider analogues of the s-curvature (cf. Definition 2.29) as follows. Assume, for simplicity, that B is torsion free, let $\mathcal{V} = \{p_1, p_2, p_3, ...\}$, and let $\delta = (\delta_p)$ be a global connection on B.

DEFINITION 3.7. Let $s \geq 2$. The *s-curvature* on B, attached to the global connection $\delta = (\delta_p)$, is the family $(\varphi_{p_{i_1}p_{i_2}...p_{i_s}})$ of elements in $\mathrm{End}(B)$,

$$\varphi_{p_{i_1}p_{i_2}...p_{i_s}} = \frac{1}{p_{i_{s-1}}p_{i_s}}\left[\phi_{p_{i_1}}, \left[\phi_{p_{i_2}}, ... \left[\phi_{p_{i_{s-1}}}, \phi_{p_{i_s}}\right]...\right]\right] : B \to B. \tag{3.10}$$

The *holonomy ring* of δ on B is the $\mathbb{Z}$-span $\mathfrak{hol} = \mathfrak{hol}(\phi)$ in $\mathrm{End}(B)$ of all endomorphisms of the form $p_{i_{s-1}}p_{i_s}\varphi_{p_{i_1}p_{i_2}...p_{i_s}}$ with $p_i \in \mathcal{V}$ and $s \geq 2$. The *holonomy $\mathbb{Q}$-algebra* of δ on B is the $\mathbb{Q}$-linear span $\mathfrak{hol}_{\mathbb{Q}}$ of $\mathfrak{hol}$ in $\mathrm{End}(B) \otimes \mathbb{Q}$.

REMARK 3.8.

(1) In the notation above $\mathfrak{hol}_{\mathbb{Q}} \simeq \mathfrak{hol} \otimes \mathbb{Q}$, $\varphi_{p_{i_1}p_{i_2}...p_{i_s}}$ are naturally elements of $\mathfrak{hol}_{\mathbb{Q}}$, and $\mathfrak{hol}_{\mathbb{Q}}$ is spanned over $\mathbb{Q}$ by all these elements (with $p_i \in \mathcal{V}$, $s \geq 2$).
(2) For B an A-algebra with A and B equipped with compatible global connections one could also consider the A-linear span $\mathfrak{hol}_A$ of $\mathfrak{hol}$ in $\mathrm{End}(B)$; but note that, in contrast with the case of classical differential geometry, $\mathfrak{hol}_A$ is not generally a Lie subring of $\mathrm{End}(B)$. So $\mathfrak{hol}_A$ is, perhaps, not a sufficiently natural object and will play no role in what follows.

If Ω is $\Omega_{B/A}$ or $\Omega_{B/A}^{for}$ one can also define higher curvatures and holonomy on Ω as in the case of classical differential geometry as follows.

DEFINITION 3.9. Let $s \geq 2$. The *s-curvature* on Ω, attached to $\delta = (\delta_p)$, is the family $(\varphi^{\Omega}_{p_{i_1}p_{i_2}\dots p_{i_s}})$ of elements in $\text{End}(\Omega)$,

$$\varphi^{\Omega}_{p_{i_1}p_{i_2}\dots p_{i_s}} = \left[\frac{\phi_{p_{i_1}}}{p_{i_1}}, \left[\frac{\phi_{p_{i_2}}}{p_{i_2}}, \dots \left[\frac{\phi_{p_{i_{s-1}}}}{p_{i_{s-1}}}, \frac{\phi_{p_{i_s}}}{p_{i_s}}\right]\dots\right]\right] : \Omega \to \Omega. \tag{3.11}$$

The *holonomy ring* of δ in Ω is the $\mathbb{Z}$-span $\mathfrak{hol}^{\Omega} = \mathfrak{hol}(\frac{\phi}{\mathbf{p}})$ in $\text{End}(\Omega)$ of all components of all s-curvatures of δ on Ω for all $s \geq 2$.

Also recall that by the general formalism of Lie rings the commutator

$$[\varphi^{\Omega}_{p_{i_1}p_{i_2}\dots p_{i_s}}, \varphi^{\Omega}_{p_{j_1}p_{j_2}\dots p_{j_t}}]$$

of any two components of the s-curvature and t-curvature respectively is a $\mathbb{Z}$-linear combination of the components

$$\varphi^{\Omega}_{p_{k_1}p_{k_2}\dots p_{k_{s+t}}}$$

of the $s+t$-curvature. In particular $\mathfrak{hol}^{\Omega}$ is a Lie subring of $\text{End}(\Omega)$. Similarly $\mathfrak{hol}$ is a Lie subring of $\text{End}(B)$ and $\mathfrak{hol}_{\mathbb{Q}}$ is a Lie $\mathbb{Q}$-subalgebra of $\text{End}(B) \otimes \mathbb{Q}$.

One can introduce an obvious analogue, in our arithmetic setting, of the concept of connection with *A-horizontal* curvature, cf. Definition 2.33.

DEFINITION 3.10. The global connection δ on B is said to have a *$\mathbb{Q}$-horizontal* curvature if one of the following equivalent conditions holds:

1. $\mathfrak{hol}_{\mathbb{Q}}$ is spanned over $\mathbb{Q}$ by the components of the curvature $(\varphi_{pp'})$.
2. All the components of the 3-curvature $(\varphi_{pp'p''})$ are $\mathbb{Q}$-linear combinations in $\mathfrak{hol}_{\mathbb{Q}}$ of the components of the curvature $(\varphi_{pp'})$.

Assume B is an A-algebra and A and B have compatible global connections in the sense of Definition 3.6. Also let $I \subset B$ be an ideal. The following gives a relation between curvature on B and curvature on Ω where the latter is $\Omega_{B/A}$ or $\Omega^{for}_{B/A}$.

LEMMA 3.11. *Assume the following hold:*

(i) *The map $A \to B/I$ is an isomorphism.*
(ii) *B/I^n is p-torsion free for all $p \in \mathcal{V}$ and $n \geq 1$.*
(iii) *The connection on A is flat (i.e. $\varphi_{pp'}(A) = 0$ for all p, p').*
(iv) *Ω is a free B-module generated by the set dI.*
(v) *$\delta_p(I) \subset I$ for all $p \in \mathcal{V}$ (equivalently $\phi_p I \subset I$ for all p).*
(vi) *For some $\nu \geq 0$ the maps $\varphi^{(\nu)}_{pp'} : I/I^{\nu+1} \to I/I^{\nu+1}$ vanish for all $p, p' \in \mathcal{V} = \{p_1, p_2, p_3, \dots\}$, in other words $\varphi_{pp'}(I) \subset I^{\nu+1}$. (This is, of course automatic for $\nu = 0$.)*

Then, for all $i_1, \dots, i_s$ with $s \geq 2$ and all $n \geq 0$ we have:

(1) $\varphi_{p_{i_1}\dots p_{i_s}}(I^n) \subset I^{n+\nu}$.
(2) $\varphi^{\Omega}_{p_{i_1}\dots p_{i_s}}(I^n\Omega) \subset I^{n+\nu}\Omega$.

PROOF. We proceed by induction on s. The induction step follows immediately from the fact that ϕ_p sends I^n and $I^n\Omega$ into themselves plus the fact that B/I^nB and $\Omega/I^n\Omega$ are p-torsion free. So we are reduced to proving the case $s = 2$. Again, since B/I^nB and $\Omega/I^n\Omega$ are p-torsion free, it is enough to prove, for $n \geq 0$ that

$$(\phi_p\phi_{p'} - \phi_{p'}\phi_p)(I^n) \subset I^{n+\nu}, \tag{3.12}$$

$$(\phi_p\phi_{p'} - \phi_{p'}\phi_p)(I^n\Omega) \subset I^{n+\nu}\Omega. \tag{3.13}$$

To check 3.12 let $f = \phi_p\phi_{p'} : B \to B$ and $g = \phi_{p'}\phi_p : B \to B$. We proceed by induction on n. For $n = 0$ we have

$$(f-g)(B) = (f-g)(A+I) = (f-g)(I) \subset I^{\nu+1} \subset I^{\nu}.$$

For $n = 1$ we are done by the condition $\varphi_{pp'}(I) \subset I^{\nu+1}$. For the induction step if $n \geq 2$, $x \in I$, $y \in I^{n-1}$ then

$$\begin{array}{rcl} f(xy) - g(xy) & = & f(x)f(y) - g(x)g(y) \\ & = & f(x)(f(y) - g(y)) + (f(x) - g(x))g(y) \\ & \in & I \cdot I^{n-1+\nu} + I^{1+\nu} \cdot I^{n-1} \subset I^{n+\nu}. \end{array}$$

Now, in order to prove 3.13 let $n \geq 0$, $x \in I^n$, $y \in I$ and note that

$$\begin{array}{rcl} (\phi_p\phi_{p'} - \phi_{p'}\phi_p)(xdy) & = & f(x)d(f(y)) - g(x)d(g(y)) \\ & = & f(x)d(f(y) - g(y)) + (f(x) - g(x))d(g(y)) \\ & \in & I^n \cdot d(I^{\nu+1}) + I^{n+\nu} \cdot dI \ \text{ by } \ 3.12 \\ & \subset & I^n \cdot I^{\nu} \cdot dI + I^{n+\nu} \cdot dI \subset I^{n+\nu}\Omega. \end{array}$$

We are done because Ω is generated as a B-module by dI. □

COROLLARY 3.12. *Under the assumptions of Lemma* 3.11 *(with $\nu \geq 0$) we have natural maps of Lie rings, for each $n \geq 1$,*

$$\mathfrak{hol} \to End(B/I^n), \tag{3.14}$$

$$\mathfrak{hol}^{\Omega} \to End(\Omega/I^n\Omega). \tag{3.15}$$

DEFINITION 3.13. Under the assumptions of Lemma 3.11 (with $\nu \geq 0$) let

$$\mathfrak{hol}_n, \ \ \mathfrak{hol}_n^{\Omega}$$

be the Lie rings, images of the maps 3.14 and 3.15, respectively. Then the Lie rings

$$\widehat{\mathfrak{hol}} := \lim_{\leftarrow} \mathfrak{hol}_n, \ \ \widehat{\mathfrak{hol}}^{\Omega} := \lim_{\leftarrow} \mathfrak{hol}_n^{\Omega}$$

are called the *completed holonomy rings* of δ on B and Ω respectively. Let us say that a Lie ring is *pro-nilpotent* if it is a projective limit of nilpotent Lie rings.

COROLLARY 3.14. *Assume the hypotheses of Lemma* 3.11 *with $\nu \geq 1$. Then the Lie rings $\mathfrak{hol}_n$ and $\mathfrak{hol}_n^{\Omega}$ are nilpotent. Hence the completed holonomy rings*

$$\widehat{\mathfrak{hol}}, \ \ \widehat{\mathfrak{hol}}^{\Omega}$$

are pro-nilpotent.

REMARK 3.15. Assume we are given a morphism of torsion free rings $A \to B$ equipped with compatible global connections $\delta^A = (\delta_p^A)$ and $\delta^B = (\delta_p^B)$ such that δ^A is flat, so the ϕ_p^A's commute on A. By analogy with 2.3 we may consider the Lie ring $\text{End}(B)$ of $\mathbb{Z}$-module endomorphisms of B and we may consider the smallest Lie subring $\mathfrak{lie}(\phi^B)$ of $\text{End}(B)$ containing all the ϕ_p^B's attached to δ_p^B. Similarly we may consider the smallest Lie subring $\mathfrak{lie}(\phi^A)$ of $\text{End}(A)$ containing all ϕ_p^A's; then

$\mathfrak{lie}(\phi^A)$ is the $\mathbb{Z}$-span of the δ_p^A's and is an Abelian Lie ring. We have at our disposal the restriction map (or *anchor map*)

$$\chi : \mathfrak{lie}(\phi^B) \to \mathfrak{lie}(\phi^A). \tag{3.16}$$

The map χ in 3.16 is a map of Lie rings. If, in addition ϕ_p^A are $\mathbb{Z}$-linearly independent in $\mathrm{End}(A)$ then χ has a section $\mathfrak{lie}(\phi^A) \to \mathfrak{lie}(\phi^B)$ in the category of $\mathbb{Z}$-modules given by $\phi_p^A \mapsto \phi_p^B$ and the kernel of χ in 3.16 equals the commutator

$$\mathfrak{hol}(\phi^B) = [\mathfrak{lie}(\phi^B), \mathfrak{lie}(\phi^B)]. \tag{3.17}$$

Linear independence of the ϕ_p^As is automatic if, for instance, $A = \mathbb{Z}[1/M, \zeta_N]$, $M \in 2\mathbb{Z}$, $N \geq 2$, and the p's in $\mathcal{V}$ do not divide MN and are distinct modulo N; cf. the independence of characters 1.2. The $\mathbb{Z}$-module section $\mathfrak{lie}(\phi^A) \to \mathfrak{lie}(\phi^B)$ given by $\phi_p^A \mapsto \phi_p^B$ is a morphism of Lie rings if and only if δ^B is flat.

REMARK 3.16. It is interesting to draw a parallel between the anchor map 2.3, its analogue 3.16, and the cyclotomic character in Galois theory. Indeed, let $K = \mathbb{Q}(\zeta_N)$, $A = \mathbb{Z}[1/M, \zeta_N]$, $M \in 2\mathbb{Z}$, $N \geq 2$, let L/K be a finite field extension with $L/\mathbb{Q}$ Galois and unramified outside the primes dividing MN. Let $B = \mathcal{O}_L[1/M]$, where $\mathcal{O}_L$ is the ring of integers of L. We have at our disposal the restriction map (the *cyclotomic character*)

$$\chi : \mathrm{Aut}(B) \to \mathrm{Aut}(A) = (\mathbb{Z}/N\mathbb{Z})^\times, \tag{3.18}$$

where Aut stands for the *group of ring automorphisms*. Assume in addition that $\mathcal{V}$ is a set of primes in $\mathbb{Z}$ not dividing MN and distinct modulo N. Let $\phi^A = (\phi_p^A)$ be the collection of Frobenius elements $\phi_p^A \in \mathrm{Aut}(A)$ corresponding to the primes $p \in \mathcal{V}$ and let $\phi^B = (\phi_p^A)$ be a collection of elements $\phi_p^B \in \mathrm{Aut}(B)$ indexed by $\mathcal{V}$ such that, for each p, ϕ_p^B is a Frobenius element of $\mathrm{Aut}(B)$ correspondding to some prime ideal $\mathfrak{P}$ of B lying above p, i.e.,

$$\phi_p^B(b) \equiv b^p \mod \mathfrak{P} \text{ for } b \in B.$$

In particular $\chi(\phi_p^B) = \phi_p^A$. (Note that ϕ_p^B is not a Frobenius lift on B but it induces a Frobenius lift on the localization $B_{\mathfrak{P}}$.) Then ϕ^B can be viewed as a Galois theoretic analogue of a connection; the failure of the ϕ_p^Bs to commute can be viewed as a Galois theoretic analogue of curvature. Let Γ^A and Γ^B be the subgroups of Aut(A) and Aut(B) generated by ϕ^A and ϕ^B respectively. Then the induced surjective group homomorphism

$$\chi : \Gamma^B \to \Gamma^B \tag{3.19}$$

can be viewed as an analogue of the maps 2.3 and 3.16. By the way the surjective map 3.19 has kernel equal to the commutator

$$[\Gamma^B, \Gamma^B]; \tag{3.20}$$

so this kernel can be viewed as a Galois theoretic analogue of holonomy. It would be interesting to see if this analogy between the maps 3.16 and 3.19 can be promoted, in special cases, to an actual link between the two maps.

We next define the first Chern form in our arithmetic context.

DEFINITION 3.17. Assume $\Omega_{B/A}$ (respectively $\Omega^{for}_{B/A}$) is a free B-module and let $\omega_1, ..., \omega_d$ be a basis. Assume (δ_p) is a global connection on B, preserving an ideal $I \subset B$, and compatible with a global connection on A. Let $\varphi^{\Omega}_{pp'}$ be the induced additive endomorphisms of $\Omega_{B/A}$ (respectively $\Omega^{for}_{B/A}$). Write

$$\varphi^{\Omega}_{pp'}\omega_i = \sum_{j=1}^{d} \mathcal{F}^j_{pp'i}\omega_j, \quad \mathcal{F}^j_{pp'i} \in B.$$

The *first Chern form* of (δ_p) (relative to the basis $\omega = (\omega_i)$) is the $\mathcal{V} \times \mathcal{V}$-matrix $(\rho_{pp'})$ defined by

$$\rho_{pp'} = \sum_{i=1}^{d} \mathcal{F}^i_{pp'i}.$$

In the superalgebra language (cf. subsection 1.3.6), for Ω equal to $\Omega_{B/A}$ or $\Omega^{for}_{B/A}$, and for

$$\frac{\phi}{\mathbf{p}} := \sum_p \frac{\phi_p}{p} \cdot dp \in \Lambda^1(\text{End}(\Omega)), \tag{3.21}$$

we have

$$\frac{1}{2}\left[\frac{\phi}{\mathbf{p}}, \frac{\phi}{\mathbf{p}}\right] = \sum_{p<p'} \varphi^{\Omega}_{pp'} \cdot dp \wedge dp' \in \Lambda^2(\text{End}(\Omega)), \tag{3.22}$$

$$c_1\left[\frac{\phi}{\mathbf{p}}, \omega\right] = \sum_{p<p'} \rho_{pp'} \cdot dp \wedge dp' \in \Lambda^2(B). \tag{3.23}$$

Then the holonomy introduced in Definition 3.9 coincides of course with the holonomy introduced in Definition 1.10.

3.1.4. Global connections on power series rings. Assume A is a ring with a flat global connection (δ_p) and assume A is p-torsion free for all $p \in \mathcal{V}$. Let $B = A[[T]]$ where T is a d-tuple of indeterminates $T_1, ..., T_d$. Assume we are given a global connection (δ_p) on B compatible with that on A and let (ϕ_p) be the attached family of Frobenius lifts on B; let $I = (T) \subset B$ be the ideal generated by T and assume that $\phi_p(I) \subset I$. Then for $p \neq p'$ the curvature $\varphi_{pp'} : B \to B$ satisfies $\varphi_{pp'}(T) = \Phi^0_{pp'}$ where $\Phi^0_{pp'}$ is the d-tuple of elements in $A[[T]]$,

$$\Phi^0_{pp'} = \frac{1}{pp'}(\phi_p(\phi_{p'}(T)) - \phi_{p'}(\phi_p(T))). \tag{3.24}$$

Clearly ϕ_p and $\phi_{p'}$ commute on B if and only if $\Phi^0_{pp'} = 0$. Furthermore the components of dT are a B-basis of $\Omega^{for}_{B/A}$ so the first Chern form $(\rho_{pp'})$ with respect to this basis is defined and $\rho_{pp'}$ are easily expressible in terms of $\Phi^0_{pp'}$; we will make this explicit later in the case of GL_n.

Let now $T^{(1)}$ and $T^{(2)}$ be d-tuples of variables and assume we are given a formal group law structure on B i.e. a d-tuple of "multiplication" series

$$\mathcal{G} \in A[[T^{(1)}, T^{(2)}]]^{\times d}$$

and an "inverse" series $\mathcal{S} \in A[[T]]$, satisfying the usual properties [**97**]. We would like to define a concept *$\mathcal{G}$-curvature* $(\varphi^{\mathcal{G}}_{pp'})$, where

$$\varphi^{\mathcal{G}}_{pp'} : B \to B$$

are ring homomorphisms (rather than merely additive group homomorphisms). This $\mathcal{G}$-curvature $\varphi^{\mathcal{G}}_{pp'}$ will appear as a "deformation" of the "usual" curvature $\varphi_{pp'}$. To proceed consider the "quotient" tuple of series

$$\mathcal{Q}(T^{(1)}, T^{(2)}) := \mathcal{G}(T^{(1)}, \mathcal{S}(T^{(2)})).$$

For any A-algebra C which is J-adically complete, where J is an ideal in C, and for any $a, b \in J^{\times d} = J \times ... \times J$ we set

$$a +_{\mathcal{G}} b = \mathcal{G}(a, b), \quad a -_{\mathcal{G}} b = \mathcal{Q}(a, b);$$

then $J^{\times d}$ is a group with respect to $+_{\mathcal{G}}$ and $(a -_{\mathcal{G}} b) +_{\mathcal{G}} b = a$. Let us say that $a \equiv b \bmod n$ for some $n \in \mathbb{Z}$ if $a - b = nc$ for some $c \in C^{\times d}$.

LEMMA 3.18. *Assume $a \equiv b$ mod m. Then $a -_{\mathcal{G}} b \equiv 0$ mod m.*

PROOF. This follows directly from the fact that $\mathcal{Q}(T, T) = 0$. □

Now consider the d-tuples

$$\Phi^{\mathcal{G}}_{pp'} = \frac{1}{pp'}(\phi_p(\phi_{p'}(T)) -_{\mathcal{G}} \phi_{p'}(\phi_p(T))). \tag{3.25}$$

By Lemma 3.18 these are tuples with coefficients in $A[[T]]$. Clearly ϕ_p and $\phi_{p'}$ commute on B if and only if $\Phi^{\mathcal{G}}_{pp'} = 0$. Consider the ring homomorphism

$$\varphi^{\mathcal{G}}_{pp'} : A[[T]] = B \to B \tag{3.26}$$

whose restriction to A is $\phi_p \phi_{p'} = \phi_{p'} \phi_p$ and such that $\varphi^{\mathcal{G}}_{pp'}(T) = \Phi^{\mathcal{G}}_{pp'}$. Also let $\nu_{pp'} : B \to B$ be the unique A-algebra homomorphism with $\nu_{pp'}(T) = pp'T$ and let

$$\mathcal{Q}^* : B \to B \widehat{\otimes}_A B \simeq A[[T^{(1)}, T^{(2)}]]$$

be the A-algebra homomorphism with $\mathcal{Q}^*(T) = \mathcal{Q}$, where the $\widehat{}$ on $\otimes$ is the completion with respect to the ideal

$$(T) \otimes B + B \otimes (T) \subset B \otimes_A B.$$

Then $\varphi^{\mathcal{G}}_{pp'}$ is the unique ring homomorphism fitting into the diagram

$$\begin{array}{ccc} B & \stackrel{\mathcal{Q}^*}{\longrightarrow} & B\widehat{\otimes}_A B \\ \nu_{pp'} \downarrow & & \downarrow \phi_p\phi_{p'} \otimes \phi_{p'}\phi_p \\ B & \stackrel{\varphi^{\mathcal{G}}_{pp'}}{\longrightarrow} & B \end{array}$$

DEFINITION 3.19. The family of ring endomorphisms of B, $(\varphi^{\mathcal{G}}_{pp'})$, is called the *$\mathcal{G}$-curvature* of (δ_p).

We note that the concept of $\mathcal{G}$-curvature does not fit into our superalgebra formalism but, rather, should be viewed as a natural "non-linear deformation" of the latter.

3.1.5. (1, 1) framework. The framework below is aimed at imitating the classical setting of hermitian geometry. Let $\delta = (\delta_p)$ and $\bar{\delta} = (\bar{\delta}_p)$ be two global connections on a ring B and let $\delta_{\bar{p}} := \bar{\delta}_p$. We have then at our disposal the curvatures $(\varphi_{pp'})$ and $(\varphi_{\overline{pp}'})$ of δ and $\overline{\delta}$ respectively. Now for $p \neq p'$ let $\varphi_{p\bar{p}'} : B \to B$ be given by

$$\varphi_{p\bar{p}'}(a) := [\delta_p, \delta_{\bar{p}'}](a) - C_{pp'}(a, \delta_p a, \delta_{\bar{p}'} a). \tag{3.27}$$

In particular $pp'\varphi_{p\bar{p}'} = [\phi_p, \phi_{\bar{p}'}]$. On the other hand for any p let $\varphi_{p\bar{p}} : B \to B$ be given by

$$\varphi_{p\bar{p}}(a) = p[\delta_p, \delta_{\bar{p}}](a) - C_{pp}(a, \delta_p a, \delta_{\bar{p}} a), \tag{3.28}$$

where

$$C_{pp}(X_0, X_1, X_2) := \frac{1}{p}((X_0^p + pX_2)^p - (X_0^p + pX_1)^p) + X_1^p - X_2^p.$$

In particular $p\varphi_{p\bar{p}} = [\phi_p, \phi_{\bar{p}}]$.

Finally let $D : B \to B$ be a derivation; it is trivial to check that $D(\varphi_{p\bar{p}}(B)) \subset pB$. So if B is p-torsion free we have a well defined map

$$\frac{1}{p} D\varphi_{p\bar{p}} : B \to B.$$

Note also that, in the situation of Definition 3.6, we have induced maps

$$\varphi^{\Omega}_{p\bar{p}'} = \left[\frac{\phi_p}{p}, \frac{\phi_{\bar{p}'}}{p'}\right] : \Omega_{B/A} \to \Omega_{B/A}, \tag{3.29}$$

and (assuming all $\delta_p, \delta_{\bar{p}}$ send I into itself) we have similar endomorphisms of $\Omega^{for}_{B/A}$.

Definition 3.20. The $\mathcal{V} \times \mathcal{V}$-matrix $(\varphi_{p\bar{p}'})$ will be called the $(1,1)$-*curvature* of δ with respect to $\bar{\delta}$. Assume in addition that $\Omega_{B/A}$ (respectively $\Omega^{for}_{B/A}$) is a free B-module and let $\omega_1, ..., \omega_d$ be a basis. Assume also that δ and $\overline{\delta}$ are compatible with corresponding global connections on A. Let $\varphi^{\Omega}_{p\bar{p}'}$ be the induced additive endomorphisms of $\Omega_{B/A}$ (respectively $\Omega^{for}_{B/A}$). Write

$$\varphi^{\Omega}_{p\bar{p}'}\omega_i = \sum_{j=1}^{d} \mathcal{F}^j_{p\bar{p}'i}\omega_j, \qquad \mathcal{F}^j_{p\bar{p}'i} \in B.$$

The *Chern* $(1,1)$-*form* of δ with respect to $\overline{\delta}$ (relative to the basis $\omega = (\omega_i)$) is the $\mathcal{V} \times \mathcal{V}$-matrix $(\rho_{p\bar{p}'})$ defined by

$$\rho_{p\bar{p}'} = \sum_{i=1}^{d} \mathcal{F}^i_{p\bar{p}'i}.$$

When $\overline{\delta}$ is understood from context we drop the reference to $\overline{\delta}$.

Remark 3.21. In the superalgebra language of Chapter 1, the above can be expressed as follows. Take as set of basic variables (cf. Definition 1.3) the set $\mathcal{U} = \mathcal{V} \cup \overline{\mathcal{V}}$; so, for any $\mathbb{Z}$-module E we have at our disposal a graded Lie superalgebra

$$\Lambda(E) = \bigoplus \Lambda^k(E), \quad \Lambda^k(E) = \bigoplus_{r+s=k} \Lambda^{r,s}(E),$$

where the elements of $\Lambda^{r,s}(E)$ are formal (infinite if $\mathcal{V}$ is infinite) sums

$$\sum_{\substack{i_1<\ldots<i_r\\ j_1<\ldots<j_s}} \eta_{i_1\ldots i_r j_1\ldots j_s}\cdot dp_{i_1}\wedge\ldots\wedge dp_{i_r}\wedge d\bar{p}_{j_1}\wedge\ldots\wedge d\bar{p}_{j_s}, \tag{3.30}$$

with the η's in E. For

$$\phi=\sum_p \phi_p\cdot dp\in\Lambda^{1,0}(\mathrm{End}(B)),\quad \bar{\phi}=\sum_p \phi_{\bar{p}}\cdot d\bar{p}\in\Lambda^{0,1}(\mathrm{End}(B)), \tag{3.31}$$

we have

$$[\phi,\bar{\phi}]=\sum_{p\neq p'} pp'\varphi_{p\bar{p}'}\cdot dp\wedge d\bar{p}'+\sum_p p\varphi_{p\bar{p}}\cdot dp\wedge d\bar{p}\in\Lambda^{1,1}(\mathrm{End}(B)). \tag{3.32}$$

Furthermore, for Ω equal to $\Omega_{B/A}$ or $\Omega^{for}_{B/A}$, and for

$$\frac{\phi}{\mathbf{p}}:=\sum_p \frac{\phi_p}{p}\cdot dp\in\Lambda^{1,0}(\mathrm{End}(\Omega)),\quad \frac{\bar{\phi}}{\mathbf{p}}:=\sum_p \frac{\phi_{\bar{p}}}{p}\cdot d\bar{p}\in\Lambda^{0,1}(\mathrm{End}(\Omega)) \tag{3.33}$$

we have

$$\begin{aligned}\frac{1}{2}\left[\frac{\phi}{\mathbf{p}}+\frac{\bar{\phi}}{\mathbf{p}},\frac{\phi}{\mathbf{p}}+\frac{\bar{\phi}}{\mathbf{p}}\right] &= \textstyle\sum_{p<p'}\varphi^{\Omega}_{pp'}\cdot dp\wedge dp'\\ &\quad+\textstyle\sum_{pp'}\varphi^{\Omega}_{p\bar{p}'}\cdot dp\wedge d\bar{p}'\\ &\quad+\textstyle\sum_{p<p'}\varphi^{\Omega}_{\bar{p}\bar{p}'}\cdot d\bar{p}\wedge d\bar{p}'\in\Lambda^2(\mathrm{End}(\Omega))\end{aligned} \tag{3.34}$$

$$\left(c_1\left[\frac{\phi}{\mathbf{p}}+\frac{\bar{\phi}}{\mathbf{p}},\omega\right]\right)^{(1,1)}=\sum_{pp'}\rho_{p\bar{p}'}\cdot dp\wedge d\bar{p}'\in\Lambda^{1,1}(B), \tag{3.35}$$

where we recall that the superscript $(1,1)$ means the $(1,1)$-component in the bigrading of our Lie superalgebra.

Next assume all ϕ_p and $\phi_{\bar{p}}$ restricted to A commute and $B=A[[T]]$ for T a d-tuple of indeterminates. Assume in addition that all ϕ_p and $\phi_{\bar{p}}$ on B send I into itself, where $I=(T)$. Then the $(1,1)$-curvature $\varphi_{p\bar{p}'}:B\to B$ satisfies $\varphi_{p\bar{p}'}(T)=\Phi^0_{p\bar{p}'}$ where

$$\Phi^0_{p\bar{p}'}=\frac{1}{pp'}(\phi_p\phi_{\bar{p}'}(T)-\phi_{\bar{p}'}\phi_p(T)),\quad \text{for}\quad p\neq p'; \tag{3.36}$$

$$\Phi^0_{p\bar{p}}=\frac{1}{p}(\phi_p\phi_{\bar{p}}(T)-\phi_{\bar{p}}\phi_p(T)). \tag{3.37}$$

Clearly ϕ_p and $\phi_{\bar{p}}$ commute on B if and only if $\Phi^0_{p\bar{p}'}=0$. Also the Chern $(1,1)$-form $(\rho_{p\bar{p}'})$ relative to dT is defined and easily expressible in terms of $\Phi^0_{p\bar{p}'}$; we will make this explicit later in the special case of GL_n.

If we assume, in addition, that we are given a formal group law $\mathcal{G}$ then, exactly as in 3.25 one can define

$$\Phi^{\mathcal{G}}_{p\bar{p}'}=\frac{1}{pp'}(\phi_p(\phi_{\bar{p}'}(T))-_{\mathcal{G}}\phi_{\bar{p}'}(\phi_p(T))),\quad \text{for}\quad p\neq p'; \tag{3.38}$$

$$\Phi^{\mathcal{G}}_{p\bar{p}}=\frac{1}{p}(\phi_p(\phi_{\bar{p}}(T))-_{\mathcal{G}}\phi_{\bar{p}}(\phi_p(T))). \tag{3.39}$$

So one can consider the ring homomorphisms $\varphi^{\mathcal{G}}_{p\bar{p}'} : B \to B$ and $\varphi^{\mathcal{G}}_{p\bar{p}} : B \to B$ whose restriction to A is $\phi_p\phi_{\bar{p}'}$ and $\phi_p\phi_{\bar{p}}$ respectively and sending T into $\Phi^{\mathcal{G}}_{p\bar{p}'}$ and $\Phi^{\mathcal{G}}_{p\bar{p}}$ respectively. Again ϕ_p and $\phi_{\bar{p}'}$ commute on B if and only if $\Phi^{\mathcal{G}}_{p\bar{p}'} = 0$.

DEFINITION 3.22. The family $(\varphi_{p\bar{p}'})$ is called the $(1,1)$-$\mathcal{G}$-curvature of the global connection δ on $B = A[[T]]$.

3.1.6. δ-rings. All of our theory will be "over a base ring," enjoying some technical assumptions, and equipped with a global flat connection. Here is our main definition.

DEFINITION 3.23. By a *δ-ring* we understand a ring A equipped with a flat global connection $\delta = (\delta_p)$ indexed by a set $\mathcal{V}$ of primes in $\mathbb{Z}$. We say a δ-ring A is *admissible* if A is a Noetherian domain of characteristic 0 and $p \notin A^\times$ for all $p \in \mathcal{V}$.

REMARK 3.24.

(1) Note that any A as above is p-torsion free and p-adically separated for all $p \in \mathcal{V}$ (by Krull's intersection theorem).
(2) The p-derivations $\delta_p : A \to A$ are viewed as an analogue of the derivations 2.1.
(3) Later, in our discussion of "$(1,1)$-case" we will also assume A comes equipped with a flat global connection $\overline{\delta} = (\overline{\delta}_p)$ on A. We sometimes write $\delta_{\overline{p}}$ instead of $\overline{\delta}_p$ and we denote by $\phi_{\overline{p}}$ the Frobenius lift attached to $\delta_{\overline{p}}$. We will always assume that δ and $\overline{\delta}$ *commute* on A in the sense that $\phi_{\overline{p}}$ and $\phi_{p'}$ commute for all p and p'. And actually, in all our examples of interest, we will have that $\phi_{\overline{p}} = \phi_p$ for all p.

We will be interested in two types of δ-rings which we call of *global type* and *local type* respectively. We start with the global ones.

DEFINITION 3.25. Let $M \neq 0$ be a supernatural number (i.e. a formal infinite product $M = \prod \ell^{e_\ell}$ where ℓ are the rational primes and $e_\ell \geq 0$), let $\mathbb{Q} \subset K$ be an abelian extension of number fields, and let $\mathcal{V}$ be a set of primes in $\mathbb{Z}$ which do not divide M and are unramified in K. Let $\mathcal{O}_K$ be the ring of integers of K. For each $p \in \mathcal{V}$ let ϕ_p be the automorphism of K that induces the p-power Frobenius automorphism on $\mathcal{O}_K/p\mathcal{O}_K$. Finally take

$$A = \mathcal{O}_K[1/M] \tag{3.40}$$

where adjoining $1/M$ means, of course, inverting all the prime divisors of M. For technical reasons we will always assume M is *even*, i.e., $2|M$. Then (ϕ_p) define a flat global connection $\delta = (\delta_p)$ on A. So A equipped with δ is a δ-ring in the sense of Definition 3.23. Such a δ-ring A will be called a δ-ring of *global type*.

EXAMPLE 3.26. The standard example we have in mind is as follows. Let $M \neq 0$ be an even supernatural number, let N be a positive integer, let $\mathcal{V}$ be a set of primes in $\mathbb{Z}$ not dividing MN, let ζ_N be a primitive root of unity, and let $K = \mathbb{Q}(\zeta_N)$. Then the ring $A = \mathcal{O}_K[1/M]$ in Definition 3.25 is

$$A = \mathbb{Z}[1/M, \zeta_N] \tag{3.41}$$

with its unique flat global connection (δ_p) indexed by $\mathcal{V}$.

Next we discuss the local case. Let $\mathcal{V} = \{p\}$ consist of one prime p only. We assume $p \neq 2$. Then we let $\mathbb{F}$ be an algebraic extension of the field $\mathbb{F}_p$ with p elements. Let

$$W(\mathbb{F}) \tag{3.42}$$

be the unique (up to isomorphism) complete discrete valuation ring with maximal ideal generated by p and residue field $\mathbb{F}$. The reduction mod p map

$$W(\mathbb{F}) \to \mathbb{F} = W(\mathbb{F})/pW(\mathbb{F})$$

has a unique multiplicative section

$$\theta : \mathbb{F} \to W(\mathbb{F}),$$

called the *Teichmüller lift*, whose image $\theta(\mathbb{F})$ consists of 0 and all the roots of unity in $W(\mathbb{F})$. Any element $a \in W(\mathbb{F})$ can be written uniquely as

$$a = \sum_{i=0}^{\infty} c_i p^i, \quad c_i \in \theta(\mathbb{F}). \tag{3.43}$$

Also $W(\mathbb{F})$ has a unique Frobenius lift $\phi = \phi_p$ given by

$$\phi_p(a) := \phi(a) := \sum_{i=0}^{\infty} c_i^p p^i. \tag{3.44}$$

Hence $W(\mathbb{F})$ has a unique p-derivation $\delta = \delta_p$,

$$\delta_p a := \delta a := \frac{\phi(a) - a^p}{p} = \frac{1}{p}\left(\left(\sum_{i=0}^{\infty} c_i^p p^i\right) - \left(\sum_{i=0}^{\infty} c_i p^i\right)^p\right). \tag{3.45}$$

Then one can take $W(\mathbb{F})$ equipped with the global connection (δ_p) indexed by $\{p\}$; it is an example of δ-ring; cf. Definition 3.23. The link between the global setting and this local setting is the following: if we consider a number field K, a supernatural number M, and if $p \nmid M$ is unramified in K then we have an isomorphism followed by an embedding

$$\mathcal{O}_K[1/M]^{\widehat{p}} \simeq \prod_{\mathfrak{P}|p} W(\mathcal{O}_K/\mathfrak{P}) \subset \prod_{\mathfrak{P}|p} R \tag{3.46}$$

where $\mathfrak{P}$ are the primes in K lying above p and

$$R := W(\mathbb{F}_p^a), \tag{3.47}$$

$\mathbb{F}_p^a$ an algebraic closure of $\mathbb{F}_p$. If K is an abelian extension of $\mathbb{Q}$ then the maps in 3.46 are compatible with the corresponding Frobenius lifts.

We will need a slightly more general local setting in which A is not necessarily complete.

Definition 3.27. Let $R := W(\mathbb{F}_p^a)$, p odd, and let $K = R[1/p]$ be the fraction field of R. By a *valuation subring* of R we will understand a ring of the form

$$A = R \cap F \subset R \tag{3.48}$$

where $F \subset K$ is a subfield of K. Any such A is a discrete valuation ring with maximal ideal generated by p, fraction field F, and residue field $\mathbb{F} = A/pA$ naturally embedded into $\mathbb{F}_p^a$. Also we have $A \cap p^n R = p^n A$ and in particular we have

$$A \subset A^{\widehat{p}} = W(\mathbb{F}) \subset R = W(\mathbb{F}_p^a). \tag{3.49}$$

A subring of R will be called a *δ-subring* if it is sent into itself by the unique p-derivation $\delta : R \to R$. A valuation subring A as in 3.48 is a δ-subring if and only if $\phi(F) \subset F$, where $\phi : K \to K$ is the natural extension of $\phi : R \to R$. A valuation subring of R which is also a δ-subring will be referred to as a *valuation δ-subring* of R. Such a valuation δ-subring A, equipped with the restriction $\delta_p : A \to A$ of the p-derivation $\delta = \delta_p : R \to R$ is, of course, an example of δ-ring (cf. Definition 3.23) with $\mathcal{V} = \{p\}$ and will be referred to as a δ-ring of *local type*.

Going back to the general case of an admissible δ-ring A (cf. Definition 3.23), if B is an A-algebra, any global connection (δ_p) on B (flat or otherwise) will be implicitly assumed to be compatible with the fixed flat global connection on A. (Similarly, if we will be considering global connections on B denoted by $(\overline{\delta}_p)$, they will be assumed to be compatible with $(\overline{\delta}_p)$ on A.) Let $\delta = (\delta_p)$ be a global connection on B and consider its curvature $(\varphi_{pp'})$. Note that the operators $\varphi_{pp'}$ vanish on A and are $\phi_p\phi_{p'}$-semi-linear maps of A-modules i.e. they are additive and

$$\varphi_{pp'}(ab) = \phi_p\phi_{p'}(a) \cdot \varphi_{pp'}(b)$$

for $a \in A$ and $b \in B$. (The fact that $\varphi_{pp'}$ are $\phi_p\phi_{p'}$-semi-linear maps is trivial if B is pp'-torsion free; in general the statement is trivially reducible to the torsion free case.)

To motivate the definitions in the next section note that global connections naturally arise from λ-structures in the sense of Grothendieck, cf. especially the work of Borger [**13**]. Indeed, by a result of Wilkerson [**122**], in the special case of torsion free rings, Grothendieck's definition reduces to the following:

DEFINITION 3.28. Let B be a torsion free ring (i.e. a ring such that $nb = 0$ with $n \in \mathbb{Z}$ and $b \in B$ implies $n = 0$ or $b = 0$). A *λ-ring* structure on B is a commuting family of Frobenius lifts (ϕ_p) indexed by the set of all primes $p \in \mathbb{Z}$.

What we called a λ-ring structure is sometimes called a *special λ-ring structure.* A λ-ring structure on a torsion free ring B is then equivalent to a flat global connection on B indexed by the set $\mathcal{V}$ of *all* primes in $\mathbb{Z}$. However in most interesting applications we have in mind global connections (flat or not) are not available directly; all that *is* available are *adelic connections*, a concept to be defined in what follows.

3.2. Adelic connections

Recall that for any Noetherian ring B, and any subset $S \subset B$ we denote by $B^{\widehat{S}} = B^{\widehat{I}}$ the completion of B with respect to the ideal $I = (S)$ generated in B by S. A similar notation is used for completions of Noetherian schemes. In particular let X be a Noetherian scheme; then for each prime $p \in \mathbb{Z}$ we denote by $X^{\widehat{p}}$ the formal scheme obtained by p-adically completing X.

DEFINITION 3.29. A p-derivation on $X^{\widehat{p}}$ is a morphism of sheaves of sets

$$\delta_p^{X^{\widehat{p}}} : \mathcal{O}_{X^{\widehat{p}}} \to \mathcal{O}_{X^{\widehat{p}}}$$

which, on each open set, is a p-derivation.

Instead of $\delta_p^{X\widehat{p}}$ we sometimes write δ_p^X or even δ_p, if no confusion may arise.

Similarly for any ring B and any p-derivation $\delta_p^{B\widehat{p}} : B^{\widehat{p}} \to B^{\widehat{p}}$ we write δ_p^B or even δ_p instead of $\delta_p^{B\widehat{p}}$.

Let X (respectively B) be a Noetherian scheme (respectively a Noetherian algebra) over an admissible δ-ring A as in Definition 3.23. We will always assume $\mathcal{V}$ is a set of odd primes.

DEFINITION 3.30. An *adelic connection* on X is a family $\delta = (\delta_p^X)$ indexed by $p \in \mathcal{V}$ where, for each p, $\delta_p = \delta_p^X$ is a p-derivation on $X^{\widehat{p}}$ compatible with the p-derivation $\delta_p^A = \delta_p$ on A.

An *adelic connection* on B is a family $\delta = (\delta_p^B)$ indexed by $p \in \mathcal{V}$ where, for each p, $\delta_p = \delta_p^B$ is a p-derivation on $B^{\widehat{p}}$ compatible with the p-derivation $\delta_p^A = \delta_p$ on A.

REMARK 3.31. Any adelic connection on B induces an adelic connection on *Spec* B; so to give an adelic connection on *Spec* B is the same as to give an adelic connection on B.

REMARK 3.32. Given an adelic connection (δ_p) on X indexed by $\mathcal{V}$ one can consider the attached Frobenius lifts

$$\phi_p : \mathcal{O}_{X^{\widehat{p}}} \to \mathcal{O}_{X^{\widehat{p}}}$$

which, in their turn, define Frobenius lifts

$$\phi_p : X^{\widehat{p}} \to X^{\widehat{p}};$$

the latter are morphisms of formal schemes compatible with the induced morphisms

$$\phi_p : Spf\ A^{\widehat{p}} \to Spf\ A^{\widehat{p}},$$

whose reduction mod p is the p-power Frobenius. Conversely given such a family (ϕ_p) of Frobenius lifts, and under the assumption that each p is a non-zero divisor on $X^{\widehat{p}}$ we have an induced adelic connection (δ_p) on X; we say that (δ_p) and (ϕ_p) are *attached* to each other.

REMARK 3.33. Any global connection δ on an affine scheme $X = Spec\ B$ of finite type over A (that is compatible with that on A) naturally induces an adelic connection on X which is said to *come from* the global connection δ.

We say that an adelic connection on B is *global* if it comes from a global connection on B; if this is the case then the global connection on B is guaranteed to be unique if one assumes, for instance, that B is a Noetherian domain in which no $p \in \mathcal{V}$ is invertible. Cf. Krull's intersection theorem.

REMARK 3.34. Any adelic connection on X induces an adelic connection on open subscheme U of X. However, even if an adelic connection on $X = Spec\ B$ comes from a global connection on B, the adelic connection induced on a principal open set *Spec* B_f may not come from a global connection on B_f. A typical example is the global connection (δ_p) on $X = Spec\ A[t]$ (where t is a variable) with $\delta_p t = 0$ for all p; this induces an adelic connection on

$$U = Spec\ A[t, (t-1)^{-1}]$$

which does not come from a global connection on U.

REMARK 3.35. Note that there is no direct way to define the curvature of an adelic connection the way this was done for global connections; indeed, the various p-derivations act on (sheaves of) rings $\mathcal{O}_{X^{\widehat{p}}}$ that vary as p varies. However there are at least 3 general methods that allow one to introduce curvature for adelic connections in various special cases:

(1) One general method applies to what we shall call *semiglobal* adelic connections. Galois theory (leading to what we shall call *Galois connections*) fits into this setting. This setting is, however, perpendicular to the situation we are most interested in which is that of adelic connections on GL_n.
(2) Another general method of attaching curvature will be referred to as *analytic continuation between primes*; cf. [**34**]. This method will work well in the case of adelic connections on GL_n.
(3) Finally there is a general method to define curvature based on what we shall call *algebraization by correspondences*. This method will, again, be applied later to the case of adelic connections on GL_n.

These 3 methods will be explained in the next 3 sections respectively.

Before proceeding to explain the above 3 methods of introducing curvature it is worth noting that the notion of adelic connection is interesting (and in particular it leads to a version of curvature) even if the set of primes $\mathcal{V}$ has one element only, $\mathcal{V} = \{p\}$. This case will be examined in detail later. In any case it is reasonable to introduce the following terminology:

DEFINITION 3.36. A *p-adic connection* on X is an adelic connection (δ_p^X) on X indexed by the set $\mathcal{V} = \{p\}$. Equivalently, a *p-adic connection* on X is a p-derivation

$$\delta^X = \delta_p^X = \delta_p : \mathcal{O}_{X^{\widehat{p}}} \to \mathcal{O}_{X^{\widehat{p}}}$$

which is compatible with the p-derivation $\delta_p^A = \delta_p$ on A.

One can introduce curvature for tuples of p-adic connections in the expected way; this can be applied to our arithmetic Levi-Cività connections to be introduced later. Indeed we have:

DEFINITION 3.37. Let X be affine and let

$$(\delta_i^X) = (\delta_1^X, ..., \delta_m^X)$$

be an m-tuple of p-adic connections on X, i.e., of p-derivations on $X^{\widehat{p}}$ compatible with the one on A. Let ϕ_i^X be the attached Frobenius lifts on $X^{\widehat{p}}$. The *curvature* of (δ_i^X) is the family (φ^{ij}), indexed by $i, j = 1, ..., m$, where φ^{ij} are the additive maps

$$\varphi^{ij} = \frac{1}{p}[\phi_i^X, \phi_j^X] : \mathcal{O}(X)^{\widehat{p}} \to \mathcal{O}(X)^{\widehat{p}}.$$

Similarly, the 3-*curvature* of (δ_i^X) is the family (φ^{ijk}), indexed by $i, j, k = 1, ..., m$, where φ^{ijk} are the additive maps

$$\varphi^{ijk} = \frac{1}{p}[\phi_i^X, [\phi_j^X, \phi_k^X]] : \mathcal{O}(X)^{\widehat{p}} \to \mathcal{O}(X)^{\widehat{p}}.$$

3.3. Semiglobal connections and their curvature; Galois connections

We introduce here a class of adelic connections which, although, not global, are close to global connections and possess curvature. They will be called *semiglobal connections* and an important example of such adelic connections will be what we will call *Galois connections*. The latter will encode information from Galois theory and will be, in some sense, perpendicular to the adelic connections on GL_n that will mainly concern us later in our book.

DEFINITION 3.38. Let $\delta = (\delta_p)$ be an adelic connection on $X = Spec\ B$ with attached Frobenius lifts $\phi_p : B^{\widehat{p}} \to B^{\widehat{p}}$. Consider the decomposition

$$Spf\ B^{\widehat{p}} = \coprod_{i \in I_p} Spf\ B_i^{\widehat{p}}$$

of $Spf\ B^{\widehat{p}}$ into connected components and consider the induced Frobenius lifts

$$\phi_p^{B_i^{\widehat{p}}} : B_i^{\widehat{p}} \to B_i^{\widehat{p}}. \tag{3.50}$$

We say that δ is *semiglobal* on X (or on B) if the maps $B \to B_i^{\widehat{p}}$ are injective for all $p \in \mathcal{V}$ and $i \in I_p$ and, in addition, the homomorphisms 3.50 send the image of B into itself. If this is the case, letting

$$\phi_{pi}^B : B \to B \tag{3.51}$$

be the ring homomorphisms induced by 3.50 we define the *mixed curvature* of δ as the family of $\mathbb{Z}$-module maps $(\varphi_{pp'}^{ii'})$ indexed by $p, p' \in \mathcal{V}$, $i \in I_p$, $i' \in I_{p'}$,

$$\varphi_{pp'}^{ii'} := [\phi_{pi}^B, \phi_{p'i'}^B] : B \to B.$$

Under quite general assumptions on our adelic connections "global" implies "semiglobal" as shown by the following:

LEMMA 3.39. *Let B be a Noetherian domain equipped with an adelic connection $\delta = (\delta_p)$ indexed by $\mathcal{V}$ and assume that no $p \in \mathcal{V}$ is invertible in B. If δ is global then it is semiglobal.*

PROOF. It is sufficient to check that, in the notation of Definition 3.38, the maps $B \to B_i^{\widehat{p}}$ are injective. For each i we have

$$B_i^{\widehat{p}} \simeq B^{\widehat{p}}/(e_i)$$

where $e_i \in B^{\widehat{p}}$ is an idempotent,

$$e_i^2 = e_i \neq 1.$$

If $e_i = 0$ we are done by Krull's intersection theorem. Assume $e_i \neq 0$ and $0 \neq b \in B$ is an element in the kernel of $B \to B_i^{\widehat{p}}$; we will derive a contradiction. By Krull's theorem we may identify B with a subring of $B^{\widehat{p}}$. So there exists $\beta \in B^{\widehat{p}}$ such that

$$b = e_i \beta. \tag{3.52}$$

Squaring we get:

$$b^2 = e_i^2 \beta^2 = e_i \beta^2 = b\beta. \tag{3.53}$$

Since $B^{\widehat{p}}$ is flat over B and b is a non-zero divisor in B we get that b is a non-zero divisor in $B^{\widehat{p}}$ so by 3.53 we get

$$b = \beta. \tag{3.54}$$

So by 3.52 and 3.54 we get

$$b = e_i b. \tag{3.55}$$

Again, since b is a non-zero divisor in $B^{\widehat{p}}$ we get $e_i = 1$, a contradiction. This ends the proof. □

REMARK 3.40. Note that we did not assume, in Definition 3.38, and it will *not* be the case in our applications, that the ring homomorphisms 3.51 are Frobenius lifts. Here is a basic example of the above; this example is, in some sense, "perpendicular" to the main examples (to be considered later) of our theory. There is a way to unify the example below with the main examples of our theory but we will not pursue this in our book.

EXAMPLE 3.41. Let $A = \mathcal{O}_K[1/M]$ be a δ-ring of global type (cf. Definition 3.25) and let L/K be a finite extension such that $L/\mathbb{Q}$ is Galois with discriminant D. Let $B = \mathcal{O}_L[1/M]$. Let $\mathcal{V}$ be a set of primes not dividing MD. Then, for $p \in \mathcal{V}$, we have

$$B^{\widehat{p}} = \prod_{\mathfrak{P}|p} B^{\widehat{\mathfrak{P}}} = \prod_{\mathfrak{P}|p} W(B/\mathfrak{P})$$

where $\mathfrak{P}$ runs through the set of primes in B lying over p. So $B^{\widehat{p}}$ comes equipped with a unique Frobenius lift $\phi_p^{B^{\widehat{p}}}$ that is clearly compatible with the one on $A^{\widehat{p}}$. Then the family $(\phi_p^{B^{\widehat{p}}})$ defines an adelic connection $\delta = (\delta_p)$ on B.

DEFINITION 3.42. The adelic connection δ^B on $B = \mathcal{O}_L[1/M]$ defined by $(\phi_p^{B^{\widehat{p}}})$ as in Example 3.41 is called the *Galois connection*; it is canonically attached to the extension L/K, to the number M, and to the set $\mathcal{V}$.

REMARK 3.43. The Galois connection is semiglobal. Indeed, clearly the maps $B \to B^{\widehat{\mathfrak{P}}}$ are injective and, since $L/\mathbb{Q}$ is Galois, $\phi_p^{B^{\widehat{p}}}$ clearly sends the image of B into itself; the induced

$$\phi_{\mathfrak{P}}^B : B \to B$$

is induced, of course, by the Frobenius element corresponding to $\mathfrak{P}$ in the Galois group $G(L/K)$, i.e. it satisfies

$$\phi_{\mathfrak{P}}^B(b) \equiv b^p \mod \mathfrak{P} \quad \text{for} \quad b \in B.$$

In particular the Galois connection on B possesses a mixed curvature. Recall that the homomorphism $\phi_{\mathfrak{P}}^B$ is generally not a Frobenius lift on B but it induces a Frobenius lift on the localization $B_{\mathfrak{P}}$.

REMARK 3.44. Let us assume in Example 3.41 that $K = \mathbb{Q}$ and that $\mathcal{V}$ is cofinite in the set of all primes (i.e. it is the complement of a finite set in the set of all primes). Let δ^B be the Galois connection in Definition 3.42. Let $\text{End}(B)$ denote, as usual, the Lie ring of $\mathbb{Z}$-module (equivalently A-module) endomorphisms of B and let us denote by $\mathfrak{lie}(\phi^B)$ the smallest Lie subring of $\text{End}(B)$ containing all the automorphisms $\phi_{\mathfrak{P}}^B$ of B. Let Γ be the group of ring automorphisms of the ring B (which is of course identifiable with the Galois group of $L/\mathbb{Q}$). Then, since $\mathcal{V}$ is cofinite in the set of all primes, by Chebotarev's density theorem, any element in the group Γ is of the form $\phi_{\mathfrak{P}}^B$ for some $\mathfrak{P}$. So, by the independence of characters 1.2 we have an isomorphism of Lie rings,

$$\mathfrak{lie}(\phi^B) \simeq \mathbb{Z}[\Gamma]. \tag{3.56}$$

In particular, if one defines, as usual, the *holonomy ring* by

$$\mathfrak{hol}(\phi^B) := [\mathfrak{lie}(\phi^B), \mathfrak{lie}(\phi^B)] \tag{3.57}$$

then we have

$$\mathfrak{hol}(\phi^B) \simeq [\mathbb{Z}[\Gamma], \mathbb{Z}[\Gamma]]. \tag{3.58}$$

Note that the Lie ring $\mathbb{Z}[\Gamma]$ is "far from solvable" in general. Indeed we have the following classical consequence of theorems of Maschke [**95**] p. 666, and Wederburn [**95**], p. 659:

THEOREM 3.45. *Assume Γ is a finite group and let $\mathbb{C}[\Gamma]$ be its complex group algebra. Then there is an associative algebra isomorphism*

$$\mathbb{C}[\Gamma] \simeq \bigoplus_{i=1}^{s} \mathfrak{gl}_{n_i}(\mathbb{C}).$$

In the above statement $n_1, ..., n_s$ are the dimensions of the complex irreducible representations of Γ.

In particular,

COROLLARY 3.46. *For Γ a finite group the Lie ring $\mathbb{Z}[\Gamma]$ is solvable if and only if Γ is abelian.*

By the way the above Corollary is still true for Γ not necessarily finite [**109**].

By Corollary 3.46 and by the equality 3.58 we get:

COROLLARY 3.47. *Let δ^B be the Galois connection in Definition 3.42 where $K = \mathbb{Q}$ and $\mathcal{V}$ is cofinite in the set of all primes. Then the Lie ring $\mathfrak{hol}(\phi^B)$ cannot be nilpotent unless it equals 0.*

This will contrast with our result of the "nilpotence of holonomy for real Chern connections attached to split metrics" which we are going to prove later; cf. Corollary 6.37.

REMARK 3.48. Here is another easy link between our adelic connections and Galois theory. Assume for simplicity that $A = \mathbb{Z}[1/M]$ and let $X = Spec\ B$ be a finite étale scheme over A. So B is a finite product

$$B = \prod_i B_i, \quad B_i = \mathcal{O}_{E_i}[1/M]$$

where $E_i/\mathbb{Q}$ are finite field extensions unramified outside the primes dividing M and $\mathcal{O}_{E_i}$ are the rings of integers of E_i. Let now $L/\mathbb{Q}$ be a Galois field extension with Galois group Γ and assume this extension is unramified outside the primes dividing M. Set $C = \mathcal{O}_L[1/M]$ where $\mathcal{O}_L$ is the ring of integers of L. Then one can consider the Galois action

$$\Gamma \times X(C) \to X(C) \tag{3.59}$$

on the set

$$X(C) = \mathrm{Hom}_{\mathrm{sch}}(Spec\ C, X) = \mathrm{Hom}_{\mathrm{ring}}(B, C) \tag{3.60}$$

given by

$$(\sigma, P) \mapsto P \circ \sigma$$

for $\sigma \in \Gamma$ viewed as $\sigma : Spec\ C \to Spec\ C$, and $P : Spec\ C \to B$. On the other hand note that, for $p \in \mathcal{V} = \{p;\ p \nmid M\}$, we have

$$X^{\widehat{p}} = \coprod_{\mathfrak{P}|p} Spf\ B^{\widehat{\mathfrak{P}}},\quad B^{\widehat{\mathfrak{P}}} = W(B/\mathfrak{P}),$$

where $\mathfrak{P} \subset B$ are the prime ideals lying above p. So the Frobebnius lifts on each $W(B/\mathfrak{P})$ define Frobenius lifts $\phi_p^X : X^{\widehat{p}} \to X^{\widehat{p}}$ and hence a unique adelic connection (δ_p) on X. Note that in general, if B is not Galois over A, this adelic connection on B is *not* semiglobal.

By the way we also have

$$C^{\widehat{p}} \simeq \prod_{\mathfrak{p}|p} C^{\widehat{\mathfrak{p}}}$$

where $\mathfrak{p} \subset C$ are the primes lying over p.

The link between our adelic connection on X and the Galois action 3.59 is as follows. Let $\mathfrak{p} \subset C$ be a prime lying above p and let $\sigma = \sigma_{\mathfrak{p}} \in \Gamma$ be the Frobenius element corresponding to $\mathfrak{p}$. Also let

$$P^{\widehat{p}} : Spf\ C^{\widehat{p}} \to X^{\widehat{p}} \quad \text{and} \quad (P^{\sigma})^{\widehat{p}} : Spf\ C^{\widehat{p}} \to X^{\widehat{p}}$$

be the morphisms naturally induced via p-adic completion from

$$P : Spec\ C \to X \quad \text{and} \quad P^{\sigma} : Spec\ C \to X.$$

Then the morphism

$$Spf\ C^{\widehat{\mathfrak{p}}} \to Spf\ C^{\widehat{p}} \xrightarrow{P^{\widehat{p}}} X^{\widehat{p}} \xrightarrow{\phi_p^X} X^{\widehat{p}}$$

coincides with the morphism

$$Spf\ C^{\widehat{\mathfrak{p}}} \to Spf\ C^{\widehat{p}} \xrightarrow{(P^{\sigma})^{\widehat{p}}} X^{\widehat{p}}.$$

This follows immediately from the fact that any ring homomorphism

$$B^{\widehat{\mathfrak{P}}} \to C^{\widehat{\mathfrak{p}}}$$

must be compatible with the Frobenius lifts on these two rings; cf. the isomorphism 1.1. The action 3.59 is, of course, the source of some basic Galois representations; think, e.g., of the case when X is the kernel of multiplication by some integer $n \in \mathbb{Z} \cap A^{\times}$ on a torus or an Abelian scheme over A.

REMARK 3.49. One can attempt to "unify" the holonomy of Galois connections 3.57 (where B is a finite algebra over A) with the holonomy rings appearing in Definition 3.7 (where B is a power series ring over A). Such a unification could be achieved by constructing exact sequences

$$0 \to \mathfrak{hol}_1 \to \mathfrak{hol}_2 \to \mathfrak{hol}_3 \to 0$$

where $\mathfrak{hol}_1$ is as in Definition 3.7 and $\mathfrak{hol}_3$ is as in 3.57. This would lead to natural homomorphisms of Lie rings,

$$\mathfrak{hol}_3 \to \mathrm{Out}(\mathfrak{hol}_1),$$

cf. 1.8, which could be analogous to the Galois representations occurring from 1.7 as in [**78**]. We will not pursue this idea in our book.

3.4. Curvature via analytic continuation between primes

In this section we will use what in [**34**] is referred to as *analytic continuation between primes* in order to define curvature for certain adelic connections on schemes X. The analytic continuation will be along certain subschemes Y of X that we call *retracts*, cf. the Definition below.

DEFINITION 3.50. Let X be a Noetherian scheme and Y a closed affine subscheme of X. We say that Y is a *retract* in X if there exists an affine open set $X_1 \subset X$ containing Y and there exists a morphism $X_1 \to Y$ such that:

(1) The composition $Y \subset X_1 \to Y$ is the identity,
(2) The kernel of $\mathcal{O}(X_1) \to \mathcal{O}(Y)$ is generated by a regular sequence.

We call X_1 a *retractable neighborhood* and we call the members of the regular sequence *retract coordinates.*

In the notation above set

$$B = \mathcal{O}(X_1), \quad \mathcal{O}(Y) = B/I, \quad I = (b_1, ..., b_r),$$

with $b_1, ..., b_r$ a regular sequence.

LEMMA 3.51. *Let $T = \{T_1, ..., T_r\}$ be variables. Then we have natural isomorphisms sending T_i into b_i:*

$$B^{\widehat{I}} \simeq (B/I)[[T]], \quad (B^{\widehat{p}})^{\widehat{I}} \simeq (B/I)^{\widehat{p}}[[T]]. \tag{3.61}$$

In particular, if I is a prime ideal in B, and p is not invertible in B/I then:

$$B^{\widehat{I}} \subset (B^{\widehat{p}})^{\widehat{I}}. \tag{3.62}$$

PROOF. By [**103**], p. 125, we have that $\bigoplus I^n/I^{n+1}$ is naturally isomorphic to the associated graded ring of $(B/I)[[T]]$ with respect to the ideal (T). Then, by induction, one gets that the map

$$(B/I)[[T]]/(T)^n \to B/I^n$$

is an isomorphism; this proves the first isomorphism in 3.61. Now $(B/I)^{\widehat{p}} = B^{\widehat{p}}/IB^{\widehat{p}}$ and, by flatness of $B \to B^{\widehat{p}}$, we have that $IB^{\widehat{p}}$ is generated by a regular sequence in $B^{\widehat{p}}$. So we may apply the same argument as above to prove the second isomorphism in 3.61. Finally, if I is a prime ideal in B, and p is not invertible in B/I then B/I is p-adically separated (by Krull's intersection theorem) so we have an inclusion $B/I \subset (B/I)^{\widehat{p}}$ and hence an inclusion 3.62. □

Let X be scheme of finite type over an admissible δ-ring A, let $\delta = (\delta_p)$ be an adelic connection on X, and let Y be a retract in X, with retractable neighborhood $X_1 = Spec\ B$ and $Y = Spec(B/I)$. Also assume, in the discussion below, that B is an integral domain, I is prime in B and p is non-zero and non-invertible in B/I for all $p \in \mathcal{V}$; in particular, by Lemma 3.51, $B^{\widehat{I}}$ is a subring of $(B^{\widehat{p}})^{\widehat{I}}$.

DEFINITION 3.52. We say that δ *fixes* Y if each $\delta_p : B^{\widehat{p}} \to B^{\widehat{p}}$ sends $IB^{\widehat{p}}$ into itself. We say that δ is *global along* Y if it fixes Y and for all p the induced p-derivation $\delta_p : (B^{\widehat{p}})^{\widehat{I}} \to (B^{\widehat{p}})^{\widehat{I}}$ sends $B^{\widehat{I}}$ into itself.

So if δ fixes Y the adelic connection on X induces an adelic connection on each infinitesimal neighborhood

$$Y_\nu = Spec\ (B/I^{\nu+1})$$

of Y in X. On the other hand if δ is global along Y we have an induced global connection $\delta = (\delta_p)$ on $B^{\widehat{I}}$.

LEMMA 3.53. *If δ is global along Y then $\delta_p : B^{\widehat{I}} \to B^{\widehat{I}}$ sends each power of $IB^{\widehat{I}}$ into itself. Hence the global connection on $B^{\widehat{I}}$ induces a global connection on each infinitesimal neighborhood Y_ν. This global connection on Y_ν, induced by that on $B^{\widehat{I}}$, is compatible (in the obvious sense) with the adelic connection on Y_ν induced from that on X.*

PROOF. The first part follows from the fact that

$$I^{\nu+1}(B^{\widehat{p}})^{\widehat{I}} \cap B^{\widehat{I}} = I^{\nu+1}B^{\widehat{I}},$$

which is a consequence of Lemma 3.51. The second part is trivially checked. □

LEMMA 3.54. *The following are equivalent:*

(1) *$\delta_p : B^{\widehat{p}} \to B^{\widehat{p}}$ sends $IB^{\widehat{p}}$ into itself;*
(2) *the Frobenius lift $\phi_p : B^{\widehat{p}} \to B^{\widehat{p}}$ attached to δ_p sends $IB^{\widehat{p}}$ into itself.*

PROOF. Of course assertion (1) implies assertion (2). Conversely, assume assertion (2), i.e.,

$$\phi_p(IB^{\widehat{p}}) \subset IB^{\widehat{p}}.$$

So we have

$$p\delta_p(IB^{\widehat{p}}) \subset IB^{\widehat{p}} \cap pB^{\widehat{p}}.$$

It is enough to prove that

$$IB^{\widehat{p}} \cap pB^{\widehat{p}} = pIB^{\widehat{p}}$$

because p is a non-zero divisor in $B^{\widehat{p}}$ (because it is a non-zero divisor in B, since B is an integral domain). So let

$$b \in IB^{\widehat{p}} \cap pB^{\widehat{p}}, \quad b = pc, \quad c \in B^{\widehat{p}}.$$

So the image

$$\overline{c} \in B^{\widehat{p}}/IB^{\widehat{p}} = (B/I)^{\widehat{p}}$$

of c satisfies $p\overline{c} = 0$. Since p is a non-zero divisor in B/I it is also a non-zero divisor in $(B/I)^{\widehat{p}}$; hence $\overline{c} = 0$, hence $c \in IB^{\widehat{p}}$, hence $b \in pIB^{\widehat{p}}$ and we are done. □

LEMMA 3.55. *The following are equivalent:*

(1) *$\phi_p : B^{\widehat{p}} \to B^{\widehat{p}}$ sends $IB^{\widehat{p}}$ into itself;*
(2) *the image of*

$$\phi_p(I) \text{ in } (B^{\widehat{p}})^{\widehat{I}} = (B/I)^{\widehat{p}}[[T]] \text{ belongs to } I(B^{\widehat{p}})^{\widehat{I}} = (T)(B/I)^{\widehat{p}}[[T]].$$

PROOF. This follows from the fact that the preimage of $I(B^{\widehat{p}})^{\widehat{I}}$ in $B^{\widehat{p}}$ equals $IB^{\widehat{p}}$. □

LEMMA 3.56. *The following are equivalent:*

(1) *$\delta_p : (B^{\widehat{p}})^{\widehat{I}} \to (B^{\widehat{p}})^{\widehat{I}}$ sends $B^{\widehat{I}}$ into itself;*
(2) *the Frobenius lift $\phi_p : (B^{\widehat{p}})^{\widehat{I}} \to (B^{\widehat{p}})^{\widehat{I}}$ attached to δ_p sends $B^{\widehat{I}}$ into itself.*

PROOF. This follows from Lemma 3.51 plus the fact that $(B/I) \cap p(B/I)^{\widehat{p}} = p(B/I)$. □

REMARK 3.57. If δ is global along Y we have an induced global connection on $B^{\widehat{I}}$; therefore we may consider the *curvature* of the latter, $\varphi = (\varphi_{pp'})$,

$$\varphi_{pp'} : B^{\widehat{I}} \to B^{\widehat{I}}. \tag{3.63}$$

This φ depends only on X, Y and the adelic connection δ on X (but not on X_1). Indeed we have an identification

$$B^{\widehat{I}} \simeq \mathcal{O}(X^{\widehat{Y}})$$

where $X^{\widehat{Y}}$ is the completion of X along Y. Also, the quotients I/I^ν depend only on X and Y and not on X_1.

REMARK 3.58. If Y is the image in X of an A-point $P \in X(A)$, $P : Spec\ A \to X$ we will simply say that Y is an *A-point.* Then $B/I = A$ hence

$$B^{\widehat{I}} \simeq A[[T]], \quad (B^{\widehat{p}})^{\widehat{I}} \simeq A^{\widehat{p}}[[T]],$$

and we write $X^{\widehat{Y}} = X^{\widehat{P}}$.

We may proceed at this point to giving our main definitions related to "curvature via analytic continuation." The hypotheses included at the beginning of our definition guarantee that our previous discussion applies.

DEFINITION 3.59. Let X be an integral scheme of finite type over an admissible δ-ring A, δ indexed by $\mathcal{V}$, and let Y be a retract in X, with retractable neighborhood $X_1 = Spec\ B$ and $Y = Spec(B/I)$. Assume Y is an integral scheme and

$$Y \otimes Spec\ \mathbb{F}_p \neq \emptyset, Y$$

for $p \in \mathcal{V}$.

(1) We denote by $\Lambda^1(X, Y)$ the set of adelic connections on X that are global along Y.
(2) We denote by $\Lambda^2(X, Y)$ the set of anti-symmetric matrices $\varphi = (\varphi_{pp'})$ with entries $\phi_p\phi_{p'}$-semi-linear maps of A-modules $\mathcal{O}(X^{\widehat{Y}}) \to \mathcal{O}(X^{\widehat{Y}})$ that send each power I^ν into itself and send A into 0.
(3) Given an adelic connection
$$\delta = (\delta_p) \in \Lambda^1(X, Y)$$
the matrix
$$\varphi = \varphi[\delta] = (\varphi_{pp'}) \in \Lambda^2(X, Y)$$
in 3.63 is called the *curvature* of δ.
(4) We say δ is *flat* if its curvature vanishes,
$$\varphi[\delta] = 0.$$
(5) Given one more adelic connection
$$\overline{\delta} = (\overline{\delta}_p) = (\delta_{\overline{p}}) \in \Lambda^1(X, Y)$$
one defines the $(1,1)$-curvature of δ with respect to $\overline{\delta}$,
$$(\varphi_{p\overline{p}'}) \in \Lambda^2(X, Y),$$
as the $(1,1)$-curvature of the global connection δ on $\mathcal{O}(X^{\widehat{Y}})$.
(6) The first Chern form $\rho[\delta] = (\rho_{pp'})$ of the global connection $\delta = (\delta_p)$ on $\mathcal{O}(X^{\widehat{Y}}) = B^{\widehat{I}}$ will be called the *first Chern form* of the adelic connection $\delta = (\delta_p)$ on X.

(7) We denote by

$$\varphi^{(\nu)}_{pp'} : I/I^{\nu+1} \to I/I^{\nu+1}, \tag{3.64}$$

the maps induced by 3.63; cf. Lemma 3.53. They are $\phi_p\phi_{p'}$-semi-linear maps of A-modules.

(8) Assume Y is the image in X of an A-point $P \in X(A)$. The holonomy rings

$$\mathfrak{hol} \subset \mathrm{End}(A[[T]]), \quad \mathfrak{hol}^{\Omega} \subset \mathrm{End}(\Omega^{for}_{A[[T]]/A})$$

of the global connection δ on $\mathcal{O}(X^{\widehat{Y}}) = A[[T]]$, and their completions

$$\widehat{\mathfrak{hol}}, \quad \widehat{\mathfrak{hol}}^{\Omega},$$

will be referred to as the *holonomy rings*, respectively the *completed holonomy rings*, of the adelic connection δ on X. Cf. Definitions 3.7, 3.9, 3.13. The connection δ on X will be said to have a *horizontal* curvature if the global connection δ on $\mathcal{O}(X^{\widehat{Y}}) = A[[T]]$ has this property, i.e., if $\mathfrak{hol}_{\mathbb{Q}}$ is spanned over $\mathbb{Q}$ by the components of the curvature $(\varphi_{pp'})$. Cf. Definition 3.10. By the way we recall that: $\mathfrak{hol}$ is the $\mathbb{Z}$-linear span in $\mathrm{End}(A[[T]])$ of all the Lie monomials

$$[\phi_{p_1}, [\phi_{p_2}, ..., [\phi_{s-1}, \phi_{p_s}]...]] : A[[T]] \to A[[T]]$$

where $s \geq 2$, $p_i \in \mathcal{V}$; $\mathfrak{hol}_{\mathbb{Q}}$ is the $\mathbb{Q}$-linear span of $\mathfrak{hol}$ in $\mathrm{End}(A[[T]]) \otimes \mathbb{Q}$; $\mathfrak{hol}^{\Omega}$ is the $\mathbb{Z}$-linear span in $\mathrm{End}(\Omega)$ of all the Lie monomials

$$[\frac{\phi_{p_1}}{p_1}, [\frac{\phi_{p_2}}{p_2}, ..., [\frac{\phi_{s-1}}{p_{s-1}}, \frac{\phi_{p_s}}{p_s}]...]] : \Omega \to \Omega$$

where $\Omega = \Omega^{for}_{A[[T]]/A}$, $s \geq 2$; and

$$\widehat{\mathfrak{hol}} = \lim_{\leftarrow} \mathfrak{hol}_n, \quad \widehat{\mathfrak{hol}}^{\Omega} = \lim_{\leftarrow} \mathfrak{hol}^{\Omega}_n$$

where $\mathfrak{hol}_n, \mathfrak{hol}^{\Omega}_n$ are the images of the maps

$$\mathfrak{hol} \to \mathrm{End}(A[[T]]/(T)^n), \quad \mathfrak{hol}^{\Omega} \to \mathrm{End}(\Omega/(T)^n\Omega) \tag{3.65}$$

respectively.

REMARK 3.60. We have the following two special cases:

(1) For Y (the image in X of) an A-point $P \in X(A)$ we write $\Lambda^1(X, P)$ and $\Lambda^2(X, P)$ in place of $\Lambda^1(X, Y), \Lambda^2(X, Y)$.
(2) For $Y = X$ the set $\Lambda^1(X, X)$ is simply the set of global connections on X; we write $\Lambda^1(X)$ in place of $\Lambda^1(X, X)$.

REMARK 3.61. If, in the above construction, $X = G$ is a smooth group scheme over A and $P = 1$ is the identity then $B^{\widehat{I}} = \mathcal{O}(G^{\widehat{1}}) = A[[T]]$ carries an induced formal group law $\mathcal{G}$. Then, by our discussion on global connections on formal groups, we have at our disposal tuples $\Phi^{\mathcal{G}}_{pp'}$ (and $\Phi^{\mathcal{G}}_{p\bar{p}'}$, in the $(1,1)$ case) with entries in $A[[T]]$ as in 3.25, 3.38, 3.39. These tuples depend, of course, on the choice of our retract coordinates T.

In what follows we define an arithmetic analogue of the *gauge action* in 2.62. Indeed assume X and Y above are given by $X = G$, a smooth affine group scheme and $Y = H$, a smooth closed subgroup scheme. In the following definition, for any $h \in H(A)$ we denote by $L_h : G \to G$ the left translation by h and, as usual, we still denote by

$$L_h : \mathcal{O}(G)^{\widehat{p}} \to \mathcal{O}(G)^{\widehat{p}}$$

the induced ring homomorphism. Also recall that for any $\delta = (\delta_p) \in \Lambda^1(G, H)$, δ_p are maps $\mathcal{O}(G)^{\widehat{p}} \to \mathcal{O}(G)^{\widehat{p}}$. So the compositions

$$L_{h^{-1}} \circ \delta_p \circ L_h : \mathcal{O}(G)^{\widehat{p}} \to \mathcal{O}(G)^{\widehat{p}}$$

define an adelic connection on G denoted by $h \star \delta$. Now since L_h sends H into itself, L_h induces an isomorphism

$$L_h : \mathcal{O}(G^{\widehat{H}}) \to \mathcal{O}(G^{\widehat{H}})$$

and hence $h \star \delta$ is global along H.

DEFINITION 3.62. Assume G is a smooth group scheme over A and H is a smooth closed subgroup scheme. Then the action

$$H(A) \times \Lambda^1(G, H) \to \Lambda^1(G, H), \quad (h, \delta) \mapsto h \star \delta, \tag{3.66}$$

is called the *gauge action* of H on $\Lambda^1(G, H)$.

Clearly the curvatures of δ and $h \star \delta$ are related as follows:

$$\varphi_{pp'}[h \star \delta] = L_{h^{-1}} \circ \varphi_{pp'}[\delta] \circ L_h : \mathcal{O}(G^{\widehat{H}}) \to \mathcal{O}(G^{\widehat{H}}). \tag{3.67}$$

Note that, for the above action to be interesting, we need that our adelic connections of interest be global along a non-trivial H. This will be often the case with our arithmetic analogues of real Chern connections (in which case H will be a torus) but not with our arithmetic analogues of Ehresmann connections. The latter will fit into a slightly different paradigm which we now set up.

Assume X is a smooth affine scheme containing an open set G which is a group scheme and assume the action of G on itself by left translations is given an extension to an action of G on X; we refer to this situation by saying that X is a *partial equivariant compactification* of G. Then for any $g \in G(A)$ we may consider the left translation

$$L_g : \mathcal{O}(X) \to \mathcal{O}(X);$$

so for any global connection $\delta = (\delta_p) \in \Lambda^1(X)$ on X, $\delta_p : \mathcal{O}(X) \to \mathcal{O}(X)$, we may consider the global connection

$$g \star \delta = (L_{g^{-1}} \circ \delta_p \circ L_g)$$

on X.

DEFINITION 3.63. Assume G is a smooth group scheme over A and X is a partial equivariant compactification of G. Then the action

$$G(A) \times \Lambda^1(X) \to \Lambda^1(X), \quad (g, \delta) \mapsto g \star \delta, \tag{3.68}$$

is called the *gauge action* of G on $\Lambda^1(X)$.

Clearly the curvatures of δ and $g \star \delta$ are related as follows:

$$\varphi_{pp'}[g \star \delta] = L_{g^{-1}} \circ \varphi_{pp'}[\delta] \circ L_g : \mathcal{O}(X) \to \mathcal{O}(X). \tag{3.69}$$

We end by introducing yet another type of curvature as follows:

DEFINITION 3.64. Let A be an admissible δ-ring, $\delta^A = (\delta_p^A)$. Let X be a smooth scheme over A and $P \in X(A)$ an A-point which is a retract in X, with retract coordinates T, hence $\mathcal{O}(X^{\widehat{P}}) = A[[T]]$. Let

$$(\delta_1^X, ..., \delta_n^X)$$

be a tuple of adelic connections on X that are global along P, $\delta_i^X = (\delta_{ip}^X)$. Let ϕ_{ip}^X be the attached Frobenius lifts on $X^{\widehat{p}}$ and let us denote by

$$\phi_{ip} : A[[T]] \to A[[T]]$$

the induced Frobenius lifts on $A[[T]]$. Then the *mixed curvature* of $(\delta_1^X, ..., \delta_n^X)$ is the family $(\varphi_{pp'}^{ij})$ indexed by $i, j = 1, ..., n$ and $p, p' \in \mathcal{V}$ where

$$\varphi_{pp'}^{ij} : A[[T]] \to A[[T]]$$

are the additive maps defined by

$$\varphi_{pp'}^{ij} = \frac{1}{pp'}[\phi_{ip}, \phi_{jp'}], \quad \text{if } p \neq p',$$

$$\varphi_{pp}^{ij} = \frac{1}{p}[\phi_{ip}, \phi_{jp}].$$

REMARK 3.65. There is a natural construction whose input is a tuple of adelic connections global along a point and whose output is an adelic connection over a larger number field which is semiglobal along a point over the larger field; in this construction the corresponding mixed curvatures in Definitions 3.64 and 3.38 correspond.

We explain our construction in a simple case. Assume the situation in Definition 3.64 with $K = \mathbb{Q}$, so $A = \mathbb{Z}[1/M]$, and let $L/\mathbb{Q}$ be a Galois extension of degree n such that all p in $\mathcal{V}$ split completely in L and do not divide M. For each p fix an ordering

$$\mathfrak{p}_1, ..., \mathfrak{p}_n$$

of the prime ideals in $C := \mathcal{O}_L[1/M]$ lying above p. Let $(\delta_1^X, ..., \delta_n^X)$ be an n-tuple of adelic connections on X that are global along $P \in X(A)$, with P a retract in X with retract coordinates T; so $\mathcal{O}(X^{\widehat{P}}) = A[[T]]$. Let (ϕ_{ip}^X) be the Frobenius lifts on $X^{\widehat{p}}$; they induce Frobenius lifts

$$\phi_{ip} : A^{\widehat{p}}[[T]] \to A^{\widehat{p}}[[T]]$$

sending $A[[T]]$ into itself. Since p splits completely in L we have natural isomorphisms,

$$A^{\widehat{p}} \simeq C^{\widehat{\mathfrak{p}}_i}.$$

So we have isomorphisms

$$C[[T]]^{\widehat{p}} \simeq C^{\widehat{p}}[[T]] \simeq \prod_{i=1}^{n} C^{\widehat{\mathfrak{p}}_i}[[T]] \simeq \prod_{i=1}^{n} A^{\widehat{p}}[[T]]. \tag{3.70}$$

Then, via the above isomorphism, the Frobenius lift

$$\prod_{i=1}^{n} \phi_{ip} : \prod_{i=1}^{n} A^{\widehat{p}}[[T]] \to \prod_{i=1}^{n} A^{\widehat{p}}[[T]]$$

induces a Frobenius lift

$$\phi_{Cp} : C[[T]]^{\widehat{p}} \to C[[T]]^{\widehat{p}}.$$

Clearly the homomorphisms

$$C[[T]] \subset C^{\widehat{\mathfrak{p}}_i}[[T]] \simeq A^{\widehat{p}}[[T]]$$

are injective. View these maps as inclusions.

Claim. Each $\phi_{ip} : A^{\widehat{p}}[[T]] \to A^{\widehat{p}}[[T]]$ sends $C[[T]]$ into itself.

To check the claim note that

$$C[[T]] = A[[T]] \otimes_A C$$

because C is a free A-module of finite rank. But by Galois theory, since p splits completely in L, the Frobenius lift on $C^{\widehat{\mathfrak{p}}_i}$ is the identity on C. Also we already know that ϕ_{ip} send $A[[T]]$ into itself. Our claim follows.

The conclusion of the above discussion is that the family (ϕ_{Cp}) defines an adelic connection (δ_{Cp}) on $C[[T]]$ which is semiglobal. The mixed curvature of the latter (in the sense of Definition 3.38) restricted to $A[[T]]$ is, of course, up to factors of the form $\frac{1}{pp'}$ and $\frac{1}{p}$, equal to the mixed curvature (in the sense of Definition 3.64) of the n-tuple of adelic connections $(\delta_1^X, \ldots, \delta_n^X)$ we started with.

Similarly, under the assumptions and notation above, we have isomorphisms

$$(X \otimes_A C)^{\widehat{p}} \simeq \prod_{i=1}^{n} X^{\widehat{p}}.$$

Then, again, via the above isomorphism, the Frobenius lift

$$\prod_{i=1}^{n} \phi_{ip}^X : \prod_{i=1}^{n} X^{\widehat{p}} \to \prod_{i=1}^{n} X^{\widehat{p}}$$

induces a Frobenius lift

$$\phi_{Cp}^X : (X \otimes_A C)^{\widehat{p}} \to (X \otimes_A C)^{\widehat{p}}.$$

The family (ϕ_{Cp}^X) induces an adelic connection (δ_{Cp}^X) on $X \otimes_A C$. This latter adelic connection is compatible, in the obvious sense, with the adelic (semiglobal) connection (δ_{Cp}) on $C[[T]]$.

3.5. Curvature via algebraization by correspondences

In the previous section we considered adelic connections on schemes X and, for those that are global along a retract Y in X, we introduced *curvature*. Many remarkable examples of adelic connections to be encountered later will indeed be global along such Y's. But, as a rule, adelic connections are only rarely global along some Y in X and for those that are not no curvature can be attached via analytic continuation between primes as in the previous section. There is, however, an alternative method to attach curvature to adelic connections: it is based on "algebraizing the Frobenius lifts." One cannot generally expect to have endomorphisms of X inducing the Frobenius lifts $\phi_p : X^{\widehat{p}} \to X^{\widehat{p}}$; rather, we will find algebraic correspondences on X that induce, via p-adic completion, our Frobenius lifts. The aim of this section is to explain this approach in the abstract case of an arbitrary scheme X. We will closely follow [**47**]. Later we will specialize to $X = GL_n$ and concrete adelic connections.

3.5.1. Correspondences on a scheme. In order to introduce our concepts we need some terminology related to correspondences on schemes. The formalism below has a motivic flavor and can be viewed as a "naive" variation on the Voevodsky formalism of finite correspondences [**121**]; this variation does not seem to naturally fit into the framework of [**121**] so our exposition will be independent of that in [**121**].

Our basic concept is the following:

DEFINITION 3.66. By a *correspondence* on a scheme X we mean a triple

$$\Gamma = (Y, \pi, \varphi)$$

where $\pi, \varphi : Y \to X$ are morphisms of schemes.

Alternatively we view Γ as a diagram

$$\begin{array}{ccccc} & & Y & & \\ & \pi \swarrow & & \searrow \varphi & \\ X & & & & X \end{array}$$

We sometimes write $Y = Y_\Gamma$ and also

$$\Gamma = \Gamma_{/X} = (Y_{/X}, \pi_{/X}, \varphi_{/X}). \tag{3.71}$$

The following technical term will also play a role later:

DEFINITION 3.67. A morphism of schemes is *totally dominant* if its image is dense and, moreover, the image of any connected component of the source is dense in the corresponding connected component of the target.

Compositions of totally dominant morphisms are totally dominant.

REMARK 3.68.

If P is a property of schemes we say Γ has P if Y_Γ has P. So we have a well defined notion, for instance, of *non-empty* correspondence, *connected* correspondence, *reduced* correspondence, etc.

If P is a property of morphisms of schemes we say Γ is *left* P (respectively *right* P) if π (respectively φ) has P. So we have a notion of *left étale* correspondence, *right étale* correspondence, *left totally dominant* correspondence, *right totally dominant* correspondence, *left finite* correspondence, *right finite* correspondence, etc.

DEFINITION 3.69. The category $\mathcal{C}(X)$ of *correspondences* on X is the category whose objects are the correspondences on X and whose morphisms

$$\Gamma' = (Y_{\Gamma'}, \pi', \varphi') \to \Gamma = (Y_\Gamma, \pi, \varphi)$$

are morphism of schemes $u : Y_{\Gamma'} \to Y_\Gamma$ such that $\pi' = \pi \circ u$ and $\varphi' = \varphi \circ u$.

REMARK 3.70. In particular there is a notion of *isomorphism* of correspondences denoted by $\Gamma \simeq \Gamma'$.

We say that a morphism of correspondences as above has property P if u has property P; in particular it is *totally dominant* if u is totally dominant.

We say that a correspondence *totally dominates* another correspondence if there is a totally dominant morphism from the first to the second.

DEFINITION 3.71. If $\Gamma = (Y, \pi, \varphi)$ is a correspondence we define its *transpose* as

$$\Gamma^t := (Y, \varphi, \pi).$$

If $\Gamma_1 \to \Gamma_2$ is a morphism of correspondences then we have an induced morphism $\Gamma_1^t \to \Gamma_2^t$ and its formation is compatible with composition of morphisms.

DEFINITION 3.72.
A correspondence $\Gamma = (Y, \pi, \varphi)$ will be called *strictly symmetric* if $\Gamma^t = \Gamma$ i.e., if $\pi = \varphi$.
A correspondence $\Gamma = (Y, \pi, \varphi)$ will be called *symmetric* if $\Gamma^t \simeq \Gamma$, i.e., if there exists an isomorphism $\sigma : Y \to Y$ such that $\varphi = \pi \circ \sigma$ and $\pi = \varphi \circ \sigma$.

REMARK 3.73. If $\Gamma \simeq \Gamma'$ and Γ is symmetric (respectively strictly symmetric) then Γ' is also symmetric (respectively strictly symmetric).

EXAMPLE 3.74. If $Y \subset X \times X$ is a closed subscheme, $\pi, \varphi : Y \to X$ are given by the two projections from $X \times X$ to X, and τ is the automorphism of $X \times X$ that permutes the factors, then the correspondence (Y, π, φ) is symmetric if and only if $\tau(Y) = Y$; this correspondence is strictly symmetric if and only if Y is contained in the diagonal.

Going back to our general discussion, there are two remarkable correspondences on any X: the *empty* correspondence,

$$\mathbb{O} := (\emptyset, \emptyset, \emptyset),$$

and the *identity* correspondence,

$$\mathbb{I} = (X, id, id).$$

A correspondence (Y, π, φ) has a morphism to $\mathbb{I}$ if and only if it is strictly symmetric.

DEFINITION 3.75. If $\Gamma = (Y_\Gamma, \pi, \varphi)$ and $\Gamma' = (Y_{\Gamma'}, \pi', \varphi')$ are two correspondences on X their *composition* is the correspondence

$$\Gamma' \circ \Gamma := (Y_{\Gamma' \circ \Gamma}, \pi \circ \pi_1', \varphi' \circ \varphi_2) \tag{3.72}$$

where the above data are defined by the following diagram in which the square is cartesian:

$$\begin{array}{ccccc} Y_{\Gamma' \circ \Gamma} & \stackrel{\varphi_2}{\to} & Y_{\Gamma'} & \stackrel{\varphi'}{\to} & X \\ \pi_1' \downarrow & & \downarrow \pi' & & \\ Y_\Gamma & \stackrel{\varphi}{\to} & X & & \\ \pi \downarrow & & & & \\ X & & & & \end{array} \tag{3.73}$$

REMARK 3.76. Here, for any two morphisms of schemes with the same target we fix once and for all a fiber product object; the indices 1 and 2 stand for first and second projection.

In general $\Gamma' \circ \Gamma$ may be empty even if Γ and Γ' are non-empty; if, however, both correspondences are, say, left and right totally dominant and left and right of finite type then their composition is non-empty.

DEFINITION 3.77. For two correspondences $\Gamma = (Y_\Gamma, \pi, \varphi)$ and $\Gamma' = (Y_{\Gamma'}, \pi', \varphi')$ their *direct sum* $\Gamma \oplus \Gamma'$ is the correspondence given by

$$Y_{\Gamma \oplus \Gamma'} := Y_\Gamma \coprod Y_{\Gamma'}$$

(disjoint union of schemes), with the obvious induced maps.

REMARK 3.78. There are natural isomorphisms of correspondences expressing the "associativity" of composition and direct sum, the "neutral element" properties of $\mathbb{O}$ and $\mathbb{I}$, and the left and right "distributivity" of composition with respect to direct sum. Cf. [**98**], pp. 158-159. All of these satisfy the standard (pentagonal and triangular) diagrams in loc. cit. plus diagrams for associativity, making $\mathcal{C}(X)$ a monoidal category with respect to both composition and direct sum, having an additional coherence structure for distributivity. Also we have canonical isomorphisms

$$(\Gamma \circ \Gamma')^t \simeq (\Gamma')^t \circ \Gamma^t, \quad \Gamma \circ \mathbb{I} \simeq \mathbb{I} \circ \Gamma \simeq \Gamma, \quad \Gamma \circ \mathbb{O} \simeq \mathbb{O} \circ \Gamma \simeq \mathbb{O}, \tag{3.74}$$

satisfying appropriate coherence diagrams. If $\Gamma' \to \Gamma''$ is a morphism of correspondences and Γ is a correspondence we have natural morphisms of correspondences

$$\Gamma' \oplus \Gamma \to \Gamma'' \oplus \Gamma, \quad \Gamma \oplus \Gamma' \to \Gamma \oplus \Gamma'', \quad \Gamma' \circ \Gamma \to \Gamma'' \circ \Gamma, \quad \Gamma \circ \Gamma' \to \Gamma \circ \Gamma'', \tag{3.75}$$

whose formation is compatible, in the obvious sense, with composition in $\mathcal{C}(X)$.

The following definition will play a key role later:

DEFINITION 3.79.

A correspondence Γ is *compatible* with a correspondence Γ' if there is an isomorphism

$$\Gamma' \circ \Gamma^t \simeq \Gamma_1 \oplus \Gamma_2$$

where Γ_1 is strictly symmetric and left (equivalently, right) totally dominant.

A correspondence Γ_1 is said to *partially commute* with Γ_2 if $\Gamma_1 \circ \Gamma_2$ is compatible with $\Gamma_2 \circ \Gamma_1$; in other words, if the correspondence

$$\Gamma_2 \circ \Gamma_1 \circ \Gamma_2^t \circ \Gamma_1^t$$

has a direct summand which is strictly symmetric and left totally dominant.

A correspondence Γ_1 is said to *commute* with Γ_2 if we have an isomorphism

$$\Gamma_1 \circ \Gamma_2 \simeq \Gamma_2 \circ \Gamma_1.$$

REMARK 3.80. If we consider the following diagram in which the square is cartesian,

$$\begin{array}{ccccc} Y_{\Gamma' \circ \Gamma^t} & \stackrel{\pi_2}{\to} & Y_{\Gamma'} & \stackrel{\varphi'}{\to} & X \\ \pi_1' \downarrow & & \downarrow \pi' & & \\ Y_\Gamma & \stackrel{\pi}{\to} & X & & \\ \varphi \downarrow & & & & \\ X & & & & \end{array} \tag{3.76}$$

the condition that Γ and Γ' be compatible is equivalent to the condition that $\varphi' \circ \pi_2$ and $\varphi \circ \pi_1'$ coincide on *some* connected component of $Y_{\Gamma' \circ \Gamma^t}$ and are totally dominant when restricted to that component.

Note that if Γ is compatible with Γ' then Γ' is compatible with Γ.

Note that, in this generality, Γ is not a priori compatible with itself; this is because $\Gamma \circ \Gamma^t$, although symmetric, does not a priori contain a strictly symmetric summand.

However, if the image of the diagonal $Y_\Gamma \to Y_{\Gamma \circ \Gamma^t}$ is open and closed then Γ is compatible with itself; this happens, for instance, if X is the spectrum of a field of characteristic zero and Γ is reduced and left finite.

Similarly commutation does not imply partial commutation in general but it does imply it if X is the spectrum of a field and we restrict ourselves to reduced, left and right finite, correspondences.

If $\sigma : X \to X$ is an automorphism of X we may consider the naturally associated correspondence $\Gamma_\sigma = (X, id, \sigma)$; we have $\Gamma_\sigma^t \simeq \Gamma_{\sigma^{-1}}$ and $\Gamma_{\sigma_1 \circ \sigma_2} \simeq \Gamma_{\sigma_1} \circ \Gamma_{\sigma_2}$.

If $X' \to X$ is a morphism of schemes then we have a natural base change functor $\mathcal{C}(X) \to \mathcal{C}(X')$, $\Gamma_{/X} \mapsto \Gamma_{/X'}$, compatible with all the operations introduced above. If E is a field of characteristic zero we simply write $\mathcal{C}(E)$ in place of $\mathcal{C}(Spec\ E)$. So, in particular, if X is an integral scheme of characteristic zero with field of rational functions E then $Spec\ E \to X$ induces a functor

$$\mathcal{C}(X) \to \mathcal{C}(E), \quad \Gamma \mapsto \Gamma \otimes E.$$

3.5.2. Correspondences on a field.

DEFINITION 3.81. Let E be a field of characteristic zero. We denote by $\mathcal{C}_0(E)$ the subcategory of $\mathcal{C}(E)$ whose objects are the non-empty, reduced, left and right finite correspondences, and whose morphisms are the totally dominant (equivalently, surjective) morphisms.

So the objects of $\mathcal{C}_0(E)$ are of the form

$$\Gamma = (Spec\ F, \pi, \varphi)$$

where F is a finite product of fields each of which is a finite extension of E via both π and φ.

DEFINITION 3.82. We define the *bidegree* of Γ as

$$\text{bideg}(\Gamma) = (\deg(\pi), \deg(\varphi)) \in \mathbb{Z}_{>0} \times \mathbb{Z}_{>0};$$

we then refer to $\deg(\pi)$ and $\deg(\varphi)$ as the *left degree* and the *right degree* of Γ.

REMARK 3.83. One immediately checks that $\mathcal{C}_0(E)$ is closed under transposition t, composition $\circ$, taking direct sums $\oplus$, and taking direct summands. Moreover, if $\Gamma' \to \Gamma''$ is a morphism in $\mathcal{C}_0(E)$ and Γ is an object in $\mathcal{C}_0(E)$ then the morphisms in 3.75 are surjective, hence they are morphisms in $\mathcal{C}_0(E)$. Any morphism $\Gamma' \to \Gamma$ in $\mathcal{C}_0(E)$ induces an injection $\mathcal{O}(Y_\Gamma) \to \mathcal{O}(Y_{\Gamma'})$. The relations of compatibility on the set of objects in $\mathcal{C}_0(E)$ are reflexive and symmetric. As noted already, symmetry follows from the fact that, in this setting, $\Gamma \circ \Gamma^t$ contains a strictly symmetric summand namely, the image of the diagonal map $Y_\Gamma \to Y_{\Gamma \circ \Gamma^t}$ which is both closed and open. Although compatibility is not transitive in general, there are cases when transitivity holds. Indeed, we have the following:

EXAMPLE 3.84. Assume that $\Gamma_i = (Y, \pi, \varphi_i)$, $i = 1, 2, 3$, are correspondences in $\mathcal{C}_0(E)$, with Y irreducible, and π Galois. It is easy to see that Γ_i and Γ_j are compatible if and only if there exists an automorphism $\sigma : Y \to Y$ with $\pi \circ \sigma = \pi$ such that $\varphi_j = \varphi_i \circ \sigma$; hence if Γ_1 if compatible with Γ_2 and Γ_2 is compatible with Γ_3 then Γ_1 is compatible with Γ_3.

As already mentioned, if two correspondences in $\mathcal{C}_0(E)$ are commuting then they are partially commuting. Commutation of correspondences is, in principle a "rare" occurrence; so it is important to put forward weaker concepts of commutation such as partial commutation (defined earlier) or the "upper $*$" and "lower $*$" commutation (to be defined presently).

Let $\mathrm{End}_{\mathrm{gr}}(E)$ be the set of all group endomorphisms (equivalently of all $\mathbb{Q}$-linear endomorphisms) of E; it is a ring with respect to addition and composition, and it is a $\mathbb{Q}$-Lie algebra in a natural way. (We used the subscript gr to avoid confusion with the monoid of field endomorphisms of E.)

DEFINITION 3.85. Given a correspondence $\Gamma = (Y, \pi, \varphi)$ in $\mathcal{C}_0(E)$, $Y_\Gamma = Spec\, F$, and denoting by $\pi, \varphi : E \to F$ the corresponding ring homomorphisms we define the map $\Gamma^* \in \mathrm{End}_{\mathrm{gr}}(E)$, as the composition

$$\Gamma^* := \mathrm{tr}_\pi \circ \varphi : E \to F \to E,$$

where $\mathrm{tr}_\pi : F \to E$ is the trace of π.

REMARK 3.86. The following are trivial to check:

(1) If $\Gamma \simeq \Gamma'$ then $\Gamma^* = (\Gamma')^*$.
(2) $(\Gamma_1 \oplus \Gamma_2)^* = \Gamma_1^* + \Gamma_2^*$ in $\mathrm{End}_{\mathrm{gr}}(E)$.
(3) $(\Gamma_1 \circ \Gamma_2)^* = \Gamma_2^* \circ \Gamma_1^*$ in $\mathrm{End}_{\mathrm{gr}}(E)$.
(4) If Γ is strictly symmetric of bidegree (m, m) then $\Gamma^* = m \cdot id$ with id the identity on E.
(5) More generally if $\Gamma' \to \Gamma$ is a morphism with Γ irreducible and the morphism has degree m then $(\Gamma')^* = m \cdot \Gamma^*$.
(6) For any Γ of bidegree $(1, n)$, we have $(\Gamma^t)^*\Gamma^* = n \cdot id$ hence Γ^* is injective.
(7) If the left degree of Γ is ≥ 2 then Γ^* is not injective, in general.
(8) If Γ is an arbitrary correspondence of left degree m and $u \in E$ is such that $\Gamma^*(u^i) = 0$ for all $1 \leq i \leq m$ then $u = 0$.

Assume (as will be the case in applications) that E comes equipped with an involution $\iota : E \to E$, i.e. a field automorphism whose square is the identity; we assume ι is not the identity. Consider the eigenspaces of ι:

$$E^+ = \{z \in E; \iota(z) = z\}, \quad E^- = \{z \in E; \iota(z) = -z\}; \tag{3.77}$$

then E has degree 2 over the field E^+ and we have an E^+-linear space decomposition $E = E^+ \oplus E^-$.

DEFINITION 3.87.

A correspondence Γ on E is *partially induced* from E^+ if $\Gamma^*(E^+) \subset E^+$ and $\Gamma^*(E^-) = 0$.

If, in addition, $\varphi(E^+) \subset \pi(E^+)$ we say that Γ is *induced* from E^+; if this is the case then, of course, the restriction of Γ^* to E^+ is an integer multiple of a field endomorphism of E^+ and, in particular, $Ker(\Gamma^*) = E^-$.

REMARK 3.88. The action of correspondences on fields extends to an action on forms as follows. Assume, for simplicity, that E is a finitely generated field (as usual, of characteristic zero). For any finite reduced E-algebra $\pi : E \to F$ let Ω_F be the F-module of Kähler differentials of F over $\mathbb{Q}$ and Ω_F^i its i-th exterior power. We have natural isomorphisms $\Omega_E^i \otimes F \simeq \Omega_F^i$ and hence naturally induced maps

$$\mathrm{tr}_\pi^i : \Omega_F^i \simeq \Omega_E^i \otimes_E F \stackrel{1\otimes tr_\pi}{\longrightarrow} \Omega_E^i \otimes_E E \simeq \Omega_E^i. \tag{3.78}$$

If $x_1, ..., x_d$ is a transcendence basis of E over $\mathbb{Q}$ and $\frac{\partial}{\partial x_i}$ are the corresponding derivations on E and F then, Ω_F is a free F-module with basis $dx_1, ..., dx_d$ and, for any $f \in F$, $df = \sum \frac{\partial f}{\partial x_i} dx_i$, hence

$$\mathrm{tr}_\pi^1(df) = \sum \mathrm{tr}_\pi\left(\frac{\partial f}{\partial x_i}\right) dx_i = \sum \frac{\partial}{\partial x_i}(\mathrm{tr}_\pi f) dx_i = d(\mathrm{tr}_\pi f). \tag{3.79}$$

(The commutation of tr_π and $\frac{\partial}{\partial x_i}$ is checked by reducing to the case when F is a Galois field extension of E in which case it follows from the commutation of $\frac{\partial}{\partial x_i}$ with the E-automorphisms of F.) Now if $\mathrm{End}(\Omega_E)$ denotes the ring of group endomorphisms (equivalently $\mathbb{Q}$-linear endomorphisms) of Ω_E and $\Gamma = (Spec\ F, \pi, \varphi)$ is a correspondence in $\mathcal{C}_0(E)$ we can define a $\mathbb{Q}$-linear map

$$\Gamma^{*i} \in \mathrm{End}(\Omega_E), \qquad \Gamma^{*i} : \Omega_E^i \to \Omega_E^i$$

by the formula

$$\Gamma^{*i} = \mathrm{tr}_\pi^i \circ \varphi^{*i} : \Omega_E^i \to \Omega_F^i \to \Omega_E^i,$$

where φ^{*i} is naturally induced from φ. One has

$$\Gamma^{*1}(df) = d(\Gamma^*(f))$$

for all $f \in F$. Also for any two correspondences Γ_1, Γ_2 we have

$$(\Gamma_1 \oplus \Gamma_2)^{*i} = \Gamma_1^{*i} \oplus \Gamma_2^{*i}, \qquad (\Gamma_1 \circ \Gamma_2)^{*i} = \Gamma_2^{*i} \circ \Gamma_1^{*i}.$$

So if Γ_1 and Γ_2 commute then Γ_1^{*i} and Γ_2^{*i} commute.

If we assume in addition that an involution $\iota : E \to E$ is given such that $\iota(x_i) = -x_i$ (which is the case in our applications) and if Γ is partially induced from E^+ (in particular $\Gamma^*(E^-) = 0$) then we have

$$\Gamma^{*1}(dx_i) = d(\Gamma^*(x_i)) = 0.$$

REMARK 3.89. There is an alternative action of correspondences on *cycles* of E as follows. Fix an algebraically closed field $\mathbb{U}$, of cardinality at least that of E, let $\mathrm{Hom}_{\text{fields}}(E, \mathbb{U})$ denote the set all field homomorphisms $\sigma : E \to \mathbb{U}$, and denote by $\mathrm{Cyc}(E)$ the free abelian group with basis $\mathrm{Hom}_{\text{fields}}(E, \mathbb{U})$. For $\sigma \in \mathrm{Hom}_{\text{fields}}(E, \mathbb{U})$ denote by $[\sigma]$ the image of σ in $\mathrm{Cyc}(E)$. The inclusion

$$\mathrm{Hom}_{\text{fields}}(E, \mathbb{U}) \to \mathrm{Hom}_{\text{gr}}(E, \mathbb{U})$$

induces a group homomorphism $\int : \mathrm{Cyc}(E) \to \mathrm{Hom}_{\text{gr}}(E, \mathbb{U})$,

$$\int \left(\sum n_i[\sigma_i]\right) := \sum n_i \sigma_i. \tag{3.80}$$

Let $\mathrm{End}(\mathrm{Cyc}(E))$ be the ring of group endomorphisms of $\mathrm{Cyc}(E)$.

DEFINITION 3.90. Given a correspondence Γ in $\mathcal{C}_0(E)$ we define the map

$$\Gamma^* \in \mathrm{End}_{\text{gr}}(\mathrm{Cyc}(E))$$

by the following recipe. Assume first $\Gamma = (Y, \pi, \varphi)$ is irreducible, so $Y_\Gamma = Spec\ F$ where F is a field and π, φ are induced by finite field homomorphisms (still denoted by) $\pi, \varphi : E \to F$. Let $\sigma : E \to \mathbb{U}$ be an embedding. Then consider all the embeddings

$$\sigma_1, \ldots, \sigma_d : F \to \mathbb{U}$$

such that $\sigma_i \circ \pi = \sigma$, where d is the degree of π, and define

$$\Gamma_*[\sigma] = \sum_{i=1}^d [\sigma_i \circ \varphi] \in \mathrm{Cyc}(E). \tag{3.81}$$

Extend Γ_* by linearity to an endomorphism Γ_* of $\mathrm{Cyc}(E)$. Finally extend the definition of Γ_* by linearity to the case when Γ is not necessarily irreducible.

REMARK 3.91.

(1) If $\Gamma \simeq \Gamma'$ then $\Gamma_* = \Gamma'_*$.
(2) $(\Gamma \oplus \Gamma')_* = \Gamma_* + \Gamma'_*$.
(3) $(\Gamma \circ \Gamma')_* = \Gamma_* \circ \Gamma'_*$.
(4) If Γ_1 and Γ_2 commute then Γ_{1*} and Γ_{2*} commute.
(5) If Γ is strictly symmetric of bidegree (d,d) then $\Gamma_* = d$, where d is the multiplication by d endomorphism of $\mathrm{Cyc}(E)$.

REMARK 3.92. The above construction

$$\Gamma \mapsto \Gamma_* \in \mathrm{End}(\mathrm{Cyc}(E))$$

is related to the previous construction

$$\Gamma \mapsto \Gamma^* \in \mathrm{End}_{\mathrm{gr}}(E) \subset \mathrm{Hom}_{\mathrm{gr}}(E, \mathbb{U})$$

as follows: for any correspondence Γ on E and any $\sigma \in \mathrm{Hom}_{\mathrm{fields}}(E, \mathbb{U})$ we have an equality

$$\int (\Gamma_*[\sigma]) = \sigma \circ \Gamma^* \tag{3.82}$$

in $\mathrm{Hom}_{\mathrm{gr}}(E, \mathbb{U})$. In particular, if Γ_1, Γ_2 are two correspondences and $\Gamma_{1*} = \Gamma_{2*}$ then $\Gamma_1^* = \Gamma_2^*$. So if Γ, Γ' are two correspondences such that Γ_* and Γ'_* commute, i.e., $[\Gamma_*, \Gamma'_*] = 0$ in $\mathrm{End}(\mathrm{Cyc}(E))$ then Γ^* and $(\Gamma')^*$ commute, i.e., $[\Gamma^*, (\Gamma')^*] = 0$ in $\mathrm{End}_{\mathrm{gr}}(E)$. We summarize the various implications between various commutation relations:

$$\begin{array}{ccc} \Gamma \text{ and } \Gamma' \text{ commute} & \Longrightarrow & \Gamma_* \text{ and } \Gamma'_* \text{ commute} \\ \Downarrow & & \Downarrow \\ \Gamma \text{ and } \Gamma' \text{ partially commute} & & \Gamma^* \text{ and } (\Gamma')^* \text{ commute} \end{array} \tag{3.83}$$

REMARK 3.93. Here is a "Grothendieck ring" theoretic formalism into which the above definitions fit; this will not be needed in the sequel but seems to provide the right context for our concepts.

Assume, as before, that E is a field of characteristic zero. We can associate to E a ring $C = C(E)$ equipped with an anti-involution as follows.

Let us denote by $C_+ = C_+(E)$ the set of isomorphism classes $[\Gamma]$ of objects Γ of $\mathcal{C}_0(E)$ to which we add one element $[\mathbb{O}]$ which we call *zero* and we denote by 0. We denote the element $[\mathbb{I}]$ by 1. The set C_+ comes equipped with the following operations:

(1) transposition: $[\Gamma]^t := [\Gamma^t]$;
(2) addition: $[\Gamma] + [\Gamma'] := [\Gamma \oplus \Gamma']$;
(3) multiplication: $[\Gamma] \cdot [\Gamma'] := [\Gamma \circ \Gamma']$.

These operations make C_+ into a *semiring with anti-involution* by which we mean that the following hold:

(i) addition and multiplication are associative and addition is commutative;
(ii) multiplication is left and right distributive;
(iii) transposition is an anti-involution (i.e. it is a homomorphism with respect to direct sum, an anti-homomorphism with respect to composition, and it is its own inverse);

(iv) 0 is a neutral element for addition, $0^t = 0$, and the product of 0 with any element is 0;

(v) 1 is a neutral element for composition and $1^t = 1$.

In addition the sum and product of two non-zero elements in C_+ is non-zero.

Let us say that an element of C_+ is *irreducible* if it is non-zero and cannot be written as a sum of two non-zero elements. Clearly $[\Gamma]$ is irreducible if and only if Y_Γ is non-empty and irreducible. Also any non-zero element of C_+ can be written as a finite sum of irreducible elements and this decomposition is unique up to the permutation of the terms. In other words C_+ with its addition is a free commutative monoid with basis the irreducible elements. In particular C_+ has the cancellation property for addition. We can now embed our semiring above into a ring as follows. Define

$$C(E) := C := (C_+ \times C_+)/\sim \tag{3.84}$$

where

$$(\gamma_1, \gamma_2) \sim (\gamma_3, \gamma_4) \quad \text{if and only if} \quad \gamma_1 + \gamma_4 = \gamma_2 + \gamma_3.$$

Due to the cancellation property $\sim$ is an equivalence relation and the map $C_+ \to C$, $\gamma \mapsto (\gamma, 0)$, is injective. Then C has a naturally induced structure of ordered ring with identity and is equipped with an induced anti-involution. Also C is a free $\mathbb{Z}$-module with respect to addition, with basis the irreducible elements of C_+.

DEFINITION 3.94. The ring C is called the *ring of correspondences* on E.

REMARK 3.95. Let $\text{End}_{\text{field}}(E)$ be the monoid of field endomorphisms of E and $\text{Aut}_{\text{field}}(E)$ be the group of field automorphisms of E. Note that one has a natural injective homomorphisms of multiplicative monoids

$$\text{End}_{\text{field}}(E) \to C(E) \tag{3.85}$$

given by

$$\sigma \mapsto [\Gamma_\sigma], \qquad \Gamma_\sigma = (Spec\ E, id, \sigma). \tag{3.86}$$

If one views this embedding as an inclusion then, for any

$$\sigma \in \text{Aut}_{\text{field}}(E) \subset \text{End}_{\text{field}}(E) \subset C(E) \tag{3.87}$$

we have $\sigma^t = \sigma^{-1}$.

By the way recall that, by the independence of characters 1.2, the natural associative ring homomorphism

$$\mathbb{Z}[\text{End}_{\text{field}}(E)] \to \text{End}_{\text{gr}}(E) \tag{3.88}$$

is injective.

One has a unique ring (anti)homomorphism

$$(\ )^* : C(E) \to \text{End}_{\text{gr}}(E) \tag{3.89}$$

such that $[\Gamma] \mapsto \Gamma^*$ for any correspondence Γ.

On the other hand the morphism 3.85 induces an associative ring homomorphism

$$\mathbb{Z}[\text{End}_{\text{field}}(E)] \to C(E) \tag{3.90}$$

and clearly the composition of 3.90 with 3.89 equals 3.88. In particular the map 3.90 is injective.

More generally one has induced ring homomorphisms

$$C(E) \to \text{End}(\Omega^i_E)$$

sending each $[\Gamma]$ into Γ^{*i}. The commutator $[\Gamma_1^*, \Gamma_2^*]$ in $\text{End}_{\text{gr}}(E)$, that will soon play a role in our theory of curvature, appears then as the image in $\text{End}_{\text{gr}}(E)$ of the negative of the commutator

$$[\Gamma_1][\Gamma_2] - [\Gamma_2][\Gamma_1]$$

in $C(E)$; and the same is the case for the commutators $[\Gamma_1^{*i}, \Gamma_2^{*i}]$ in $\text{End}(\Omega^i_E)$.

On the other hand there is a natural ring homomorphism

$$(3.91) \qquad (\ )_* : C(E) \to \text{End}(\text{Cyc}(E))$$

sending any $[\Gamma]$ into Γ_* hence the commutator $[\Gamma_{1*}, \Gamma_{2*}]$ in $\text{End}_{\text{gr}}(\text{Cyc}(E))$ is the image of the commutator $[\Gamma_1][\Gamma_2] - [\Gamma_2][\Gamma_1]$ in $C(E)$. Note that by 3.82 we have, for any embedding $\sigma : E \to \mathbb{U}$, a commutative diagram,

$$(3.92) \qquad \begin{array}{ccc} C(E) & \overset{(\ )^*}{\longrightarrow} & \text{End}_{\text{gr}}(E) \\ (\ )_* \downarrow & & \downarrow \sigma\circ \\ \text{End}(\text{Cyc}(E)) & \overset{\int((\ )[\sigma])}{\longrightarrow} & \text{Hom}_{\text{gr}}(E, \mathbb{U}). \end{array}$$

We end by noting that it would be interesting to answer the following questions.

(1) Is the subset of $C(E)$ consisting of all elements of the form $[\Gamma]$ with Γ strictly symmetric definable in $C(E)$ in terms of the ring operations and the transposition operation only?
(2) Is partial commutation in $C(E)$ definable in terms of these operations only?
(3) What are the kernels of the ring homomorphisms $(\ )^*$ and $(\ )_*$?

Since the right vertical homomorphism in the above diagram is injective we get, of course, that

$$Ker((\ )_*) \subset Ker((\ )^*).$$

3.5.3. Correspondences compatible with Frobenius lifts. Let us assume X is an affine smooth connected scheme over a Noetherian regular integral domain A. Note that X is then integral; we denote by K the fraction field of A and we denote by E the fraction field of X. A correspondence on X will be said to be *over* A if the morphisms in the correspondence are over A.

Let $\mathcal{V}$ be a fixed collection of primes in $\mathbb{Z}$ and let $p \in \mathcal{V}$.

DEFINITION 3.96. We say that a Frobenius lift ϕ_p on $X^{\widehat{p}}$ and a correspondence $\Gamma_p = (Y_p, \pi_p, \varphi_p)$ in $\mathcal{C}(E)$ are *compatible* if there exists a correspondence $\Gamma_{p/X} = (Y_{p/X}, \pi_{p/X}, \varphi_{p/X})$ on X over A such that the following properties are satisfied.

(1) $\Gamma_p \simeq \Gamma_{p/X} \otimes E$;
(2) $\Gamma_{p/X}$ is affine and left étale;
(3) $\pi^{\widehat{p}}_{p/X} : Y^{\widehat{p}}_{p/X} \to X^{\widehat{p}}$ is an isomorphism;
(4) $\varphi^{\widehat{p}}_{p/X} = \phi_p \circ \pi^{\widehat{p}}_{p/X} : Y^{\widehat{p}}_{p/X} \to X^{\widehat{p}}$.

If ϕ_p is compatible with a correspondence Γ_p then the correspondence is, of course, not unique; but it is "essentially unique" in the following sense.

LEMMA 3.97. *Assume that ϕ_p is compatible with a correspondence Γ_p in $\mathcal{C}(E)$. Then Γ_p is in $\mathcal{C}_0(E)$. Assume in addition that $X \otimes \mathbb{F}_p$ is connected for all $p \in \mathcal{V}$ and ϕ_p is also compatible with a correspondence Γ'_p in $\mathcal{C}(E)$; then Γ_p and Γ'_p are compatible.*

REMARK 3.98. In our applications

$$A = \mathbb{Z}[1/M, \zeta_N], \quad X = GL_{n/A} := Spec\ A[x, \det(x)^{-1}].$$

So, for $p \nmid MN$, the condition that $X \otimes \mathbb{F}_p$ be connected is equivalent to p being inert in $\mathbb{Q}(\zeta_N)$.

PROOF OF LEMMA 3.97. Let

$$\Gamma_{p/X} = (Y_{p/X}, \pi_{p/X}, \varphi_{p/X})$$

be as in Definition 3.96 and let Y be a connected component of $Y_{p/X}$ that meets $Y_{p/X} \otimes \mathbb{F}_p$. Then Y is smooth over A, hence Y is a regular scheme, hence an integral scheme (because it is connected), hence $\mathcal{O}(Y)$ is p-adically separated (by Krull's intersection theorem, because p is not invertible on Y) and therefore $\mathcal{O}(Y)$ embeds into $\mathcal{O}(Y)^{\widehat{p}}$. But the maps

$$\pi^{\widehat{p}}_{p/X}, \varphi^{\widehat{p}}_{p/X} : \mathcal{O}(X)^{\widehat{p}} \to \mathcal{O}(Y)^{\widehat{p}}$$

are injective (because their reductions mod p are injective and the rings in question are p-adically complete with p a non-zero divisor in them). So the maps

$$\pi_{p/X}, \varphi_{p/X} : \mathcal{O}(X) \to \mathcal{O}(Y)$$

are injective. This immediately implies that Γ_p is non-empty. Since $\pi_{p/X}$ is étale it is generically finite so $\mathcal{O}(Y_p)$ is a product of fields which are finite extensions of E. Due to considerations of transcendence degrees we get that φ_p is finite. Hence Γ_p is non-empty, reduced, and left and right finite.

Assume now ϕ_p is also compatible with a correspondence Γ'_p and let

$$\Gamma'_{p/X} = (Y'_{p/X}, \pi'_{p/X}, \varphi'_{p/X})$$

be as in Definition 3.96. Let Y be a connected component of $Y_{\Gamma'_{p/X} \circ \Gamma^t_{p/X}}$ that meets $Y_{\Gamma'_{p/X} \circ \Gamma^t_{p/X}} \otimes \mathbb{F}_p$ (the latter being non-empty because $X \otimes \mathbb{F}_p$ is irreducible). Then, as before, $\mathcal{O}(X)$ and $\mathcal{O}(Y)$ embed into their p-adic completions. On the other hand, with notation as in diagram 3.76, with

$$\pi = \pi_{p/X}, \quad \pi' = \pi'_{p/X}, \quad \varphi = \varphi_{p/X}, \quad \varphi' = \varphi'_{p/X},$$

we have that the p-adic completions of $\varphi' \circ \pi_2$ and $\varphi \circ \pi'_1$ are equal. So $\varphi' \circ \pi_2$ and $\varphi \circ \pi'_1$ are equal on Y hence $\Gamma_{p/X}$ and $\Gamma'_{p/X}$ are compatible. Since, as before,

$$\pi' \circ \pi_2 : \mathcal{O}(X) \to \mathcal{O}(Y)$$

is injective it follows that $Y \otimes E$ is non-empty, hence Γ_p and Γ'_p are compatible. □

3.5.4. $*$-curvatures for adelic connections. Assume we are given an adelic connection (δ_p) on an affine smooth connected scheme X over a regular Noetherian domain A. Let E be the fraction field of X and let (ϕ_p) be the family of Frobenius lifts attached to (δ_p).

DEFINITION 3.99. A *correspondence structure* for $\delta = (\delta_p)$ is a collection (Γ_p) of correspondences on E such that, for each p, $\Gamma_p = (Y_p, \pi_p, \varphi_p)$ is *compatible* with the corresponding Frobenius lift ϕ_p.

DEFINITION 3.100. Let (Γ_p) be a *correspondence structure* for (δ_p). We define the *upper* $*$*-curvature* of (Γ_p) as the matrix $(\varphi^*_{pp'})$,

$$\varphi^*_{pp'} = \frac{1}{pp'}[\Gamma^*_{p'}, \Gamma^*_p] \in \mathrm{End}_{\mathrm{gr}}(E). \tag{3.93}$$

Given one more adelic connection $(\delta_{\overline{p}})$ with correspondence structure $(\Gamma_{\overline{p}})$ we define the *upper* $*$*-$(1,1)$-curvature* of (Γ_p) with respect to $(\Gamma_{\overline{p}})$ as the matrix $(\varphi^*_{p\overline{p}'})$,

$$\varphi^*_{p\overline{p}'} = \frac{1}{pp'}[\Gamma^*_{\overline{p}'}, \Gamma^*_p] \in \mathrm{End}_{\mathrm{gr}}(E) \quad \text{for } p \neq p', \tag{3.94}$$

$$\varphi^*_{p\overline{p}} = \frac{1}{p}[\Gamma^*_{\overline{p}}, \Gamma^*_p] \in \mathrm{End}_{\mathrm{gr}}(E). \tag{3.95}$$

Similarly we define the *lower* $*$*-curvature*

$$\varphi_{pp'*} = \frac{1}{pp'}[\Gamma_{p'*}, \Gamma_{p*}] \in \mathrm{End}_{\mathrm{gr}}(\mathrm{Cyc}(E)) \tag{3.96}$$

and the *lower* $*$*-$(1,1)$-curvature* $(\varphi_{p\overline{p}'*})$,

$$\varphi_{p\overline{p}'*} = \frac{1}{pp'}[\Gamma_{\overline{p}'*}, \Gamma_{p*}] \in \mathrm{End}_{\mathrm{gr}}(\mathrm{Cyc}(E)) \quad \text{for } p \neq p', \tag{3.97}$$

$$\varphi_{p\overline{p}*} = \frac{1}{p}[\Gamma_{\overline{p}*}, \Gamma_{p*}] \in \mathrm{End}_{\mathrm{gr}}(\mathrm{Cyc}(E)). \tag{3.98}$$

Note that, by 3.82, the non-vanishing of the upper $*$-curvature, respectively $*$-$(1,1)$-curvature, implies the non-vanishing of the corresponding lower $*$-curvature. Note also that these $*$-curvatures are attached to the correspondence structure (Γ_p) and *not* to the adelic connection (δ_p) directly. A priori these $*$-curvatures will depend on the choice of the correspondence structure. Later we will prove the existence of correspondence structures for some remarkable classes of adelic connections; we will then prove vanishing/non-vanishing results for the $*$-curvatures of these correspondence structures.

EXAMPLE 3.101. Here is an example of a canonical correspondence structure for the Galois connection; cf Example 3.41 and Definition 3.42. Recall the setting of the latter definition in a simple case. Let $A = \mathbb{Z}[1/M]$ and let $E/\mathbb{Q}$ be a finite Galois extension with discriminant D. Let $\mathcal{O}_E$ be the ring of integers of E, set $B = \mathcal{O}_E[1/M]$, $X = Spec\ B$, and let $\mathcal{V}$ be a set of primes in $\mathbb{Z}$ not dividing MD. Let $\delta = (\delta_p)$ be the Galois connection on B (equivalently, on X) in the sense of Definition 3.42. Recall that this is the adelic connection indexed by $\mathcal{V}$ whose Frobenius lifts (ϕ_p) are given by

$$\phi_p = \prod_{\mathfrak{P}|p} \phi_{\mathfrak{P}} : B^{\widehat{p}} = \prod_{\mathfrak{P}|p} B^{\widehat{\mathfrak{P}}} \to B^{\widehat{p}} = \prod_{\mathfrak{P}|p} B^{\widehat{\mathfrak{P}}},$$

where $\phi_{\mathfrak{P}} : B^{\widehat{\mathfrak{P}}} \to B^{\widehat{\mathfrak{P}}}$ is the unique Frobenius lift on $B^{\widehat{\mathfrak{P}}} = W(B/\mathfrak{P})$ and $\mathfrak{P} \subset B$ runs through the set of primes of B lying above p. Let Γ be the Galois group of $E/\mathbb{Q}$. Then recall that $\phi_{\mathfrak{P}}$ sends B into B and induces the Frobenius element $\sigma_{\mathfrak{P}} \in \Gamma$ corresponding to $\mathfrak{P}$. Consider the correspondences

$$\Gamma_{\sigma_{\mathfrak{P}}} := (E, id, \sigma_{\mathfrak{P}}), \tag{3.99}$$

cf. 3.86, and consider the correspondences

$$\Gamma_p := \bigoplus_{\mathfrak{P}|p} \Gamma_{\sigma_{\mathfrak{P}}}. \tag{3.100}$$

We have the following:

LEMMA 3.102.

(1) *The family* (Γ_p) *is a correspondence structure for the Galois connection* (δ_p).
(2) *The upper and lower* $*$*-curvatures of* (Γ_p) *vanish.*

PROOF. We prove assertion (1). For each $\mathfrak{P}|p$ the Chinese remainder theorem gives an element $b_{\mathfrak{P}} \in B$ satisfying:

$$\begin{aligned} b_{\mathfrak{P}} &\equiv 0 \mod \mathfrak{P}' \quad \text{for all} \quad \mathfrak{P}'|p \quad \text{with} \quad \mathfrak{P}' \neq \mathfrak{P}, \\ b_{\mathfrak{P}} &\equiv 1 \mod \mathfrak{P}. \end{aligned}$$

Define then the correspondences

$$\Gamma_{p/X} = (Y_{p/X}, \pi_{p/X}, \varphi_{p/X})$$

on X, where

$$Y_{p/X} := Spec\ C_p, \quad C_p := \left(\prod_{\mathfrak{P}|p} B[1/b_{\mathfrak{P}}] \right),$$

the map

$$\pi_{p/X} : Spec\ C_p \to Spec\ B$$

is given by the diagonal embedding $B \to C_p$, and

$$\varphi_{p/X} : Spec\ C_p \to Spec\ B$$

is induced by

$$\prod_{\mathfrak{P}|p} \sigma_{\mathfrak{P}} : E \to E \times \ldots \times E.$$

Note that, by our choice of the $b_{\mathfrak{P}}$'s we have

$$C^{\widehat{p}} = \prod_{\mathfrak{P}|p} B[1/b_{\mathfrak{P}}]^{\widehat{p}} \simeq \prod_{\mathfrak{P}|p} B^{\widehat{\mathfrak{P}}} \simeq B^{\widehat{p}}.$$

Using this isomorphism it is easy to see that $\Gamma_{p/X}$ satisfies the conditions in Definition 3.96; hence our assertion (1) is proved.

We prove now assertion (2). Note that the image $[\Gamma_p]$ of Γ_p in the ring of correspondences $C(E)$ equals the image in $C(E)$ of the element

$$\sum_{\mathfrak{P}|p} \sigma_{\mathfrak{P}} \in \mathbb{Z}[\mathrm{Aut}_{\mathrm{field}}(E)] \tag{3.101}$$

via the inclusion 3.90. Note on the other hand that, for each p, the set

$$\{\sigma_{\mathfrak{P}};\ \mathfrak{P}|p\}$$

is a conjugacy class in the group $\mathrm{Aut}_{\mathrm{field}}(E)$. So the elements 3.101 belong to the center of the group ring $\mathbb{Z}[\mathrm{Aut}_{\mathrm{field}}(E)]$. In particular for any p, p' the classes $[\Gamma_p]$, $[\Gamma_{p'}]$ are pairwise commuting in $C(E)$. Hence their images Γ_p^* and $\Gamma_{p'}^*$ in $\mathrm{End}_{\mathrm{gr}}(E)$ commute; similarly the images Γ_{p*} and $\Gamma_{p'*}$ in $\mathrm{End}_{\mathrm{gr}}(\mathrm{Cyc}(E))$ commute. □

REMARK 3.103.

(1) Note that $\Gamma_{p/X}$ is not canonically attached to our Galois connection because the $b_{\mathfrak{P}}$'s are not canonical; but the correspondence structure (Γ_p) is, of course, canonically attached to our situation.
(2) A version of this example can be developed for the case when X is an arbitrary finite étale scheme over A (whose connected components are not necessarily Galois over A). The example presented above and its non-Galois version are, however, in some sense, perpendicular to the examples of correspondence structures we will later consider for our Chern connections. For one thing the $*$-curvatures of some of our main examples of correspondence structures for real Chern connections will *not* vanish.

3.6. Arithmetic jet spaces and the Cartan connection

We would like to proceed by introducing arithmetic analogues of differential equations, solutions, prime integrals, Lie theory, logarithmic derivatives, and various remarkable connections (Ehresmann, Chern, Levi-Cività, Fedosov, Lax). In order to achieve this we need to further develop our technical apparatus. In particular we need to introduce arithmetic jet spaces [**23**] and to systematically discuss arithmetic differential equations [**35**]. This can be done by concentrating on each prime at a time.

So for the rest of this chapter we will concentrate on the case of one fixed prime p only. Therefore we denote the p-adic completion of a ring B or scheme X by $\widehat{B}$ and $\widehat{X}$ respectively (rather than by $B^{\widehat{p}}$ and $X^{\widehat{p}}$). We assume p is odd, we let $\mathbb{F}$ an algebraic extension of $\mathbb{F}_p$, we denote by $R = W(\mathbb{F}^a)$ the unique (up to isomorphism) complete discrete valuation ring with maximal ideal generated by p and residue field $k = \mathbb{F}^a$, and we let $A \subset R$ be a δ-ring of local type i.e., a valuation δ-subring of R with residue field $\mathbb{F}$ (cf. Definition 3.27). Hence

$$A \subset \widehat{A} = W(\mathbb{F}) \subset R = W(\mathbb{F}^a); \tag{3.102}$$

note that A can be $W(\mathbb{F})$, and even R itself. Recall that A and R come equipped with unique Frobenius lifts $\phi = \phi_p$ and hence with unique p-derivations $\delta = \delta_p$ compatible with the inclusion $A \subset R$. For most applications we are only interested in the case $A = \widehat{A} = W(\mathbb{F})$; the case when A is not necessarily complete will become important, however, when we later discuss Galois theory.

We review in what follows some of the material in [**23, 35, 42**]. For convenience we start by recalling (and enhancing) some of the terminology in section 3.2.

DEFINITION 3.104.

A *p-formal scheme* (respectively a p-formal scheme of finite type over a base ring) is a formal scheme locally isomorphic to the p-adic completion of a Noetherian scheme (respectively of a scheme of finite type over the base ring).

A *group p-formal scheme* is a group object in the category of p-formal schemes.

Recall that if $u : A \to B$ is a ring homomorphism we usually still denote by $u : Spec\ B \to Spec\ A$ the induced map; and conversely if $f : X \to Y$ is a morphism of schemes we still denote by $f : \mathcal{O}_Y \to f_*\mathcal{O}_X$ the induced morphism on functions.

DEFINITION 3.105. A *Frobenius lift* on a scheme (or p-formal scheme) X is a morphism of schemes (or p-formal schemes respectively)

$$\phi^X : X \to X$$

whose reduction mod p is the p-power Frobenius.

REMARK 3.106. If X is a p-formal scheme over A and $\mathcal{O}_X$ is p-torsion free the standard construction globalizes to yield an operator (a map of sheaves of sets)

$$\delta^X : \mathcal{O}_X \to \mathcal{O}_X$$

referred to as the *p-derivation* on X *attached* to ϕ^X. If no confusion arises we will sometimes write ϕ and δ instead of ϕ^X and δ^X. If X is a scheme over A and $\phi^{\widehat{X}}$ is a Frobenius lift on $\widehat{X}$ we usually denote $\phi^{\widehat{X}}$ and $\delta^{\widehat{X}}$ simply by ϕ^X and δ^X respectively or simply by ϕ and δ if there is no danger of confusion; in this case we refer to δ as a *p-derivation* on $\widehat{X}$. This is the same as what we earlier called an *adelic connection* indexed by the family $\{p\}$ consisting of one prime p only or a *p-adic connection* on X.

DEFINITION 3.107. If $\pi : X \to Y$ is a morphism of schemes (or p-formal schemes) and X, Y are given Frobenius lifts ϕ^X, ϕ^Y we say that π is *horizontal* with respect to ϕ^X, ϕ^Y if the following diagram commutes

$$\begin{array}{ccc} X & \xrightarrow{\phi^X} & X \\ \pi \downarrow & & \downarrow \pi \\ Y & \xrightarrow{\phi^Y} & Y \end{array}$$

We also say that ϕ^X is *π-horizontal* with respect to ϕ^Y.

If $Z \subset X$ is a closed subscheme and we are given a Frobenius lift ϕ^X on $\widehat{X}$ we say Z is *ϕ^X-horizontal* if there exists a (necessarily unique) Frobenius lift ϕ^Z on $\widehat{Z}$ such that $\widehat{Z} \subset \widehat{X}$ is horizontal with respect to ϕ^Z, ϕ^X; this is equivalent to the condition that ϕ^X sends the defining ideal of $\widehat{Z}$ into itself. If $\mathcal{O}_X$ and $\mathcal{O}_Z$ are p-torsion free then we have at our disposal attached p-derivations and we sometimes say *δ^X-horizontal* instead of *ϕ^X-horizontal*.

REMARK 3.108. If X is a scheme (or a p-formal scheme) over A then Frobenius lifts $\phi^X : X \to X$ will always be assumed to be such that $X \to Spec\ A$ is horizontal with respect to ϕ^X, ϕ^A; in particular ϕ^X in *not* a morphism of A-schemes unless $A \subset \mathbb{Z}_p$. Nevertheless there is a naturally induced map between R-points,

$$\phi^X_* : X(R) \to X(R)$$

that sends any R-point $P \in X(R)$,

$$P : Spec\ R \to X$$

into the R-point $\phi^X_*(P) \in X(R)$ defined by

$$\phi^X_*(P) := \phi^X \circ P \circ (\phi^R)^{-1} : Spec\ R \to Spec\ R \to X \to X. \tag{3.103}$$

We will usually simply write ϕ^X instead of ϕ^X_* in what follows.

The map $\phi^X = \phi^X_* : X(R) \to X(R)$ is, of course, *not* induced, in general, by a map of formal schemes and it is *not* generally a *δ-map* in the sense of Definition 3.119 below.

Indeed, in case, say,

$$X = \mathbb{A}^1 = Spec\ A[x],$$

if $\phi^X(x) = \Phi(x) \in A[x]\hat{}$ and $P \in \mathbb{A}^1(R)$ corresponds to the map

$$A[x] \to R, \quad x \mapsto a \in R$$

then

$$\phi^X_*(P) \in \mathbb{A}^1(R)$$

corresponds to the map

$$A[x] \to R, \quad x \mapsto (\phi^R)^{-1}(\Phi(a)) \in R.$$

So note, in particular, that, in this example, the map

$$\phi^X = \phi^X_* : R \to R, \quad a \mapsto \phi^X(a) = (\phi^R)^{-1}(\Phi(a))$$

induced by the map $\phi^X : X(R) \to X(R)$ via the natural identification

$$X(R) = \mathbb{A}^1(X) \simeq R$$

is *not* the same as the map

$$\phi^R : R \to R, \quad a \mapsto \phi^R(a).$$

This is a distinction that will be important to keep in mind.

More generally if X, Y are schemes (or p-formal schemes) over A then a morphism of schemes over $\mathbb{Z}$ (or p-formal schemes over $\mathbb{Z}_p$) $f : X \to Y$ will be called *ϕ^r-linear* if the following diagram is commutative

$$\begin{array}{ccc} X & \xrightarrow{f} & Y \\ \downarrow & & \downarrow \\ Spec\ R & \xrightarrow{\phi^r} & Spec\ R \end{array} \tag{3.104}$$

Any such f induces a map of sets

$$f_* : X(R) \to Y(R), \quad f_*(P) = f \circ P \circ (\phi^R)^{-r}. \tag{3.105}$$

Again, for simplicity, we will write f instead of f_*.

We introduce now the following abuse of notation. For any smooth scheme (or p-formal scheme) X over A we usually continue to denote by X the set $X(R)$ of R-points of X. If $Y \subset X$ is a closed smooth subscheme (respectively p-formal scheme), both over R, then it is trivial to check that Y is ϕ^X-horizontal if and only if for all R-points $P \in Y$ one has $\phi^X(P) \in Y$. (The latter is an easy exercise using Nullstellensatz over k and the surjectivity of $Y(R) \to Y(k)$.) Also two Frobenius lifts on a smooth $\widehat{X}$ coincide if and only if the corresponding maps on points coincide.

LEMMA 3.109. *Let $Z \subset Y \subset X$ be closed embeddings of smooth schemes over R and let ϕ_0 and ϕ_1 be two Frobenius lifts on $\widehat{X}$. Assume that Z is ϕ_0-horizontal and that ϕ_0 and ϕ_1 coincide on the set $Y \cap \phi_0^{-1}(Y)$. Then Z is also ϕ_1-horizontal and the restrictions of ϕ_0 and ϕ_1 to Z coincide.*

PROOF. It is enough to show that for any point $P \in Z$ we have $\phi_1(P) \in Z$ and $\phi_1(P) = \phi_0(P)$. Since Z is ϕ_0-horizontal we have $\phi_0(P) \in Z$. So $P \in Y \cap \phi_0^{-1}(Y)$. Hence $\phi_1(P) = \phi_0(P) \in Z$ and we are done. □

Recall the following:

LEMMA 3.110. *Let $X \to Y$ be an étale morphism of affine schemes of finite type over R. Then any p-derivation $\mathcal{O}(\widehat{Y}) \to \mathcal{O}(\widehat{X})$ lifts to a unique p-derivation $\mathcal{O}(\widehat{X}) \to \mathcal{O}(\widehat{X})$.*

PROOF. Cf. [**35**], Lemma 3.14, p. 76. $\square$

The following is also useful and trivial to check:

LEMMA 3.111. *Let U be an affine open subscheme of the affine space $\mathbb{A}^n$ over R and let X be a closed p-formal subscheme of $\widehat{U}$. Then any Frobenius lift on X can be extended to a Frobenius lift on $\widehat{U}$.*

Next we review p-jet spaces by closely following the discussion in [**23, 35**].

DEFINITION 3.112. A *prolongation sequence* $S^* = (S^n)$ is a sequence

$$S^0 \overset{\varphi_0}{\to} S^1 \overset{\varphi_1}{\to} S^2 \overset{\varphi_2}{\to} \dots$$

of ring homomorphisms together with a sequence of maps

$$S^0 \overset{\delta_0}{\to} S^1 \overset{\delta_1}{\to} S^2 \overset{\delta_2}{\to} \dots$$

such that each δ_n is a $p-$derivation (where S^{n+1} is viewed as a S^n-algebra via φ_n) and each δ_n prolongs δ_{n-1}, i.e., $\delta_n \circ \varphi_{n-1} = \varphi_n \circ \delta_{n-1}$.

DEFINITION 3.113. By a morphism of prolongation sequences $S^* \to U^*$ we understand a sequence of ring homomorphisms

$$u_n : S^n \to U^n$$

such that

$$u_n \circ \varphi_{n-1} = \varphi_{n-1} \circ u_{n-1}$$

and

$$u_n \circ \delta_{n-1} = \delta_{n-1} \circ u_{n-1}$$

for all $n \geq 1$. A prolongation sequence *over* A will mean a prolongation sequence all of whose ring homomorphisms are $A-$algebra maps and all of whose δ_n are compatible with the δ on A. Prolongation sequences form, in an obvious way, a category.

Let now x be a tuple of indeterminates over A and consider tuples of indeterminates $x', \dots, x^{(n)}, \dots$ of the same cardinality as x. We consider the ring of polynomials in infinitely many variables

$$A\{x\} = A[x, x', x'', \dots]$$

and we still denote by $\phi : A\{x\} \to A\{x\}$ the unique Frobenius lift extending ϕ on A such that

$$\phi(x) = x^p + px', \quad \phi(x') = x^p + px'', \quad \text{e.t.c.}$$

Then we consider the p-derivation $\delta : A\{x\} \to A\{x\}$ defined by

$$\delta f = \frac{\phi(f) - f^p}{p}.$$

DEFINITION 3.114. For any affine scheme of finite type

$$X = Spec \frac{A[x]}{(f)}$$

over A, where f is a tuple of polynomials, we define the *p-jet spaces* of X as being the p-formal schemes

$$J^n(X) = Spf \frac{A[x, x', ..., x^{(n)}]\widehat{}}{(f, \delta f, ..., \delta^n f)}. \tag{3.106}$$

So $J^0(X) = \widehat{X}$. Clearly we have an induced structure of prolongation sequence on the sequence

$$\mathcal{O}(J^*(X)) := (\mathcal{O}(J^n(X))) \tag{3.107}$$

where

$$\varphi_n : \mathcal{O}(J^n(X)) \to \mathcal{O}(J^{n+1}(X))$$

are induced by the identity on $A\{x\}$ and

$$\delta_n : \mathcal{O}(J^n(X)) \to \mathcal{O}(J^{n+1}(X)) \tag{3.108}$$

are induced by δ on $A\{x\}$. We refer to the δ_n's in 3.108 as the *universal p-derivations*. Set

$$\mathcal{O}^n(X) := \mathcal{O}(J^n(X)), \quad \mathcal{O}^\infty(X) = \lim_{\to} \mathcal{O}^n(X)$$

where the limit is taken with respect to the homomorphisms φ_n. Then we get an induced p-derivation

$$\delta : \mathcal{O}^\infty(X) \to \mathcal{O}^\infty(X). \tag{3.109}$$

The latter is, of course, an arithmetic analogue of the *Cartan connection* (or *total derivative*, or *Cartan distribution*) in classical differential geometry; cf. 2.232.

DEFINITION 3.115. The map 3.109 is called the *Cartan connection.*

REMARK 3.116. The prolongation sequence 3.107 has the following universality property: for any prolongation sequence S^* over A with S^n p-adically complete and for any A-algebra homomorphism $u : \mathcal{O}(\widehat{X}) \to S^0$ there exists a unique morphism of prolongation sequences $\mathcal{O}(J^*(X)) \to S^*$ over A such that $u^0 = u$.

DEFINITION 3.117. For X of finite type but not necessarily affine we define the *p-jet space* of X by

$$J^n(X) = \bigcup J^n(X_i), \tag{3.110}$$

where $X = \bigcup X_i$ is an affine cover.

The fact that $J^n(X_i)$ naturally glue together follows from the fact that we have natural isomorphisms

$$J^n(X_j) \times_{X_j} (X_i \cap X_j) \simeq J^n(X_i \cap X_j) \simeq J^n(X_i) \times_{X_i} (X_i \cap X_j); \tag{3.111}$$

this was proved in [**23, 35**].

REMARK 3.118.

(1) The spaces $J^n(X)$ only depend on $\widehat{X}$ and not on X; so they can be defined for X a p-formal scheme of finite type as well.

(2) There are natural A-morphisms of p-formal schemes

(3.112) $$\pi : J^n(X) \to J^{n-1}(X)$$

induced by the identity of $A\{x\}$ and also ϕ-linear morphisms of p-formal schemes

(3.113) $$\pi_\phi : J^n(X) \to J^{n-1}(X)$$

and homomorphisms

(3.114) $$\phi : \mathcal{O}(J^{n-1}(X)) \to \mathcal{O}(J^n(X))$$

induced by the morphism $\phi : A\{x\} \to A\{x\}$.

(3) If X/A is smooth then $J^n(X)$ is a smooth p-formal scheme over A; more precisely, if X possesses an étale map

$$X \to \mathbb{A}^d = Spec\ A[T], \quad T = \{T_1, ..., T_d\},$$

then the natural morphism

(3.115) $$\mathcal{O}(X)[T', ..., T^{(n)}]\hat{} \to \mathcal{O}(J^n(X)), \ T' \mapsto \delta T, \ T'' \mapsto \delta^2 T, \ \text{etc.}$$

is an isomorphism.

(4) The universality property yields natural maps on sets of R-points

(3.116) $$\nabla^n : X(R) \to J^n(X)(R).$$

For $X = \mathbb{A}^m$ (the affine space), $J^n(X) = \mathbb{A}^{m(n+1)}$ and

$$\nabla^n(a) = (a, \delta a, ..., \delta^n a).$$

(5) Let X be smooth over $A = R$ and write

$$X_0^n := J^n(X) \otimes k, \quad X_0^\infty = \varprojlim X_0^n.$$

Then we have a natural bijection induced by the ∇^n's

(3.117) $$X(R) \simeq X_0^\infty(k).$$

By the way X_0^∞ coincides with the *Greenberg transform* of X. So our p-jet spaces $J^n(X)$ can be viewed as lifts to characteristic zero of the Greenberg transforms; the p-derivations on our p-jet spaces "do not survive," of course, after reduction mod p. For all of this we refer to [**24, 35**].

DEFINITION 3.119. Let X, Y be schemes of finite type over A. By a *δ-map* of order n, denoted by

$$f : X \to Y,$$

we understand a map of p-formal schemes

$$J^n(X) \to J^0(Y) = \widehat{Y}.$$

Two δ-maps $X \to Y$ and $Y \to Z$ of orders n and m respectively can be composed to yield a δ-map of order $n+m$: indeed one can consider the composition

$$J^{n+m}(X) \to J^m(Y) \to \widehat{Z}$$

where $J^{n+m}(X) \to J^m(Y)$ is induced from $J^n(X) \to \widehat{Y}$ via the universality property.

Any δ-map $f : X \to Y$ induces a set theoretic map

$$f_* : X(R) \to Y(R)$$

defined by

$$f_*(P) = f(\nabla^n(P)).$$

If X, Y are smooth the map f_* uniquely determines the map f and, in this case, we simply write f instead of f_* (and X, Y instead of $X(R), Y(R)$).

REMARK 3.120.

(1) A δ-map $X \to Y$ of order zero is nothing but a map of p-formal schemes

$$\widehat{X} \to \widehat{Y}.$$

(2) The functors J^n commute with products and send group objects into group objects.

DEFINITION 3.121. A *δ-homomorphism* $f : G \to H$ between two group schemes (or group p-formal schemes) is a group homomorphism

$$J^n(G) \to J^0(H) = \widehat{H}.$$

We end by discussing the formalism of flows and prime integrals, following [**42**]. Let X be a an affine smooth scheme over A.

DEFINITION 3.122.

(1) A system of *arithmetic differential equations* of order r is simply a subset $\mathcal{E}$ of $\mathcal{O}(J^r(X))$. We denote by $Z^r(\mathcal{E})$ the closed formal subscheme of $J^r(X)$ whose ideal is generated by $\mathcal{E}$.
(2) By a *solution* of $\mathcal{E}$ we mean an R-point $P \in X$ such that $f(P) = 0$ for all $f \in \mathcal{E}$. We denote by $Sol(\mathcal{E}) \subset X$ the set of solutions of $\mathcal{E}$.
(3) By a *prime integral* of $\mathcal{E}$ we mean a function $\mathcal{H} \in \mathcal{O}(\widehat{X})$ such that

$$\delta(\mathcal{H}(P)) = 0, \quad \text{for all} \quad P \in Sol(\mathcal{E}). \tag{3.118}$$

(Intuitively $\mathcal{H}$ is "constant along the solutions of $\mathcal{E}$.")
(4) Let δ^X be a p-derivation on $\mathcal{O}(\widehat{X})$. The *δ^A-flow* (or simply *δ-flow*) associated to δ^X is the ideal $\mathcal{E}(\delta^X)$ in $\mathcal{O}(J^1(X))$ generated by all elements of the form

$$\delta f - \delta^X f$$

where $f \in \mathcal{O}(\widehat{X})$. Here δf is the image of f under the "universal" p-derivation

$$\delta : \mathcal{O}(\widehat{X}) \to \mathcal{O}(J^1(X))$$

while $\delta^X f$ is the image of f under the p-derivation

$$\delta^X : \mathcal{O}(\widehat{X}) \to \mathcal{O}(\widehat{X}) \subset \mathcal{O}(J^1(X)).$$

(5) By a *δ-flow* on X we will understand an ideal in $\mathcal{O}(J^1(X))$ of the form $\mathcal{E}(\delta^X)$ where δ^X is some p-derivation on $\mathcal{O}(\widehat{X})$.

REMARK 3.123. If we denote by $\mathcal{F}(X)$ the set of δ-flows on X then we have natural bijections

$$\begin{array}{rcl} \mathcal{F}(X) & := & \{\delta\text{-flows on } X\} \\ & \simeq & \{p\text{-derivations on } \widehat{X}\} \\ & \simeq & \{p\text{-adic connections on } X\} \\ & \simeq & \{\text{Frobenius lifts on } \widehat{X}\} \\ & \simeq & \{\text{sections of } J^1(X) \to \widehat{X}\} \end{array}$$

If δ^X is a p-derivation on $\mathcal{O}(\widehat{X})$ and

$$\sigma : \widehat{X} \to J^1(X)$$

is the corresponding section then the ideal $\mathcal{E}(\delta^X)$ equals the ideal of the image of σ.

LEMMA 3.124. *Let $\mathcal{O}(X) = (R[x_1, ..., x_n]/I)_g$ for indeterminates $x_1, ..., x_n$ and some $g \in R[x_1, ..., x_n]$ and let us continue to denote by $x_i \in \mathcal{O}(\widehat{X})$ the images of x_i. Then the elements*

$$\delta x_i - \delta^X x_i, \quad i = 1, ..., n \tag{3.119}$$

generate the ideal $\mathcal{E}(\delta^X)$. In particular

$$Sol(\mathcal{E}(\delta^X)) = Sol(\delta x_1 - \delta^X x_1, ..., \delta x_n - \delta^X x_n).$$

PROOF. Denote by $\mathcal{I}$ the ideal of $\mathcal{O}(J^1(X))$ generated by the elements 3.119 and consider the set

$$\mathcal{A} = \{f \in \mathcal{O}(\widehat{X}); \delta f - \delta^X f \in \mathcal{I}\}.$$

Then $\mathcal{A}$ is easily checked to be a subring of $\mathcal{O}(\widehat{X})$ and it clearly contains R and x_i for all i. It also contains the image of $1/g$ because

$$\delta\left(\frac{1}{g}\right) - \delta^X\left(\frac{1}{g}\right) = -\frac{\delta g - \delta^X g}{(g^p + p\delta g)(g^p + p\delta^X g)}.$$

By Noetherianity $\mathcal{I}$ is p-adically closed in $\mathcal{O}(J^1(X))$ so $\mathcal{A}$ is p-adically closed in $\mathcal{O}(X)$; so it must coincide with $\mathcal{O}(\widehat{X})$. □

DEFINITION 3.125. An element $\mathcal{H} \in \mathcal{O}(\widehat{X})$ is a *prime integral* for a p-derivation δ^X on $\widehat{X}$ if $\delta^X \mathcal{H} = 0$. Alternatively we say $\mathcal{H}$ us a δ^X-*constant.*

LEMMA 3.126. *If $\mathcal{H} \in \mathcal{O}(\widehat{X})$ is a prime integral for δ^X then $\mathcal{H}$ is a prime integral for $\mathcal{E}(\delta^X)$.*

PROOF. Indeed if $P \in Sol(\mathcal{E}(\delta^X))$ then

$$0 = (\delta\mathcal{H} - \delta^X\mathcal{H})(P) = (\delta\mathcal{H})(P) - (\delta^X\mathcal{H})(P) = (\delta\mathcal{H})(P).$$

□

In what follows, in order to model arithmetic analogues of order 2 systems of differential equations, we choose a smooth affine scheme Y and apply some of the constructions discussed above to $X := J^1(Y)$. In this case one can define a special class of δ-flows on $J^1(Y)$ which will be called *canonical δ-flows.*

DEFINITION 3.127. A *canonical δ-flow* on $J^1(Y)$ is a δ-flow $\mathcal{E}(\delta^{J^1(Y)})$ with the property that the composition of $\delta^{J^1(Y)} : \mathcal{O}(J^1(Y)) \to \mathcal{O}(J^1(Y))$ with the pull back map $\varphi_0 = \pi : \mathcal{O}(Y) \to \mathcal{O}(J^1(Y))$ equals the universal p-derivation $\delta : \mathcal{O}(Y) \to \mathcal{O}(J^1(Y))$.

REMARK 3.128. Notice that in view of the universality property of p-jet spaces, one gets a natural closed embedding

$$\iota : J^2(Y) \to J^1(J^1(Y)).$$

This induces an injective map (which we view as an identification) from the set of sections of

$$J^2(Y) \to J^1(Y)$$

to the set of sections of

$$J^1(J^1(Y)) \to J^1(Y).$$

The sections of

$$J^2(Y) \to J^1(Y)$$

are in a natural bijection with canonical δ-flows on $J^1(Y)$ whereas the sections of

$$J^1(J^1(Y)) \to J^1(Y)$$

are in a bijection with (not necessarily canonical) δ-flows on $J^1(Y)$.

DEFINITION 3.129. Consider a system of arithmetic differential equations of order 2, $\mathcal{F} \subset \mathcal{O}(J^2(Y))$. We say $\mathcal{F}$ *defines a δ-flow* on $J^1(Y)$ if the map

$$Z^2(\mathcal{F}) \to J^1(Y)$$

is an isomorphism.

REMARK 3.130. In this case then $Z^2(\mathcal{F}) \to J^1(Y)$ defines a section of

$$J^2(Y) \to J^1(Y)$$

and hence a canonical δ-flow $\mathcal{E}(\delta^{J^1(Y)})$ on $J^1(Y)$ such that

$$\iota(Z^2(\mathcal{F})) = Z^1(\mathcal{E}(\delta^{J^1(Y)})).$$

REMARK 3.131.

(1) The differential algebra counterpart of the above definition yields the natural concept of flow on the (co)tangent space defining a second order differential equation.
(2) For our applications to arithmetic Painlevé VI equations the above concept of *canonical flow* will have to be slightly generalized.

3.7. Arithmetic Lie algebras and arithmetic logarithmic derivative

In this section we introduce arithmetic analogues of some basic Lie theoretic concepts in the case of "abstract" group schemes. The "concrete" case of classical groups will be analyzed later. We continue to observe the conventions at the beginning of section 3.6: we only consider one prime p, we use the symbol $\widehat{\ }$ for p-adic completion, and we are working with a δ-ring of local type $A \subset R$ as in 3.102.

DEFINITION 3.132. Let G be a smooth group scheme (or a group p-formal scheme) over A. The *δ-Lie algebra* of G is the group p-formal scheme $L_\delta(G)$ over A given by:

$$L_\delta(G) := Ker(J^1(G) \to J^0(G)). \tag{3.120}$$

We usually denote by

$$(a, b) \mapsto ab, \quad a \mapsto a^{-1}$$

the multiplication and the inverse in G and $J^n(G)$ and by

$$(a, b) \mapsto a +_\delta b, \quad a \mapsto -_\delta a$$

the multiplication and the inverse in $L_\delta(G)$. So

$$a +_\delta b = ab, \quad -_\delta a = a^{-1} \quad \text{for} \quad a, b \in L_\delta(G).$$

REMARK 3.133. This construction is obviously functorial in the following sense: if $f : G \to G'$ is a homomorphism of group schemes (or group p-formal schemes) then we have an induced homomorphism of group p-formal schemes

$$L_\delta(f) : L_\delta(G) \to L_\delta(G'). \tag{3.121}$$

REMARK 3.134. Recall that the *Lie algebra* $L(G)$ of G is the vector group scheme

$$L(G) = \text{Ker}(TG \to G)$$

where TG is the tangent bundle of G, $(TG)(C) = G(C[\epsilon])$, $\epsilon^2 = 0$, C any A-algebra. Then we have a (non-canonical!) isomorphism of p-formal schemes

$$L_\delta(G) \simeq \widehat{L(G)};$$

cf. [**23**]. This is not an isomorphism of groups in general; indeed $L(G)$ is always commutative whereas $L_\delta(G)$ is commutative if and only if G is commutative.

In what follows we continue to freely use our convention that if X is a scheme (or a p-formal scheme) of finite type over A we denote its set $X(R)$ of R-points simply by X; in particular we write $a \in X$ instead of $a \in X(R)$. We also identify, as usual, the δ-maps $X \to Y$ with the maps between the sets of R-points. Note that if X is a smooth R-scheme then

$$\widehat{X}(R) \simeq X(R).$$

So, according to our conventions, both these sets will sometimes be denoted simply by X.

In what follows we define the *δ-adjoint action.* Consider the morphism

$$G \times G \to G \tag{3.122}$$

which on R-points acts as

$$(g, h) \mapsto ghg^{-1}, \quad g, h \in G.$$

We get an obvious induced morphism of p-formal schemes (i.e. δ-map of order 0)

$$J^1(G) \times J^1(G) \to J^1(G) \tag{3.123}$$

which, by restriction, yields a δ-map of order 0

$$\star : J^1(G) \times L_\delta(G) \to L_\delta(G). \tag{3.124}$$

The latter is, of course, an action that can be referred to as the *adjoint action.* Now the identity $id : J^1(G) \to J^1(G)$ defines a δ-homomorphism of order 1, $\nabla^1 = id_*$,

$$\nabla^1 : G \to J^1(G). \tag{3.125}$$

We get an induced δ-map of order 1,

$$\nabla^1 \times id : G \times L_\delta(G) \to J^1(G) \times L_\delta(G). \tag{3.126}$$

DEFINITION 3.135. The δ-map of order 1

$$\star_\delta : G \times L_\delta(G) \to L_\delta(G) \tag{3.127}$$

obtained by composing 3.126 with 3.124 is called the *δ-adjoint action.*

This map is indeed an action. By the way one has a similar construction in classical differential geometry; in this case the resulting action has order 0 and is the usual adjoint action.

Here are two other basic actions. Denote by $\mathcal{F}(G)$ the set of δ-flows on G; we identify δ-flows on G with sections $\sigma : \widehat{G} \to J^1(G)$ in the category of p-formal schemes of the projection $J^1(G) \to \widehat{G}$. (N.B. σ are not group homomorphisms in general!) Then $\mathcal{F}(G)$ is acted upon, via left multiplication, by the group

$$L_\delta(G)(\widehat{G})$$

of p-formal scheme maps $\widehat{G} \to L_\delta(G)$; of course $\mathcal{F}(G)$ is a principal homogeneous space for this action so it is either empty or in bijection with $L_\delta(G)(\widehat{G})$. On the other hand the group $L_\delta(G) = L_\delta(G)(R)$ acts via left multiplication on $\mathcal{F}(G)$: if $\alpha \in L_\delta(G)$ and $\sigma \in \mathcal{F}(G)$, $\sigma : \widehat{G} \to J^1(G)$, then the action is denoted by

$$(\alpha, \sigma) \mapsto \sigma^\alpha$$

and is defined by

$$\sigma^\alpha(v) = \alpha \cdot \sigma(v).$$

So if one fixes, once and for all, a section $\sigma^G \in \mathcal{F}(G)$ then we have a natural identification

$$L_\delta(G)\backslash L_\delta(G)(\widehat{G}) \simeq L_\delta(G)\backslash \mathcal{F}(G). \tag{3.128}$$

The orbits of $L_\delta(G)$ on $L_\delta(G)(\widehat{G})$ and $\mathcal{F}(G)$ respectively will be referred to as *linear equivalence classes.*

There is also a right action of $G = G(R)$ on $\mathcal{F}(G)$ defined as follows: for any $u \in G$ and any $\sigma \in \mathcal{F}(G)$, $\sigma : \widehat{G} \to J^1(G)$, the action is denoted by

$$(u, \sigma) \mapsto \sigma^{\star u}$$

and is given by

$$\sigma^{\star u}(v) = \nabla^1(u)^{-1} \cdot \sigma(uv).$$

Assume now in addition that G is affine. Then, by smoothness, there exists a Frobenius lift $\phi^G : \mathcal{O}(\widehat{G}) \to \mathcal{O}(\widehat{G})$ extending $\phi = \phi^A : A \to A$ (and not necessarily compatible with the comultiplication) hence a Frobenius lift ϕ^G on $\widehat{G}$; one can attach to ϕ^G the p-derivation $\delta^G : \mathcal{O}(\widehat{G}) \to \mathcal{O}(\widehat{G})$ defined by

$$\delta^G f = p^{-1}(\phi^G(f) - f^p).$$

The latter induces a section $\sigma^G : \widehat{G} \to J^1(G)$ of the projection $J^1(G) \to \widehat{G}$ in the category of p-formal schemes. Fix from now on ϕ^G (equivalently δ^G or σ^G.) We may then consider the δ-map of order 0,

$$\{\ ,\ \}_G : \widehat{G} \times \widehat{G} \to L_\delta(G)$$

which on R-points is given by:

$$\{\ ,\ \}_G : G \times G \to L_\delta(G), \quad \{a, b\}_G = \sigma^G(a)\sigma^G(b)\sigma^G(ab)^{-1}.$$

Recall that by a δ^G-*horizontal* (or ϕ^G-*horizontal*) subgroup scheme of G we understand a smooth closed subgroup scheme $H \subset G$ such that the ideal I_H of $\widehat{H}$ in $\mathcal{O}(\widehat{G})$ satisfies $\delta^G(I_H) \subset I_H$; the latter is equivalent to $\phi^G(I_H) \subset I_H$ (due to the flatness of H). The indices G in $\phi^G, \delta^G, \sigma^G, \{\ ,\ \}_G$ are not meant to indicate that these objects are attached in any natural way to G; rather the index is meant to suggest that these objects are to be considered as part of the data whenever G is being considered. Dropping the index G from ϕ^G, δ^G may create confusion even if the reference to G is clear, because ϕ and δ have other meanings; so we will drop the index G from ϕ^G, δ^G only reluctantly; on the other hand we will usually drop the index G from $\sigma^G, \{\ ,\ \}_G$.

DEFINITION 3.136. A δ-map $f : G \to L_\delta(G)$ is called a *skew δ-cocycle* if it satisfies

$$f(ab) = (a \star_\delta f(b)) \cdot f(a) \cdot \{a, b\}$$

for all $a, b \in G$. We say f is *δ-coherent* if for any δ^G-horizontal closed subgroup scheme $H \subset G$ we have that $f(H) \subset L_\delta(H)$.

If f is a fixed skew δ-cocycle then we set

$$G^f = \{b \in G; f(b) = 1\}.$$

This is not a subgroup of G in general. For all $a \in G$ and all $b \in G^f$ with $\{a, b\} = 1$ we have $f(ab) = f(a)$.

DEFINITION 3.137. The map $l\delta = l\delta^G : G \to L_\delta(G)$, given on R-points by

$$l\delta(a) = \nabla^1(a)\sigma(a)^{-1}, \ a \in G$$

(operations performed in $J^1(G)$) is called the *arithmetic logarithmic derivative* attached to ϕ^G (or to δ^G or to σ^G).

The map $l\delta$ above is a δ-map of order 1 and is easily seen to be a skew δ-coherent δ-cocycle; we view it as an analogue of the logarithmic derivative 2.108. By the way for any $b \in G^{l\delta}$ with $\{a, b\} = 1$ we have $l\delta(ab) = l\delta(a)$. Note also that the map of p-formal schemes

$$l\delta^1 : J^1(G) \to L_\delta(G)$$

inducing the δ-map $l\delta : G \to L_\delta(G)$ is given by

$$l\delta^1(g) = g \cdot \sigma(\pi(g))^{-1}.$$

Indeed, clearly, $l\delta^1(\nabla^1(a)) = l\delta(a)$ i.e. $(l\delta^1)_* = l\delta$.

DEFINITION 3.138. Let G be a smooth affine group scheme equipped with a Frobenius lift ϕ^G and let $\alpha \in L_\delta(G)$. Then the equality

$$l\delta(u) = \alpha$$

will be referred to as a δ^G-*linear equation* with unknown $u \in G$.

REMARK 3.139. The arithmetic logarithmic derivatives

$$l\delta : G \to L_\delta(G)$$

attached to various Frobenius lifts ϕ^G all derive from a certain δ-map

$$l^1\delta : J^1(G) \to L_\delta(G)$$

which we describe in what follows. Let G be any smooth group scheme over A and denote by

$$\pi : J^1(G) \to G$$

the canonical projection. One can consider the δ-map, of order 1,

$$l^1\delta : J^1(G) \to J^1(G)$$

defined on points by

$$l^1\delta a := \nabla^1(\pi(a)) \cdot a^{-1}, \quad \text{for} \quad a \in J^1(G).$$

Clearly this map takes values in $L_\delta(G)$ so it induces a δ-map of order 1

$$l^1\delta : J^1(G) \to L_\delta(G).$$

The following are trivial to check:

(1) The map $l^1\delta : J^1(G) \to L_\delta(G)$ is a δ-cocycle for the $\star$-action in the sense that

$$l^1\delta(ab) = (l^1\delta a) \cdot (a \star (l^1\delta b)), \quad a, b \in J^1(G).$$

(2) For $a, b \in J^1(G)$ one has $l^1\delta a = l^1\delta b$ if and only if there exists $c \in G$ such that $ab^{-1} = \nabla^1 c$. In particular the composition of

$$l^1\delta : J^1(G) \to L_\delta(G)$$

with the map

$$\nabla^1 : G \to J^1(G)$$

is the constant map with value $1 \in L_\delta(G)$.

(3) The composition of

$$l^1\delta : J^1(G) \to L_\delta(G)$$

with any section

$$s : G \to J^1(G)$$

of the projection

$$\pi : J^1(G) \to G$$

equals the arithmetic logarithmic derivative

$$l\delta : G \to L_\delta(G)$$

attached to the Frobenius lift defined by s.

(4) The composition of

$$l^1\delta : J^1(G) \to L_\delta(G)$$

with the inclusion $L_\delta(G) \subset J^1(G)$ is the antipode

$$L_\delta(G) \to L_\delta(G), \qquad a \mapsto a^{-1}.$$

(5) The map of formal schemes

$$l^1\delta^1 : J^1(J^1(G)) \to L_\delta(G)$$

defining the δ-map

$$l^1\delta : J^1(G) \to L_\delta(G)$$

(i.e. with the property that $l^1\delta = (l^1\delta^1)_*$) can be described as follows. Let

$$\pi_G^1 : J^1(G) \to J^0(G)$$

denote the natural projection. Then one may consider the homomorphism

$$\pi_G^{11} := \pi_{J^1(G)}^1 \times J^1(\pi_G^1) : J^1(J^1(G)) \to J^1(G) \times_{J^0(G)} J^1(G). \tag{3.129}$$

On the other hand the quotient map

$$J^1(G) \times J^1(G) \to J^1(G), \quad (a,b) \mapsto ba^{-1}$$

induces a morphism

$$J^1(G) \times_{J^0(G)} J^1(G) \to L_\delta(G). \tag{3.130}$$

Then the map $l^1\delta^1$ is obtained by composing the map 3.130 above with the morphism π_G^{11} defined in 3.129. It was checked in [**41**] that

$$(l^1\delta^1)^{-1}(1) = J^2(G).$$

Moreover the restriction of $l^1\delta^1$ to $L_\delta(J^1(G))$ is obtained by composing the map

$$L_\delta(\pi_G^1) : L_\delta(J^1(G)) \to L_\delta(G)$$

(induced by $J^1(\pi_G^1)$) with the antipode map

$$L_\delta(G) \to L_\delta(G).$$

(6) For any section

$$s : \widehat{G} \to J^1(G)$$

of the projection

$$\pi_G^1 : J^1(G) \to \widehat{G}$$

the corresponding arithmetic logarithmic derivative

$$l\delta : G \to L_\delta(G)$$

is induced by the map

$$l^1\delta^1 \circ J^1(s) : J^1(G) \to J^1(J^1(G)) \to L_\delta(G);$$

in other words,

$$(l^1\delta^1 \circ J^1(s))_* = l\delta : G \to L_\delta(G).$$

3.8. Compatibility with translations and involutions

In this section we continue our investigation of the one prime case; we continue to use the notation introduced in the beginning of section 3.6. In particular $A \subset R$ is a δ-ring of local type as in 3.102. For smooth schemes X over A we write X instead of $X(R)$. We discuss here compatibility of Frobenius lifts with:

1. left and right translation by subgroups;
2. "outer" automorphisms and their symmetric spaces;
3. "inner" automorphisms and conjugacy classes.

We start with the first type of compatibility.

DEFINITION 3.140. Let G be a smooth group scheme over A and let $H \subset G$ be a smooth closed subgroup scheme. Let δ^G be a p-derivation on $\widehat{G}$ with attached Frobenius lift ϕ^G. We say that ϕ^G (or δ^G) is *left invariant* with respect to H if H is δ^G-horizontal and the following diagram is commutative:

$$\begin{array}{ccc} \widehat{H}\times\widehat{G} & \to & \widehat{G} \\ \phi^H\times\phi^G\downarrow & & \downarrow\phi^G \\ \widehat{H}\times\widehat{G} & \to & \widehat{G} \end{array} \tag{3.131}$$

where $\phi^H : \widehat{H} \to \widehat{H}$ is induced by ϕ^G and the horizontal maps are given by multiplication. We say that ϕ^G (or δ^G) is *right invariant* with respect to H if H is δ^G-horizontal and the following diagram is commutative:

$$\begin{array}{ccc} \widehat{G}\times\widehat{H} & \to & \widehat{G} \\ \phi^G\times\phi^H\downarrow & & \downarrow\phi^G \\ \widehat{G}\times\widehat{H} & \to & \widehat{G} \end{array} \tag{3.132}$$

We say that ϕ^G is *bi-invariant* with respect to H if it is both left and right invariant with respect to H. Note that the vertical maps in the diagrams 3.131 and 3.132 are morphisms of formal schemes over $\mathbb{Z}_p$ and not necessarily over A!

REMARK 3.141. By a *split torus* over A we will understand a group scheme of the form

$$T = Spec\ A[t_1, t_1^{-1}, ..., t_n, t_n^{-1}] \tag{3.133}$$

with t_i group like elements, i.e. elements defining multiplicative characters. There is exactly one Frobenius lift ϕ^T on $\widehat{T}$ which is bi-invariant with respect to T. This ϕ^T acts as $\phi^T(t_i) = t_i^p$. Indeed this can be proved as follows. Any Frobenius lift on $\widehat{T}$ which is bi-invariant with respect to T can be easily shown to send the identity of T into itself and hence it induces an endomorphism of the formal group law $\mathcal{G}$ of T over A, $\mathcal{G} = (\mathcal{G}_i)$,

$$\mathcal{G}_i(t,s) = t_i + s_i + t_i s_i.$$

On the other hand, if $\overline{\mathcal{F}}$ over $\overline{A} = A/pA$ is the reduction mod p of a one dimensional formal group $\mathcal{F}$ over A and $\overline{\mathcal{F}}$ is not $\overline{A}$-isomorphic to the additive formal group then the ring homomorphism

$$\mathrm{End}_A(\mathcal{F}) \to \mathrm{End}_{\overline{A}}(\overline{\mathcal{F}})$$

is known to be injective [**97**]; in particular, for $\mathcal{F}$ the multiplicative one dimensional formal group over A and $\mathcal{G} = \mathcal{F}^n$ we have that

$$\mathrm{End}_A(\mathcal{G}) \to \mathrm{End}_{\overline{A}}(\overline{\mathcal{G}})$$

is injective and we are done.

REMARK 3.142. By a *finite constant group scheme* over A we will understand a group scheme which is a disjoint union W of schemes *Spec* A indexed by some finite abstract group, with the obvious structure of group scheme. For any finite constant group scheme W there is exactly one Frobenius lift on

$$\mathcal{O}(\widehat{W}) = \widehat{A} \times ... \times \widehat{A}, \tag{3.134}$$

hence on $\widehat{W}$. Clearly this Frobenius lift is bi-invariant with respect to W.

REMARK 3.143. More generally one can consider semidirect products N of a split torus T and a finite constant group scheme W acting on T. Then there is exactly one Frobenius lift on $\widehat{N}$ that is bi-invariant with respect to N.

LEMMA 3.144. *Assume one of the following holds:*

(1) ϕ^G *is left invariant with respect* H, $a \in H$, $b \in G$;
(2) ϕ^G *is right invariant with respect to* H, $a \in G$, $b \in H$;

Then $\{a, b\}_G = 1$ *and, in particular, if* $l\delta : G \to L_\delta(G)$ *is the logarithmic derivative attached to* δ^G *then*

$$l\delta(ab) = (a \star_\delta l\delta(b)) \cdot l\delta(a).$$

PROOF. Clear from the definitions. □

Next we discuss compatibility with automorphisms.

Recall the following setting from section 2.3. Let G be an affine smooth group scheme over A. For a group scheme automorphism $\tau : G \to G$, written on points as $x \mapsto x^\tau$, one can attach the closed subgroup scheme

$$G^+ \subset G,$$

of fixed points of $x \mapsto x^\tau$ and the closed subscheme

$$G^- \subset G,$$

of fixed points of $x \mapsto x^{-\tau} := (x^\tau)^{-1}$. Cf. 2.92, 2.93.

Also one may consider the *quadratic map*

$$\mathcal{H} : G \to G, \quad \mathcal{H}(x) = x^{-\tau}x$$

and the *bilinear map*

$$\mathcal{B} : G \times G \to G, \quad \mathcal{B}(x, y) = x^{-\tau}y$$

defined by τ. Cf. 2.94, 2.96.

Finally, for $g \in G(A)$, one may consider the maps

$$\mathcal{H}_g : G \to G, \quad \mathcal{H}_g(x) = gx^{-\tau}x$$

and

$$\mathcal{B}_g : G \times G \to G, \quad \mathcal{B}_g(x, y) = gx^{-\tau}y.$$

Cf. 2.95 and 2.97 respectively.

The identity component

$$S := (G^+)^\circ$$

is called the *subgroup defined by* τ. On the other hand G^- is not a subgroup of G; recall, however, that G^- is stable under the maps

$$G \to G, \quad a \mapsto a^\nu$$

for all $\nu \in \mathbb{Z}$.

Consider now Frobenius lifts ϕ^G and ϕ_0^G on $\widehat{G}$.

DEFINITION 3.145. We say that ϕ^G is $\mathcal{H}_g$*-horizontal* with respect to ϕ_0^G if the following diagram is commutative:

$$
\begin{array}{ccc}
\widehat{G} & \stackrel{\phi^G}{\longrightarrow} & \widehat{G} \\
\mathcal{H}_g \downarrow & & \downarrow \mathcal{H}_g \\
\widehat{G} & \stackrel{\phi_0^G}{\longrightarrow} & \widehat{G}
\end{array}
\tag{3.135}
$$

We say that ϕ^G is $\mathcal{B}_g$*-symmetric* with respect to ϕ_0^G if the following diagram is commutative:

$$
\begin{array}{ccc}
\widehat{G} & \stackrel{\phi_0^G\times\phi^G}{\longrightarrow} & \widehat{G}\times\widehat{G} \\
\phi^G\times\phi_0^G \downarrow & & \downarrow \mathcal{B}_g \\
\widehat{G}\times\widehat{G} & \stackrel{\mathcal{B}_g}{\longrightarrow} & \widehat{G}
\end{array}
\tag{3.136}
$$

We say that ϕ^G is $(\mathcal{H}_g,\mathcal{B}_g)$*-horizontal* with respect to ϕ_0^G if the following diagram is commutative:

$$
\begin{array}{ccc}
\widehat{G} & \stackrel{\phi_0^G\times\phi^G}{\longrightarrow} & \widehat{G}\times\widehat{G} \\
\mathcal{H}_g \downarrow & & \downarrow \mathcal{B}_g \\
\widehat{G} & \stackrel{\phi_0^G}{\longrightarrow} & \widehat{G}
\end{array}
\tag{3.137}
$$

The issue of existence of Frobenius lifts with the above properties will be addressed later. In what follows we make a series of remarks on these properties.

REMARK 3.146. Note that if ϕ^G is $\mathcal{H}$-horizontal with respect to ϕ_0^G and $\phi_0^G(1) = 1$ then $S := (G^+)^\circ$ is ϕ^G-horizontal and $\phi^G(1) \in S$; in particular there is an induced Frobenius lift ϕ^S on $\widehat{S}$.

REMARK 3.147. Assume in this remark that τ is an involution and $A = R$. Then recall that one gets an induced morphism $\mathcal{H} : G \to G^-$. Denote by an upper bar the functor $\otimes_{\mathbb{Z}}\mathbb{F}_p$. Assume in addition that G/A has connected geometric fibers and G^+ is smooth over A. Then by 2.101 we have an induced dominant morphism

$$\overline{\mathcal{H}} : \overline{G} \to (\overline{G^-})^\circ_s,$$

where the upper $\circ$ means "connected component passing through 1" and the lower s means "smooth locus." Since $(\overline{G^-})^\circ_s$ is an integral scheme the map

$$\mathcal{O}((\overline{G^-})^\circ_s) \to \mathcal{O}(\overline{G})$$

induced by $\overline{\mathcal{H}}$ is injective. Hence the map

$$\mathcal{O}((\widehat{G^-})^\circ_s) \to \mathcal{O}(\widehat{G})$$

induced by

$$\widehat{\mathcal{H}} : \widehat{G} \to (\widehat{G^-})^\circ_s$$

is injective. In particular, if there exists ϕ^G that is $\mathcal{H}$-horizontal with respect to ϕ_0^G then ϕ_0^G sends $(\widehat{G^-})^\circ_s$ into itself.

Here is an alternative approach to the commutativity of diagrams 3.135 and 3.136. Consider the map

$$\mathcal{Q} : G \times G \to G, \quad \mathcal{Q}(x,y) = x^{-1}y. \tag{3.138}$$

For any two ϕ-linear morphisms $f_1, f_2 : \widehat{G} \to \widehat{G}$ of p-formal schemes one can consider the ϕ-linear morphism $f_1 \times f_2 : \widehat{G} \to \widehat{G} \times \widehat{G}$ (where $\times = \times_A$) and hence the composition

$$f_1^{-1} \cdot f_2 := \widehat{\mathcal{Q}} \circ (f_1 \times f_2) : \widehat{G} \to \widehat{G} \times \widehat{G} \to \widehat{G}.$$

Similarly if $\mathcal{Q}' : G \times G \to G$, $\mathcal{Q}'(x, y) = xy^{-1}$ then we set

$$f_1 \cdot f_2^{-1} := \widehat{\mathcal{Q}'} \circ (f_1 \times f_2) : \widehat{G} \to \widehat{G} \times \widehat{G} \to \widehat{G}.$$

We have the following:

LEMMA 3.148. *Let G be an affine smooth group scheme over A and $g \in G(A)$. Let τ be an involution and let ϕ_0^G, ϕ^G be Frobenius lifts on $\widehat{G}$.*

(1) *ϕ^G is $\mathcal{B}_g$-symmetric with respect to ϕ_0^G if and only if*

$$(\phi^G \cdot (\phi_0^G)^{-1})^{-\tau} = \phi^G \cdot (\phi_0^G)^{-1}. \tag{3.139}$$

(2) *Assume ϕ^G is $\mathcal{H}_g$-horizontal with respect to ϕ_0^G. Then ϕ^G is $\mathcal{B}_g$-symmetric with respect to ϕ_0^G if and only if:*

$$((\phi_0^G)^{-1} \cdot \phi^G)^2 = (\mathcal{H}_g \circ \phi_0^G)^{-1} \cdot (\phi_0^G \circ \mathcal{H}_g). \tag{3.140}$$

(3) *ϕ^G is $(\mathcal{H}_g, \mathcal{B}_g)$-horizontal with respect to ϕ_0^G if and only if*

$$(\phi_0^G)^{-1} \cdot \phi^G = (\mathcal{H}_g \circ \phi_0^G)^{-1} \cdot (\phi_0^G \circ \mathcal{H}_g). \tag{3.141}$$

(4) *Assume ϕ_1^G is $(\mathcal{H}_g, \mathcal{B}_g)$-horizontal with respect to ϕ_0^G and assume ϕ_2^G is $\mathcal{H}_g$-horizontal and $\mathcal{B}_g$-symmetric with respect to ϕ_0^G. Then*

$$(\phi_0^G)^{-1} \cdot \phi_1^G = ((\phi_0^G)^{-1} \cdot \phi_2^G)^2. \tag{3.142}$$

REMARK 3.149. In the above statement the equalities in the displayed equations are equalities of maps $\widehat{G} \to \widehat{G}$. The symbol $(\phi^G \cdot (\phi_0^G)^{-1})^{-\tau}$ denotes, of course, the composition of

$$\phi^G \cdot (\phi_0^G)^{-1} : \widehat{G} \to \widehat{G}$$

with the map

$$\widehat{G} \to \widehat{G}, \quad x \mapsto x^{-\tau}.$$

Similarly

$$((\phi_0^G)^{-1} \cdot \phi^G)^2 = (\phi_0^G)^{-1} \cdot \phi^G \cdot (\phi_0^G)^{-1} \cdot \phi^G$$

is the composition of

$$(\phi_0^G)^{-1} \cdot \phi^G : \widehat{G} \to \widehat{G}$$

with the squaring map

$$\widehat{G} \to \widehat{G}, \quad x \mapsto x^2.$$

We use similar notation for the rest of the expressions.

PROOF OF LEMMA 3.148. To check assertion (1) note that diagram 3.136 is equivalent to

$$(\phi^G)^{-\tau} \cdot \phi_0^G = (\phi_0^G)^{-\tau} \cdot \phi^G,$$

which in its turn is equivalent to 3.139.

To check the rest of the assertions note first that we have

$$\mathcal{B}_g(x, y) = \mathcal{H}_g(x) \cdot \mathcal{Q}(x, y).$$

Hence, in particular,

$$\mathcal{B}_g \circ (\phi_0^G \times \phi^G) = (\mathcal{H}_g \circ \phi_0^G) \cdot ((\phi_0^G)^{-1} \cdot \phi^G). \tag{3.143}$$

To check assertion (2) note that

$$(3.144)\quad \mathcal{B}_g \circ (\phi^G \times \phi_0^G) = (\mathcal{H}_g \circ \phi^G) \cdot ((\phi^G)^{-1} \cdot \phi_0^G) = (\phi_0^G \circ \mathcal{H}_g) \cdot ((\phi^G)^{-1} \cdot \phi_0^G).$$

So, by 3.143 and 3.144, the diagram 3.136 is commutative if and only if equality 3.140 holds.

Assertion (3) follows directly from equation 3.143.

Assertion (4) follows from assertions (2) and (3). □

By the way, assertion (3) in Lemma 3.148 implies:

COROLLARY 3.150. *Let $\phi_0^G, \phi_1^G, \phi_2^G$ be Frobenius lifts on $\widehat{G}$ such that ϕ_1^G and ϕ_2^G are $(\mathcal{H}_g, \mathcal{B}_q)$-horizontal with respect to ϕ_0^G. Then $\phi_1^G = \phi_2^G$.*

We will next use Lemma 3.148 in combination with the following:

LEMMA 3.151. *Let $\mathcal{A}$ be a p-adically separated flat A-algebra, let $\nu \geq 1$ be an integer not divisible by p, and let $M_1, M_2 \in GL_n(\mathcal{A})$ two matrices such that $M_1^\nu = M_2^\nu$ and $M_1 \equiv M_2 \equiv 1$ mod p. Then $M_1 = M_2$.*

PROOF. We prove by induction on n that $M_1 \equiv M_2$ mod p^n. Indeed assume the latter and write $M_1 = M_2 + p^n M$. Then

$$M_2^\nu = M_1^\nu \equiv M_2^\nu + p^n \left(\sum_{i+j=\nu-1} M_2^i M M_2^j \right) \quad mod \quad p^{n+1}.$$

Since $M_2 \equiv 1$ mod p we get $p^n \nu M \equiv 0$ mod p^{n+1}. Since $\mathcal{A}$ is A-flat we get $\nu M \equiv 0$ mod p and hence, since $\nu \in R^\times$, we get $M \equiv 0$ mod p so $M_1 \equiv M_2$ mod p^{n+1}. □

In the next corollary recall that a group scheme is called *linear* if it is isomorphic to a closed subgroup scheme of GL_n.

COROLLARY 3.152. *Let G be an smooth linear group scheme over A. Let $\phi_0^G, \phi_1^G, \phi_2^G$ be Frobenius lifts on $\widehat{G}$ such that ϕ_1^G and ϕ_2^G are $\mathcal{H}_g$-horizontal and $\mathcal{B}_g$-symmetric with respect to ϕ_0^G. Then $\phi_1^G = \phi_2^G$.*

PROOF. By Lemma 3.148 we have

$$((\phi_0^G)^{-1} \cdot \phi_1^G)^2 = ((\phi_0^G)^{-1} \cdot \phi_2^G)^2.$$

We also have

$$(\phi_0^G)^{-1} \cdot \phi_1^G \equiv (\phi_0^G)^{-1} \cdot \phi_2^G \equiv 1 \quad mod \quad p.$$

View G as embedded into some GL_n. Interpreting the latter congruence as a congruence between matrices with entries in the ring $\mathcal{O}(\widehat{G})$ it follows from Lemma 3.151 that

$$(\phi_0^G)^{-1} \cdot \phi_1^G = (\phi_0^G)^{-1} \cdot \phi_2^G$$

and we are done. □

Here is another consequence of Lemma 3.148. Assume τ is an involution and consider the naturally induced involution $L_\delta(\tau)$ on $J^1(G)$; it preserves $L_\delta(G)$ so it induces an involution $L_\delta(\tau)$ on $L_\delta(G)$. When there is no danger of confusion we continue to denote by τ the involution $L_\delta(\tau)$. We also identify $L_\delta(G)$ with its set of R-points. Set, as usual,

$$L_\delta(G)^+ = \{b \in L_\delta(G); b^\tau = b\},$$
$$L_\delta(G)^- = \{b \in L_\delta(G); b^\tau = b^{-1}\}.$$

Again $L_\delta(G)^+$ is a subgroup of $L_\delta(G)$ but $L_\delta(G)^-$ is not a subgroup of $L_\delta(G)$ in general; however $L_\delta(G)^-$ is, again, stable under the maps

$$L_\delta(G) \to L_\delta(G), \quad b \mapsto \nu_\delta b := b^\nu$$

for all $\nu \in \mathbb{Z}$. Note also that $L_\delta(G^+)$ is a subgroup of $L_\delta(G)^+$. Let

$$\sigma^G, \sigma_0^G : G \to J^1(G)$$

be the sections defined by the Frobenius lifts ϕ^G, ϕ_0^G and let

$$l\delta, l\delta_0 : G \to L_\delta(G)$$

be the corresponding arithmetic logarithmic derivatives.

COROLLARY 3.153. *The following are equivalent:*

(1) ϕ^G *is* $\mathcal{B}$*-symmetric with respect to* ϕ_0^G.
(2) *The image of the map* $\phi^G \cdot (\phi_0^G)^{-1} : G \to G$ *is contained in* G^-.
(3) *The image of the map* $\sigma^G \cdot (\sigma_0^G)^{-1} : G \to L_\delta(G)$ *is contained in* $L_\delta(G)^-$.
(4) *The image of the* δ*-map* $(l\delta_0)^{-1} \cdot (l\delta) : G \to L_\delta(G)$ *is contained in* $L_\delta(G)^-$.

PROOF. The equivalence of (1) and (2) is a rephrasing of assertion (1) in Lemma 3.148. The rest of the Corollary is a trivial exercise. □

On the other hand here is a consequence of Lemma 3.151:

COROLLARY 3.154. *Let* G *be a smooth linear group scheme. The following hold:*

(1) *For all* $\nu \in \mathbb{Z}$ *not divisible by* p *the following maps are injective:*
$$L_\delta(G) \to L_\delta(G), \quad a \mapsto \nu_\delta a = a^\nu.$$
(2) *If* τ *is an involution of* G *then the multiplication map below is injective:*
$$L_\delta(G)^+ \times L_\delta^-(G) \to L_\delta(G), \quad (b, c) \mapsto b +_\delta c = bc.$$

PROOF. To check assertion (1) embed G into some GL_n. Then there is a natural injective homomorphism between groups of points, $\epsilon : L_\delta(G) \to GL_n$; it is given by the map

$$\mathcal{O}(\widehat{GL_n}) \to \mathcal{O}(\widehat{G}) \xrightarrow{\phi} \mathcal{O}(J^1(G)) \to \mathcal{O}(L_\delta(G)),$$

where $\phi(a) = a^p + p\delta a$, δ the universal p-derivation. Any matrix in the image of ϵ is $\equiv 1 \bmod p$. So if $a^\nu = b^\nu$ with $a, b \in L_\delta(G)$ then $\epsilon(a)^\nu = \epsilon(b)^\nu$. By Lemma 3.151 we get $\epsilon(a) = \epsilon(b)$ and hence $a = b$.

To check assertion (2) assume $b_1 c_1 = b_2 c_2$ with $b_i^\tau = b_i$ and $c_i^\tau = c_i^{-1}$. Set $b = b_2^{-1} b_1$. Then $b^\tau = b$ and $b = c_2 c_1^{-1}$. Hence

$$c_1 c_2^{-1} = b^{-1} = b^{-\tau} = (c_2 c_1^{-1})^{-\tau} = c_1^\tau c_2^{-\tau} = c_1^{-1} c_2.$$

Hence $c_1^2 = c_2^2$. By assertion (1) we get $c_1 = c_2$. Hence $b = 1$ and hence $b_1 = b_2$. □

DEFINITION 3.155. In the notation of Corollary 3.154, for any $a \in L_\delta(G)$ in the image of the map

$$L_\delta(G)^+ \times L_\delta^-(G) \to L_\delta(G) \tag{3.145}$$

we have a unique representation

$$a = a^+ +_\delta a^- := a^+ a^-$$

with $a^{\pm} \in L_{\delta}(G)^{\pm}$ which we shall refer to as the *Cartan decomposition* of a with respect to τ.

We will later show that for $G = GL_n$ and τ certain remarkable involutions the map 3.145 is a bijection; so in that case we will have that any element of $L_{\delta}(G)$ has a Cartan decomposition.

Next we discuss compatibility of Frobenius lifts with conjugation.

DEFINITION 3.156. Let G be a smooth affine group scheme over A and let

$$\mathcal{C} : G \times G \to G, \quad \mathcal{C}(h,g) = g^{-1}hg \tag{3.146}$$

be the conjugation map. Let H be a smooth closed subgroup scheme, and $G^* \subset G$ an open set which is invariant under the action of G on G by conjugation. Let $H^* = H \cap G^*$ and let

$$\mathcal{C} : H^* \times G \to G^* \tag{3.147}$$

be the induced map. Let ϕ_0^G be a Frobenius lift on $\widehat{G}$ and ϕ^{G^*} a Frobenius lift on $\widehat{G^*}$. We say that ϕ_0^G is $\mathcal{C}$*-horizontal* with respect to ϕ^{G^*} if H is ϕ_0^G-horizontal and, upon denoting by $\phi_0^{H^*}$ the Frobenius lift on $\widehat{H^*}$ induced by ϕ_0^G, the following diagram is commutative:

$$\begin{array}{ccc} \widehat{H^*} \times \widehat{G} & \stackrel{\phi_0^{H^*} \times \phi_0^G}{\longrightarrow} & \widehat{H^*} \times \widehat{G} \\ \mathcal{C} \downarrow & & \downarrow \mathcal{C} \\ \widehat{G^*} & \stackrel{\phi^{G^*}}{\longrightarrow} & \widehat{G^*} \end{array} \tag{3.148}$$

There is an easy relation between $\mathcal{C}$-horizontality and $\mathcal{H}$-horizontality as follows.

LEMMA 3.157. *Assume in Definition 3.156 that ϕ_0^G is $\mathcal{C}$-horizontal with respect to ϕ^{G^*}. Let $h \in H^*$ be ϕ_0^G-horizontal and let V_h be the p-adic completion of the Zariski closure of the image of $\mathcal{C}_h : G \to G^*$, $g \mapsto \mathcal{C}_h(g) = g^{-1}hg$. The following hold:*

1. *V_h is ϕ^{G^*}-horizontal.*
2. *Assume $G = GL_n$ and let τ is the automorphism $x^{\tau} = h^{-1}xh$ of G. Assume furthermore that V_h is closed in $\widehat{G}$. Then there exists a Frobenius lift ϕ^G on $\widehat{G}$ such that ϕ_0^G is $\mathcal{H}$-horizontal with respect to ϕ^G.*

PROOF. To prove assertion (1) note that by combining the commutative diagram 3.148 with the the commutative diagram

$$\begin{array}{ccc} \widehat{G} & \stackrel{\phi_0^G}{\longrightarrow} & \widehat{G} \\ h \times id \downarrow & & \downarrow h \times id \\ \widehat{H^*} \times \widehat{G} & \stackrel{\phi_0^{H^*} \times \phi_0^G}{\longrightarrow} & \widehat{H^*} \times \widehat{G} \end{array} \tag{3.149}$$

one gets a commutative diagram

$$\begin{array}{ccc} \widehat{G} & \stackrel{\phi_0^G}{\longrightarrow} & \widehat{G} \\ \mathcal{C}_h \downarrow & & \downarrow \mathcal{C}_h \\ \widehat{G^*} & \stackrel{\phi^{G^*}}{\longrightarrow} & \widehat{G^*} \end{array} \tag{3.150}$$

which immediately implies assertion (1). Let us check assertion (2). Putting together the commutative diagram 3.150 and the diagram

$$\begin{array}{ccc} \widehat{G^*} & \stackrel{\phi^{G^*}}{\longrightarrow} & \widehat{G^*} \\ L_h^{-1} \downarrow & & \downarrow L_h^{-1} \\ h^{-1}\widehat{G^*} & \stackrel{L_h^{-1}\phi^{G^*}L_h}{\longrightarrow} & h^{-1}\widehat{G^*} \end{array}$$

and noting that $L_h^{-1}\phi_{G^*}L_h$ is a Frobenius lift we get a commutative diagram

$$\begin{array}{ccc} \widehat{G} & \stackrel{\phi_0^G}{\longrightarrow} & \widehat{G} \\ \downarrow & & \downarrow \\ h^{-1}V_h & \stackrel{L_h^{-1}\phi^{G^*}L_h}{\longrightarrow} & h^{-1}V_h \end{array} \tag{3.151}$$

Now since V_h is closed in $\widehat{G}$ so is $h^{-1}V_h$ hence by Lemma 3.111 the bottom arrow of 3.151 extends to a Frobenius lift ϕ^G on $\widehat{G}$ and we are done. □

3.9. Arithmetic Lie brackets and exponential

We continue our discussion of arithmetic Lie theory by discussing brackets. We continue to work over a δ-ring of local type $A \subset R$. Let G be a smooth group scheme or group p-formal scheme over A.

DEFINITION 3.158. The *order r δ-Lie algebra* of G is the group p-formal scheme,

$$L_\delta^r(G) := Ker(\pi : J^r(G) \to J^{r-1}(G)) \tag{3.152}$$

where π is as in 3.112. In particular $L_\delta^1(G) = L_\delta(G)$.

We denote by $+_{\delta,r}$ the multiplication on $L_\delta^r(G)$. When we do not want to specify r we write $+_{\delta,*}$ instead of $+_{\delta,r}$. Note that, with previous notation, $+_{\delta,1} = +_\delta$.

Also we consider the composition

$$ex^r : L_\delta^r(G) \subset J^r(G) \stackrel{\pi_\phi}{\longrightarrow} J^{r-1}(G) \stackrel{\pi_\phi}{\longrightarrow} \dots \stackrel{\pi_\phi}{\longrightarrow} J^1(G) \stackrel{\pi_\phi}{\longrightarrow} \widehat{G}, \tag{3.153}$$

where π_ϕ is as in 3.113; ex^r can be thought of as an "exponential" and is a ϕ^r-linear morphism of p-formal schemes. It is clearly functorial in G with respect to homomorphisms of group (p-formal) schemes. As usual we still denote by ex^r : $L_\delta^r(G) \to G$ the map induced on sets of R-points. We also set $ex = ex^1$. In the next chapter we will prove the following:

PROPOSITION 3.159. *Let G be a smooth linear group scheme over A with commutator map $[\ ,\ ] : G \times G \to G$, $[a, b] = aba^{-1}b^{-1}$ and let $r, s \geq 1$.*

(1) *There exists a unique δ-map*

$$[\ ,\ ]_\delta : L_\delta^r(G) \times L_\delta^s(G) \to L_\delta^{r+s}(G)$$

of order $\max\{r, s\}$ *making the following diagram of sets of R-points commutative*

$$\begin{array}{ccc} L_\delta^r(G) \times L_\delta^s(G) & \stackrel{[\ ,\ ]_\delta}{\longrightarrow} & L_\delta^{r+s}(G) \\ ex^r \times ex^s \downarrow & & \downarrow ex^{r+s} \\ G \times G & \stackrel{[\ ,\ ]}{\longrightarrow} & G \end{array} \tag{3.154}$$

(2) *The δ-map $[\ ,\ ]_\delta$ is functorial in G with respect to homomorphisms $f : G \to G'$ of group schemes, i.e., for the induced maps $L^i_\delta(f) : L^i_\delta(G) \to L^i_\delta(G)$, $i = r, s$, we have*

$$[L^r_\delta(f)(\alpha), L^s_\delta(f)(\beta)]_\delta = L^{r+s}_\delta(f)([\alpha,\beta]_\delta).$$

(3) *If $\alpha_1, \alpha_2 \in L^r_\delta(G)$ and $\beta_1, \beta_2 \in L^s_\delta(G)$ are such that $\alpha_1 \equiv \alpha_2$ mod p^m and $\beta_1 \equiv \beta_2$ mod p^m then*

$$[\alpha_1, \beta_1]_\delta \equiv [\alpha_2, \beta_2]_\delta \quad mod\ \ p^m.$$

(4) *(Linearity) If $\alpha_1, \alpha_2 \in L^r_\delta(G)$, $\beta \in L^s_\delta(G)$ then*

$$[\alpha_1 +_{\delta,*} \alpha_2, \beta]_\delta \equiv [\alpha_1, \beta]_\delta +_{\delta,*} [\alpha_2, \beta]_\delta \quad mod\ \ p.$$

(5) *(Antisymmetry) If $\alpha \in L^r_\delta(G)$, $\beta \in L^s_\delta(G)$ then*

$$[\alpha, \beta]_\delta +_{\delta,*} [\beta, \alpha]_\delta \equiv 0 \ \ mod\ \ p.$$

(6) *(Jacobi identity) If $\alpha \in L^r_\delta(G)$, $\beta \in L^s_\delta(G)$, $\gamma \in L^t_\delta(G)$ then*

$$[[\alpha,\beta]_\delta, \gamma]_\delta +_{\delta,*} [[\beta,\gamma]_\delta, \alpha]_\delta +_{\delta,*} [[\beta,\gamma]_\delta, \alpha]_\delta \equiv 0 \ \ mod\ \ p.$$

(7) *Let $G = GL_n$ and let $[\ ,\ ] : \mathfrak{gl}_n \times \mathfrak{gl}_n \to \mathfrak{gl}_n$ be the commutator map $[a, b] = ab - ba$. Then for any $\alpha \in L^r_\delta(GL_n)$ and $\beta \in L^s_\delta(GL_n)$ we have a congruence in $\mathfrak{gl}_n$:*

$$[\alpha, \beta]_\delta \equiv [\alpha^{(p^s)}, \beta^{(p^r)}] \ \ mod\ \ p.$$

As a matter of notation used in the statement above if a and b are two R-points of an R-scheme we write $a \equiv b$ mod p^m if the R/p^mR-points induced by a and b coincide. Note also that the map of sets $ex^r \times ex^s$ is not generally induced by a morphism of p-formal schemes if $r \neq s$; but it is induced by a ϕ^r-linear morphism of p-formal schemes if $r = s$. On the other hand the map ex^{r+s} is induced by a ϕ^{r+s}-linear morphism of p-formal schemes.

REMARK 3.160. The higher order δ-Lie algebras $L^r_\delta(G)$ have an alternative description as follows. Let us define by induction $L^{\circ 1}_\delta(G) := L_\delta(G)$ and

$$L^{\circ r}_\delta(G) := L_\delta(L^{\circ(r-1)}_\delta(G)).$$

By the above discussion we have natural morphisms

$$ex^{\circ r} : L^{\circ r}_\delta(G) \xrightarrow{ex} L^{\circ(r-1)}_\delta \xrightarrow{ex} \dots \xrightarrow{ex} \widehat{G}.$$

Then we claim there exist unique isomorphisms $L^{\circ r}_\delta(G) \simeq L^r_\delta(G)$ of group p-formal schemes fitting into commutative diagrams of abstract groups

$$\begin{array}{ccc} L^{\circ r}_\delta(G) & \simeq & L^r_\delta(G) \\ ex^{\circ r} \downarrow & & \downarrow ex^r \\ \widehat{G} & = & \widehat{G} \end{array}$$

This is an easy consequence of [**23**], Proposition 2.2.

3.10. Hamiltonian formalism and Painlevé

In this section we consider the case of the ring $R = W(k)$, $k = \mathbb{F}_p^a$. Our aim here is to develop an arithmetic analogue of the Hamiltonian formalism and to apply it to an arithmetic analogue of the Painlevé VI equation. We follow **[23]** and **[42]** closely. Recall that by an elliptic curve over R we mean a smooth projective curve of genus 1 equipped with an R-point 0; such an elliptic curve has a unique group scheme structure with 0 as the identity. Recall that by Grothendieck's existence theorem in formal geometry $\widehat{E}$ has a Frobenius lift if and only if E has a Frobenius lift, i.e., a morphism of $\mathbb{Z}$-schemes $E \to E$ whose reduction mod p is the absolute Frobenius.

Recall a basic result from **[23]**:

THEOREM 3.161. **[23]** *Let E be an elliptic curve over R that has no Frobenius lift. Then there exists an element $\psi \in \mathcal{O}(J^2(E))$, $\psi \not\equiv 0$ mod p, with the property that the corresponding morphism*

$$\psi : J^2(E) \to \widehat{\mathbb{G}}_a$$

is a group homomorphism. Moreover ψ is unique up to multiplication by an element of $R^\times$.

DEFINITION 3.162. ψ above is called the *canonical δ-character* of order 2 on E; it is defined up to a unit in R.

Also we have the following:

THEOREM 3.163. **[41]** *Let E be an elliptic curve over R that has no Frobenius lift. Then the following hold:*

(1) $\mathcal{O}(J^1(E)) = R$.
(2) *$J^1(E)$ has no Frobenius lift.*

REMARK 3.164. Assertion (2) in Theorem 3.163 is equivalent to saying that the projection $J^1(J^1(E)) \to J^1(E)$ does not admit a section in the category of p-formal schemes; in particular the projection $J^2(E) \to J^1(E)$ does not have a section. This justifies the Definition 3.166 below.

Let V be an affine p-formal scheme over R. Modules of vertical differential forms on V are defined as

$$\Omega_V = \lim_{\leftarrow} \Omega_{V_n/R_n} \tag{3.155}$$

where $R_n = R/p^{n+1}R$, $V_n = V \otimes_R R_n$.

Given now a system of arithmetic differential equations

$$\mathcal{F} \subset \mathcal{O}(J^r(Y)), \tag{3.156}$$

denote by

$$Z := Z^r(\mathcal{F}) \tag{3.157}$$

the closed formal subscheme of $J^r(Y)$ with ideal $I_Z \subset \mathcal{O}(J^r(Y))$ generated by $\mathcal{F}$. In particular one can consider the module

$$\Omega_Z = \lim_{\leftarrow} \Omega_{Z_n/R_n} \tag{3.158}$$

and the module

$$\Omega_J = \lim_{\leftarrow} \Omega_{J_n/R_n}, \tag{3.159}$$

where $J := J^r(Y)$. On the other hand put

$$\Omega'_Z := \frac{\Omega_J}{\langle I_Z \Omega_J, dI_Z \rangle} \tag{3.160}$$

Moreover define Ω'^i_Z to be the i-th wedge power $\wedge^i \Omega'_Z$.

Under quite general hypotheses the modules 3.155 and 3.160 coincide; indeed, with $\mathcal{F}_n$ the image of $\mathcal{F}$ in $\mathcal{O}(J_n)$ we have:

LEMMA 3.165. *Assume the following hold:*

(1) *Y is smooth over R.*
(2) *Z_n is smooth over R_n for all n.*
(3) *$\mathcal{F}$ is a regular sequence in $\mathcal{O}(J)$.*
(4) *$\mathcal{F}_n$ is a regular sequence in $\mathcal{O}(J_n)$ for all n.*

Then we have a natural isomorphism

$$\Omega'_Z \to \Omega_Z.$$

PROOF. One has

$$\mathcal{O}(Z_n) = \frac{\mathcal{O}(J_n)}{I_Z \mathcal{O}(J_n)}$$

hence by condition (2) we have a split exact sequence

$$0 \to \frac{I_Z \mathcal{O}(J_n)}{I_Z^2 \mathcal{O}(J_n)} \to \Omega_{J_n/R_n} \otimes \mathcal{O}(Z_n) \to \Omega_{Z_n/R_n} \to 0.$$

Since the sequence is split, taking projective limits, one gets a commutative diagram with exact rows:

$$\begin{array}{ccccccccc} 0 & \to & \lim \left(\frac{I_Z \mathcal{O}(J_n)}{I_Z^2 \mathcal{O}(J_n)} \right) & \to & \lim \left(\Omega_{J_n/R_n} \otimes \mathcal{O}(Z_n) \right) & \to & \Omega_Z & \to & 0 \\ & & \alpha \uparrow & & \beta \uparrow & & \gamma \uparrow & & \\ & & \frac{I_Z}{I_Z^2} & \to & \Omega_J \otimes \mathcal{O}(Z) & \to & \Omega'_Z & \to 0 & \end{array}$$

By assumption 1 we may assume Ω_{J_n/R_n} is a free $\mathcal{O}(J_n)$-module; this immediately implies that β is an isomorphism. Also by assumptions 3 and 4 we have that the source and the target of α are free $\mathcal{O}(Z)$-modules of the same rank and α sends the basis induced by $\mathcal{F}$ on the source into the corresponding basis of the target; so α is an isomorphism. We conclude that γ is an isomorphism. □

Going back to $\mathcal{F}$ and Z as in 3.156 and 3.157, for each $s \leq r$, there is a natural map

$$\pi_{r,s} : Z \to J^s(Y).$$

We also have natural maps

$$\frac{\phi^*}{p^i} : \Omega^i_{J^{r-1}(Y)} \to \Omega^i_{J^r(Y)}.$$

For $\mathcal{F} \subset \mathcal{O}(J^2(Y))$ and $Z = Z^2(\mathcal{F})$, they induce maps which we will denote by

$$\frac{\phi_Z^*}{p^i} : \Omega^i_{J^1(Y)} \to \Omega'^i_Z.$$

DEFINITION 3.166. We say that $\mathcal{F} \subset \mathcal{O}(J^2(Y))$ defines a *generalized δ-flow* on $J^1(Y)$, if the induced map

$$\pi_{2,1}^* \Omega_{J^1(Y)} \to \Omega'_Z$$

is injective, and its cokernel is annihilated by a power of p.

REMARK 3.167. By Lemma 3.165 we have that, under the assumptions of that Lemma, if $\mathcal{F}$ defines a δ-flow on $J^1(Y)$ in the sense of Definition 3.129 then $\mathcal{F}$ defines a generalized δ-flow on $J^1(Y)$ in the sense of Definition 3.166.

DEFINITION 3.168. Let X be a smooth surface (i.e. a smooth scheme of relative dimension 2) over R; alternatively let X be the p–adic completion of such a surface.

(a) A *symplectic form* on X is an invertible 2-form on X over R.
(b) A *contact form* on X is a 1-form on X over R such that $d\nu$ is symplectic.
(c) Let Y be a smooth curve over R. A 1-form ν on $X := J^1(Y)$ is called *canonical* if $\nu = f\beta$, where $f \in \mathcal{O}(X)$ and β is an 1-form lifted from Y.

Notice that any closed canonical 1-form on $X = J^1(Y)$ is lifted from Y.

DEFINITION 3.169. Let Y be a smooth affine curve over R and let $f \in \mathcal{O}(J^2(Y))$ be a function defining a generalized δ-flow on $J^1(Y)$.

(a) The respective generalized δ-flow is called *Hamiltonian* with respect to the symplectic form η on $J^1(Y)$, if

$$\phi_Z^* \eta = \mu \cdot \eta$$

in $\Omega_Z'^2$ for some $\mu \in pR$ called the *eigenvalue.*

(b) Assume that moreover $\eta = d\nu$ for some canonical 1-form ν on $J^1(Y)$. Then we call

$$\epsilon := \frac{1}{p}(\phi_Z^* \nu - \mu\nu) := \frac{\phi_Z^*}{p}\nu - \frac{\mu}{p}\nu \in \Omega'_Z \tag{3.161}$$

the *Euler-Lagrange form* attached to ν.

Now we pass to our arithmetic analogue of Painlevé VI.

Let again E be an elliptic curve over R that does not admit a Frobenius lift and let $\psi \in \mathcal{O}(J^2(E))$ be the canonical δ-character of order 2; cf. Definition 3.162. Also consider an invertible 1-form ω on E. Denote by

$$\phi^i = \pi_\phi \circ \ldots \circ \pi_\phi : J^i(E) \to J^0(E)$$

the natural map induced by ϕ^i on the corresponding rings and put

$$\omega^{(i)} := \frac{(\phi^i)^*(\omega)}{p^i}, \ i = 0, 1, 2. \tag{3.162}$$

Then in view of [**35**], p. 203 (where our $\omega^{(i)}$ were denoted ω_i), after replacing ψ by a $R^\times$-multiple of it, we have

$$d\psi = p\omega^{(2)} + \lambda_1\omega^{(1)} + \lambda_0\omega^{(0)} \tag{3.163}$$

with $\lambda_1 \in R$, $\lambda_1 \in R^\times$ depending on (E, ω). Notice that here d means the usual differential taken in the "geometric", or "vertical" direction, that is $dc = 0$ for any $c \in R$. Moreover, in [**25**] it was proved that if E is defined over $\mathbb{Z}_p \subset R$, then $\lambda_0 = 1$, $\lambda_1 = -a_p$ where a_p is the trace of Frobenius on the reduction of E mod p.

Consider the symplectic form

$$\eta = \omega^{(0)} \wedge \omega^{(1)}$$

on $J^1(E)$: cf. 3.162. Let $Y \subset E$ be an affine open set and let $r \in \mathcal{O}(Y)$. Assume in addition that Y has an étale coordinate. (The basic example is E with sections of the second order removed). Denoting such an étale coordinate by T, put

$$\mathcal{A}_2 = K[[T, T']], \quad \mathcal{A}_3 = K[[T, T', T'']],$$

where K is the quotient field of R. As usual, for $\mathcal{A}$ one of the rings $\mathcal{A}_i$, $i = 2, 3$, we set

$$\Omega_{\mathcal{A}} = \bigoplus_{j=0}^{i-1} \mathcal{A} \cdot dT^{(j)}.$$

THEOREM 3.170. *The following assertions hold:*

(1) *The function*

$$f = \psi - \phi(r)$$

defines a generalized δ-flow on $J^1(Y)$ which is Hamiltonian with respect to η.

(2) *There exists a canonical 1-form ν on X such that $d\nu = \eta$; in particular the symplectic form η is exact and if*

$$\epsilon := \frac{1}{p}(\phi_Z^* \nu - \mu\nu)$$

is the Euler-Lagrange form then $p\epsilon$ is closed.

(3) *Let ϵ be Euler-Lagrange form and $f = \psi - \phi(r)$. Then we have the following equality in $\Omega_{\mathcal{A}_3}$:*

$$\epsilon = f\omega^{(1)} - \frac{1}{p}(\phi^* - \mu)\nu.$$

REMARK 3.171. The function f in assertion (1) of the Theorem is our analogue of the Painlevé VI equation; cf. 2.235. By the Theorem it defines a generalized δ-flow; however, it does not define a δ-flow in the sense of Definition 3.122 which is our motivation for generalizing the definition of δ-flow.

PROOF. From 3.162, we get the following equality in $\Omega^2_{J^2(Y)}$:

$$\frac{\phi^*\eta}{p^2} = \frac{\phi^*\omega^{(0)}}{p} \wedge \frac{\phi^*\omega^{(1)}}{p} = \omega^{(1)} \wedge \omega^{(2)}.$$

We have the following equality:

$$df = p\omega^{(2)} + (\lambda_1 - p\phi(\partial r))\omega^{(1)} + \lambda_0\omega^{(0)}$$

in $\Omega_{J^2(E)}$ where $\lambda_1 \in R$, $\lambda_0 \in R^\times$. Set, as usual, $Z = Z^2(f)$. Hence, if we keep notation $\omega^{(0)}, \omega^{(1)}$ also for the images of the respective forms in Ω'_Z, in view of $\omega^{(1)} \wedge df = 0$ in Ω'_Z, we have the following equality in Ω'^2_Z:

$$\phi_Z^*\eta = -p \cdot \omega^{(1)} \wedge ((\lambda_1 - p\phi(dr))\omega^{(1)} + \lambda_0\omega^{(0)}) = p\lambda_0\eta,$$

This completes the proof of (1).

Write now

$$\omega = dL = \frac{dL}{dT} dT$$

where $L = L(T) \in TK[[T]]$. (For instance, we can take L to be the formal logarithm of E). Then

$$\mathcal{A}_3 = K[[T, \phi(T), \phi^2(T)]] = K[[L, \phi(L), \phi^2(L)]].$$

So the image of ψ in $\mathcal{A}_3$ is

$$\psi = \frac{1}{p}(\phi^2(L) + \lambda_1\phi(L) + p\lambda_0 L) + \lambda_{-1}$$

for some $\lambda_{-1} \in R$. Therefore the maps

$$\Omega_{\mathcal{A}_2} \to \frac{\Omega_{\mathcal{A}_3}}{\langle f\Omega_{\mathcal{A}_3}, df\rangle}, \qquad \Omega^2_{\mathcal{A}_2} \to \frac{\Omega^2_{\mathcal{A}_3}}{\langle f\Omega^2_{\mathcal{A}_3}, df \wedge \Omega_{\mathcal{A}_3}\rangle}$$

are isomorphisms and so we have induced Frobenii maps

$$\phi_f^* : \Omega_{\mathcal{A}_2} \to \Omega_{\mathcal{A}_2} \quad \text{and} \quad \phi_f^* : \Omega^2_{\mathcal{A}_2} \to \Omega^2_{\mathcal{A}_2}.$$

Since T is étale the Frobenius lift $T \mapsto T^p$ on $\mathbb{A}^1$ extends to a Frobenius lift $\phi_0 : \widehat{Y} \to \widehat{Y}$ of the p-adic completion of Y. Also the derivation $\frac{d}{dT}$ on $R[T]$ extends to a derivation still denoted by $\frac{d}{dT}$ on $\mathcal{O}(\widehat{Y})$. We claim that

$$\frac{\phi(L) - \phi_0(L)}{p},$$

which a priori is an element of $\mathcal{A}_2$, actually belongs to $\mathcal{O}(J^1(Y))$. Indeed we have the following expansion in $\mathcal{A}_2$:

$$\begin{aligned}
\tfrac{\phi(L)-\phi_0(L)}{p} &= \tfrac{L^{(\phi)}(T^p+pT')-L^{(\phi)}(T^p)}{p} \\
&= \textstyle\sum_{i=1}^{\infty} \frac{p^{i-1}}{i!}\frac{d^i L^{(\phi)}}{dT^i}(T^p)(T')^i \\
&= \textstyle\sum_{i=1}^{\infty} \frac{p^{i-1}}{i!}\phi_0\left(\frac{d^i L}{dT^i}\right)(T')^i \\
&= \textstyle\sum_{i=1}^{\infty} \frac{p^{i-1}}{i!}\phi_0\left(\left(\frac{d}{dT}\right)^{i-1}\left(\frac{\omega}{dT}\right)\right)(T')^i \\
&\in \mathcal{O}(J^1(Y)),
\end{aligned}$$

where the superscript (ϕ) means twisting coefficients by ϕ. The latter inclusion follows because

$$T' = \delta T \in \mathcal{O}(J^1(Y)), \qquad \frac{\omega}{dT} \in \mathcal{O}(Y) \subset \mathcal{O}(\widehat{Y}).$$

Now set

$$\nu := -\frac{\phi(L) - \phi_0(L)}{p}\,\omega \in \mathcal{O}(J^1(Y))\,\omega \subset \Omega_{J^1(Y)}.$$

Then

$$\begin{aligned}
d\nu &= -d\left(\tfrac{\phi(L)-\phi_0(L)}{p}\right) \wedge \omega \\
&= -d\left(\tfrac{\phi(L)}{p}\right) \wedge \omega + d\left(\tfrac{\phi_0(L)}{p}\right) \wedge \omega \\
&= -\omega^{(1)} \wedge \omega^{(0)} = \eta.
\end{aligned}$$

This completes the proof of (2).

Next for $f=\psi-\phi(r)$ we have the following computation in $\Omega_{\mathcal{A}_3}$:

$$\begin{aligned}
p\epsilon &= \phi_f^*\nu-\mu\nu\\
&= -\phi_f^*\left(\tfrac{\phi(L)-\phi_0(L)}{p}\omega\right)+\mu\tfrac{\phi(L)-\phi_0(L)}{p}\omega\\
&= -\phi_f\phi(L)\omega^{(1)}+\phi\phi_0(L)\omega^{(1)}+\mu\tfrac{\phi(L)-\phi_0(L)}{p}\omega\\
&= (\lambda_1\phi(L)+p\lambda_0L+\phi\phi_0(L)-p\phi(r)+p\lambda_{-1})\omega^{(1)}+\mu\tfrac{\phi(L)-\phi_0(L)}{p}\omega^{(0)}\\
&= (\lambda_1\phi(L)+p\lambda_0L+\phi\phi_0(L)-p\phi(r)+p\lambda_{-1})\omega^{(1)}\\
&\quad +\phi^*\left(\tfrac{\phi(L)-\phi_0(L)}{p}\omega^{(0)}\right)-(\phi^*-\mu)\nu\\
&= \left(\phi^2(L)+\lambda_1\phi(L)+p\lambda_0L-p\phi(r)+p\lambda_{-1}\right)\omega^{(1)}-(\phi^*-\mu)\nu\\
&= pf\omega^{(1)}-(\phi^*-\mu)\nu.
\end{aligned}$$

This ends the proof of assertion (3). □

REMARK 3.172.

(a) Some of our arguments above break down if $\psi-\phi(r)$ is replaced by $\psi-r$. This is our main motivation for studying $\psi-\phi(r)$ rather than $\psi-r$.

(b) Assertion (3) of the Theorem implies that if

$$[\epsilon]_2,\ \ [f\omega^{(1)}]_2\in\frac{\Omega_{\mathcal{A}_3}}{(\phi^*-\mu)\Omega_{\mathcal{A}_2}}$$

are the images of $\epsilon, f\omega^{(1)}$ then

$$[\epsilon]_2=[f\omega^{(1)}]_2.$$

This can be viewed as an analogue of the "integration by parts" appearing in the classical theory of Euler-Lagrange equations and justifies our suggestion that f is the "Euler-Lagrange equation attached to our Hamiltonian data."

3.11. p-adic connections on curves: Weierstrass and Riccati

At this point one can ask for arithmetic analogues of connections on smooth projective curves; cf. Example 2.169. Let X be a smooth projective curve over $R=W(\mathbb{F}_p^a)$, of genus g. Then the following hold:

(1) If $g\geq 2$ then X does not possess a p-adic connection; this is a reformulation of a result of Raynaud [**111**].

(2) If $g=1$ then X possess a p-adic connection if and only if it has a Frobenius lift. The set of p-adic connections is in bijection with R and can be described as follows. The p-adic connections are in bijection with sections of $J^1(X)\to\widehat{X}$. The sections of the latter are a principal homogeneous space for the group Γ of morphism (of p-formal schemes) $X\to N$ where N is the kernel of $J^1(X)\to\widehat{X}$. But $N\simeq\widehat{\mathbb{G}}_a$ [**23**]. So any morphism $X\to N$ must be constant and the group Γ is identified with the group of R-points of N which is R. The p-adic connections on X that we just described

can be viewed as an arithmetic analogue of the Weierstrass connections in Example 2.169.

(3) If $g = 0$ then $X = \mathbb{P}^1 := Proj\ R[x_0, x_1]$ and the p-adic connections on X are in bijection with pairs (F, G) of homogeneous polynomials of degree p, $F, G \in R[x_0, x_1]$ as follows. The p-adic connections on X are in bijection with Frobenius lifts on $\widehat{X}$. Since X is projective and R is complete, by Grothendieck's existence theorem in formal geometry, the Frobenius lifts on $\widehat{X}$ are in bijection with the Frobenius lifts on X itself. By the general description of morphisms to projective spaces it follows that the latter are all given by

$$x_0 \mapsto x_0^p + pF, \qquad x_1 \mapsto x_1^p + pG$$

where $F, G \in R[x_0, x_1]$ are homogeneous of degree p. The connections on X we just described can be viewed as arithmetic analogues of the Riccati connections in Example 2.169.

CHAPTER 4

Arithmetic differential geometry: the case of GL_n

In the previous chapter we presented some of our main concepts of arithmetic differential geometry in the "abstract setting" of arbitrary schemes and group schemes. In this chapter we specialize our discussion to the case of the group scheme GL_n over a δ-ring A. We will start by considering the case when our δ-ring A is of local type i.e., $A \subset R$ is a valuation δ-subring of $R = W(\mathbb{F}_p^a)$; cf. Definition 3.27. In this setting we will explore the arithmetic Lie theory, arithmetic logarithmic derivative, and compatibilities with outer and inner automorphisms respectively. In particular we will prove the existence of Chern and Levi-Cività connections. We will also prove the existence/non-existence of various Lax-type connections, the existence of Fedosov connections in dimension 2, and the non-existence of Fedosov connections in dimension ≥ 4. Next we will consider A a δ-ring of global type, $A = \mathcal{O}_K[1/M]$, cf. Definition 3.25, and we will explore, in this setting, the various notions of curvature and gauge action. We end by proving some more negative results; these results play a key motivational role in our theory. The first of these results shows that the classical concept of logarithmic derivative from classical differential geometry does not have a direct analogue in arithmetic differential geometry; this is, of course, our main motivation for introducing arithmetic logarithmic derivatives in the first place. Another negative result shows that simple algebraic groups do not contain proper *Zariski dense mod p δ-subgroups*, an analogue of the classical result of Cassidy in differential algebra. As a consequence we show that GL_n does not possess Frobenius lifts compatible with its group structure; this justifies our search for the "next best" Frobenius lifts in our theory.

4.1. Arithmetic logarithmic derivative and Ehresmann connections

In this section we continue to assume $A \subset R = W(\mathbb{F}^a)$ is a δ-ring of local type i.e., a valuation δ-subring of R as in Definition 3.27 and we continue to denote by $\widehat{\ }$ (rather than by $\widehat{p}$) the operation of p-adic completion. For any smooth R-scheme X we continue to often denote the set $X(R)$ of R-points by X. We let

$$G = GL_n = Spec\ A[x, \det(x)^{-1}] \tag{4.1}$$

be the general linear group scheme over A, where $x = (x_{ij})$ is an $n \times n$ matrix of indeterminates. Consider an arbitrary Frobenius lift

$$\phi^G : A[x, \det(x)^{-1}]\widehat{\ } \to A[x, \det(x)^{-1}]\widehat{\ }, \tag{4.2}$$

$$\phi^G(x_{ij}) = \Phi_{ij}(x) \in A[x, \det(x)^{-1}]\widehat{\ }, \tag{4.3}$$

$$\delta^G(x_{ij}) = \Delta_{ij}(x) \in A[x, \det(x)^{-1}]\widehat{\ }. \tag{4.4}$$

So $\Phi_{ij}(x) = x_{ij}^p + p\Delta_{ij}(x)$. Write $\Phi(x)$ and $\Delta(x)$ for the matrices $(\Phi_{ij}(x))$ and $(\Delta_{ij}(x))$ respectively. We also write $\Phi(x) = x^{(p)}\Lambda(x)$; hence

$$\Delta(x) = \frac{1}{p}x^{(p)}(\Lambda(x) - 1) \tag{4.5}$$

DEFINITION 4.1. The matrix $\Delta(x)^t$ is called the *Christoffel symbol of the* 2nd *kind* of δ^G (or of ϕ^G). We write

$$\Gamma_j^k := \Delta_{kj} := \delta^G(x_{kj}). \tag{4.6}$$

We will later introduce the *Christoffel symbol of the* 1st *kind* (relative to a symmetric matrix, used to "lower the indices").

Conversely any choice of a matrix Δ defines a unique Frobenius lift ϕ^G by the above formulas. As usual, we view the Lie algebra

$$L(G) = \mathfrak{g} = \mathfrak{gl}_n = Spec\ A[x'] \tag{4.7}$$

as a vector group over A where $x' = (x'_{ij})$ is a matrix of indeterminates. We identify

$$J^1(G) = Spf\ A[x, x', \det(x)^{-1}]\hat{}, \quad \delta x = x'. \tag{4.8}$$

If u, v are two matrices of indeterminates which are coordinates on $G \times G$ then the multiplication $G \times G \to G$ is given, of course, by the map $x \mapsto uv$. This map induces, by universality, a multiplication map on $J^1(G)$ induced by the map

$$\mu : A[x, x', \det(x)^{-1}]\hat{} \to A[u, v, u', v', \det(u)^{-1}, \det(v)^{-1}]\hat{}$$

with $\mu(x) = uv$ and $\mu(x') = \delta(uv)$. If $u = (u_{ij})$ then we set $u^{(p)} = (u_{ij}^p)$ the matrix whose entries are the p-th powers of the entries of u, we set $\phi(u) = (\phi(u_{ij}))$, and we set $\delta u = u' = (\delta u_{ij})$ the matrix whose entries are obtained from the entries of u by applying δ; hence we have

$$\phi(u) = u^{(p)} + p\delta u;$$

and similarly for v. Also, using the above notation we may write

$$\Phi(x) = x^{(p)} + p\Delta(x). \tag{4.9}$$

In particular we have $\phi^G(x) = \phi(x) = \Phi(x)$. From the identity of matrices $\phi(uv) = \phi(u)\phi(v)$ we get:

$$\delta(uv) = u^{(p)}v' + u'v^{(p)} + pu'v' + p^{-1}(u^{(p)}v^{(p)} - (uv)^{(p)}), \tag{4.10}$$

$$\delta(u^{-1}) = -\phi(u)^{-1}(u'(u^{-1})^{(p)} + p^{-1}(u^{(p)}(u^{-1})^{(p)} - 1)). \tag{4.11}$$

Recall that, by our definition of the δ-Lie algebra $L_\delta(G)$ we have

$$L_\delta(G) := Ker(J^1(G) \to \widehat{G}) = Spf\ A[x']\hat{} \tag{4.12}$$

and consequently we have (canonical !) identifications as p-formal schemes (but not as groups)

$$J^1(G) \simeq \widehat{G} \times \widehat{\mathfrak{g}}, \quad L_\delta(G) \simeq \widehat{\mathfrak{g}}.$$

(For groups other than GL_n there is still an identification as above but it is not canonical in general !) The set of R-points of $J^1(G)$ will be identified with pairs $(a_0, a_1) \in G \times \mathfrak{g}$; the set of R-points of $L_\delta(G)$ will be identified with the set of pairs $(1, a)$ with $a \in \mathfrak{g}$ and hence with $\mathfrak{g}$. More generally, and similarly, we have a natural identification of p-formal schemes

$$J^r(G) \simeq Spf\ A[x, x', ..., x^{(r)}, \det(x)^{-1}]\hat{} \simeq \widehat{G} \times \widehat{\mathfrak{g}}^r, \tag{4.13}$$

$$L_\delta^r(G) \simeq \widehat{\mathfrak{g}}. \tag{4.14}$$

Under these identifications, and freely using the notation of section 3.7 we have:

• the group structure on $J^1(G)$ induces a group structure on the set $G \times \mathfrak{g}$; we denote the multiplication and inverse on $G \times \mathfrak{g}$ by $\circ$ and ι.

• the action $\star_\delta : G \times L_\delta(G) \to L_\delta(G)$ induces an action (still denoted by)

$$\star_\delta : G \times \mathfrak{g} \to \mathfrak{g}.$$

• the group structure $+_{\delta,r}$ on $L_\delta^r(G)$ induces a group structure on $\mathfrak{g}$,

$$+_{\delta,r} : \mathfrak{g} \times \mathfrak{g} \to \mathfrak{g}.$$

• the inverse map $-_\delta$ on $L_\delta(G)$ induces a map

$$-_\delta : \mathfrak{g} \to \mathfrak{g}.$$

• $\nabla^1 : \widehat{G} \to J^1(G)$ induces a homomorphism (still denoted by)

$$\nabla^1 : G \to G \times \mathfrak{g}$$

• the map $\sigma : \widehat{G} \to J^1(G)$ induces a map (still denoted by)

$$\sigma : G \to G \times \mathfrak{g}$$

• for $\sigma : \widehat{G} \to J^1(G)$ as above, and $\alpha \in L_\delta(G) = \mathfrak{g}$, the map $\sigma^\alpha := \alpha \cdot \sigma : \widehat{G} \to J^1(G)$ induces a map (still denoted by)

$$\sigma^\alpha : G \to G \times \mathfrak{g}.$$

• for $\sigma : \widehat{G} \to J^1(G)$ as above, and $u \in G$, the map $\sigma^{\star u} : \widehat{G} \to J^1(G)$ defined by $\sigma^{\star u}(v) = \nabla(u)^{-1} \cdot \sigma(uv)$ induces a map (still denoted by)

$$\sigma^{\star u} : G \to G \times \mathfrak{g}.$$

• the arithmetic logarithmic derivative $l\delta : G \to L_\delta(G)$ induces a map (still denoted by)

$$l\delta : G \to \mathfrak{g}$$

• the map $\{\ ,\ \} : G \times G \to L_\delta(G)$ induces a map (still denoted by)

$$\{\ ,\ \} : G \times G \to \mathfrak{g}.$$

• The map $ex^r : L_\delta^r(G) \to G$ induces a group homomorphism

$$ex^r : \mathfrak{g} \to G.$$

The following proposition provides a list of formulas computing the maps above. In the proposition below $+$ and $\cdot$ (or simply juxtaposition) in $\mathfrak{g}$ denote addition and multiplication in the associative ring $\mathfrak{g}$ of $n \times n$ matrices.

PROPOSITION 4.2.

(1) *The multiplication* $\circ$ *on* $G \times \mathfrak{g}$ *is given by:*

$$(a_0, a_1) \circ (b_0, b_1) = (a_0 b_0, a_0^{(p)} b_1 + a_1 b_0^{(p)} + p a_1 b_1 + p^{-1}(a_0^{(p)} b_0^{(p)} - (a_0 b_0)^{(p)})).$$

(2) *The identity on* $G \times \mathfrak{g}$ *is the pair*

$$(1, 0).$$

(3) *The inverse* ι *map on* $G \times \mathfrak{g}$ *is given by*

$$\iota(a_0, a_1) = (a_0^{-1}, -(a_0^{(p)} + p a_1)^{-1}(a_1 (a_0^{-1})^{(p)} + p^{-1}(a_0^{(p)}(a_0^{-1})^{(p)} - 1))).$$

(4) *The homomorphism* $\nabla^1 : G \to G \times \mathfrak{g}$ *is given by*
$$\nabla^1(a) = (a, \delta a).$$
(5) *The multiplication* $+_{\delta,r}$ *on* $\mathfrak{g}$ *is given by*
$$a +_{\delta,r} b := a + b + p^r ab.$$
(6) *The action* $\star_\delta : G \times \mathfrak{g} \to \mathfrak{g}$ *is given by*
$$a \star_\delta b = \phi(a) \cdot b \cdot \phi(a)^{-1}.$$
(7) *The arithmetic logarithmic derivative* $l\delta : G \to \mathfrak{g}$ *satisfies*
$$l\delta(ab) = (\phi(a) \cdot l\delta(b) \cdot \phi(a)^{-1}) +_\delta l\delta(a) +_\delta \{a, b\}.$$
(8) *The map* $\sigma : G \to G \times \mathfrak{g}$ *is given by*
$$\sigma(a) = (a, \Delta(a)).$$
(9) *The map* $l\delta : G \to \mathfrak{g}$ *is given by*
$$l\delta a := \frac{1}{p}\left(\phi(a)\Phi(a)^{-1} - 1\right) = (\delta a - \Delta(a))(a^{(p)} + p\Delta(a))^{-1}.$$
(10) *The map* $\{\ ,\ \} : G \times G \to \mathfrak{g}$ *is given by*
$$\{a, b\} = p^{-1}(\Phi(a)\Phi(b)\Phi(ab)^{-1} - 1).$$
(11) *If* $\epsilon = 1 + p\alpha$ *then the* δ^G*-linear equation* $l\delta(u) = \alpha$ *is equivalent to the equation*
$$\delta u = \Delta^\alpha(u) \tag{4.15}$$
and also to the equation
$$\phi(u) = \Phi^\alpha(u), \tag{4.16}$$
where $\Delta^\alpha(x) := \alpha \cdot \Phi(x) + \Delta(x)$, $\Phi^\alpha(x) = \epsilon \cdot \Phi(x)$, $\epsilon = 1 + p\alpha$. *(We shall refer to Equation 4.15 as a* Δ*-linear equation.)*

(12) *For* $\alpha \in \mathfrak{g}$ *the map* $\sigma^\alpha : G \to G \times \mathfrak{g}$ *is given by*
$$\sigma^\alpha(a) = (a, \Delta^\alpha(a)).$$
(So we can refer to $\{\Delta^\alpha(x); \alpha \in \mathfrak{g}\} \subset \mathfrak{g}(R[x, \det(x)^{-1}]\hat{\ })$ *as the linear equivalence class of* $\Delta(x)$*.)*

(13) *For* $u \in G$ *the map* $\sigma^{\star u} : G \to G \times \mathfrak{g}$ *is given by*
$$\sigma^{\star u}(a) = (a, \Delta^{\star u}(a)),$$
where $\Delta^{\star u}(x) = p^{-1}(\Phi^{\star u}(x) - x^{(p)})$, $\Phi^{\star u}(x) = \phi(u)^{-1}\Phi(ux)$. *(So in particular if* $\delta u = \Delta^\alpha(u)$ *then* $(\Phi^\alpha)^{\star u}(x) = \Phi(u)^{-1}\Phi(ux)$*.)*

(14) *The group homomorphism* $ex^r : \mathfrak{g} \to G$ *is given by*
$$ex^r(a) = 1 + p^r\phi^{-r}(a).$$

PROOF. Assertions (1), (2), (3) follow directly from 4.10 and 4.11. Assertion (4) follows directly from definitions. Assertions (12), (13) follow from assertion (1). Assertion (5) for $r = 1$ follows from assertion (1); for arbitrary r assertion (5) follows from [**23**], Proposition 2.2, which gives the expression for $+_{\delta,r}$ in terms of the formal group law. Assertion (7) follows from assertion (6). Assertion (8) is clear. Assertion (12) is an easy exercise. Assertion (11) follows from assertion (9). Assertion (14) follows from the formula
$$\phi^r(x)_{|x=1, x'=\ldots=x^{(r-1)}=0} = 1 + p^r x^{(r)}.$$

To prove the rest of the assertions one uses the following adaptation of a standard trick in Witt vector theory. Define the "ghost map"

$$w : G \times \mathfrak{g} \to G \times \mathfrak{g}, \quad w(a_0, a_1) = (a_0, a_0^{(p)} + pa_1).$$

Then w is a homomorphism of monoids with identity where the source $G \times \mathfrak{g}$ is the monoid (actually group) equipped with multiplication $\circ$ and the target $G \times \mathfrak{g}$ is equipped with the structure of direct product of multiplicative monoids (where $\mathfrak{g}$ is viewed as a monoid with respect to multiplication in the associative algebra $\mathfrak{g}$). The homomorphism w injective. So in order to prove the identities in assertions (6), (9), (10) it is enough to prove them after composition with w; this however follows trivially using the equalities:

$$w(\nabla^1(a)) = (a, \phi(a)), \quad w(\sigma(a)) = (a, \Phi(a)).$$

□

REMARK 4.3. In the terminology above, any Δ-linear equation $\delta u = \Delta^\alpha(u)$, cf. 4.15, may be identified with the system of arithmetic differential equations on GL_n,

$$\delta x_{ij} - \Delta^\alpha_{ij}(x), \quad i, j = 1, ..., n, \tag{4.17}$$

where $(\Delta^\alpha_{ij}(x)) = \Delta^\alpha(x)$, and hence, by Lemma 3.124, it has the same solution set as the δ-flow $\mathcal{E}(\delta^{GL_n,\alpha})$ on GL_n, where $\delta^{GL_n,\alpha}(x) = \Delta^\alpha(x)$. Also note that, by Lemma 3.126, for any $\mathcal{H} \in \mathcal{O}(GL_n)\widehat{}$ with $\delta^{GL_n,\alpha}\mathcal{H} = 0$ we have that $\mathcal{H}$ is a prime integral of the system 4.17; in particular for all solutions u of $\delta u = \Delta^\alpha(u)$ we have $\delta(\mathcal{H}(u)) = 0$.

DEFINITION 4.4. Denote by $T \subset G = GL_n$ the split torus of diagonal matrices, by N the normalizer of T in GL_n, and by $W \subset GL_n$ the group of permutation matrices. Then $N = WT = TW$ and W is isomorphic to the *Weyl group* N/T. In what follows we will view T, N, W either as abstract subgroups of the abstract group $G = GL_n$ of R-points of GL_n or as subgroup schemes over A of the group scheme GL_n over A; all three group schemes are smooth over A. This ambiguity in our notation should create no confusion; it will be clear from context which of the two meanings is being assumed. Finally one can consider the involution $x^\tau = x$ on GL_n; the group defined by τ is, of course, GL_n itself and we refer to τ as the canonical involution defining GL_n. Note that the associated maps $\mathcal{H}$ and $\mathcal{B}$ are given by $\mathcal{H}(x) = 1$ and $\mathcal{B}(x, y) = x^{-1}y$.

REMARK 4.5. Writing $G = GL_n$ we have

$$N = \{b \in G; (ab)^{(p)} = a^{(p)}b^{(p)} \text{ for all } a \in G\},$$

$$N = \{b \in G; (ba)^{(p)} = b^{(p)}a^{(p)} \text{ for all } a \in G\}.$$

Also note that for any $w \in W$ we have $w^t = w^{-1}$, $w^{(p)} = w$, $\phi(w) = w$. Also, for any $d \in T$, $d^t = d$ and $d^{(p)} = d^p$. A Frobenius lift on GL_n is (left, right, bi-) invariant with respect to N if and only if it is (left, right, bi-) invariant with respect to both T and W.

REMARK 4.6. The letter T has been used above to denote the diagonal torus in GL_n; it was also previously used to denote a set (or a matrix) of variables. The two uses of the letter T should not lead to any confusion.

In the Theorem below $\widehat{GL_n}$ is viewed as an open subset of $\widehat{\mathfrak{gl}_n}$.

THEOREM 4.7. *The Frobenius lift $\phi_0^{GL_n}(x)=x^{(p)}$ on $\widehat{GL_n}$ is the unique Frobenius lift that is bi-invariant with respect to N and extends to a Frobenius lift on $\widehat{\mathfrak{gl}_n}$.*

PROOF. Set $G=GL_n=Spec\ A[x,\det(x)^{-1}]$. Let

$$T=Spec\ A[s_1,s_1^{-1},...,s_n,s_n^{-1}]$$

and consider another copy of T given by

$$T=Spec\ A[t_1,t_1^{-1},...,t_n,t_n^{-1}].$$

The multiplication map $\mu:T\times G\times T\to G$ is given by the map

$$\mu:A[x,\det(x)^{-1}]\to A[x,\det(x)^{-1},s_1,s_1^{-1},...,s_n,s_n^{-1},t_1,t_1^{-1},...,t_n,t_n^{-1}]$$

$$\mu x=sxt$$

where $s=\text{diag}(s_1,...,s_n)$, $t=\text{diag}(t_1,...,t_n)$. Let ϕ^{GL_n} be a Frobenius lift on $\widehat{GL_n}$ that is bi-invariant with respect to T and W and extends to a Frobenius lift on $\widehat{\mathfrak{gl}_n}$. Then $\phi^{GL_n}(x)=\Phi(x)$, $\Phi(x)=(\Phi_{ij}(x))$, $\Phi_{ij}(x)\in A[x]\hat{}$. Bicompatibility with T implies

$$\Phi(sxt)=s^p\Phi(x)t^p.$$

For any $n\times n$ matrix $M=(m_{ij})$ with entries $m_{ij}\in\mathbb{Z}_+$ (non-negative integers) we write $x^M=\prod_{ij}x_{ij}^{m_{ij}}$. We say that x^M has *weight*

$$(u,v)\in\mathbb{Z}_+^n\times\mathbb{Z}_+^n,\quad u=(u_1,...,u_n),\quad v=(v_1,...,v_n),$$

if $u_i=\sum_j m_{ij}$ for all i and $v_j=\sum_i m_{ij}$ for all j. An element in $A[x]\hat{}$ is said to have weight (u,v) if it is a (possibly infinite) sum of monomials λx^M with $\lambda\in A$ and x^M of weight (u,v). Write

$$\Phi_{ij}(x)=\sum_{u,v\in\mathbb{Z}_+^n}\Phi_{ijuv}$$

with Φ_{ijuv} of weight (u,v). Write $s^u=s_1^{u_1}...s_n^{u_n}$ and similarly for t^v. Then we have

$$\mu\Phi_{ij}=\sum_{u,v}s^ut^v\Phi_{ijuv}.$$

On the other hand

$$\mu\Phi_{ij}=s_i^pt_j^p\Phi_{ij}=\sum_{u,v}s_i^pt_j^p\Phi_{ijuv}.$$

This implies that Φ_{ij} has weight (pe_i^t,pe_j^t) where $e_1,...,e_n$ are the columns of the identity matrix $1_n=[e_1,...,e_n]$. There is, however, only one monomial x^M with weight (pe_i^t,pe_j^t), namely $x^M=x_{ij}^p$. So $\Phi_{ij}=\lambda_{ij}x_{ij}^p$ for some $\lambda_{ij}\in A$. Now we use the bicompatibility with W to show that all λ_{ij} are equal; this will end the proof because the common value of the λ_{ij}s must be 1 (which follows by looking at the restriction of ϕ^{GL_n} to T). To use bicompatibility with W denote by $w_\sigma\in W$ the matrix

$$w_\sigma=[e_{\sigma(1)},...,e_{\sigma(n)}]$$

obtained by permuting the columns of the identity matrix according to the permutation σ. For any matrix $a=(a_{ij})$ we have the formulae

$$(aw_\sigma)_{ij}=a_{i\sigma(j)},\qquad(w_\sigma a)_{\sigma(i)j}=a_{ij}.$$

Bicompatibility with W yields

$$\Phi(xw_\sigma) = \Phi(x)w_\sigma \quad \text{and} \quad \Phi(w_\sigma x) = w_\sigma \Phi(x). \tag{4.18}$$

The first of these equalities yields

$$(\Phi(xw_\sigma))_{ij} = (\Phi(x)w_\sigma)_{ij}.$$

Now

$$(\Phi(xw_\sigma))_{ij} = \Phi_{ij}(xw_\sigma) = \lambda_{ij}((xw_\sigma)_{ij})^p = \lambda_{ij}x^p_{i\sigma(j)};$$

on the other hand

$$(\Phi(x)w_\sigma)_{ij} = \Phi_{i\sigma(j)}(x) = \lambda_{i\sigma(j)}x^p_{i\sigma(j)}.$$

We conclude that $\lambda_{ij} = \lambda_{i\sigma(j)}$. Similarly, using the second of the equalities in 4.18 we get $\lambda_{ij} = \lambda_{\sigma(i)j}$. This ends the proof. □

REMARK 4.8. Theorem 4.7 fails if one removes the condition that the Frobenius lift extend to one on $\widehat{\mathfrak{gl}_n}$. Indeed assume $n = 2$ and consider

$$u(x) \in A[x, \det(x)^{-1}]\hat{}$$

defined by

$$u(x) = \frac{x_{11}x_{12}x_{21}x_{22}}{\det(x)^2}.$$

Note that $u(sxt) = u(x)$ for all $s, t \in T$. Define the Frobenius lift ϕ^{GL_2} on $\widehat{GL_2}$ by

$$\phi^{GL_2}(x) = (\Phi_{ij}(x)), \qquad \Phi_{ij}(x) = u(x) \cdot x^p_{ij}.$$

Then clearly ϕ^{GL_2} is bi-invariant with respect to T and W.

DEFINITION 4.9. We let $\phi_0^{GL_n}$ be the Frobenius lift on $\widehat{GL_n}$ given by $\phi_0^{GL_n}(x) = x^{(p)}$ and we denote by $\delta_0^{GL_n}$ the corresponding p-derivation on $\widehat{GL_n}$; so $\delta_0^{GL_n}x = 0$. We refer to $\phi_0^{GL_n}$ and $\delta_0^{GL_n}$ as the *trivial* Frobenius lift, respectively the *trivial* p-derivation, on $\widehat{GL_n}$.

PROPOSITION 4.10. *Let $x^\tau = x$ on GL_n. The following hold.*

(i) $\phi_0^{GL_n}$ *is the unique Frobenius lift that is $\mathcal{H}$-horizontal and $\mathcal{B}$-symmetric with respect to* $\phi_0^{GL_n}$.

(ii) *The arithmetic logarithmic derivative $l\delta : GL_n \to \mathfrak{gl}_n$ attached to $\delta_0^{GL_n}$ is given by*

$$l\delta a := \delta a \cdot (a^{(p)})^{-1}.$$

(iii) *The map $\{\ ,\ \} : GL_n \times GL_n \to \mathfrak{gl}_n$ attached to $\delta_0^{GL_n}$ is given by*

$$\{a, b\} = p^{-1}(a^{(p)} \cdot b^{(p)} \cdot ((ab)^{(p)})^{-1} - 1).$$

(iv) *If $\epsilon = 1 + p\alpha$ then the equation $l\delta(u) = \alpha$ is equivalent to either of the following equations:*

$$\delta u = \alpha \cdot u^{(p)}, \tag{4.19}$$

$$\phi(u) = \epsilon \cdot u^{(p)}. \tag{4.20}$$

(v) *For all $a, b \in GL_n$ with either a or b in N we have*

$$l\delta(ab) = (\phi(a) \cdot l\delta(b) \cdot \phi(a)^{-1}) +_\delta l\delta(a).$$

PROOF. Uniqueness in (i) follows from Corollary 3.152. (ii), (iii), (iv) are a specialization of Proposition 4.2. (v) follows from Lemma 3.144 □

DEFINITION 4.11. For any $\alpha \in \mathfrak{gl}_n(A)$ the p-derivation $\delta^{GL_n,\alpha}$ on $\widehat{GL_n}$ defined by

$$\delta^{GL_n,\alpha}x = \alpha \cdot x^{(p)}$$

is called the *Ehresmann connection* attached to α.

The arithmetic differential equation corresponding to an Ehresmann connection as above is, of course, the equation 4.19, or equivalently 4.20.

REMARK 4.12. Consider a p-adically complete ring $\mathcal{A}$, a matrix $U \in GL_n(\mathcal{A})$ with $U \equiv 1 = 1_n \bmod p$, and an integer $\nu \in \mathbb{Z}$ not divisible by p. Then there exists a unique matrix $U_p^{1/\nu} \in GL_n(\mathcal{A})$ whose ν-th power is U and which is $\equiv 1 \bmod p$. The uniqueness follows from Lemma 3.151. For the existence let $V = \frac{U-1}{p}$ and set

$$U_p^{1/\nu} := (1+pV)_p^{1/\nu} := \sum_{m=0}^{\infty} \left(\begin{array}{c} 1/\nu \\ m \end{array} \right) p^m V^m \in GL_n(\mathcal{A}). \tag{4.21}$$

DEFINITION 4.13. We refer to $U_p^{1/\nu}$ as the *principal* $1/\nu$-root of U.

COROLLARY 4.14. *Let $G = GL_n$. Then, for any integer $\nu \in \mathbb{Z}$ coprime to p, the map $L_\delta(G) \to L_\delta(G)$, $b \mapsto \nu_\delta b := b^\nu$ is bijective.*

PROOF. Injectivity follows from Corollary 3.154. Surjectivity follows from Remark 4.12. □

REMARK 4.15. Let $G \subset GL_n$ be a smooth closed subgroup scheme and let ϕ^{GL_n} be a Frobenius lift on $\mathcal{O}(\widehat{GL_n}) = A[x, \det(x)^{-1}]\widehat{\ }$ such that G is δ^{GL_n}-horizontal. (Such a ϕ^{GL_n} always exists due to the smoothness of G and Lemma 3.111.) Let ϕ^G be the induced Frobenius lift on $\mathcal{O}(\widehat{G})$. Then the embedding $G \subset GL_n$ induces a closed embedding $J^1(G) \subset J^1(GL_n)$ and hence a closed embedding $L_\delta(G) \subset L_\delta(GL_n)$. Composing the latter with the natural identification $L_\delta(GL_n) = \mathfrak{gl}_n$ we get a natural embedding of groups $L_\delta(G) \subset \mathfrak{gl}_n$ where $\mathfrak{gl}_n$ is viewed as a group with respect to $+_\delta$. Note that the usual Lie algebra $L(G)$ is also embedded as a subgroup into $L(GL_n) = \mathfrak{gl}_n$ where this time $\mathfrak{gl}_n$ is equipped with the usual addition $+$ of matrices; but note that, as subsets of $\mathfrak{gl}_n$ we have $L_\delta(G) \neq L(G)$ in general! Indeed it is trivial to check that if

$$G = Spec\ A[x, \det(x)^{-1}]\widehat{\ }/(f_1(x), ..., f_m(x))$$

and if

$$\delta_1 f_i(x') := p^{-1} f_i^{(\phi)}(1 + px')$$

where $f_i^{(\phi)}$ is obtained from f_i by applying ϕ to the coefficients then

$$L_\delta(G) = Spf\ A[x']\widehat{\ }/(\delta_1 f_1(x'), ..., \delta_1 f_m(x'))$$

hence the R-points of the latter are the zeroes in $\mathfrak{gl}_n$ of the polynomials $\delta_1 f_i(x')$. On the other hand the Lie algebra $L(G)$ is given by

$$L(G) = Spec\ A[x']/(d_1 f_1(x'), ..., d_1 f_m(x'))$$

where

$$d_1 f_i(x') = \epsilon^{-1} f_i(1 + \epsilon x') := \sum_{j,k=1}^{n} \frac{\partial f_i}{\partial x_{jk}}(1) x'_{jk}.$$

So the set of R-points of $L(G)$ is the zero locus in $\mathfrak{gl}_n$ of the polynomials $d_1 f_i$. These sets of R-points of $L(G)$ and $L_\delta(G)$ do not coincide in general; cf. our discussion of examples in the next section. Going back to our discussion it follows from horizontality that

$$l\delta : G \to L_\delta(G)$$

is the restriction of the map

$$l\delta : GL_n \to L_\delta(GL_n) = \mathfrak{gl}_n.$$

Similarly the map

$$\{\ ,\ \}_G : G \times G \to L_\delta(G)$$

is the restriction of the map

$$\{\ ,\ \}_{GL_n} : GL_n \times GL_n \to L_\delta(GL_n) = \mathfrak{gl}_n.$$

In particular the formulas involving $l\delta$ and $\{\ ,\ \}$ in Proposition 4.2 continue to be valid if one replaces $G = GL_n$ by $G \subset GL_n$ an arbitrary smooth closed subgroup of GL_n, provided G is δ^{GL_n}-horizontal and provided we make the identifications explained above. Finally note that under the identifications above we have

$$L^r_\delta(GL_n) = Spf\ A[x^{(r)}]\hat{\ }, \tag{4.22}$$

$$L^r_\delta(G) = Spf\ A[x^{(r)}]\hat{\ }/(p^{-r}f_1^{(\phi^r)}(1+p^r x^{(r)}), ..., p^{-r}f_m^{(\phi^r)}(1+p^r x^{(r)})). \tag{4.23}$$

In particular the map $ex^r : L^r_\delta(GL_n) \to GL_n$, $ex^r(a) = 1 + p^r\phi^{-r}(a)$, satisfies:

$$(ex^r)^{-1}(G) = L^r_\delta(G). \tag{4.24}$$

We are ready to give the:

PROOF OF PROPOSITION 3.159. For $r, s \geq 1$ define the map

$$[\ ,\ ]_\delta : L^r_\delta(GL_n) \times L^s_\delta(GL_n) \to L^{r+s}_\delta(GL_n) \tag{4.25}$$

by the formula

$$[\alpha, \beta]_\delta := \frac{(1+p^r\phi^s(\alpha))(1+p^s\phi^r(\beta))(1+p^r\phi^s(\alpha))^{-1}(1+p^s\phi^r(\beta))^{-1} - 1}{p^{r+s}}. \tag{4.26}$$

The right hand side of 4.26 is trivially seen to be a restricted power series in $\phi^s(\alpha)$ and $\phi^r(\beta)$ so 4.25 is a δ-map. Clearly the following diagram of sets is commutative:

$$\begin{array}{ccc} L^r_\delta(GL_n) \times L^s_\delta(GL_n) & \stackrel{[\ ,\ ]_\delta}{\longrightarrow} & L^{r+s}_\delta(GL_n) \\ ex^r \times ex^s \downarrow & & \downarrow ex^{r+s} \\ GL_n \times GL_n & \stackrel{[\ ,\ ]}{\longrightarrow} & GL_n \end{array} \tag{4.27}$$

By 4.24 the upper horizontal arrow in 4.27 sends $L^r_\delta(G) \times L^s_\delta(G)$ into $L^{r+s}_\delta(G)$ so the δ-map 4.25 induces a δ-map as in assertion (1) of the proposition. (Uniqueness follows from the injectivity of the ex maps.) Assertion (2) of the proposition is clear from the functoriality of $L^r_\delta(G)$ and ex^r. Assertion (7) is a direct computation. Now to prove assertions (3), (4), (5), (6) it is enough to prove them in the case $G = GL_n$. Assertion (3) clearly holds in this case. Assertion (5) for $G = GL_n$ follows directly from assertion (7). Assertion (6) follows from assertions (3) and (7) because of the usual Jacobi identity plus the following computation:

$$\begin{aligned}
[[\alpha,\beta]_\delta,\gamma]_\delta &\equiv [[\alpha^{(p^s)},\beta^{(p^r)}],\gamma]_\delta && \bmod p\\
&\equiv [[\alpha^{(p^s)},\beta^{(p^r)}]^{(p^t)},\gamma^{(p^{r+s})}] && \bmod p\\
&\equiv [[\alpha^{(p^{s+t})},\beta^{(p^{r+t})}],\gamma^{(p^{r+s})}] && \bmod p.
\end{aligned}$$

Assertion (4) can be checked in the same way. □

4.2. Existence of Chern connections

In this section we prove our main existence results for arithmetic analogues of real and complex Chern connections; the real case is more subtle and we will mostly concentrate on this case.

We continue to assume here that A is a δ-ring of local type i.e., that $A \subset R$ is a valuation δ-subring of R as in Definition 3.27.

Let again $G = GL_n$ and consider the *trivial* Frobenius lift $\phi_0^{GL_n}(x) = x^{(p)}$ on $\widehat{G}$. Let x^t be the transpose of x and let $q \in GL_n(A)$ be either symmetric or anti-symmetric, i.e. $q^t = \pm q$. Recall that q is called a *metric* if $q^t = q$ and it is called a *2-form* if $q^t = -q$. For such a q the formula $x^\tau = q^{-1}(x^t)^{-1}q$ defines an involution on GL_n. The associated quadratic map $\mathcal{H}$, normalized quadratic map $\mathcal{H}_q$, bilinear map $\mathcal{B}$, and normalized bilinear map $\mathcal{B}_q$ are

$$\mathcal{H}(x) = q^{-1}x^tqx, \quad \mathcal{H}_q(x) = x^tqx, \quad \mathcal{B}(x,y) = q^{-1}x^tqy, \quad \mathcal{B}_q(x,y) = x^tqy. \tag{4.28}$$

Let $SO(q) = (G^+)^\circ$ be the group defined by τ; we refer to τ as the *canonical involution* defining $SO(q)$. The group $SO(q)$ is trivially seen to be smooth over A. Also

$$G^- = q^{-1}G_\pm, \quad G_\pm = \{x; x^t = \pm x\}; \tag{4.29}$$

so G^- is smooth (an open set of an affine space). If $T \subset GL_n$ is the torus of diagonal matrices in GL_n and $N \subset GL_n$ is the normalizer of T in GL_n we set

$$T_{SO(q)} := T \cap SO(q) \tag{4.30}$$

$$N_{SO(q)} := N \cap SO(q). \tag{4.31}$$

It is easy to see that $T_{SO(q)}$ and $N_{SO(q)}$ are smooth over A. Note that the (set of R-points of the) δ-Lie algebra $L_\delta(SO(q))$ identifies with the set of matrices $a \in \mathfrak{gl}_n$ satisfying

$$a^t\phi(q) + \phi(q)a + pa^t\phi(q)a = 0. \tag{4.32}$$

On the other hand the Lie algebra $L(SO(q))$ identifies with the set of matrices $a \in \mathfrak{gl}_n$ satisfying

$$a^tq + qa = 0. \tag{4.33}$$

DEFINITION 4.16. We say q is *split* if q is one of the matrices

$$\begin{pmatrix} 0 & 1_r \\ -1_r & 0 \end{pmatrix}, \begin{pmatrix} 0 & 1_r \\ 1_r & 0 \end{pmatrix}, \begin{pmatrix} 1 & 0 & 0 \\ 0 & 0 & 1_r \\ 0 & 1_r & 0 \end{pmatrix}. \tag{4.34}$$

REMARK 4.17. If q is split then

$$qq^t = 1, \quad q^t = \pm q, \quad q \in N, \quad q^{(p)} = q, \quad \phi(q) = q, \quad \delta q = 0,$$

and, for all $b \in G$,

$$(qb)^{(p)} = qb^{(p)}, \quad (bq)^{(p)} = b^{(p)}q.$$

LEMMA 4.18. *G^- is $\phi_0^{GL_n}$-horizontal.*

PROOF. Since G^- is smooth it is enough to check that if $a \in G^-$ then $a^{(p)} \in G^-$ which is a trivial exercise. □

REMARK 4.19. Let us assume that q is split. In this case $SO(q)$ is denoted by Sp_{2r}, SO_{2r}, SO_{2r+1} respectively. The groups $T_{SO(q)}$ are then the *split maximal tori*:

$$\begin{aligned} T_{Sp_{2r}} &= \{d \in T; d = \text{diag}(d_1, ..., d_r, d_1^{-1}, ..., d_r^{-1})\} \\ T_{SO_{2r}} &= \{d \in T; d = \text{diag}(d_1, ..., d_r, d_1^{-1}, ..., d_r^{-1})\} \\ T_{SO_{2r+1}} &= \{d \in T; d = \text{diag}(1, d_1, ..., d_r, d_1^{-1}, ..., d_r^{-1})\}. \end{aligned}$$

One also has that $N_{SO(q)}$ is the normalizer of $T_{SO(q)}$ in $SO(q)$; this follows from the fact that if $a \in SO(q)$ normalizes $T_{SO(q)}$ and $d \in T_{SO(q)}$ is a regular matrix (by which we mean d has distinct mod p elements on the diagonal) then $a^{-1}da = wdw^{-1}$ for some $w \in W$ hence aw is in the centralizer of d; by regularity of d we get $aw \in T$, hence $a \in N$, hence $a \in N_{SO(q)}$.

Note now that $N_{SO(q)}$ contains all the matrices of the form

$$\left(\begin{array}{cc} w_r & 0 \\ 0 & w_r \end{array}\right), \left(\begin{array}{cc} w_r & 0 \\ 0 & w_r \end{array}\right), \left(\begin{array}{ccc} 1 & 0 & 0 \\ 0 & w_r & 0 \\ 0 & 0 & w_r \end{array}\right), \tag{4.35}$$

respectively, where $w_r \in W_r$ is an arbitrary permutation matrix in GL_r. The matrices 4.35 modulo $T_{SO(q)}$ do not exhaust the *Weyl group* $W_{SO(q)} := N_{SO(q)}/T_{SO(q)}$. A system of representatives of the Weyl group can be obtained by adding to the matrices in 4.35 the matrices in the set W' which we describe below. Let $e_1, ..., e_n$ be the columns of the identity matrix 1_n. Then, for Sp_{2r}, W' consists of all products $w'_{i_1}...w'_{i_s}$ where $s \geq 1$ and

$$w'_i := [e_1, ..., e_{i-1}, e_{r+i}, e_{i+1}, ..., e_{r+i-1}, -e_i, e_{r+i+1}, ...].$$

For SO_{2r}, W' consists of all products $w'_{i_1}...w'_{i_s}$ where $s \geq 2$ is even and

$$w'_i := [e_1, ..., e_{i-1}, e_{r+i}, e_{i+1}, ..., e_{r+i-1}, e_i, e_{r+i+1}, ...].$$

For SO_{2r+1}, W' consists of all products $w'_{i_1}...w'_{i_s}$ where $s \geq 1$ and

$$w'_i := [-e_1, e_2, ..., e_i, e_{r+i+1}, e_{i+2}, ..., e_{r+i}, e_{i+1}, e_{r+i+2}, ...].$$

(Cf. [**67**], section 2.5.1; the forms of the classical groups used in loc. cit. are different from ours; however, in order to conclude our claims, the only information needed from loc. cit. is the cardinality of the Weyl groups.) The bottom-line of the above description of the Weyl groups is the following:

LEMMA 4.20. *Assume q is split. Then for any $v \in N_{SO(q)}$ we have $v^{(p)} \in N_{SO(q)}$.*

PROOF. Indeed any element w_i' in the description above can be written as $w_i' = rw$ with $r \in T$, $r^2 = 1$, $w \in W$. So

$$(w_i')^{(p)} = (rw)^{(p)} = r^{(p)}w^{(p)} = rw = w_i'.$$

Now if $v \in N_{SO(q)}$ then either $v = tw$ or $v = tw_{i_1}'...w_{i_s}'$ with $t \in T_{SO(q)}$, $w \in W$, hence either

$$v^{(p)} = t^{(p)}w^{(p)} = t^p w = t^{p-1}v \in N_{SO(q)}$$

or

$$v^{(p)} = t^{(p)}(w_{i_1}')^{(p)}...(w_{i_s}')^{(p)} = t^p w_{i_1}'...w_{i_s}' = t^{p-1}v \in N_{SO(q)}.$$

□

REMARK 4.21. We review here well know facts, to be used later, about characters and roots of the split classical groups. The groups of characters of $T_{Sp_{2r}}$ and $T_{SO_{2r}}$ have bases $\chi_1, ..., \chi_r$ where

$$\chi_i(\text{diag}(d_1, ..., d_r, d_1^{-1}, ..., d_r^{-1})) = d_i.$$

The group of characters of $T_{SO_{2r+1}}$ has basis $\chi_1, ..., \chi_r$ where

$$\chi_i(\text{diag}(1, d_1, ..., d_r, d_1^{-1}, ..., d_r^{-1})) = d_i.$$

The *roots* of Sp_{2r} are

$$\pm 2\chi_i, \quad \pm\chi_i \pm \chi_j, \quad i \neq j.$$

The *roots* of SO_{2r} are

$$\pm\chi_i \pm \chi_j, \quad i \neq j.$$

The *roots* of SO_{2r+1} are

$$\pm\chi_i, \quad \pm\chi_i \pm \chi_j, \quad i \neq j.$$

Recall that if G is any of the groups $Sp_{2r}, SO_{2r}, SO_{2r+1}$ and if χ is a root of G then there is an attached *root subgroup* U_χ of G over the algebraic closure K^a of the field of fractions K of R; it is the unique subgroup isomorphic to $\mathbb{G}_a$ over K^a normalized by the corresponding maximal torus on which this maximal torus acts via χ. In all these cases U_χ comes from an embedding of $\mathbb{G}_a$ into G over R and has the following explicit description which we need to review. We consider $r \times r$ matrices as follows. For $i = 1, ..., r$ let F_{ii} be the matrix with 1 on position (i, i) and 0 everywhere else. Let now $i \neq j$ between 1 and r. Let E_{ij} be the matrix with 1s on the diagonal, 1 on position (i, j) and 0 everywhere else. Let

$$F_{ij} = E_{ij} + E_{ji}, \qquad G_{ij} = E_{ij} - E_{ji}.$$

Finally let e_i be the columns of the identity matrix 1_r. Then the groups U_χ are given as follows (where $\mu \in R$).

For Sp_{2r} we have:

$$\begin{aligned}
U_{2\chi_i} &= \{\begin{pmatrix} 1 & \mu F_{ii} \\ 0 & 1 \end{pmatrix}\}, \ U_{-2\chi_i} = U_{2\chi_i}^t \\
U_{\chi_i - \chi_j} &= \{\begin{pmatrix} 1 + \mu E_{ij} & 0 \\ 0 & 1 - \mu E_{ji} \end{pmatrix}\} \\
U_{\chi_i + \chi_j} &= \{\begin{pmatrix} 1 & \mu F_{ij} \\ 0 & 1 \end{pmatrix}\}, \ U_{-\chi_i - \chi_j} = U_{\chi_i + \chi_j}^t
\end{aligned}$$

For SO_{2r} we have:

$$U_{\chi_i+\chi_j} = \{\begin{pmatrix} 1 & \mu G_{ij} \\ 0 & 1 \end{pmatrix}\}, \quad U_{-\chi_i-\chi_j} = U^t_{\chi_i+\chi_j}$$

$$U_{\chi_i-\chi_j} = \{\begin{pmatrix} 1+\mu E_{ij} & 0 \\ 0 & 1-\mu E_{ji} \end{pmatrix}\}$$

For SO_{2r+1} we have:

$$U_{\pm\chi_i\pm\chi_j} = \{\begin{pmatrix} 1 & 0 \\ 0 & \star \end{pmatrix}\}, \quad \text{where } \star \text{ is as for } SO_{2r}$$

$$U_{\chi_i} = \{\begin{pmatrix} 1 & \mu e_i^t & 0 \\ 0 & 1 & 0 \\ -\mu e_i & -\frac{\mu^2}{2}e_ie_i^t & 1 \end{pmatrix}\}, \quad U_{-\chi_i} = U^t_{\chi_i}.$$

Remark that if G is Sp_{2r} and χ is any root or if G is SO_{2r} and χ is any root or if G is SO_{2r+1} and $\chi \neq \pm\chi_i$ we have that U_χ is $\phi_0^{GL_n}$-horizontal. On the other hand if $G = SO_{2r+1}$ and $\chi = \pm\chi_i$ then U_χ is not $\phi_0^{GL_n}$-horizontal (because it is not stable under the map $u \mapsto u^{(p)}$). For obvious reasons we shall refer to $\pm\chi_i$ as the *short roots* of SO_{2r+1}. A root χ of $SO(q)$ for a split q will be called *abnormal* if $SO(q) = SO_{2r+1}$ and χ is a short root of the latter.

We are ready to prove our main results. We begin with a result on the existence of Cartan decompositions. Recall that $+_\delta$ and $-_\delta$ denote the multiplication and inverse on $L_\delta(G) = \mathfrak{gl}_n$.

THEOREM 4.22. *Assume q is split and consider the involution $x^\tau = q^{-1}(x^t)^{-1}q$ on $G = GL_n$ and the induced involution $b \mapsto b^{L_\delta(\tau)}$ on $L_\delta(G) = \mathfrak{gl}_n$. Then the following hold:*

1. *For all $b \in L_\delta(G)$ we have $b^{L_\delta(\tau)} = -_\delta(q^{-1}b^tq)$.*
2. *The multiplication map $L_\delta(G)^+ \times L_\delta(G)^- \to L_\delta(G)$ is a bijection. In particular any element $a \in L_\delta(G)$ has a Cartan decomposition*

$$a = a^+ +_\delta a^-$$

with respect to τ.

PROOF. To prove assertion (1) note that if $-\tau$ is the map $x \mapsto x^{-\tau}$ then the action of $L_\delta(-\tau)$ on $L_\delta(G) = Spf\ A[x']\hat{}$ is given by the following computation, where we use the fact that $x^{t(p)} = x^{(p)t}$, $q \in N$, and $q^{(p)} = q$:

$$\begin{aligned} (x')^{L_\delta(-\tau)} &= \delta(x^{-\tau})_{|x=1} = \delta(q^{-1}x^tq)_{|x=1} \\ &= p^{-1}(\phi(q^{-1}x^tq) - (q^{-1}x^tq)^{(p)})_{|x=1} \\ &= p^{-1}(q^{-1}(x^{(p)} + px')^tq - q^{-1}x^{t(p)}q) \\ &= q^{-1}(x')^tq. \end{aligned}$$

We conclude that

$$(x')^{L_\delta(\tau)} = -_\delta((x')^{L_\delta(-\tau)}) = -_\delta(q^{-1}(x')^tq).$$

Then assertion (1) follows because x' are the coordinates used in the identification $L_\delta(G) = \mathfrak{gl}_n$.

Let us prove assertion (2). We already know this map is injective; cf. Corollary 3.154. To prove surjectivity let $a \in L_\delta(G) = \mathfrak{gl}_n$ and set $A = 1 + pa$.

We claim first that

$$A = UV$$

with

$$U, V \in GL_n, \quad U^\tau = U, \quad V^\tau = V^{-1}.$$

This is standard and analogous to the argument for the existence of the "polar decomposition." Indeed if we set

$$V = \{A^{-\tau} A\}_p^{1/2}$$

and

$$U = AV^{-1}$$

then the series expression of V shows that $V^\tau = V^{-1}$; on the other hand

$$U^{-\tau} U = (V^{-\tau})^{-1} A^{-\tau} A V^{-1}$$

and the latter equals 1 if and only if

$$A^{-\tau} A = V^{-\tau} V.$$

But

$$V^{-\tau} V = V^2 = A^{-\tau} A$$

so we conclude that $U^{-\tau} U = 1$ and our claim is proved.

Going back to the proof of the theorem set

$$u = \frac{U - 1}{p}, \quad v = \frac{V - 1}{p}.$$

We get that

$$1 + pa = (1 + pu)(1 + pv)$$

hence

$$a = u +_\delta v.$$

On the other hand, by assertion (1),

$$u^{L_\delta(-\tau)} = q^{-1} u^t q = q^{-1} \frac{U^t - 1}{p} q = \frac{U^{-\tau} - 1}{p} = \frac{U^{-1} - 1}{p} = -_\delta u.$$

Hence $u^{L_\delta(\tau)} = u$. Similarly $v^{L_\delta(\tau)} = -_\delta v$ which ends our proof. □

In the next statement fractional powers are, again, principal roots as in Definition 4.13. Here is our main existence result for arithmetic analogues of real Chern connections.

THEOREM 4.23. *Assume $q^t = \pm q \in GL_n(A)$ and consider the involution $x^\tau = q^{-1}(x^t)^{-1} q$ on GL_n. Consider the associated maps $\mathcal{H}_q : GL_n \to GL_n$, $\mathcal{H}_q(x) = x^t q x$, and $\mathcal{B}_q : GL_n \times GL_n \to GL_n$, $\mathcal{B}_q(x, y) = x^t q y$. Let $\phi_1^{GL_n}$ be a Frobenius lift on $\widehat{GL_n}$ which commutes with the automorphism of $\widehat{GL_n}$ induced by $x \mapsto \pm x^t$ and let $\phi_1^{GL_n}(x) =: \Phi_1(x)$.*

Then there exists a unique Frobenius lift ϕ^{GL_n} on $\widehat{GL_n}$ that is $\mathcal{H}_q$-horizontal and $\mathcal{B}_q$-symmetric with respect to $\phi_1^{GL_n}$. It is given by $\phi^{GL_n}(x) = \Phi(x)$, where

$$\Phi(x) = \Phi_1(x) \cdot \{(\Phi_1(x)^t \phi(q) \Phi_1(x))^{-1} \Phi_1(x^t q x)\}_p^{1/2}.$$

REMARK 4.24. In particular one can take, in the above theorem $\phi_1^{GL_n} = \phi_0^{GL_n}$ in which case $\Phi_1(x) = x^{(p)}$ and hence $\phi^{GL_n}(x) = \Phi(x)$,

$$\Phi(x) = x^{(p)} \cdot \{(x^{(p)t}\phi(q)x^{(p)})^{-1}(x^tqx)^{(p)}\}_p^{1/2}. \tag{4.36}$$

By the way, setting $x = 1$, the identity matrix, one gets, in this case,

$$\Phi(1) = \{\phi(q)^{-1}q^{(p)}\}_p^{1/2}. \tag{4.37}$$

DEFINITION 4.25. Let $q^t = \pm q \in GL_n(A)$ and consider the involution $x^\tau = q^{-1}(x^t)^{-1}q$ on GL_n. Let ϕ^{GL_n} be the unique Frobenius lift on $\widehat{GL_n}$ that is $\mathcal{H}_q$-horizontal and $\mathcal{B}_q$-symmetric with respect to $\phi_0^{GL_n}$; cf. Theorem 4.23. Then the p-derivation δ^{GL_n} on $\widehat{GL_n}$ attached to ϕ^{GL_n} is called the *real Chern connection* on GL_n attached to q.

REMARK 4.26. The above real Chern connection is the analogue of real Chern connection in classical differential geometry. Note that, unlike in classical differential geometry, Chern connections in the sense of Definition 4.25 are *not* a priori Ehresmann connections in the sense of Definition 4.11 if $n \geq 2$. (On the other hand, for $n = 1$, real Chern connections are trivially seen to be Ehresmann connections.)

REMARK 4.27. Note the following property of real Chern connections which is similar to that in Remark 2.81. The property referred to is that the matrices $\phi^{GL_n}(x)$ are obtained from the matrices $q, \delta q, x, x^t$ via "size independent" *matrix* operations, namely:

$$+\ , \times,\ (\ )^{-1},\ (\ )_p^{1/2},\ (\ )^{(p)}.$$

This is, again, in the spirit of free non-commutative function theory [**85**] and is different from saying that the entries of $\phi^{GL_n}(x)$ are obtained from the entries of $q, \delta q, x$ via *scalar* operations on these entries.

For the next statement recall from 2.74 that we denote by GL_r^c the centralizer of $q_0 := \begin{pmatrix} 0 & 1 \\ -1 & 0 \end{pmatrix}$ in GL_{2r}. Recall also from 2.82 that we denoted by U_r^c the complexified unitary group. Recall that a metric (respectively a 2-form) $q \in GL_{2r}(A)$ is called *hermitian* if it commutes with q_0, i.e., if $q \in GL_r^c(A)$. For a hermitian metric q the matrix $\check{q} := -qq_0$ is a hermitian form. Finally we consider the A-homomorphism

$$c : GL_r^c = Spec\ A[a, b, \det(x)^{-1}] \to GL_r = Spec\ A[v, \det(v)^{-1}]$$

given by $v \to x^c := a + \sqrt{-1} \cdot b$, where $x = \begin{pmatrix} a & b \\ -b & a \end{pmatrix}$.

THEOREM 4.28. *Let $q^t = q \in GL_r^c(A) \subset GL_n(A)$, $n = 2r$, and let $\check{q} = -qq_0$. Let δ^{GL_n} be the real Chern connection on GL_n attached to q, with attached Frobenius lift ϕ^{GL_n}, and let $\check{\delta}^{GL_n}$ be the real Chern connection on GL_n attached to $\check{q}$, with attached Frobenius lift $\check{\phi}^{GL_n}$. Then:*

(i) *GL_r^c is both ϕ^{GL_n}-horizontal and $\check{\phi}^{GL_n}$-horizontal. Moreover the induced Frobenius lifts $\phi^{GL_r^c}$ and $\check{\phi}^{GL_r^c}$ on $\widehat{GL_r^c}$ coincide.*

(ii) *Let $q = 1_{2r}$. Then ϕ^{GL_n} and $\check{\phi}^{GL_n}$ induce equal Frobenius lifts $\phi^{U_r^c}$ and $\check{\phi}^{U_r^c}$ on the complexified unitary group $\widehat{U_r^c}$.*

(iii) *Let $q = 1_{2r}$. Then there is no Frobenius lift ϕ^{GL_r} on $\widehat{GL_r}$ such that $\phi^{GL_r^c}$ is c-horizontal with respect to ϕ^{GL_r}.*

The coincidence of connections in assertion (i) of Theorem 4.28 is reminiscent of the coincidence of connections 2.149 in classical differential geometry.

On the other hans the situation in assertion (iii) of Theorem 4.28 contrasts with the situation in classical differential geometry where one has the diagram 2.152.

Also the following Theorem 4.29 contrasts with the situation in classical differential geometry; cf. Proposition 2.83.

THEOREM 4.29. *Let $q \in GL_n(A)$, be split with $q^t = q$, $n = 2r \geq 2$, and let δ^{GL_n} be the real Chern connection attached to q. Then SL_n is not δ^{GL_n}-horizontal.*

Note that if δ^{GL_2} is the real Chern connection attached to the split anti-symmetric matrix in GL_2 then $Sp_2 = SL_2$ is δ^{GL_2}-horizontal; cf also assertion (1) of Remark 4.36 below. It is not clear at this point if SL_{2r} is $\delta^{GL_{2r}}$-horizontal for $\delta^{GL_{2r}}$ the real Chern connection attached to the split anti-symmetric matrix in GL_{2r}, $r \geq 2$.

The following Theorem is largely in analogy with classical differential geometry.

THEOREM 4.30. *Let $q^t = \pm q \in GL_n(A)$ be split and let δ^{GL_n} be the real Chern connection attached to q, with attached Frobenius lift ϕ^{GL_n}. Then:*

(i) *ϕ^{GL_n} is the unique Frobenius lift that is $\mathcal{H}$-horizontal and $\mathcal{B}$-symmetric with respect to $\phi_0^{GL_n}$.*

(ii) *ϕ^{GL_n} is right invariant with respect to N and also left invariant (and hence bi-invariant) with respect to $N_{SO(q)}$.*

(iii) *ϕ^{GL_n} and $\phi_0^{GL_n}$ coincide on the set*

$$SO(q) \cap \phi_{GL_n,0}^{-1}(SO(q));$$

in particular if χ is not abnormal then the root subgroup U_χ is ϕ^{GL_n}-horizontal and the Frobenius lift on $\widehat{\mathbb{G}_a} = Spf\ A[z]\widehat{\ }$ induced by ϕ^{GL_n} is given by $z \mapsto z^p$.

(iv) *If $l\delta : GL_n \to \mathfrak{gl}_n$ is the arithmetic logarithmic derivative attached to ϕ^{GL_n} then for all $a \in N_{SO(q)}$ and $b \in GL_n$ (alternatively for all $a \in GL_n$ and $b \in N$) we have*

$$l\delta(ab) = (\phi(a) \cdot l\delta(b) \cdot \phi(a)^{-1}) +_\delta l\delta(a).$$

(v) *If $l\delta_0 : GL_n \to \mathfrak{gl}_n$ is the arithmetic logarithmic derivative attached to $\phi_0^{GL_n}$ then for any $a \in GL_n$ we have*

$$l\delta(a) \in l\delta_0(a) +_\delta \mathfrak{gl}_n^-.$$

In particular if $a \in SO(q)$ and

$$l\delta_0(a) = (l\delta_0(a))^+ +_\delta (l\delta_0(a))^-$$

is the Cartan decomposition of $l\delta_0(a)$ then we have

$$l\delta(a) = (l\delta_0(a))^+.$$

(vi) *Let $\alpha \in L_\delta(SO(q))$, let $\phi^{GL_n,\alpha}$ be the Frobenius lift on $\mathcal{O}(\widehat{GL_n})$ defined by $\phi^{GL_n,\alpha}(x) = \epsilon \cdot \phi^{GL_n}(x)$, $\epsilon = 1 + p\alpha$, and let $\delta^{GL_n,\alpha}$ be the p-derivation on $\mathcal{O}(\widehat{GL_n})$ attached to $\phi^{GL_n,\alpha}$. Then $\phi^{GL_n,\alpha}$ is $\mathcal{H}$-horizontal with respect to $\phi_0^{GL_n}$; equivalently,*

$$\delta^{GL_n,\alpha}(\mathcal{H}) = 0.$$

Finally, the following theorem gives an arithmetic analogue of the complex Chern connection:

THEOREM 4.31. *Assume $q \in GL_n(A)$, $q^t = \pm q$, and let $\phi_0^{GL_n}$ be the Frobenius lift on $\widehat{GL_n}$ given by $\phi_0^{GL_n}(x) = x^{(p)}$. Then there is a unique Frobenius lift ϕ^{GL_n} on $\widehat{GL_n}$ that is $(\mathcal{H}_q, \mathcal{B}_q)$-horizontal with respect to $\phi_0^{GL_n}$. It is given by $\phi^{GL_n}(x) = \Phi(x)$ where*

$$\Phi(x) = x^{(p)}(x^{(p)t}\phi(q)x^{(p)})^{-1}(x^t qx)^{(p)} = \phi(q)^{-1}(x^{(p)t})^{-1}(x^t qx)^{(p)}. \tag{4.38}$$

REMARK 4.32. Note that the expression $(x^{(p)})^{-1}\Phi(x)$ in 4.38 is the square of the expression $(x^{(p)})^{-1}\Phi(x)$ in 4.36. This is analogous to the fact that the Christoffel symbols of the complex Chern connection in classical differential geometry are equal to twice the Christoffel symbols of the real Chern connection; cf. equations 2.132 and 2.133.

DEFINITION 4.33. Let $q^t = \pm q \in GL_n(A)$ and let ϕ^{GL_n} be the unique Frobenius lift on $\widehat{GL_n}$ in Theorem 4.31. Then the p-derivation δ^{GL_n} on $\widehat{GL_n}$ attached to ϕ^{GL_n} is called the *complex Chern connection* on GL_n attached to q.

In what follows we concentrate on proving our Theorems.

PROOF OF THEOREM 4.23. The condition that $\phi_1^{GL_n}$ commutes with $x \mapsto \pm x^t$ translates into the identity:

$$\Phi_1(\pm x^t) = \pm\Phi_1(x)^t. \tag{4.39}$$

Consider the matrices

$$A := \Phi_1(x)^t \cdot \phi(q) \cdot \Phi_1(x), \qquad B := \Phi_1(x^t qx). \tag{4.40}$$

Clearly $A^t = \pm A$ according as $q^t = \pm q$. Also, by 4.39,

$$B^t = \pm(\pm\Phi_1(x^t qx)^t) = \pm\Phi_1(\pm(x^t qx)^t) = \pm\Phi_1(x^t qx) = \pm B,$$

according as $q^t = \pm q$. Note that $A \equiv B$ mod p and set

$$C = p^{-1}(B - A).$$

Clearly $C^t = \pm C$. Define the matrix

$$\Lambda = \Lambda(x) := \{A^{-1}B\}_p^{1/2} := \{1 + pA^{-1}C\}_p^{1/2} := \sum_{i=0}^{\infty} \binom{1/2}{i} p^i (A^{-1}C)^i. \tag{4.41}$$

Clearly $\Lambda \equiv 1$ mod p. Note that

$$A\Lambda = \Lambda^t A, \tag{4.42}$$

equivalently, $(A\Lambda)^t = \pm A\Lambda$; this follows because

$$(A(A^{-1}C)^i)^t = ((CA^{-1})^{i-1}C)^t = \pm C(A^{-1}C)^{i-1} = \pm A(A^{-1}C)^i.$$

On the other hand we claim that

$$\Lambda^t A\Lambda = B. \tag{4.43}$$

Indeed note first that the identity

$$1 + t = (1+t)^{1/2}(1+t)^{1/2} = \sum_{n \geq 0}\left(\sum_{i+j=n}\binom{1/2}{i}\binom{1/2}{j}\right)t^n$$

yields

$$\sum_{i+j=n} \binom{1/2}{i}\binom{1/2}{j} = 0 \quad \text{for} \quad n \geq 2. \tag{4.44}$$

Then 4.43 follows from the following computation:

$$\begin{aligned}
\Lambda^t A \Lambda &= \left(\sum_{i=0}^{\infty} \binom{1/2}{i} p^i (CA^{-1})^i\right) A \left(\sum_{j=0}^{\infty} \binom{1/2}{j} p^j (A^{-1}C)^i\right) \\
&= A + \sum_{n=1}^{\infty} p^n (CA^{-1})^{n-1} C \sum_{i+j=n} \binom{1/2}{i}\binom{1/2}{j} \\
&= A + pC, \quad \text{by } 4.44 \\
&= B.
\end{aligned}$$

To prove the existence part of our theorem define the Frobenius lift ϕ^{GL_n} by

$$\phi^{GL_n}(x) := \Phi(x) := \Phi_1(x) \cdot \Lambda(x).$$

The commutativity of 3.135 for $g = q$ is equivalent to the condition

$$\Phi(x)^t \cdot \phi(q) \cdot \Phi(x) = \Phi_1(x^t q x), \tag{4.45}$$

which is trivially equivalent to equation 4.43. On the other hand the commutativity of 3.136 is equivalent to

$$\Phi_1(x)^t \cdot \phi(q) \cdot \Phi(x) = \Phi(x)^t \cdot \phi(q) \cdot \Phi_1(x), \tag{4.46}$$

which is trivially equivalent to equality 4.42. This ends the proof of the existence part.

The uniqueness part of our theorem follows from Corollary 3.152. □

PROOF OF ASSERTIONS (i) AND (ii) IN THEOREM 4.28 Assertion (ii) follows from assertion (i). Assertion (i) follows from the fact that, if $q_0 q q_0^{-1} = q$ then $\phi^{GL_{2r}}$ and $\breve{\phi}^{GL_{2r}}$ send any R-point of GL_r^c into the same point of GL_r^c, as can be seen directly by using the formula 4.36 and the fact that conjugation by q_0 commutes with the maps $GL_{2r} \to GL_{2r}$ given by

$$u \mapsto u^{(p)}, \quad u \mapsto u^t, \quad u \mapsto u^{-1}, \quad u \mapsto \phi(u).$$

□

PROOF OF THEOREM 4.30. Consider the notation in the proof of Theorem 4.23 with $\Phi_1(x) = x^{(p)}$. Let us consider the right action of $SO(q)$ on $A[x, \det(x)^{-1}]\hat{}$

$$F(x) \bullet c := F(xc), \quad F \in A[x, \det(x)^{-1}]\hat{}, \quad c \in SO(q),$$

and the left action of $SO(q)$ on $A[x, \det(x)^{-1}]\hat{}$

$$c \bullet F(x) := F(cx), \quad F \in R[x, \det(x)^{-1}]\hat{}, \quad c \in SO(q).$$

Let u be in N and let v be in $N_{SO(q)}$. Then, using Lemma 4.20 one checks that:

$$A \bullet u = (u^{(p)})^t A u^{(p)}, \quad v \bullet A = A, \quad B \bullet u = (u^{(p)})^t B u^{(p)}, \quad v \bullet B = B, \tag{4.47}$$

$$A(u) = B(u); \tag{4.48}$$

hence

$$C \bullet u = (u^{(p)})^t C u^{(p)}, \quad v \bullet C = C, \tag{4.49}$$

$$\Lambda \bullet u = (u^{(p)})^{-1} \Lambda u^{(p)}, \quad v \bullet \Lambda = \Lambda, \tag{4.50}$$

$$C(u) = 0, \quad \Lambda(u) = 1. \tag{4.51}$$

We proceed now to proving the various assertions of our theorem.

Assertion (i) is clear.

To check assertion (ii) we check first that N is ϕ^{GL_n}-horizontal. For this it is enough to show that $\phi^{GL_n}(u) \in N$ if $u \in N$; but this follows from the fact that $\Lambda(u) = 1$ in 4.51.

In order to check right invariance of ϕ^{GL_n} with respect to N and left invariance of ϕ^{GL_n} with respect to $N_{SO(q)}$ it is sufficient to notice that, by 4.50, we get

$$\begin{aligned}
\phi^{GL_n}(xu) &= \phi^{GL_n}(x) \bullet u = (x^{(p)}\Lambda) \bullet u) \\
&= (xu)^{(p)}(\Lambda \bullet u) = (x^{(p)}u^{(p)})((u^{(p)})^{-1}\Lambda u^{(p)}) \\
&= x^{(p)}\Lambda u^{(p)} = \phi^{GL_n}(x) \cdot \phi^N(u)
\end{aligned}$$

and

$$\begin{aligned}
\phi^{GL_n}(vx) &= v \bullet \phi^{GL_n}(x) = v \bullet (x^{(p)}\Lambda) \\
&= (vx)^{(p)}(v \bullet \Lambda) = v^{(p)}x^{(p)}\Lambda \\
&= \phi^{N_{SO(q)}}(v) \cdot \phi^{GL_n}(x).
\end{aligned}$$

To check assertion (iii) let

$$g \in SO(q) \cap \phi^{-1}_{GL_n,0}(SO(q)).$$

Then $g^t q g = q$ hence

$$B(g) = q^{(p)} = q;$$

and also

$$\phi_{GL_n,0}(g)^t q \phi_{GL_n,0}(g) = q,$$

hence

$$q = \phi(q) = \phi(\phi_{GL_n,0}(g)^t q \phi_{GL_n,0}(g)) = (g^{(p)})^t \phi(q) g^{(p)} = (g^{(p)})^t q g^{(p)},$$

so $A(g) = q$. Hence $C(g) = 0$, hence $\Lambda(g) = 1$, and hence

$$\phi_{GL_n}(g) = \phi^{-1}(x^{(p)})\phi^{-1}(\Lambda(g)) = \phi_{GL_n,0}(g).$$

This ends the proof of the first assertion in (iii). The second assertion in (iii) follows from Lemma 3.109.

Assertion (iv) follows from Lemma 3.144.

Assertion (v) follows from Corollary 3.153 plus the fact that if $a \in SO(q)$ then

$$l\delta(a) \in L_\delta(SO(q)) \subset \mathfrak{gl}_n^+.$$

To check assertion (vi) note that, by 4.45, and since q is split, we have

$$q^{-1}\Phi(x)^t q \Phi(x) = (q^{-1}x^t q x)^{(p)}. \tag{4.52}$$

We conclude by the following computation (where we use $\epsilon \in SO(q)$):

$$\begin{aligned}\phi^{GL_n,\alpha}(q^{-1}x^tqx) &= q^{-1}(\phi^{GL_n,\alpha}(x))^tq\phi^{GL_n,\alpha}(x) = q^{-1}(\epsilon\Phi(x))^tq\epsilon\Phi(x)\\ &= q^{-1}\Phi(x)^t\epsilon^tq\epsilon\Phi(x) = q^{-1}\Phi(x)^tq\Phi(x)\\ &= q^{-1}(x^tqx)^{(p)} = (q^{-1}x^tqx)^{(p)}.\end{aligned}$$

□

PROOF OF THEOREM 4.31. The argument is analogous to that in the proof of Theorem 4.23 but, this time, it is completely trivial. □

REMARK 4.34. In contrast with assertion (iii) in Theorem 4.30 above if χ is an abnormal root (i.e. $SO(q) = SO_{2r+1}$ and $\chi = \pm\chi_i$) then one can check (by direct computation involving the formula 4.36) that U_χ is not ϕ_{GL_n}-horizontal.

REMARK 4.35. Let us assume $n = 1$ in Theorem 4.23. Furthermore assume

$$A = \mathbb{Z}_p = W(\mathbb{F}_p), \quad q \in \mathbb{Z}_p^\times, \quad \phi_1^{GL_1}(x) = \phi_0^{GL_1}(x) = x^p.$$

Then the Frobenius lift ϕ^{GL_1} on $\widehat{GL_1}$ in Theorem 4.23 is given by:

$$\phi^{GL_1}(x) = \left(\frac{q}{p}\right)\cdot q^{(p-1)/2}\cdot x^p, \tag{4.53}$$

where

$$\left(\frac{-}{p}\right) : \mathbb{Z}_p^\times \to \{\pm 1\}$$

is the Legendre symbol. This follows from the obvious fact that

$$\left(\frac{q}{p}\right)\cdot q^{(p-1)/2}$$

is the principal square root of $\phi(q)^{-1}q^p = q^{p-1}$ in $\mathbb{Z}_p^\times$.

REMARK 4.36. Here are some computations for the case $n = 2$ where, in Theorem 4.23, we assume $\phi_1^{GL_1}(x) = \phi_0^{GL_1}(x) = x^{(p)}$.

(1) Let

$$q = \begin{pmatrix} 0 & 1\\ -1 & 0\end{pmatrix}.$$

Then the Frobenius lift ϕ^{GL_2} on $\widehat{GL_2}$ in Theorem 4.23 is given by:

$$\phi^{GL_2}(x) = x^{(p)}\{((x^{(p)})^tqx^{(p)})^{-1}(x^tqx)^{(p)})\}_p^{1/2} = \left(\frac{\det(x^{(p)})}{(\det(x))^p}\right)_p^{-1/2} x^{(p)}. \tag{4.54}$$

Note that

$$\begin{aligned}\phi^{GL_2}(\det(x)) &= \det(\phi^{GL_2}(x))\\ &= \det\left(\left(\frac{\det(x^{(p)})}{(\det(x))^p}\right)_p^{-1/2} x^{(p)}\right)\\ &= \left(\frac{\det(x^{(p)})}{(\det(x))^p}\right)^{-1}\cdot\det(x^{(p)})\\ &= (\det(x))^p.\end{aligned}$$

Hence

$$\delta^{GL_2}(\det(x)) = 0. \tag{4.55}$$

In particular SL_2 is δ^{GL_2}-horizontal.

(2) Let

$$x = \begin{pmatrix} a & b \\ c & d \end{pmatrix}, \quad q = \begin{pmatrix} 0 & 1 \\ 1 & 0 \end{pmatrix}.$$

Then the Frobenius lift ϕ^{GL_2} on $\widehat{GL_2}$ in Theorem 4.23 is given by

$$\phi_{GL_2}(x) = \begin{pmatrix} a^p & b^p \\ c^p & d^p \end{pmatrix} \begin{pmatrix} u & v \\ w & u \end{pmatrix}_p^{1/2}$$

where

$$\begin{array}{rcl} u & = & \frac{(a^pd^p+b^pc^p)(ad+bc)^p-2^{p+1}a^pb^pc^pd^p}{(a^pd^p-b^pc^p)^2}, \\ v & = & b^pd^p\cdot\frac{2^p(a^pd^p+b^pc^p)-2(ad+bc)^p}{(a^pd^p-b^pc^p)^2}, \\ w & = & a^pc^p\cdot\frac{2^p(a^pd^p+b^pc^p)-2(ad+bc)^p}{(a^pd^p-b^pc^p)^2}. \end{array} \tag{4.56}$$

(3) Let $A = R$ and consider an arbitrary $q \in GL_2(R)$ with $q^t = q$, write

$$q = \begin{pmatrix} \alpha & \beta \\ \beta & \gamma \end{pmatrix}$$

where $\alpha, \beta, \gamma \in R$, and consider the Frobenius lift $\phi^{GL_n}(x) = x^{(p)}\Lambda(x)$ in Theorem 4.23. We would like to "compute" $\Lambda(1)$. First note that we have

$$\Lambda(1) = \left(1 + \frac{p}{\alpha^p\gamma^p - \beta^{2p}}V\right)^{-1/2}, \quad V := \begin{pmatrix} \gamma^p\delta\alpha - \beta^p\delta\beta & \gamma^p\delta\beta - \beta^p\delta\gamma \\ \alpha^p\delta\beta - \beta^p\delta\alpha & \alpha^p\delta\gamma - \beta^p\delta\beta \end{pmatrix}.$$

Set

$$\begin{array}{rcl} \{\alpha, \beta\}_\delta & := & \alpha^p\delta\beta - \beta^p\delta\alpha, \\ \{\alpha, \beta, \gamma\}_\delta & := & \frac{1}{2}(\alpha^p\delta\gamma - 2\beta^p\delta\beta + \gamma^p\delta\alpha), \\ D & := & \{\alpha, \gamma\}_\delta^2 - 4\{\alpha, \beta\}_\delta\{\beta, \gamma\}_\delta. \end{array}$$

Assume $D \not\equiv 0 \bmod p$. Then the eigenvalues λ_1, λ_2 of V are in R and are non-congruent modulo p; they are given by

$$\lambda_{1,2} = \{\alpha, \beta, \gamma\}_\delta \pm \frac{\sqrt{D}}{2}.$$

Let u_i be an eigenvector of V with coefficients in R for the eigenvalue λ_i; dividing u_i by appropriate powers of p we may assume $u_i \not\equiv 0 \bmod p$. Since λ_1, λ_2 are non-congruent modulo p it follows that the reductions mod p of u_1, u_2 are linearly independent; in other words the 2×2 matrix $U = [u_1, u_2]$ is in $GL_2(R)$ and in particular, $V = U \begin{pmatrix} \lambda_1 & 0 \\ 0 & \lambda_2 \end{pmatrix} U^{-1}$. We conclude that

$$\Lambda(1) = U \begin{pmatrix} \{\varphi_1\}_p^{1/2} & 0 \\ 0 & \{\varphi_2\}_p^{1/2} \end{pmatrix} U^{-1}, \quad \varphi_i := (1 + \frac{p\lambda_i}{\alpha^p\gamma^p - \beta^{2p}})^{-1}, \tag{4.57}$$

where the fractional powers are principal roots.

Note that if we assume α, β, γ are algebraic over $\mathbb{Q}$ then the coefficients of V are also algebraic over $\mathbb{Q}$ and hence so are λ_1, λ_2. Then one can arrange U to have coefficients algebraic over $\mathbb{Q}$ and, hence, $\Lambda(1)$ has coefficients algebraic over $\mathbb{Q}$.

Let us assume, on the other hand, that $\alpha, \beta, \gamma \in \mathbb{Z}_p$, denote again by $\left(\frac{}{p}\right) : \mathbb{Z}_p^\times \to \{\pm 1\}$ the Legendre symbol, and assume that $\left(\frac{D}{p}\right) = 1$. Then $\lambda_1, \lambda_2 \in \mathbb{Z}_p$ and, in particular, if $\epsilon_i \in \mathbb{Z}_p$ are *any* elements satisfying $\varphi_i = \epsilon_i^{p-1}$, then

$$\{\varphi_i\}_p^{1/2} = \left(\frac{\epsilon_i}{p}\right) \cdot \epsilon_i^{(p-1)/2}.$$

As a special case of the above discussion let us assume that $\alpha, \beta \in \mathbb{Z}_{(p)}$, $\gamma = \alpha$, $\alpha \not\equiv \pm\beta \bmod p$, and $\{\alpha, \beta\}_\delta \not\equiv 0 \bmod p$. Then one can take, in 4.57,

$$\varphi_1 = \frac{\alpha^p - \beta^p}{\alpha - \beta}, \quad \varphi_2 = \frac{\alpha^p + \beta^p}{\alpha + \beta}, \quad U = \begin{pmatrix} 1 & 1 \\ -1 & 1 \end{pmatrix}.$$

(4) Let, again, $A = \mathbb{Z}_{(p)}$, $q \in GL_2(A)$, $q^t = q$,

$$q = \begin{pmatrix} \alpha & \beta \\ \beta & \gamma \end{pmatrix},$$

and consider the Frobenius lift $\phi^{GL_n}(x) = x^{(p)}\Lambda(x)$ in Theorem 4.23. We have:

$$\Lambda(1) = (q^{-1}q^{(p)})_p^{1/2} \in GL_2(\mathbb{Z}_p).$$

If $\beta = 0$ the matrix $(q^{-1}q^{(p)})_p^{1/2}$ is expressible in terms of the Legendre symbols of α, γ, and, in particular, still belongs to $GL_2(A)$. However, for $\beta \neq 0$, the matrix $(q^{-1}q^{(p)})_p^{1/2}$ does not necessarily belong to $GL_2(A)$ and indeed its determinant does not generally belong to A because the determinant

$$\frac{\alpha^p\gamma^p - \beta^{2p}}{\alpha\gamma - \beta^2}$$

of $q^{-1}q^{(p)}$ is not generally a square in $\mathbb{Q}$.

For instance take $\alpha = \gamma = 2$, $\beta = 1$. Then the determinant above is

$$\frac{4^p - 1}{3} = \frac{(2^p + 1)(2^p - 1)}{3}. \tag{4.58}$$

We claim that 4.58 is never a square in $\mathbb{Q}$ (equivalently not a square in $\mathbb{Z}$). Indeed assume the contrary. Since

$$2^p + 1 \equiv (-1)^p + 1 \equiv 0 \mod 3$$

it follows that

$$\frac{2^p + 1}{3} \in \mathbb{Z}.$$

Since $2^p + 1$ and $2^p - 1$ are coprime, it follows that $2^p - 1$ is the square of an integer $m \in \mathbb{Z}$. Hence

$$2^p = m^2 + 1 = (m + i)(m - i) \tag{4.59}$$

in $\mathbb{Z}[i]$, $i = \sqrt{-1}$. Hence

$$m + i = i^a \pi^b,$$

where $\pi = 1 + i$ and a, b are integers. So

$$m - i = \bar{i}^a \bar{\pi}^b = i^c \pi^b$$

for some integer c. By 4.59 $b = p$. Hence

$$2i = (m+i) - (m-i) = (i^a - i^c)\pi^p.$$

Taking norms we get $4 \geq 2 \cdot 2^p$, a contradiction. This proves our claim.

(5) Assume $q = 1 \in GL_2(A)$ is the identity matrix and let $\phi^{GL_1^c}$ be the Frobenius lift on

$$\mathcal{O}(\widehat{GL_1^c}) = A[a, b, \frac{1}{a^2 + b^2}]^{\widehat{}}$$

induced by the Frobenius lift ϕ^{GL_2} on $\widehat{GL_2}$ attached to q; here, as usual, we set

$$x = \begin{pmatrix} a & b \\ c & d \end{pmatrix}$$

and we view

$$\mathcal{O}(GL_1^c) = A[a, b, c, d, \det(x)^{-1}]/(a - d, b + c).$$

A trivial computation gives

$$\phi^{GL_1^c}\begin{pmatrix} a & b \\ -b & a \end{pmatrix} = \begin{pmatrix} U(a,b) & V(a,b) \\ -V(a,b) & U(a,b) \end{pmatrix}, \tag{4.60}$$

$$U(a,b) = a^p K(a,b), \quad V(a,b) = b^p K(a,b), \tag{4.61}$$

$$K(a,b) := \left(\frac{(a^2 + b^2)^p}{a^{2p} + b^{2p}}\right)_p^{1/2}. \tag{4.62}$$

PROOF OF ASSERTION (iii) IN 4.28 We first prove this assertion in case $r = 1$. In this case

$$\mathcal{O}(GL_1^c) = A[a, b, \frac{1}{a^2 + b^2}], \qquad \mathcal{O}(GL_1) = A[v, v^{-1}],$$

where a, b, v are 3 variables. Assume there is a Frobenius lift ϕ^{GL_1} such that $\phi^{GL_1^c}$ is c-horizontal with respect to ϕ^{GL_1} and seek a contradiction. Set

$$\phi^{GL_1}(v) = F(v) \in A[v, v^{-1}]^{\widehat{}}.$$

Then the horizontality condition gives (with notation as in 4.60),

$$F(a + \sqrt{-1} \cdot b) = U(a,b) + \epsilon \cdot \sqrt{-1} \cdot V(a,b)$$

where $\phi(\sqrt{-1}) = \epsilon \cdot \sqrt{-1}$, $\epsilon = \pm 1$. Taking $\frac{\partial}{\partial a}$ and $\frac{\partial}{\partial b}$ in the previous equality we get

$$\frac{\partial F}{\partial v}(a + \sqrt{-1} \cdot b) = \frac{\partial U}{\partial a} + \epsilon \cdot \sqrt{-1} \cdot \frac{\partial V}{\partial a},$$

$$\frac{\partial F}{\partial v}(a + \sqrt{-1} \cdot b) \times \sqrt{-1} = \frac{\partial U}{\partial b} + \epsilon \cdot \sqrt{-1} \cdot \frac{\partial V}{\partial b},$$

hence

$$\frac{\partial U}{\partial a} = \epsilon \cdot \frac{\partial V}{\partial b}, \quad \frac{\partial V}{\partial a} = -\epsilon \cdot \frac{\partial U}{\partial b}. \tag{4.63}$$

Now by 4.61, 4.62, one trivially gets:

$$\frac{\partial U}{\partial a} = pa^{p-1}K - pa^{p+1}b^2K^{-1}L, \tag{4.64}$$

$$\frac{\partial V}{\partial b} = pb^{p-1}K + pb^{p+1}a^2K^{-1}L, \tag{4.65}$$

$$L := \frac{(a^2+b^2)^{p-1}(a^{2p-2} - b^{2p-2})}{(a^{2p}+b^{2p})^2}. \tag{4.66}$$

Taking the image of 4.63 via the map

$$R[a, b, \frac{1}{a^2+b^2}]\hat{} \to R[b, b^{-1}]\hat{}, \quad a \mapsto 0,$$

one gets, by 4.64, that $pb^{p-1} = 0$ in $R[b, b^{-1}]\hat{}$, a contradiction. This ends the proof of the case $r = 1$ of assertion (iii) in our theorem.

Assume now that $r \geq 2$ and assume there is a Frobenius lift ϕ^{GL_r} such that $\phi^{GL_r^c}$ is c-horizontal with respect to ϕ^{GL_r}; again, we seek a contradiction. Consider the commutative diagram

$$\begin{array}{ccc} GL_1^c & \xrightarrow{i} & GL_r^c \\ c_1 \downarrow & & \downarrow c_r \\ GL_1 & \xrightarrow{j} & GL_r \end{array}$$

where c_1, c_r are the natural projections and

$$j(z) = z \cdot 1_r, \quad i\begin{pmatrix} a & b \\ -b & a \end{pmatrix} = \begin{pmatrix} a \cdot 1_r & b \cdot 1_r \\ -b \cdot 1_r & a \cdot 1_r \end{pmatrix}.$$

Let $\phi^{GL_1^c}$ be the Frobenius lift on $\widehat{GL_1^c}$ induced by the real Chern connection attached to $1 \in GL_{2r}$. Note that i is horizontal with respect to $\phi^{GL_1^c}$ and $\phi^{GL_r^c}$; this follows directly from the formula 4.36, giving $\phi^{GL_r^c}$, and from the fact that the group $i(GL_1^c)$ is invariant under the operations

$$x \mapsto x^{(p)}, \quad x \mapsto x^t, \quad (1+py) \mapsto \{1+py\}_p^{1/2}.$$

Since $\phi^{GL_r^c}$ sends $i(GL_1^c)$ into itself and since c_1 is surjective it follows that ϕ^{GL_r} sends $j(GL_1)$ into itself so it induces a Frobenius lift ϕ^{GL_1} on $\widehat{GL_1}$. Clearly $\phi^{GL_1^c}$ is then c_1-horizontal with respect to ϕ^{GL_1}, contradicting the case $r = 1$ of our proof. This ends our proof of the case $r \geq 2$. □

PROOF OF THEOREM 4.29. The theorem will be proved if we can find a matrix $x_0 \in SL_n(A)$ such that $\Phi(x_0) \notin SL_n(A)$, where Φ is as in 4.36. Pick $a, b, c, d \in R$ and consider the $2r \times 2r$ matrix

$$x_0 = \begin{pmatrix} a \cdot 1_r & b \cdot 1_r \\ c \cdot 1_r & d \cdot 1_r \end{pmatrix}.$$

Then

$$\det(x_0) = (ad - bc)^r. \tag{4.67}$$

By the way, in particular, $\det(q) = (-1)^r$. We assume

$$\det(\Phi(x_0)) = 1 \tag{4.68}$$

for all a, b, c, d with $ad - bc = 1$ and we seek a contradiction. Squaring the equality 4.68 and explicitly computing using the formulas 4.36 and 4.67 we get

$$((ad+bc)^{2p} - 4a^p b^p c^p d^p)^r = 1. \tag{4.69}$$

Now let $z \in R$ be arbitrary and set, in 4.69,

$$a = 1+z, \quad b = z, \quad c = -z, \quad d = 1-z.$$

We get

$$((1-2z^2)^{2p} + 4^p(1-z^2)z^{2p})^r = 1. \tag{4.70}$$

Since 4.70 is true for infinitely many z's it is valid as an equality of polynomials in z; but the coefficient of z^2 in

$$(1-2z^2)^{2p} + 4^p(1-z^2)z^{2p}$$

is $-4p \neq 0$ so the left hand side of 4.70 has degree > 0 in z, a contradiction. □

REMARK 4.37. The Frobenius lift ϕ^{GL_n} in Theorem 4.23 depends, of course, on q; we would like to discuss, in this remark, the nature of this dependence. Assume $A = R$ and $\phi_1^{GL_n}(x) = \phi_0^{GL_n}(x) = x^{(p)}$. Let $G_\pm$ be the locus of all $q \in G = GL_n$ such that $q^t = \pm q$ and for each $q \in G_\pm$ let

$$\sigma_q : \widehat{G} \to J^1(G)$$

be the section of $J^1(G) \to \widehat{G}$ corresponding to the Frobenius lift ϕ^{GL_n} attached to q in Theorem 4.23. Then by formula 4.36 the map of sets

$$G_\pm \times G \to J^1(G), \quad (q, a) \mapsto \sigma_q(a)$$

is a δ-map of order 1 and actually comes from a morphism of p-formal schemes

$$J^1(G_\pm) \times \widehat{G} \to J^1(G). \tag{4.71}$$

Explicitly, if y is a matrix of indeterminates, and we view

$$G_\pm = Spec\ R[y]/(y^t \mp y),$$

then the map 4.71 is given by the map of $R[x, \det(x)^{-1}]\widehat{\ }$-algebras

$$R[x, x', \det(x)^{-1}]\widehat{\ } \to R[x, y, y', \det(x)^{-1}]\widehat{\ }/(y^t \mp y, (y')^t \mp y')$$

sending

$$x' \mapsto \frac{1}{p} x^{(p)} \{\{(x^{(p)t}(y^{(p)} + py')x^{(p)})^{-1}(x^t y x)^{(p)}\}_p^{1/2} - 1\}.$$

4.3. Existence of Levi-Cività connections

The discussion below is devoted to an analogue of the Levi-Cività connection. The classical Levi-Cività connection is a connection on the frame bundle (of the tangent bundle) of a manifold hence, in particular the size, n, of the matrices of Christoffel symbols equals the number, m, of derivations on the ground ring of functions on the manifold. Our arithmetic analogue of the Levi-Cività connection *will not* exhibit this property: rather, we will assume we have *one* prime p only whereas the size, n, of the matrices will be arbitrary. A classical analogue of what follows is the case of cohomogeneity one metrics; cf. Example 2.53. Indeed for such metrics local coordinates ξ_i on the manifold can be chosen such that the partial derivatives $\frac{\partial}{\partial \xi_i}$ induce one and the same derivation δ of the ring generated by the components of our metrics. This single derivation δ corresponds, in our arithmetic setting, to our considering one prime only; nevertheless the size of the Christoffel symbols matrices will still be n.

So, in this section, we continue to consider a δ-ring of local type A.

Here is our existence and uniqueness result in this setting:

THEOREM 4.38. *Assume $q^t = q \in GL_n(A)$ and consider the involution $x^\tau = q^{-1}(x^t)^{-1}q$ on GL_n and the associated quadratic map $\mathcal{H}_q : GL_n \to GL_n$, $\mathcal{H}_q(x) = x^t qx$. Let $\phi_0^{GL_n}$ be the trivial Frobenius lift on $\widehat{GL_n}$ defined by $\phi_0^{GL_n}(x) = x^{(p)}$. Then there exists a unique n-tuple $(\phi_1^{GL_n}, ..., \phi_n^{GL_n})$ of Frobenius lifts on $\widehat{GL_n}$ such that if $\delta_i^{GL_n}$ are the p-derivations attached to $\phi_i^{GL_n}$, $\delta_i^{GL_n}(x) = \Delta_i(x) = (\Delta_{ijk})$, and if $\Gamma_{ij}^k = \Delta_{ikj}$ are the corresponding Christoffel symbols of the 2nd kind then, for $i, j, k = 1, ..., n$, we have:*

(1) *$\phi_i^{GL_n}$ is $\mathcal{H}_q$-horizontal with respect to $\phi_0^{GL_n}$;*
(2) *$\Gamma_{ij}^k = \Gamma_{ji}^k$.*

In addition we have:

(3) *If $\delta q = 0$ then $\delta_i^{GL_n}$ fixes the identity 1 and is left invariant with respect to $N_{SO(q)}$.*

The symmetry in assertion (2) of the theorem is, of course, similar to the "vanishing of the torsion" condition 2.159 in the definition of the classical Levi-Cività connection; it reads:

$$\delta_i^{GL_n}(x_{kj}) = \delta_j^{GL_n}(x_{ki}).$$

Compare also with 2.160.

DEFINITION 4.39. The tuple

$$(\delta_1^{GL_n}, ..., \delta_n^{GL_n}) \tag{4.72}$$

in Theorem 4.38 is called the *Levi-Cività connection* on GL_n attached to q.

PROOF OF THEOREM 4.38. We prove the existence of the tuple 4.72; its uniqueness is proved similarly. Set

$$A = x^{(p)t} \cdot \phi(q) \cdot x^{(p)}, \quad B = (x^t qx)^{(p)}.$$

We will construct by induction a sequence of n-tuples

$$(\Lambda_1^{(\nu)}, ..., \Lambda_n^{(\nu)}), \quad \nu \geq 1$$

of $n \times n$ matrices in entries in $\mathcal{A} := \mathcal{O}(\widehat{GL_n})$ with the following properties:

(i) $\Lambda_i^{(1)} = 1$,
(ii) $\Lambda_i^{(\nu+1)} \equiv \Lambda_i^{(\nu)} \bmod p^\nu$,
(iii) $\Lambda_i^{(\nu)t} A \Lambda_i^{(\nu)} \equiv B \bmod p^\nu$,
(iv) If $\Delta_i^{(\nu)} := \frac{1}{p} x^{(p)}(\Lambda_i^{(\nu)} - 1) = (\Delta_{ijk}^{(\nu)})$ then $\Delta_{ikj}^{(\nu)} = \Delta_{jki}^{(\nu)}$,
(v) In case $\delta q = 0$ we have

$$\Lambda_i^{(\nu)} \equiv 1 \mod (x-1) \quad \text{and} \quad \Lambda_i^{(\nu)}(ux) = \Lambda_i^{(\nu)}(x) \quad \text{for} \quad u \in N_{SO(q)}.$$

Then we are done by setting

$$\Lambda_i = \lim_{\nu \to \infty} \Lambda_i^{(\nu)}$$

and

$$\phi_i^{GL_n}(x) = x^{(p)} \Lambda_i.$$

Indeed condition (iii) implies assertion (1) of the Theorem; condition (iv) implies the symmetry in assertion (2) of the Theorem; and condition (v) implies assertion (3) of the Theorem.

To construct of our sequence of n-tuples define the n-tuple for $\nu = 1$ by condition (i), assume the n-tuple was constructed for some ν and seek the n-tuple for $\nu + 1$ in the form

$$\Lambda_i^{(\nu+1)} = \Lambda_i^{(\nu)} + p^\nu Z_i. \tag{4.73}$$

Write

$$\Lambda_i^{(\nu)t} A \Lambda_i^{(\nu)} = B - p^\nu C_i. \tag{4.74}$$

Then

$$\begin{aligned} \Lambda_i^{(\nu+1)t} A \Lambda_i^{(\nu+1)} &\equiv \Lambda_i^{(\nu)t} A \Lambda_i^{(\nu)} + p^\nu (\Lambda_i^{(\nu)t} A Z_i + Z_i^t A \Lambda_i^{(\nu)}) \mod p^{\nu+1} \\ &\equiv B + p^\nu(-C_i + AZ_i + Z_i^t A) \mod p^{\nu+1}. \end{aligned} \tag{4.75}$$

Now $A^t = A$ and $B^t = B$ so $C_i^t = C_i$ so if $C_i = (C_{ijk})$ then $C_{ijk} = C_{ikj}$. So there exists a (unique) family (D_{ijk}) of elements in $\mathcal{A}$ such that

$$D_{ijk} = D_{jik} \tag{4.76}$$

and

$$D_{ijk} + D_{ikj} = C_{ijk}; \tag{4.77}$$

explicitly,

$$D_{ijk} = \frac{1}{2}(C_{ijk} + C_{jik} - C_{kij}). \tag{4.78}$$

Cf. Remark 2.113. Now define the matrices $D_i = (D_{ijk})$ hence

$$D_i + D_i^t = C_i.$$

Setting

$$Z_i := A^{-1} D_i^t \tag{4.79}$$

we get $D_i^t = AZ_i$ and $D_i = Z_i^t A$ so, by 4.75,

$$\Lambda_i^{(\nu+1)t} A \Lambda_i^{(\nu+1)} \equiv B \mod p^{\nu+1}.$$

So condition (iii) holds for $\Lambda_i^{(\nu+1)}$ mod $p^{\nu+1}$. Also note that, by 4.76, we have

$$(D_i E)_{jk} = (D_j E)_{ik} \tag{4.80}$$

for any square matrix E; cf. Remark 2.114.

To check condition (iv) for $\Delta_{ikj}^{(\nu+1)}$ we note that

$$\begin{aligned} \Delta_i^{(\nu+1)t} &= \tfrac{1}{p}(\Lambda_i^{(\nu+1)t} - 1)x^{(p)t} \\ &= \tfrac{1}{p}(\Lambda_i^{(\nu)t} + p^\nu Z_i^t - 1)x^{(p)t} \\ &= \Delta_i^{(\nu)t} + p^{\nu-1} Z_i^t x^{(p)t} \\ &= \Delta_j^{(\nu)t} + p^{\nu-1} D_i A^{-1} x^{(p)t}. \end{aligned}$$

Using 4.80 we get

$$\begin{aligned}\Delta_{ikj}^{(\nu+1)} &= (\Delta_i^{(\nu+1)t})_{jk}\\ &= (\Delta_i^{(\nu)t})_{jk} + p^{\nu-1}(D_i A^{-1} x^{(p)t})_{jk}\\ &= (\Delta_j^{(\nu)t})_{ik} + p^{\nu-1}(D_j A^{-1} x^{(p)t})_{ik}\\ &= \Delta_{jki}^{(\nu+1)}.\end{aligned}$$

Finally, in order to check condition (v) note that if $\delta q = 0$ we have

$$A \equiv B \mod (x-1)$$

hence, by 4.74,

$$C_i \equiv 0 \mod (x-1)$$

hence, by 4.78

$$D_{ijk} \equiv 0 \mod (x-1)$$

hence, by 4.79,

$$Z_i \equiv 0 \mod (x-1)$$

so, by 4.73,

$$\Lambda_i^{(\nu+1)} \equiv 1 \mod (x-1).$$

Also the condition that $\delta q = 0$ is trivially seen to imply that A and B are invariant under $x \mapsto ux$ for $u \in N_{SO(q)}$, hence so are C_i, D_i, Z_i, and $\Lambda_i^{(\nu+1)}$, respectively. This ends our induction step in the construction of the sequence $(\Lambda_i^{(\nu)})_{\nu \geq 1}$ and hence ends our proof. □

We would like to compare the real Chern connection δ^{GL_n} and Levi-Cività connection $(\delta_1^{GL_n}, ..., \delta_n^{GL_n})$ attached to a symmetric matrix $q \in GL_n(A)$; cf. Definitions 4.25 and 4.39.

DEFINITION 4.40. Let δ^{GL_n} be the real Chern connection attached to the symmetric matrix q, let $\Delta(x) := \delta^{GL_n}(x)$, $A := x^{(p)t} \cdot \phi(q) \cdot x^{(p)}$, and let Δ^t be the Christoffel symbol of the 2nd kind. Then the matrix

$$\Gamma := \Delta^t \cdot (x^{(p)t})^{-1} \cdot A = \Delta^t \cdot \phi(q) \cdot x^{(p)} \tag{4.81}$$

is called the *Christoffel symbol of the* 1st *kind* of the real Chern connection. Similarly, let

$$(\delta_1^{GL_n}, ..., \delta_n^{GL_n})$$

be the Levi-Cività connection attached to q, $\Delta_i(x) := \delta_i^{GL_n}(x)$, and let Δ_i^t be the Christoffel symbols of the 2nd kind. Then the matrices

$$\Gamma_i := \Delta_i^t \cdot (x^{(p)t})^{-1} \cdot A = \Delta_i^t \cdot \phi(q) \cdot x^{(p)} \tag{4.82}$$

are called the *Christoffel symbols of the* 1st *kind* of the Levi-Cività connection. We write

$$\Gamma = (\Gamma_{jk}), \quad \Gamma_i = (\Gamma_{ijk}). \tag{4.83}$$

REMARK 4.41. Note that one passes from 2nd kind to 1st kind Christoffel symbols via right multiplication by $\phi(q)x^{(p)}$; this is an analogue of the "lowering of the indices" in classical differential geometry. Note also that assertion (2) in Theorem 4.38 immediately implies (by Remark 2.114) the following symmetry for the Christoffel symbols of the 1st kind:

$$\Gamma_{ijk} = \Gamma_{jik}. \tag{4.84}$$

Here is a corollary to the proofs of Theorems 4.23 and 4.38.

COROLLARY 4.42.

$$\Gamma \equiv \frac{1}{2p}\{(x^t q x)^{(p)} - x^{(p)t}\phi(q)x^{(p)}\} \quad mod \ \ p; \tag{4.85}$$

in particular

$$\Gamma_{jk} \equiv \Gamma_{kj} \quad mod \ \ p \tag{4.86}$$

and

$$\Gamma_{jk} \equiv -\frac{1}{2}\delta q_{jk} \quad mod \ \ (p, x-1). \tag{4.87}$$

Moreover,

$$\Gamma_{ijk} \equiv \Gamma_{jk} + \Gamma_{ij} - \Gamma_{ik} \quad mod \ \ p, \tag{4.88}$$

hence

$$\Gamma_{ijk} \equiv -\frac{1}{2}(\delta q_{jk} + \delta q_{ij} - \delta q_{ik}) \quad mod \ \ (p, x-1). \tag{4.89}$$

The above congruences are, of course, analogous to the corresponding formulas in classical differential geometry. There are also some differences. For instance there are sign differences coming from slightly different conventions in the two situations. Also the operators δ have no indices in the arithmetic case; this makes our arithmetic Levi-Cività paradigm an analogue of the Levi-Cività paradigm for cohomogeneity one metrics in the case of classical differential geometry. Indeed in classical differential geometry all the partial derivative operators act identically on the δ-ring generated by the coefficients of a cohomogeneity one metric. Finally note that, as expected, we only have congruences mod p in the arithmetic case, and not equalities; this is an instance of the principle that the arithmetic analogues of classical differential geometric objects are non-linear objects which, mod p, behave like the classical, linear ones.

PROOF OF COROLLARY 4.42. With

$$A = x^{(p)t}\phi(q)x^{(p)}, \quad B = (x^t q x)^{(p)}, \quad C = \frac{1}{p}(B - A),$$

as in the proof of Theorem 4.23 we have, by the proof of that theorem, that $\Phi(x) = x^{(p)}\Lambda(x)$ with

$$\Lambda(x) = (1 + pA^{-1}C)_p^{1/2} \equiv 1 + \frac{p}{2}A^{-1}C \quad mod \ \ p^2,$$

hence

$$\begin{aligned}
\Gamma^t &= A \cdot (x^{(p)})^{-1} \cdot \frac{1}{p} \cdot (x^{(p)}\Lambda - x^{(p)}) \\
&= \frac{1}{p}(A\Lambda - A) \\
&\equiv \frac{1}{p}(A(1 + \frac{p}{2}A^{-1}C) - A) \quad mod \ \ p \\
&= \frac{1}{2}C \\
&= \frac{1}{2p}(B - A),
\end{aligned}$$

which proves 4.85.

Similarly, with the notation in the proof of Theorem 4.38 we have

$$\Lambda_i \equiv 1 + pZ_i \quad mod \ \ p^2$$

where $D_i := Z_i^t A$ satisfies

$$D_{ijk} = \frac{1}{2}(C_{jk} + C_{ik} - C_{ij}).$$

Hence

$$\Gamma_i^t \equiv \frac{1}{p}(A\Lambda_i - A) \equiv D_i^t \quad mod \ \ p.$$

This plus the previously checked equality $\Gamma^t = \frac{1}{2}C$ ends the proof of 4.88. The rest of the congruences then follow. □

4.4. Existence/non-existence of Fedosov connections

In contrast with our existence theorem for Levi-Cività connections we will prove here that a corresponding natural arithmetic analogue of Fedosov connection does not exist for $n \geq 4$ but it exists for $n = 2$.

We consider a δ-ring of local type A and $p \neq 2, 3$.

THEOREM 4.43. *Assume $q^t = -q \in GL_n(A)$ and consider the associated map $\mathcal{H}_q : GL_n \to GL_n$, $\mathcal{H}_q(x) = x^t q x$. Let $\phi_0^{GL_n}$ be the trivial Frobenius lift on $\widehat{GL_n}$ defined by $\phi_0^{GL_n}(x) = x^{(p)}$. For any n-tuple $(\phi_1^{GL_n}, ..., \phi_n^{GL_n})$ of Frobenius lifts on $\widehat{GL_n}$ denote by $\delta_i^{GL_n}$ the p-derivations attached to $\phi_i^{GL_n}$, $\delta_i^{GL_n}(x) = \Delta_i(x) = (\Delta_{ijk})$, and let $\Gamma_{ij}^k = \Delta_{ikj}$, be the corresponding Christoffel symbols of the 2nd kind. Consider the following conditions:*

1. *$\phi_i^{GL_n}$ is $\mathcal{H}_q$-horizontal with respect to $\phi_0^{GL_n}$ for all i;*
2. *$\Gamma_{ij}^k \equiv \Gamma_{ji}^k$ mod $(p, x-1)$ for all i, j, k.*
3. *$\Gamma_{ij}^k \equiv \Gamma_{ji}^k$ mod p for all i, j, k.*
4. *$\Gamma_{ij}^k = \Gamma_{ji}^k$ for all i, j, k.*
5. *$\delta q_{jk} + \delta q_{ki} + \delta q_{ij} \equiv 0$ mod p for all i, j, k.*

Then the following hold:

- (I) *Conditions (1) and (2) imply condition (5).*
- (II) *If q is split and $n \geq 4$ then conditions (1) and (3) cannot hold simultaneously.*
- (III) *If $n = 2$ there exists $(\phi_1^{GL_2}, \phi_2^{GL_2})$ satisfying conditions (1) and (4).*

DEFINITION 4.44. Assume $(\phi_1^{GL_2}, ..., \phi_n^{GL_n})$ is a tuple satisfying conditions (1) and (4) in Theorem 4.43. Then the corresponding tuple of p-derivations

$$(\delta_1^{GL_2}, ..., \delta_n^{GL_n})$$

is called a *Fedosov connection* attached to q.

REMARK 4.45. Of course we have the following trivial implications between the conditions in the Theorem:

$$4 \Rightarrow 3 \Rightarrow 2.$$

So Theorem 4.43 shows that Fedosov connections do not exist for $n \geq 4$ and q split but they exist for $n = 2$ and any q. Of course, this leaves open the possibility that weaker arithmetic analogues of Fedosov connections exist for all n; it would be interesting to find such a weaker concept. By the way, condition (5) in our Theorem is an analogue of the condition defining a *symplectic form* in classical differential geometry; note that δ in condition (5) does not possess an index which makes condition (5) correspond, as in the discussion of Levi-Cività connection, to the classical case of *cohomogeneity one* metrics/forms. Finally let us note that conditions (1) and (2) are sometimes simultaneously satisfied for all even n: this is the case, for instance, if all $\phi_i^{GL_n}$ are equal to the real Chern connection attached to a q for which $\delta q = 0$; in this case condition (5) is, of course, also satisfied. It is not clear at this point whether condition (5) in the theorem guarantees the existence of a tuple of Frobenius lifts satisfying conditions (1) and (2).

PROOF OF THEOREM 4.43. Assume first that conditions (1) and (2) in the Theorem hold. Consider again the matrices

$$A = x^{(p)t}\phi(q)x^{(p)}, \quad B = (x^t qx)^{(p)}, \quad C = \frac{1}{p}(B - A). \tag{4.90}$$

Also set $\phi_i(x) = x^{(p)}\Lambda_i(x)$, $\Lambda_i = 1+pM_i$, hence $\Delta_i = x^{(p)}M_i$. Recall that condition 1 is equivalent to

$$\Lambda_i^t \cdot A \cdot \Lambda_i = B,$$

from which we get

$$M_i^t A + AM_i \equiv C \mod p. \tag{4.91}$$

Set $M_i = (M_{ijk})$. Condition (2) reads

$$\Delta_{ikj} \equiv \Delta_{jki} \mod (p, x - 1)$$

which immediately implies

$$M_{ikj} \equiv M_{jki} \mod (p, x - 1); \tag{4.92}$$

cf. Remark 2.114. Equation 4.91 implies

$$\sum_l M_{ilj}A_{lk} + \sum_l M_{ilk}A_{jl} \equiv C_{jk} \mod p. \tag{4.93}$$

Permuting the indices cyclically we get:

$$\sum_l M_{jlk}A_{li} + \sum_l M_{jli}A_{kl} \equiv C_{ki} \mod p, \tag{4.94}$$

$$\sum_l M_{kli}A_{lj} + \sum_l M_{klj}A_{il} \equiv C_{ij} \mod p. \tag{4.95}$$

Adding 4.93, 4.94, 4.95, and using 4.92 plus the fact that $A_{ij} = -A_{ji}$ we get

$$C_{jk} + C_{ki} + C_{ij} \equiv 0 \mod (p, x-1). \tag{4.96}$$

On the other hand clearly

$$C_{jk} \equiv -\delta q_{jk} \mod (x-1). \tag{4.97}$$

Now 4.96 and 4.97 imply condition (5) in the Theorem; this ends the proof of assertion (I) in the Theorem.

Assume now conditions (1) and (3) in the Theorem hold for q split and $n = 2r \geq 4$ and seek a contradiction; this will end the proof of assertion (II) in the Theorem. As above condition (3) implies

$$M_{ikj} \equiv M_{jki} \mod p. \tag{4.98}$$

Adding 4.93, 4.94, 4.95, and using 4.98 plus the fact that $A_{ij} = -A_{ji}$ we get

$$C_{jk} + C_{ki} + C_{ij} \equiv 0 \mod p. \tag{4.99}$$

We may view 4.99 as a congruence between polynomials in the entries of the $r \times r$ matrices a, b, c, d, where $x = \begin{pmatrix} a & b \\ c & d \end{pmatrix}$. Set, in this congruence, $a = d = 1$, $c = 0$, and take $i = 2$, $j = r+1$, and $k = r+2$. We get a congruence

$$\gamma_{r+1,r+2} + \gamma_{r+2,2} + \gamma_{2,r+1} \equiv 0 \mod p, \tag{4.100}$$

where

$$\gamma = \frac{1}{p}\begin{pmatrix} 0 & 0 \\ 0 & (b^t - b)^{(p)} - b^{(p)t} + b^{(p)} \end{pmatrix}.$$

Since $\gamma_{r+2,2} = \gamma_{2,r+1} = 0$ and $\gamma_{r+1,r+2} \not\equiv 0$ mod p, we obtain a contradiction.

Finally let us prove assertion (III) of the Theorem. Again consider the matrices A and B in 4.90. We will construct by induction a sequence of pairs

$$(\Lambda_1^{(\nu)}, \Lambda_2^{(\nu)}), \quad \nu \geq 1$$

of 2×2 matrices in entries in $\mathcal{A} := \mathcal{O}(\widehat{GL_2})$ with the following properties:

(i) $\Lambda_i^{(1)} = 1$,
(ii) $\Lambda_i^{(\nu+1)} \equiv \Lambda_i^{(\nu)} \bmod p^\nu$,
(iii) $\Lambda_i^{(\nu)t} A \Lambda_i^{(\nu)} \equiv B \bmod p^\nu$,
(iv) If $\Delta_i^{(\nu)} := \frac{1}{p} x^{(p)} (\Lambda_i^{(\nu)} - 1) = (\Delta_{ijk}^{(\nu)})$ then $\Delta_{ikj}^{(\nu)} = \Delta_{jki}^{(\nu)}$,

Then we are done by setting

$$\Lambda_i = \lim_{\nu \to \infty} \Lambda_i^{(\nu)}, \quad \phi_i^{GL_n}(x) = x^{(p)} \Lambda_i.$$

Indeed condition (iii) implies condition (1) of the Theorem; condition (iv) implies the symmetry in condition (4) of the Theorem. We proceed as in the proof of Theorem 4.38. To construct of our sequence of n-tuples define the n-tuple for $\nu = 1$ by condition (i), assume the n-tuple was constructed for some ν and seek the n-tuple for $\nu + 1$ in the form

$$\Lambda_i^{(\nu+1)} = \Lambda_i^{(\nu)} + p^\nu Z_i. \tag{4.101}$$

Write

$$\Lambda_i^{(\nu)t} A \Lambda_i^{(\nu)} = B - p^\nu C_i. \tag{4.102}$$

Then

$$\begin{aligned}\Lambda_i^{(\nu+1)t}A\Lambda_i^{(\nu+1)} &\equiv \Lambda_i^{(\nu)t}A\Lambda_i^{(\nu)}\\ &\quad +p^\nu(\Lambda_i^{(\nu)t}AZ_i+Z_i^tA\Lambda_i^{(\nu)}) \mod p^{\nu+1} \qquad (4.103)\\ &\equiv B+p^\nu(-C_i+AZ_i+Z_i^tA) \mod p^{\nu+1}.\end{aligned}$$

Now $A^t=-A$ and $B^t=-B$ so $C_i^t=-C_i$ so if $C_i=(C_{ijk})$ then

$$C_{ijk}=-C_{ikj}. \tag{4.104}$$

So there exists a (unique) family (D_{ijk}) of elements in $\mathcal{A}$ such that

$$D_{ijk}=D_{jik} \tag{4.105}$$

and

$$D_{ikj}-D_{ijk}=C_{ijk}; \tag{4.106}$$

explicitly,

$$D_{ijk}=-\frac{1}{3}(C_{ijk}+C_{jik}). \tag{4.107}$$

Indeed we have

$$D_{ikj}-D_{ijk}=C_{ijk}-\frac{1}{3}(C_{ijk}+C_{jki}+C_{kij}). \tag{4.108}$$

But

$$C_{ijk}+C_{jki}+C_{kij}=0 \tag{4.109}$$

because $n=2$ so at least 2 of the indices i,j,k are equal and we have that 4.104 holds. (This is the only place where we use the hypothesis $n=2$ in our proof.) Now define the matrices $D_i=(D_{ijk})$ hence

$$D_i^t-D_i=C_i.$$

Setting

$$Z_i:=A^{-1}D_i^t \tag{4.110}$$

we get $D_i^t=AZ_i$ and $D_i=-Z_i^tA$ so, by 4.103,

$$\Lambda_i^{(\nu+1)t}A\Lambda_i^{(\nu+1)}\equiv B \mod p^{\nu+1}.$$

So condition (iii) holds for $\Lambda_i^{(\nu+1)}$ mod $p^{\nu+1}$. Also note that, by 4.105, we have

$$(D_iE)_{jk}=(D_jE)_{ik} \tag{4.111}$$

for any square matrix E; cf. Remark 2.114.

To check condition (iv) for $\Delta_{ikj}^{(\nu+1)}$ we note that

$$\begin{aligned}\Delta_i^{(\nu+1)t} &= \tfrac{1}{p}(\Lambda_i^{(\nu+1)t}-1)x^{(p)t}\\ &= \tfrac{1}{p}(\Lambda_i^{(\nu)t}+p^\nu Z_i^t-1)x^{(p)t}\\ &= \Delta_i^{(\nu)t}+p^{\nu-1}Z_i^tx^{(p)t}\\ &= \Delta_j^{(\nu)t}-p^{\nu-1}D_iA^{-1}x^{(p)t}.\end{aligned}$$

Using 4.111 we get

$$\begin{aligned}\Delta_{ikj}^{(\nu+1)} &= (\Delta_i^{(\nu+1)t})_{jk}\\ &= (\Delta_i^{(\nu)t})_{jk} - p^{\nu-1}(D_iA^{-1}x^{(p)t})_{jk}\\ &= (\Delta_j^{(\nu)t})_{ik} - p^{\nu-1}(D_jA^{-1}x^{(p)t})_{ik}\\ &= \Delta_{jki}^{(\nu+1)}.\end{aligned}$$

This ends the proof. □

REMARK 4.46. Assume $n = 2$. Then the Fedosov connections attached to a given q, whose existence was proved in Theorem 4.43, are not unique; it is actually an easy exercise to parameterize all such Fedosov connections in terms of "symmetric 3-tensors" with components in $\mathcal{A} = \mathcal{O}(\widehat{GL_2})$; cf. 2.174.

EXAMPLE 4.47. Let $n = 2$ and let

$$q = \begin{pmatrix} 0 & 1 \\ -1 & 0 \end{pmatrix}, \quad x = \begin{pmatrix} a & b \\ c & d \end{pmatrix}, \quad \mu := \frac{(ad-bc)^p - (a^pd^p - b^pc^p)}{3p(ad-bc)^{2p}}.$$

Then the Fedosov connection $(\delta_1^{GL_2}, \delta_2^{GL_2})$ attached to q, constructed in the proof of Theorem 4.43, is trivially seen to satisfy the congruences:

$$\Delta_1 \equiv \mu \cdot \begin{pmatrix} 0 & a^p + b^p \\ 0 & c^p + d^p \end{pmatrix} \mod p, \tag{4.112}$$

$$\Delta_2 \equiv \mu \cdot \begin{pmatrix} a^p + b^p & 0 \\ c^p + d^p & 0 \end{pmatrix} \mod p, \tag{4.113}$$

where we recall that $\Delta_i = \delta_i^{GL_2}(x)$, $i = 1, 2$.

4.5. Existence/non-existence of Lax-type connections

In this section we continue to assume $A \subset R$ is a δ-ring of local type. We will state and prove here our main existence/non-existence results for "Lax-type connections," by which we loosely mean at this point connections compatible (in various ways) with inner automorphisms of GL_n. (Later we will define *Lax connections* in a precise way.) Throughout this section we continue to consider the Frobenius lift $\phi_0^{GL_n}(x) = x^{(p)}$ on $\widehat{GL_n}$. We continue to denote by T and N the maximal diagonal torus and its normalizer in $G = GL_n$.

Let us start by considering an inner involution $x^\tau = q^{-1}xq$, $q \in T(A) \subset GL_n(A)$ with $q^2 = 1$, and let $\mathcal{H}(x) = q^{-1}x^{-1}qx$ and $\mathcal{B}(x,y) = q^{-1}x^{-1}qy$ be the attached quadratic and bilinear maps. For such an involution the group G^+ is the centralizer of q and the corresponding factor $G^+\backslash G$ can be identified with the conjugacy class of q. One can ask if, in analogy with the case of outer involutions (Theorem 4.23), there exists a Frobenius lift ϕ^{GL_n} on $\widehat{GL_n}$ that is $\mathcal{H}$-horizontal and $\mathcal{B}$-symmetric with respect to $\phi_0^{GL_n}$. In the trivial case $q = \pm 1$ the answer is yes (and then the necessarily unique ϕ^{GL_n} equals $\phi_0^{GL_n}$ itself). But, except for this case, the answer is *no*; cf. the next Theorem.

THEOREM 4.48. *Assume* $q \in T(A) \subset GL_n(A)$, $\phi(q) = q$, $q^2 = 1$, $q \neq \pm 1$. *Consider the involution* $x^\tau = q^{-1}xq$ *on* GL_n, *the associated quadratic map* $\mathcal{H}(x) = q^{-1}x^{-1}qx$, *and the Frobenius lift* $\phi_0^{GL_n}(x) = x^{(p)}$ *on* $\widehat{GL_n}$. *Then there is no Frobenius lift* ϕ^{GL_n} *on* $\widehat{GL_n}$ *that is* $\mathcal{H}$*-horizontal with respect to* $\phi_0^{GL_n}$.

PROOF. Let A and B be defined by

$$A = (x^{(p)})^{-1}qx^{(p)}, \quad B = (x^{-1}qx)^{(p)}. \tag{4.114}$$

Assume $\phi^{GL_n}(x) = x^{(p)} \cdot \Lambda(x)$ is a Frobenius lift on $\widehat{GL_n}$ that is $\mathcal{H}$-horizontal with respect to $\phi_0^{GL_n}$ and seek a contradiction. By $\mathcal{H}$-horizontality we get

$$q^{-1}\Phi(x)^{-1}q\Phi(x) = (q^{-1}x^{-1}qx)^{(p)},$$

hence $\Lambda^{-1}A\Lambda = B$. Taking determinants we get

$$\det(B) = \det(q). \tag{4.115}$$

In order to get a contradiction it is enough to show that 4.115 fails for some value of x. In order to check the latter we may assume

$$q = \begin{pmatrix} q' & 0 \\ 0 & q'' \end{pmatrix}, \quad q' \in GL_{n-2}, \quad q'' = \begin{pmatrix} 1 & 0 \\ 0 & -1 \end{pmatrix}.$$

Let us assume 4.115 holds for all matrices x of the form

$$x = \begin{pmatrix} 1_{n-2} & 0 & 0 \\ 0 & a & b \\ 0 & c & d \end{pmatrix} \in GL_n(R)$$

and seek a contradiction. Then 4.115 yields the equality

$$(ad+bc)^{2p} - 2^{2p}a^pb^pc^pd^p = (ad-bc)^{2p}$$

valid for all $a, b, c, d \in R$ with $ad - bc \in R^\times$. Since R/pR is algebraically closed the above equality is valid as an equality between polynomials in 4 variables a, b, c, d. Picking out the coefficients of $a^{2p-1}d^{2p-1}bc$ we get a contradiction. □

REMARK 4.49. The above theorem shows that inner automorphisms $x^\tau = q^{-1}xq$ do not fit into the paradigm of Theorem 4.23. The next theorem shows that such inner automorphisms are, on the other hand, naturally compatible with $\mathcal{C}$-horizontality; cf. Definition 3.156.

So let, in what follows, $D^*(x) \in A[x]$ be the discriminant of the characteristic polynomial $\det(s \cdot 1 - x) \in A[x][s]$ and let $GL_n^* \subset GL_n$ be the open subset where $D^*(x)$ is invertible, i.e.

$$GL_n^* = Spec\ A[x, \det(x)^{-1}, D^*(x)^{-1}].$$

We refer to GL_n^* as the locus of *regular matrices*. Then $T^* = T \cap GL_n^*$ is the locus of *regular diagonal matrices*. If $T = Spec\ A[t, \det(t)^{-1}]$, with $t = \text{diag}(t_1, ..., t_n)$ then $\det(t) = t_1...t_n$ and

$$T^* = Spec\ A[t, \det(t), D(t)^{-1}], \quad D(t) = \prod_{i<j}(t_i - t_j)^2.$$

Consider the morphism

$$\mathcal{C} : T^* \times GL_n \to GL_n^*, \quad \mathcal{C}(d, a) = a^{-1}da. \tag{4.116}$$

It is defined by the map

$$A[x, \det(x)^{-1}, D^*(x)^{-1}] \to A[t, \det(t)^{-1}, D^*(t)^{-1}, x, \det(x)^{-1}], \quad x \mapsto x^{-1}tx.$$

Note that, under this map, $\det(x) \mapsto \det(t)$, $D^*(x) \mapsto D^*(t)$. Also consider, as usual, the Frobenius lift $\phi_0^{GL_n}(x) = x^{(p)}$ on $\widehat{GL_n}$. Recall that $\phi_0^G = \phi_0^{GL_n}$ is called $\mathcal{C}$*-horizontal* with respect to a Frobenius lift $\phi^{G^*} = \phi^{GL_n^*}$ on $\widehat{G^*} = \widehat{GL_n^*}$ if the following diagram is commutative.

$$\begin{array}{ccc} \widehat{T^*} \times \widehat{G} & \stackrel{\phi_0^{T^*} \times \phi_0^G}{\longrightarrow} & \widehat{T^*} \times \widehat{G} \\ \mathcal{C} \downarrow & & \downarrow \mathcal{C} \\ \widehat{G^*} & \stackrel{\phi^{G^*}}{\longrightarrow} & \widehat{G^*} \end{array} \tag{4.117}$$

Then we will prove:

THEOREM 4.50. *There exists a unique Frobenius lift ϕ^{G^*} on $\widehat{G^*}$ such that ϕ_0^G is $\mathcal{C}$-horizontal with respect to ϕ^{G^*}.*

By Theorem 4.50 and Lemma 3.157 one trivially gets:

THEOREM 4.51. *Let $q \in T(A)$ be regular (i.e. $q \in T^*(A)$) with $q^{(p)} = q$, let τ be the automorphism of $G = GL_n$ defined by $x^\tau = q^{-1}xq$, and consider the associated quadratic map $\mathcal{H}(x) = q^{-1}x^{-1}qx$. Consider the Frobenius lift $\phi_0^G(x) = x^{(p)}$ on $\widehat{G}$. Then there exists a Frobenius lift ϕ^G on $\widehat{G}$ such that ϕ_0^G is $\mathcal{H}$-horizontal with respect to ϕ^G.*

To prove Theorem 4.50 we need some preliminaries. In the discussion below we will identify k-varieties ($k = R/pR$) with their sets of k-points. Furthermore we let $GL_{n,k}$, T_k, N_k be the corresponding algebraic groups over k. Note that N acts on the left on $T^* \times GL_n$ by the rule

$$n \cdot (d, a) = (ndn^{-1}, na), \quad n \in N,\ d \in T^*,\ a \in GL_n. \tag{4.118}$$

In particular N_k acts on

$$\mathcal{O}(T_k^* \times GL_{n,k}) = k[t, \det(t)^{-1}, D^*(t)^{-1}, x, \det(x)^{-1}].$$

LEMMA 4.52. *The ring of N_k-invariant elements in $\mathcal{O}(T_k^* \times GL_{n,k})$ is given by*

$$k[t, \det(t)^{-1}, D^*(t)^{-1}, x, \det(x)^{-1}]^{N_k} = k[x^{-1}tx, \det(t)^{-1}, D^*(t)^{-1}].$$

In particular the same holds with k replaced by $\mathbb{F}_p$.

PROOF. The variety $GL_{n,k}$ is a Zariski locally trivial principal homogeneous space for T_k over the affine variety U_k obtained from $(\mathbb{P}_k^{n-1})^n$ by removing the zero locus of the determinant. (Here the elements of $\mathbb{P}_k^{n-1}$ correspond to the rows of our matrices.) Then the variety

$$X_k := T_k^* \times GL_{n,k}$$

is a Zariski locally trivial principal homogeneous space over

$$Y_k := T_k^* \times U_k.$$

We have that T_k acts on X_k via its left action on $GL_{n,k}$, Y_k is set theoretically the quotient

$$Y_k = T_k \backslash X_k,$$

and

$$\mathcal{O}(X)^{T_k} = \mathcal{O}(Y_k).$$

Consider further the quotient

$$Z_k = W_k \backslash Y_k$$

of Y_k by the finite permutation group W_k acting via the action induced by that of N_k; so

$$\mathcal{O}(Z_k) = \mathcal{O}(Y_k)^{W_k}.$$

Hence

$$Z_k = N_k \backslash X_k$$

as sets and

$$\mathcal{O}(Z_k) = \mathcal{O}(X_k)^{N_k}.$$

Set

$$V_k = GL^*_{n,k}.$$

The morphism $\mathcal{C}_k : X_k \to V_k$ induced by $\mathcal{C}$ is surjective and its set-theoretic fibers coincide with the orbits of the N_k-action on X_k. Since the entries of the matrix $x^{-1}tx$ are N_k-invariant it follows that the map $\mathcal{C}_k$ factors as

$$\mathcal{C}_k : X_k \xrightarrow{\alpha} Y_k \xrightarrow{\beta} Z_k \xrightarrow{\gamma} V_k$$

and γ is a bijection.

Claim 1. The scheme theoretic fiber of $\mathcal{C}_k$ through any point of the form $(d,1) \in X_k$ is smooth at $(d,1)$.

Indeed the tangent space at 1 of the above scheme theoretic fiber identifies with the linear space of all pairs (a_1, d_1) of $n \times n$ matrices with entries in k such that

$$(1 + \epsilon a_1)^{-1}(d + \epsilon d_1)(1 + \epsilon a_1) = d,$$

where $\epsilon^2 = 0$; the above condition reads $da_1 - a_1 d + d_1 = 0$ which implies that $d_1 = 0$ and a_1 is diagonal; so the dimension of this linear space is n, equal to the dimension of the fiber of $\mathcal{C}_k$ through $(d,1)$. This ends the proof of Claim 1.

Claim 2. γ is birational.

Indeed assume γ is not birational. Then all the scheme theoretic fibers of γ are non-reduced. Hence all scheme theoretic fibers of $\mathcal{C}_k$ above points in a Zariski open set of V_k must be non-reduced. But this contradicts Claim 1 and ends the proof of Claim 2.

By Claim 2 and Zariski's Main Theorem it follows that γ is an isomorphism. So $\mathcal{O}(V_k) = \mathcal{O}(X_k)^{N_k}$ and we are done. □

PROOF OF THEOREM 4.50. Uniqueness is clear because the map

$$\mathcal{O}(GL_n^*)\widehat{\ } \to \mathcal{O}(T^* \times GL_n)\widehat{\ }$$

induced by $\mathcal{C}$ is injective. Let's prove the existence part. We may assume $A = \mathbb{Z}_p$. Let $\phi^{GL_n^*}$ be any Frobenius lift on $\widehat{GL_n^*}$, $\phi^{GL_n^*}(x) = x^{(p)}\Lambda(x)$ where Λ is an $n \times n$ matrix with entries in $A[x, \det(x)^{-1}, D^*(x)^{-1}]\widehat{\ }$, $\Lambda \equiv 1 \bmod p$. Set

$$A(t,x) = (x^{(p)})^{-1} t^{(p)} x^{(p)}, B(t,x) = (x^{-1}tx)^{(p)} \in \mathcal{O}(T^* \times GL_n).$$

The condition that $\phi^{GL_n^*}$ is $\mathcal{C}$-horizontal with respect to $\phi_0^{GL_n}$ translates into the equality

$$\Lambda(x^{-1}tx) = B(t,x)^{-1} A(t,x) \tag{4.119}$$

in the ring $A[t, \det(t)^{-1}, D^*(t)^{-1}, x, \det(x)^{-1}]\hat{}$. Note that $A(t,x)$ and $B(t,x)$ are N-invariant for the action 4.118. We shall construct a sequence of matrices $\Lambda_\nu(x)$ with entries in $A[x, \det(x)^{-1}, D^*(x)^{-1}]\hat{}$, with $\Lambda_0 = 1$, satisfying the following properties:

(1) $\Lambda_{\nu+1}(x) \equiv \Lambda_\nu(x) \bmod p^{\nu+1}$ for $\nu \geq 0$;
(2) $\Lambda_\nu(x^{-1}tx) \equiv B(t,x)^{-1}A(t,x) \bmod p^{\nu+1}$ for $n \geq 0$.

This will end the proof by taking

$$\Lambda(x) = \lim \Lambda_\nu(x). \tag{4.120}$$

To construct Λ_ν we proceed by induction. Assume Λ_ν has been constructed for some $\nu \geq 0$. Write

$$\Lambda_\nu(x^{-1}tx) = B(t,x)^{-1}A(t,x) + p^{\nu+1}C(t,x).$$

Then $C(t,x)$ is N-invariant. By Lemma 4.52 there exists $F \in A[x, \det(x)^{-1}, D^*(x)^{-1}]$ such that

$$C(x) \equiv F(x^{-1}qx) \mod p.$$

Setting $\Lambda_{\nu+1} = \Lambda_\nu(x) - p^{\nu+1}F(x)$ one immediately checks that

$$\Lambda_{\nu+1}(x^{-1}qx) \equiv B(x)^{-1}A(x) \mod p^{\nu+2}$$

which ends our construction and hence the proof of the theorem. □

In contrast with the situation for regular matrices described in Theorem 4.51, we have the following results for non-regular matrices:

THEOREM 4.53. *Let $q \in T(A)$ with $q^{(p)} = q$, let τ be the automorphism of GL_n defined by $x^\tau = q^{-1}xq$, and let $\mathcal{H}(x) = q^{-1}x^{-1}qx$. Consider the Frobenius lift $\phi_0^{GL_n}(x) = x^{(p)}$ on $\widehat{GL_n}$. Assume q is not regular (i.e. $q \notin T^*$) and not scalar. Then there exists no Frobenius lift ϕ^{GL_n} on $\widehat{GL_n}$ such that $\phi_0^{GL_n}$ is $\mathcal{H}$-horizontal with respect to ϕ^{GL_n}.*

PROOF. We may assume $A = R$. Let ϕ^{GL_n} be any Frobenius lift on $\widehat{GL_n}$, $\phi^{GL_n}(x) = x^{(p)}\Lambda(x)$ and set

$$A(x) = (x^{(p)})^{-1}qx^{(p)}, \quad B(x) = (x^{-1}qx)^{(p)}.$$

The condition that $\phi_0^{GL_n}$ is $\mathcal{H}$-horizontal with respect to ϕ^{GL_n} means by definition that the following diagram is commutative:

$$\begin{array}{ccc} \widehat{GL_n} & \stackrel{\phi_0^{GL_n}}{\longrightarrow} & \widehat{GL_n} \\ \mathcal{H}\downarrow & & \downarrow\mathcal{H} \\ \widehat{GL_n} & \stackrel{\phi^{GL_n}}{\longrightarrow} & \widehat{GL_n} \end{array}$$

Hence this condition translates into the following equality:

$$\Lambda(q^{-1}x^{-1}qx) = B(x)^{-1}A(x). \tag{4.121}$$

Let $q \in T$ be not regular and not scalar, with $q^{(p)} = q$. Let us assume 4.121 and seek a contradiction. We may assume

$$q = \text{diag}(q_1 \cdot 1_{r_1}, \ldots, q_s \cdot 1_{r_s})$$

with $q_1, \ldots, q_s$ distinct and $s \geq 2$, $r_1 \geq 2$. It is enough then to obtain a contradiction in the case $n = 3$, $r_1 = 2$, $r_2 = 1$. Let $H \subset GL_3$ be the group of matrices of the form $\begin{pmatrix} a & b \\ c & d \end{pmatrix}$ with $a \in GL_2$; hence $q \in T \cap H$ and the centralizer of q in GL_3 is H. Note that $B(hx) = B(x)$ and

$$\Lambda(q^{-1}(hx)^{-1}q(hx)) = \Lambda(q^{-1}x^{-1}qx)$$

for all $h \in H$. From 4.121 it follows that $A(hx) = A(x)$ for $h \in H$, hence

$$((hx)^{(p)})^{-1}q(hx)^{(p)} = (x^{(p)})^{-1}qx^{(p)}, \quad h \in H. \tag{4.122}$$

This implies that $(hx)^{(p)}(x^{(p)})^{-1}$ is in the centralizer of q and hence

$$(hx)^{(p)}(x^{(p)})^{-1} = \begin{pmatrix} w_{11} & 0 \\ 0 & w_{22} \end{pmatrix} \tag{4.123}$$

with w_{ij} functions of x, w_{11} a 2×2 matrix. Write

$$h = \begin{pmatrix} h_{11} & 0 \\ 0 & h_{22} \end{pmatrix}, \quad x = \begin{pmatrix} x_{11} & x_{12} \\ x_{21} & x_{22} \end{pmatrix},$$

with h_{11}, x_{11} 2×2 matrices. Then 4.123 yields:

$$((h_{11}x_{11})^{(p)})^{-1}(h_{11}x_{12})^{(p)} = (x_{11}^{(p)})^{-1}x_{12}^{(p)}$$

for any $h_{11} \in GL_2$. Let x_{12}^1, x_{12}^2 be column vectors of indeterminates and let x_{12}^* be the 2×2 matrix with columns x_{12}^1, x_{12}^2. Also let y_{11} be a 2×2 matrix of indeterminates. Then we get

$$((y_{11}x_{11})^{(p)})^{-1}(y_{11}x_{12}^*)^{(p)} = (x_{11}^{(p)})^{-1}(x_{12}^*)^{(p)}.$$

Setting $x_{12}^* = y_{11}^{-1}$ and $x_{11} = 1$ we get

$$((x_{12}^*)^{-1})^{(p)})^{-1} = (x_{12}^*)^{(p)},$$

which is easily seen to be a contradiction. □

By Theorem 4.53 and (the proof of) Theorem 4.50 we will get the following:

THEOREM 4.54. *Assume $n \geq 3$. Then the Frobenius lift $\phi^{GL_n^*}$ on $\widehat{GL_n^*}$ in* Theorem 4.50 *does not extend to a Frobenius lift on $\widehat{GL_n}$.*

PROOF. Assume the Frobenius lift $\phi^{GL_n^*}$ on $\widehat{GL_n^*}$ extends to a Frobenius lift ϕ^{GL_n} on $\widehat{GL_n}$ and seek a contradiction. We get that the matrix $\Lambda(x)$ in 4.120 has coefficients in $A[x, \det(x)^{-1}]\hat{}$. Hence the equation 4.119 holds in the ring $A[t, \det(t)^{-1}, x, \det(x)^{-1}]\hat{}$ because the map

$$A[t, \det(t)^{-1}, x, \det(x)^{-1}]\hat{} \to A[t, \det(t)^{-1}, D^*(t)^{-1}, x, \det(x)^{-1}]\hat{}$$

is injective. Now pick any $q \in T(A)$ with $q^{(p)} = q$ which is not regular and not scalar; this is possible because $n \geq 3$. Let τ be the automorphism of GL_n defined by $x^\tau = q^{-1}xq$. Setting $t = q$ in equality 4.119 we get an equality

$$\Lambda(x^{-1}qx) = B(q, x)^{-1}A(q, x)$$

in the ring $A[x, \det(x)^{-1}]\hat{}$. But this is immediately seen to imply that the Frobenius lift ϕ^{GL_n} on $\widehat{GL_n}$ defined by $\phi_{GL_n}(x) = x^{(p)}\Lambda(qx)$ has the property that $\phi_0^{GL_n}$ is $\mathcal{H}$-horizontal with respect to ϕ_{GL_n}. This, however, contradicts Theorem 4.53, which ends our proof. □

We examine now the interaction between Frobenius lifts and the map

$$\mathcal{P} : GL_n \to \mathbb{A}^n = Spec\ A[z_1, ..., z_n]$$

with coordinates $\mathcal{P}_1, ..., \mathcal{P}_n \in A[x]$ given by the formula

$$\det(s \cdot 1 - x) = \sum_{j=0}^{n} (-1)^j \mathcal{P}_j(x) s^{n-j}; \tag{4.124}$$

so

$$\mathcal{P}_1(x) = \text{tr}(x), ..., \mathcal{P}_n(x) = \det(x).$$

Consider the Frobenius lift $\phi_0^{\mathbb{A}^n}$ on $\widehat{\mathbb{A}^n}$ defined by $\phi_0^{\mathbb{A}^n}(z_i) = z_i^p$.

Definition 4.55. A Frobenius lift $\phi^{G^{**}}$ on an open set $\widehat{G^{**}}$ of $\widehat{G} = \widehat{GL_n}$ is *$\mathcal{P}$-horizontal* with respect to $\phi_0^{\mathbb{A}^n}$ if the diagram

$$\begin{array}{ccc} (G^{**})^{\widehat{p}} & \stackrel{\phi^{G^{**}}}{\longrightarrow} & (G^{**})^{\widehat{p}} \\ \mathcal{P} \downarrow & & \downarrow \mathcal{P} \\ (\mathbb{A}^n)^{\widehat{p}} & \stackrel{\phi_0^{\mathbb{A}^n}}{\longrightarrow} & (\mathbb{A}^n)^{\widehat{p}}, \end{array} \tag{4.125}$$

is commutative.

We say that $\phi^{G^{**}}$ is a *T-deformation* of $\phi_0^G(x) = x^{(p)}$ if there is a commutative diagram

$$\begin{array}{ccc} \widehat{G^{**}} & \stackrel{\phi_0^{G^{**}} \times \phi^{G^{**}}}{\longrightarrow} & \widehat{G^{**}} \times \widehat{G^{**}} \\ \downarrow & & \downarrow \mathcal{Q} \\ \widehat{T} & \subset & \widehat{G} \end{array} \tag{4.126}$$

where $\mathcal{Q}(x, y) = x^{-1}y$ and $\phi_0^{G^{**}}$ is induced by ϕ_0^G; the left vertical map in 4.126 is then necessarily unique.

If G^{**} is invariant under conjugation by N we say $\phi^{G^{**}}$ is *$\mathcal{C}$-compatible* with N if the following diagram is commutative:

$$\begin{array}{ccc} \widehat{G^{**}} \times \widehat{N} & \stackrel{\mathcal{C}}{\longrightarrow} & \widehat{G^{**}} \\ \phi^{G^{**}} \times \phi_0^N \downarrow & & \downarrow \phi^{G^{**}} \\ \widehat{G^{**}} \times \widehat{N} & \stackrel{\mathcal{C}}{\longrightarrow} & \widehat{G^{**}} \end{array} \tag{4.127}$$

where ϕ_0^N is induced by $\phi_0^G(x) = x^{(p)}$.

THEOREM 4.56.

(1) *There exists an open set* G^{**} *of* $G = GL_n$ *and a Frobenius lift* $\phi^{G^{**}}$ *on* $\widehat{G^{**}}$ *such that* G^{**} *is invariant under conjugation by* N, $G^{**} \cap T = T^*$, $\phi^{G^{**}}$ *is* $\mathcal{P}$*-horizontal with respect to* $\phi_0^{\mathbb{A}^n}$, *and* $\phi^{G^{**}}$ *is a* T*-deformation of* ϕ_0^G.

(2) *For any* G^{**} *as above,* $\phi^{G^{**}}$ *is unique with the above properties, it is* $\mathcal{C}$*-compatible with* N, *and, if* $p \equiv 1$ *mod* n, *it fixes individually all the* n*-torsion points of* $T(A)$ *contained in* $T^*(A)$.

(3) *For any matrix* $\alpha(x)$ *with coefficients in* $\mathcal{O}(\widehat{G^{**}})$ *the Frobenius lift* $\phi^{G^{**}(\alpha)}$ *on* $\widehat{G^{**}}$ *defined by*

$$\phi^{G^{**}(\alpha)}(x) = \epsilon(x) \cdot \phi^{G^{**}}(x) \cdot \epsilon(x)^{-1}, \quad \epsilon(x) := 1 + p\alpha(x)$$

is $\mathcal{P}$*-horizontal with respect to* $\phi_0^{\mathbb{A}^n}$. *So if* $\delta^{G^{**}(\alpha)}$ *is the* p*-derivation on* $\mathcal{O}(\widehat{G^{**}})$ *attached to* $\phi^{G^{**}(\alpha)}$ *then*

$$\delta^{G^{**}(\alpha)}\mathcal{P}_i = 0, \quad i = 1, ..., n.$$

Furthermore, if $p \equiv 1$ *mod* n *and* P *is an* n*-torsion point of* $T(A)$ *contained in* $T^*(A)$ *such that* $\alpha(P)$ *is diagonal then* $\phi^{G^{**}(\alpha)}$ *fixes* P.

PROOF. Consider the action of N on $A[x, \det(x)^{-1}]\widehat{}$ by conjugation, $n \star x = nxn^{-1}$, $n \in N$. Let $e_1, ..., e_n$ be the columns of the $n \times n$ identity matrix, i.e., $1_n = [e_1, ..., e_n]$, and let

$$1_{n,j} = [e_1, ..., e_{j-1}, 0, e_{j+1}, ..., e_n],$$

$$\det(s \cdot 1_{n,j} - x) = \sum_{i=0}^{n-1} (-1)^i \mathcal{P}_{ij}(x) s^{n-1-i} \in A[x][s],$$

$$D^{**}(x) := \det(\mathcal{P}_{ij}(x)) \in A[x].$$

Clearly

$$d \star \mathcal{P}_{ij}(x) = \mathcal{P}_{ij}(x), \quad d \star D^{**}(x) = D^{**}(x) \tag{4.128}$$

for $d \in T$. One also easily checks that

$$w_\sigma \star \mathcal{P}_{ij}(x) = \mathcal{P}_{i\sigma^{-1}(j)}(x), \quad w_\sigma \star D^{**}(x) = \pm D^{**}(x), \tag{4.129}$$

for all $w_\sigma = [e_{\sigma(1)}, ..., e_{\sigma(n)}] \in W$. Define G^{**} as the locus in G where D^{**} is invertible; so G^{**} is invariant under conjugation by N. Note that if $\mu = \text{diag}(\mu_1, ..., \mu_n) \in T$ then $D^{**}(\mu) \in R^\times$ if and only if the polynomials

$$\prod_{i \neq 1}(s - \mu_i), ..., \prod_{i \neq n}(s - \mu_i)$$

are linearly independent mod p which holds if and only if $\mu_1, ..., \mu_n$ are distinct mod p; this shows that $G^{**} \cap T = T^*$. Let $\phi^{G^{**}}$ be a Frobenius lift on $\widehat{G^{**}}$, $\phi^{G^{**}}(x) = x^{(p)}\Lambda(x)$, where $\Lambda(x)$ is a matrix $\equiv 1$ mod p with coefficients in

$$A[x, \det(x)^{-1}, D^{**}(x)^{-1}]\widehat{}.$$

The commutativity of 4.125 is equivalent to the equalities

$$\mathcal{P}_i(x^{(p)}\Lambda(x)) = \mathcal{P}_i(x)^p \tag{4.130}$$

for all $i = 1, ..., n$, hence to the equality

$$V(x,s) := \sum_{i=0}^{n} (-1)^i \mathcal{P}_i(x)^p s^{n-i} = \sum_{i=0}^{n} (-1)^i \mathcal{P}_i(x^{(p)}\Lambda(x)) s^{n-i},$$

hence to the equality

$$V(x,s^p) = \sum_{i=0}^{n} (-1)^i \mathcal{P}_i(x^{(p)}\Lambda(x)) s^{p(n-i)}.$$

Note the following:

(i) If $\phi^{G^{**}}$ is $\mathcal{P}$-horizontal with respect to $\phi_0^{\mathbb{A}^n}$, then for any matrix $\alpha(x)$ with coefficients in $\mathcal{O}(\widehat{G^{**}})$ the Frobenius lift $\phi^{G^{**}(\alpha)}$ in the statement of the Theorem is $\mathcal{P}$-horizontal with respect to $\phi_0^{\mathbb{A}^n}$.

(ii) $\phi^{G^{**}}$ is a T-deformation of ϕ_0^G if and only if $\Lambda(x)$ is a diagonal matrix.

(iii) $\phi^{G^{**}}$ is $\mathcal{C}$-compatible with N if and only if $d \star \Lambda(x) = \Lambda(x)$ for $d \in T$ and $w \star \Lambda(x) = w\Lambda(x)w^{-1}$ for $w \in W$.

(iv) If $p \equiv 1 \mod n$ then $\phi^{G^{**}}$ fixes individually all the n-torsion points of $T(A)$ contained in $T^*(A)$ if and only if $\Lambda(d) = 1$ for all $d \in T^*(A)$ with $d^n = 1$.

In what follows we prove the existence of a diagonal matrix $\Lambda(x)$ as above. The uniqueness part of the theorem is proved via a similar argument. To prove the existence of $\Lambda(x)$ it is enough to construct a sequence $\Lambda_\nu(x)$ of diagonal matrices satisfying the following properties:

(1) $\det(s^p \cdot 1_n - x^{(p)}\Lambda_\nu(x)) \equiv V(x,s^p) \mod p^{\nu+1}$, $\nu \geq 0$.
(2) $\Lambda_{\nu+1}(x) \equiv \Lambda_\nu(x) \mod p^{\nu+1}$, $\nu \geq 0$.
(3) $d \star \Lambda_\nu(x) = \Lambda_\nu(x)$ for $d \in T$.
(4) $w \star \Lambda_\nu(x) = w\Lambda_\nu(x)w^{-1}$ for $w \in W$.
(5) If $p \equiv 1 \mod n$ then $\Lambda_\nu(d) = 1$ for all $d \in T^*$ with $d^n = 1$.

We take $\Lambda_0(x) = 1$. Assume now $\Lambda_\nu(x)$ was constructed and seek $\Lambda_{\nu+1}(x)$ of the form $\Lambda_\nu(x) - p^{\nu+1}Z(x)$ where $Z(x)$ is a diagonal matrix. Write

$$\begin{array}{rcl} A(x) & = & s^p \cdot 1_n - x^{(p)}\Lambda_\nu(x) = (a_{ij}), \\ \det(A(x)) & = & V(x,s^p) + p^{\nu+1}U(x,s^p), \\ U(x,s) & = & \sum_{i=0}^{n-1} U_i(x) s^{n-1-i}, \\ Z & = & \text{diag}(z_1, ..., z_n), \\ B & = & x^{(p)}Z(x) = (b_{ij}) = (z_j x_{ij}^p). \end{array}$$

If $d^n = 1$, $d \in T^*$, $p \equiv 1 \mod n$ we have that the entries on the diagonal of d are precisely the n-th roots of unity in some order; hence $\det(s \cdot 1_n - d) = s^n - 1$. By induction

$$A(d) = s^p \cdot 1_n - d^{(p)}\Lambda_\nu(d) = s^p \cdot 1_n - d$$

for $d \in T^*$, $d^n = 1$, $p \equiv 1 \mod n$. Since, in this case,

$$V(d,s^p) = \sum_{i=0}^{n} (-1)^i \mathcal{P}_i(d)^p s^{p(n-i)} = \sum_{i=0}^{n} (-1)^i \mathcal{P}_i(d) s^{p(n-i)} = \det(A(d)),$$

we get $U(d, s^p) = 0$ hence, for $d \in T^*$, $d^n = 1$, $p \equiv 1 \bmod n$, we have

$$U_i(d) = 0. \tag{4.131}$$

Since $\mu \star V(x,s) = V(x,s)$ and $\mu \star A(x) = A(x)$ for $\mu \in N$ it follows that

$$\mu \star U_i(x) = U_i(x) \tag{4.132}$$

for all i and $\mu \in N$. Now

$$\det(s^p \cdot 1_n - x^{(p)}\Lambda_{\nu+1}(x)) = \det \begin{pmatrix} a_{11} + p^{\nu+1}b_{11} & a_{12} + p^{\nu+1}b_{12} & * \\ a_{21} + p^{\nu+1}b_{21} & a_{22} + p^{\nu+1}b_{22} & * \\ * & * & * \end{pmatrix}$$

which is $\equiv \bmod p^{\nu+2}$ to

$$\det(A) + p^{\nu+1}W(x, s^p)$$

where

$$W(x, s^p) = \det \begin{pmatrix} b_{11} & a_{12} & * \\ b_{21} & a_{22} & * \\ * & * & * \end{pmatrix} + \det \begin{pmatrix} b_{11} & a_{12} & * \\ b_{21} & a_{22} & * \\ * & * & * \end{pmatrix} + \ldots$$

So, for property (1) above, it is enough to find $Z(x)$ such that

$$U(x, s^p) + W(x, s^p) \equiv 0 \mod p. \tag{4.133}$$

On the other hand we have

$$A(x) \equiv s^p \cdot 1_n - x^{(p)} \mod p$$

so $W(x, s^p)$ is congruent mod p to

$$-z_1 \det \begin{pmatrix} -x_{11} & -x_{12} & * \\ -x_{21} & s - x_{22} & * \\ * & * & * \end{pmatrix}^p - z_2 \det \begin{pmatrix} s - x_{11} & -x_{12} & * \\ -x_{21} & -x_{22} & * \\ * & * & * \end{pmatrix}^p - \ldots$$

Now 4.133 is satisfied if the following system can be solved:

$$\sum_{j=1}^{n} \mathcal{P}_{ij}(x)^p z_j = (-1)^i U_i(x), \quad i = 0, \ldots, n-1. \tag{4.134}$$

This system has determinant $\equiv D^{**}(x)^p \bmod p$ so it has a (unique) solution $(z_1, \ldots, z_n)$ in the ring $A[x, D^{**}(x)^{-1}]\widehat{\ }$. By Cramer's rule and 4.128, 4.132, we have that $d \star z_j = z_j$ for all j and $d \in T$. On the other hand by applying w_σ to 4.134 we get

$$\sum_{j=1}^{n} \mathcal{P}_{i\sigma^{-1}(j)}(x)^p (w_\sigma \star z_j) = (-1)^i U_i(x)$$

hence $w_\sigma \star z_j = z_{\sigma^{-1}(j)}$, hence $w_\sigma \star Z = w_\sigma Z w_\sigma^{-1}$, hence

$$w_\sigma \star \Lambda_{\nu+1}(x) = w_\sigma \Lambda_{\nu+1}(x) w_\sigma^{-1}.$$

Finally, by 4.131 and Cramer's rule again, if $d \in T^*$, $d^n = 1$, $p \equiv 1 \bmod n$, we get that z_j vanish for $x = d$ so $\Lambda_{\nu+1}(d) = 1$. This ends our construction and hence ends our proof. $\square$

REMARK 4.57. By assertion (3) in Theorem 4.56 and Lemma 3.126 it follows that $\mathcal{P}_i$ are prime integrals for the system of arithmetic differential equations

$$\delta x_{ij} - \Delta_{ij}^{**(\alpha)}(x),$$

where $\Delta_{ij}^{**(\alpha)}(x) = \delta^{G^{**}(\alpha)}(x_{ij})$; in particular if $\Delta^{**(\alpha)} = (\Delta_{ij}^{**(\alpha)})$ then for any solution $u \in G^{**}$ of the equation

$$\delta u = \Delta^{**(\alpha)}(u) \tag{4.135}$$

we have $\delta(\mathcal{P}_i(u)) = 0$ for all i. On the other hand we make the:

Claim. Equation 4.135 is equivalent to

$$l\delta u = -_\delta (u \star_\delta \alpha(u)) +_\delta \alpha(u), \tag{4.136}$$

where $l\delta : G^{**} \to \mathfrak{gl}_n$ is defined by

$$l\delta u := \frac{1}{p}(\phi(u)\Phi^{**}(u)^{-1} - 1), \quad \Phi^{**}(x) := \phi^{G^{**}}(x).$$

This makes assertion (3) in our Theorem 4.56 an analogue of equation 2.210 and hence of the equation 2.209 corresponding to Lax connections in classical differential geometry. To check the claim note that Equation 4.136 is equivalent to

$$1 + pl\delta u = (1 + p\phi(u)\alpha(u)\phi(u)^{-1})^{-1}(1 + p\alpha(u))$$

hence to

$$\Phi^{**}(u) = (1 + p\alpha(u))^{-1}\phi(u)(1 + p\alpha(u)) = \epsilon(u)^{-1}\phi(u)\epsilon(u),$$

hence to Equation 4.135, and our claim is proved. In view of the analogy between 4.135 and Lax connections we make the following definition.

DEFINITION 4.58. Let $\phi^{G^{**}}$ be as in assertion (1) of Theorem 4.56 and let $\delta^{G^{**}}$ be the attached p-derivation on $\widehat{G^{**}}$. Then $\delta^{G^{**}}$ is called the *canonical isocharacteristic Lax connection* on G^{**}. More generally any p-derivation on $\widehat{G^{**}}$ whose Frobenius lift is $\mathcal{P}$-horizontal with respect to $\phi_0^{\mathbb{A}^n}$ is called an *isocharateristic Lax connection* on G^{**}. So the p-derivation $\delta^{G^{**}(\alpha)}$ on $\widehat{G^{**}}$ attached to the $\phi^{G^{**}(\alpha)}$ in assertion (3) of Theorem 4.56 is an isocharacteristic Lax connection called the *isocharacteristic Lax connection* on G^{**} attached to α.

In the same spirit we make the following:

DEFINITION 4.59. Let ϕ^{G^*} be as in Theorem 4.50 and let δ^{G^*} be the attached p-derivation on $\widehat{G^*}$. Then δ^{G^*} is called the *canonical isospectral Lax connection* on G^*. More generally if a matrix $\alpha(x) \in \mathfrak{gl}_n(\widehat{\mathcal{O}(G^*)})$ is given then the p-derivation $\delta^{G^*(\alpha)}$ on $\widehat{G^*}$ attached to the Frobenius lift on $\widehat{G^*}$,

$$\phi^{G^*(\epsilon)}(x) := \Phi^{(\epsilon)}(x) := \epsilon(x) \cdot \phi(x) \cdot \epsilon(x)^{-1}, \quad \epsilon(x) = 1 + p\alpha(x),$$

is referred to as the *isospectral Lax connection* on G^* attached to $\alpha(x)$.

Recall that $\delta^{G^{**}(\alpha)}$ enjoys an "isospectrality-like" property which we referred to as the *isocharacteristic* property; cf. Theorem 4.56. In a similar way $\delta^{G^*(\alpha)}$ also enjoys an "isospectrality-like" property which justifies our *isospectral Lax* terminology. Indeed we have:

THEOREM 4.60. *Let $u = a^{-1}ba$, $a \in G(R)$, $b \in T^*(R)$, $b = diag(b_1, ..., b_n)$, be such that $\phi(u) = \Phi^{(\epsilon)}(u)$ where $\phi^{G^*(\epsilon_p)}(x) = \Phi^{(\epsilon)}(x)$ corresponds to the isospectral Lax connection attached to α, and $\epsilon(x) = 1+p\alpha(x)$. Then $\delta b_i = 0$ for all $i = 1, ..., n$.*

The difference between the two forms of "isospectrality" in Remark 4.57 and Theorem 4.60 is this: in Theorem 4.60 the eigenvalues of u are δ-constant (i.e. roots of unity) while in Theorem 4.56 the coefficients of the characteristic polynomial of u are δ-constant (i.e., roots of unity or 0). In usual calculus the corresponding two properties are equivalent but in our arithmetic calculus these two properties are quite different.

PROOF OF THEOREM 4.60. We have a commutative diagram

$$\begin{array}{ccc} T^*(R) \times G(R) & \stackrel{\phi_{0*}}{\longrightarrow} & T^*(R) \times G(R) \\ \mathcal{C}_* \downarrow & & \downarrow \mathcal{C}_* \\ G^*(R) & \stackrel{\phi_*^{G^*}}{\longrightarrow} & G^*(R) \end{array}$$

Let the canonical isospectral Lax connection be given by $\phi(x) = x^{(p)}\Lambda(x)$ hence

$$\Phi^{(\epsilon)}(x) = \epsilon(x)^{-1} \cdot x^{(p)} \cdot \Lambda(x) \cdot \epsilon(x).$$

From $\phi(u) = \Phi^{(\epsilon)}(u)$ we get

$$\phi^{-1}(\epsilon(u)^{-1}u^{(p)}\Lambda(u)\epsilon(u)) = u$$

or, equivalently, setting $v = \phi^{-1}(\epsilon(u))$,

$$\begin{aligned} a^{-1}ba &= v^{-1}\phi_*^{G^*}(u)v \\ &= v^{-1}\phi_*^{G^*}\mathcal{C}_*(b,a)v \\ &= v^{-1}\mathcal{C}_*\phi_{0*}(b,a)v \\ &= v^{-1}\mathcal{C}_*(\phi^{-1}(b^p), \phi^{-1}(a^{(p)}))v \\ &= v^{-1}(\phi^{-1}(a^{(p)}))^{-1}\phi^{-1}(b)^p\phi^{-1}(a^{(p)})v. \end{aligned}$$

So b and $\phi^{-1}(b)^p$ are conjugate, hence $\phi(b)$ and b^p are conjugate, hence the latter two matrices have the same characteristic polynomial, hence we have an equality of sets

$$\{\phi(b_1), ..., \phi(b_n)\} = \{b_1^p, ..., b_n^p\}.$$

Hence there exists a permutation σ of $\{1, ..., n\}$ such that $\phi(b_i) = b_{\sigma(i)}^p$ for all i. Pick an integer $k \geq 1$ such that $\sigma^k = 1$, the identity. Then $\phi^k(b_i) = b_i^{p^k}$ for all $i = 1, ..., n$. This implies that b_i are all roots of unity. Indeed, if some b_i is not a root of unity it can be written as $b_i = \zeta + p^m u$ where $\zeta \in R$ is a root of unity and $u \in R^\times$. But then $\phi^k(b_i) = \zeta^{p^k} + p^m\phi(u)$ and $b_i^{p^k} = (\zeta + p^m u)^{p^k} = \zeta^{p^k} + p^{m+1}v$ for some $v \in R$ hence $\phi(u) \in pR$ which is a contradiction. This ends the proof. $\square$

REMARK 4.61. We end by summarizing (and comparing) the various existence and non-existence results that we proved in this section. Here are the results:

(1) (Compatibility with conjugation) There exists a unique lift of Frobenius ϕ^{G^*} on $\widehat{G^*}$ such that ϕ_0^G is $\mathcal{C}$-horizontal with respect to ϕ^{G^*}.
(2) (Singularity along the discriminant) For $n \geq 3$ the lift of Frobenius ϕ^{G^*} in assertion (1) does not extend to a lift of Frobenius on the whole of $\widehat{G}$.
(3) (Compatibility with characteristic polynomial) There exists a Frobenius lift $\phi^{G^{**}}$ on $\widehat{G^{**}}$ such that $\phi^{G^{**}}$ is $\mathcal{P}$-horizontal with respect to $\phi_0^{\mathbb{A}^n}$ and $\phi^{G^{**}}$ is a T-deformation of ϕ_0^G. Moreover $\phi^{G^{**}}$ is unique with the above properties and it is $\mathcal{C}$-compatible with N.

   Furthermore let $q \in G$ be such that $q^{(p)} = q$ and consider the involution $x^\tau = q^{-1}xq$ on G. Then the following hold.
(4) (Inner involutions) If $q \in T$, $q^2 = 1$, q non-scalar, then there is no lift of Frobenius ϕ^G on $\widehat{G}$ that is $\mathcal{H}$-horizontal with respect to ϕ_0^G.
(5) (Regular inner automorphisms) If $q \in T^*$ then there exists a lift of Frobenius ϕ^G on $\widehat{G}$ such that ϕ_0^G is $\mathcal{H}$-horizontal with respect to ϕ^G.
(6) (Non-regular inner automorphisms) If $q \in T \backslash T^*$ and q is non-scalar then there is no lift of Frobenius ϕ^G on $\widehat{G}$ such that ϕ_0^G is $\mathcal{H}$-horizontal with respect to ϕ^G.

4.6. Existence of special linear connections

In the previous sections we investigated Frobenius lifts compatible with outer and inner automorphisms respectively; those compatible with outer involutions lead to natural Frobenius lifts on $SO(q)$. One can ask if there are natural Frobenius lifts on SL_n constructed in a similar way. We construct such Frobenius lifts in this section; there will be, by the way, a "simplest" one which we call the *special linear connection.* Recall that, in contrast with the case of classical differential geometry (Proposition 2.83), SL_{2r} is *never* horizontal with respect to the real Chern connection attached to a split symmetric q; "fixing" this anomaly is one of the main motivations for the present section.

Again, we set $G = GL_n$ over $A \subset R$, a δ-ring of local type, and we let δ_0^G be the trivial p-derivation on $\widehat{G}$, $\delta_0^G x = 0$ with attached Frobenius lift ϕ_0^G on $\widehat{G}$ given by $\phi_0^G(x) = x^{(p)}$. Let $\det : G \to \mathbb{G}_m = GL_1$ be the determinant homomorphism and let $\mathbb{G}_m \simeq Z \subset G$ be the center of G.

DEFINITION 4.62. A Frobenius lift ϕ^G on $\widehat{G}$ is det-*horizontal* with respect to ϕ_0^G if the following diagram is commutative:

$$
\begin{array}{ccc}
\widehat{G} & \xrightarrow{\phi^G} & \widehat{G} \\
\det \downarrow & & \downarrow \det \\
\mathbb{G}_m & \xrightarrow{\phi_0^{\mathbb{G}_m}} & \mathbb{G}_m
\end{array}
\tag{4.137}
$$

The above condition is equivalent to the condition that $\delta^G(\det(x)) = 0$, i.e., that $\det(x)$ is a prime integral for δ^G.

DEFINITION 4.63. A Frobenius lift ϕ^G on $\widehat{G}$ is a *Z-deformation* of ϕ_0^G if there is a commutative diagram

$$\begin{array}{ccc} \widehat{G} & \stackrel{\phi_0^G \times \phi^G}{\longrightarrow} & \widehat{G} \times \widehat{G} \\ \downarrow & & \downarrow \mathcal{Q} \\ \widehat{Z} & \subset & \widehat{G} \end{array} \tag{4.138}$$

where $\mathcal{Q}(x,y) = x^{-1}y$; the left vertical map in 4.138 is then necessarily unique. The above condition is equivalent to the condition that $\phi^G(x) = \lambda(x) \cdot x^{(p)}$ for some $\lambda(x) \in \mathcal{O}(\widehat{G})^\times$.

THEOREM 4.64. *Assume $p \nmid n$. Then there exists a unique Frobenius lift ϕ^G on $\widehat{G}$ which is* det*-horizontal with respect to ϕ_0^G and is a Z-deformation of ϕ_0^G; it is given by $\phi^G(x) = \lambda(x) \cdot x^{(p)}$ where $\lambda(x)$ is given by the principal root:*

$$\lambda(x) = \left(\frac{\det(x^{(p)})}{(\det(x))^p} \right)_p^{-1/n}. \tag{4.139}$$

Moreover ϕ^G is bi-invariant with respect to N.

PROOF. A computation similar to that in assertion (1) of Remark 4.36. □

DEFINITION 4.65. The p-derivation δ^G on $\widehat{G}$ attached to the Frobenius lift ϕ^G in Theorem 4.64 is called the *special linear* connection on G (or the *p-adic connection on G attached to SL_n*); it induces a p-derivation on $\widehat{SL_n}$.

4.7. Existence of Euler connections

Our aim here is to give an arithmetic analogue of the Euler equations 2.221. We refer to [**48**] for proofs.

We continue to consider a δ-ring $A \subset R$ of local type.

We consider quadratic forms $H_1, H_2 \in A[x_1, x_2, x_3]$,

$$H_1 := \sum_{i=1}^{3} a_i x_i^2, \quad H_2 := \sum_{i=1}^{3} x_i^2, \tag{4.140}$$

where $a_1, a_2, a_3 \in A$ are such that

$$(a_1 - a_2)(a_2 - a_3)(a_3 - a_1) \in A^\times. \tag{4.141}$$

We consider the affine spaces

$$\mathbb{A}^2 = Spec\ A[z_1, z_2], \quad \mathbb{A}^3 = Spec\ A[x_1, x_2, x_3] \tag{4.142}$$

and the morphism

$$\mathcal{H} : \mathbb{A}^3 \to \mathbb{A}^2 \tag{4.143}$$

defined by

$$z_1 \mapsto H_1, \quad z_2 \mapsto H_2. \tag{4.144}$$

For any $c = (c_1, c_2) \in A^2 = \mathbb{A}^2(A)$ let

$$E_c = Spec\ A[x_1, x_2, x_3]/(H_1 - c_1, H_2 - c_2) \tag{4.145}$$

be fiber of $\mathcal{H}$ at c. Moreover let

$$N(z_1, z_2) = \prod_{i=1}^{3}(z_1 - a_i z_2) \in A[z_1, z_2]. \tag{4.146}$$

If

$$N(c_1, c_2) \not\equiv 0 \mod p,$$

E_c is smooth and has a smooth projective compactification equipped with a global 1-form whose restriction to E_c is given by

$$\omega_c = \frac{dx_1}{(a_2 - a_3)x_2x_3} = \frac{dx_2}{(a_3 - a_1)x_3x_1} = \frac{dx_3}{(a_1 - a_2)x_1x_2}. \tag{4.147}$$

The following was proved in [**48**]:

THEOREM 4.66. *Given H_1, H_2 as above, there exist:*

(i) *open sets $Y \subset \mathbb{A}^2$ and $X \subset \pi^{-1}(Y) \subset \mathbb{A}^3$, with $X \otimes \mathbb{F}_p \neq \emptyset$, $N \in \mathcal{O}(Y)^\times$;*
(ii) *a function $A_{p-1} \in \mathcal{O}(X)^\times$;*
(iii) *a p-derivation δ^X on $\widehat{X}$ with attached Frobenius lift ϕ^X on $\widehat{X}$;*

such that:

(1) *The following equalities hold in $\mathcal{O}(\widehat{X})$:*
$$\delta^X H_1 = \delta^X H_2 = 0;$$
(2) *For any point $c = (c_1, c_2) \in Y(A)$ with $\delta c_1 = \delta c_2 = 0$ the Frobenius lift ϕ_c on $\widehat{E_c} \cap \widehat{X}$ induced by ϕ^X and the 1-form ω_c on $E_c \cap X$ satisfy the following congruence on $\widehat{E}_c \cap \widehat{X}$:*

$$\frac{\phi_c^*}{p}\omega_c \equiv A_{p-1}(c_1, c_2)^{-1} \cdot \omega_c \quad mod \quad p. \tag{4.148}$$

The congruence in assertion (2) above is a congruence in the module of 1-forms on $\widehat{E}_c \cap \widehat{X}$ defined by:

$$\Omega_{\widehat{E}_c \cap \widehat{X}} = \lim_{\leftarrow} \Omega_{(E_c \cap X)_n/A}$$

where

$$(E_c \cap X)_n := (E_c \cap X) \otimes (A/p^{n+1}A).$$

DEFINITION 4.67. A p-derivation δ^X as in Theorem 4.66 is called an *Euler connection* on $\widehat{X}$, attached to the data H_1, H_2, X, Y, A_{p-1}.

REMARK 4.68.

(1) Conditions (1) and (2) in the above theorem are analogous to the conditions 2.222 and 2.226 satisfied by the classical Euler equations. Condition (1) says that H_1 and H_2 are *prime integrals* for the δ-flow attached to δ^X; this condition is equivalent to saying that ϕ^X is $\mathcal{H}$-horizontal with respect to ϕ_0^Y, where $\mathcal{H} : \widehat{X} \to \widehat{Y}$ is the induced map and ϕ_0^Y is the Frobenius lift on $\widehat{Y}$ induced by the trivial p-adic connection $\phi_0^{\mathbb{A}^2}$ on $\widehat{\mathbb{A}^2}$, $\phi_0^{\mathbb{A}^2}(z_i) = z_i^p$. Recall that ϕ^X is *$\mathcal{H}$-horizontal* with respect to ϕ_0^Y if the following diagram is commutative:

$$\begin{array}{ccc} \widehat{X} & \xrightarrow{\phi^X} & \widehat{X} \\ \mathcal{H} \downarrow & & \downarrow \mathcal{H} \\ \widehat{Y} & \xrightarrow{\phi_0^Y} & \widehat{Y} \end{array}$$

Condition (2) is analogous to the classical condition that the flow be "linearized" on the elliptic curves appearing as level sets of the prime integrals.

(2) Euler connections are, of course, not unique. By the way one can choose δ^X in Theorem 4.66 such that $\delta^X x_3$ is in $\mathcal{O}(X)$ (rather than in $\mathcal{O}(\widehat{X})$), and moreover $\delta^X x_1, \delta^X x_2$ are algebraic over $\mathcal{O}(X)$; this can be used to construct correspondence structures for our Euler connections. We will not pursue this in our book.

(3) As the notation suggests the function A_{p-1} is related to the Hasse invariant of E_c; roughly speaking $A_{p-1}(c_1, c_2)$ is the Hasse invariant of a natural quartic defining an elliptic curve isogenous to the elliptic curve E_c; see [**48**] for details.

(4) It is not clear at this point if a strengthened form of the above theorem continues to hold in which one replaces the congruence 4.148 by an equality and one replaces A_{p-1} by an appropriate p-adic modular form.

By the way Euler connections are "forced to have singularities" on the affine 3-space at least for a_i's "generic". Indeed the following non-existence result was proved in [**48**]:

THEOREM 4.69. *Assume that one of the elements* $\delta a_1, \delta a_2, \delta a_3 \in A$ *is in* $A^\times$. *Then there is no p-derivation* $\delta^{\mathbb{A}^3}$ *on* $\mathcal{O}(\widehat{\mathbb{A}^3})$ *such that*

$$\delta^{\mathbb{A}^3} H_1 = \delta^{\mathbb{A}^3} H_2 = 0.$$

It is possible to give a parametrization of Euler connections "mod p." This parameterization will involve the prime integrals mod p of the *classical* Euler equations! Indeed let us say that two p-derivations on $\mathcal{O}(\widehat{X})$ are congruent mod p if their values at any element of $\mathcal{O}(\widehat{X})$ are congruent mod p. Also let $\mathbb{F} = A/pA$ be the residue field of A and let $\overline{X} = X \otimes_A \mathbb{F}$. More generally, in what follows, an overline will denote "reduction mod p." Let us denote now by $\overline{\delta}^{cl}$ the derivation on $\mathcal{O}(\overline{X})$ defined by the *classical Euler formulas*:

$$\begin{array}{rcl} \overline{\delta}^{cl} x_1 & = & (a_2 - a_3) x_2 x_3, \\ \overline{\delta}^{cl} x_2 & = & (a_3 - a_1) x_3 x_1, \\ \overline{\delta}^{cl} x_3 & = & (a_1 - a_2) x_1 x_2. \end{array} \tag{4.149}$$

Then the following was proved in [**48**]:

THEOREM 4.70. *Assume* $\mathbb{F}$ *is infinite. Then there exist* X, Y, A_{p-1} *as in Theorem* 4.66 *such that the quotient set*

$$\frac{\{\textit{Euler connections on } \widehat{X} \textit{ attached to } H_1, H_2, X, Y, A_{p-1}\}}{\{\textit{congruence mod } p\}}$$

is a principal homogeneous space for the $\mathbb{F}$*-linear space*

$$\{\overline{K} \in \mathcal{O}(\overline{X});\ \overline{\delta}^{cl}\overline{K} = 0\}.$$

REMARK 4.71. Recall that a set is called a *principal homogeneous space* for a group if it has a simply transitive action of that group. A principal homogeneous space is either empty or in bijection with the group. The principal homogeneous space in Theorem 4.70 is non-empty due to Theorem 4.66.

REMARK 4.72. The $\mathbb{F}$-linear space in Theorem 4.70 can be viewed as the space of *prime integrals mod p* of the classical Euler equations. It is, of course, rather large: it contains the $\mathbb{F}$-subalgebra of $\mathcal{O}(\overline{X})$ generated by $\overline{H}_1$, $\overline{H}_2$, and by all the p-powers of elements of $\mathcal{O}(\overline{X})$.

4.8. Curvature formalism and gauge action on GL_n

4.8.1. Notation. In the previous sections we concentrated on the case when A is a δ-ring of local type and developed some of the theory in that case. Beginning with this subsection we revert to case when A is a δ-ring of global type, $A = \mathcal{O}_K[1/M]$, cf. Definition 3.25. Our aim here is to make some of our abstract definitions of connection and curvature explicit for the case of GL_n over such a global A. For the abstract discussion of connection and curvature we refer to section 3.4. So recall that, for us, GL_n is the group scheme

$$G := GL_n = Spec\ A[x, \det(x)^{-1}] \tag{4.150}$$

and we set

$$\mathfrak{g} = \mathfrak{gl}_n = Spec\ A[x].$$

For the A-point $P = 1$ of G given by the identity matrix we consider the retract coordinates

$$T = x - 1; \tag{4.151}$$

cf. Definition 3.50. For each odd p we have $\mathcal{O}(G)^{\widehat{p}} = A[x, \det(x)^{-1}]^{\widehat{p}}$. Also we have $\mathcal{O}(G^{\widehat{P}}) = A[[T]]$. Given an adelic connection $\delta = (\delta_p)$ on G, the p-derivations $\delta_p : A[x, \det(x)^{-1}]^{\widehat{p}} \to A[x, \det(x)^{-1}]^{\widehat{p}}$ should be viewed as an analogue of the derivations in 2.8.

DEFINITION 4.73. The collection of matrices

$$(\Delta_p^t)$$

with entries in $A[x, \det(x)^{-1}]^{\widehat{p}}$, where $\delta_p x = \Delta_p$, $\Delta_p = \Delta_p(x)$, will be referred to as the *Christoffel symbol of the 2nd kind* of (δ_p).

Let (ϕ_p) be the family of Frobenius lifts,

$$\phi_p : A[x, \det(x)^{-1}]^{\widehat{p}} \to A[x, \det(x)^{-1}]^{\widehat{p}}, \tag{4.152}$$

attached to an adelic connection (δ_p) on G. Set $\phi_p(x) = \Phi_p$, $\Phi_p = \Phi_p(x)$, which has entries in $A[x, \det(x)^{-1}]^{\widehat{p}}$; hence $\Phi_p(x) = x^{(p)} + p\Delta_p(x)$.

4.8.2. Adelic connections on GL_n global along 1. An adelic connection δ on G fixes 1 if and only if $\Phi_p(1) = 1$; cf. Lemma 3.54. Assume this is the case. Then the p-derivations $\delta_p : A^{\widehat{p}}[[T]] \to A^{\widehat{p}}[[T]]$ are attached to the Frobenius lift $\phi_p : A^{\widehat{p}}[[T]] \to A^{\widehat{p}}[[T]]$ defined by

$$\phi_p(T) = \Phi_p^0, \quad \Phi_p^0(T) = \Phi_p(1+T) - 1, \tag{4.153}$$

$$\Phi_p^0 = (\Phi_{pij}^0), \quad \Phi_{pij}^0 = \phi_p(T_{ij}) \in A^{\widehat{p}}[[T]].$$

Moreover δ is global along 1 if, and only if, all entries of Φ_{pij}^0 are series in $(T)A[[T]]$; cf. Lemma 3.56. If this is the case we still denote by $\phi_p : A[[T]] \to A[[T]]$ the induced Frobenius lifts. Finally, the curvature of δ satisfies $\varphi_{pp'}[d] = 0$ for some pair (p, p') if and only if ϕ_p and $\phi_{p'}$ commute on $A[[T]]$.

Recall our basic definitions/notation in this setting:

DEFINITION 4.74. Let $\delta = \delta^G$ be an adelic connection on G that is global along 1, i.e. $\delta \in \Lambda^1(G, 1)$, and let $\varphi[\delta] = (\varphi_{pp'})$ be its curvature, so

$$\varphi_{pp'} : A[[T]] \to A[[T]] \tag{4.154}$$

are $\phi_p\phi_{p'}$-semi-linear maps of A-modules sending each ideal $(T)^\nu$ into itself. One also has the induced truncations

$$\varphi_{pp'}^{(\nu)} : I/I^{\nu+1} \to I/I^{\nu+1}, \tag{4.155}$$

where $I = (T)$. Also, by Definition 3.59, we have at our disposal the holonomy rings

$$\mathfrak{hol} \subset \text{End}(A[[T]]), \quad \mathfrak{hol}^\Omega \subset \text{End}(\Omega^{for}_{A[[T]]/A}) \tag{4.156}$$

of δ^G and their completions

$$\widehat{\mathfrak{hol}}, \quad \widehat{\mathfrak{hol}}^\Omega. \tag{4.157}$$

Consider the matrices of series

$$\begin{aligned} \Phi^0_{pp'} &= \varphi_{pp'}(T) = \frac{1}{pp'}(\phi_p\phi_{p'}(T) - \phi_{p'}\phi_p(T)), \\ \Phi^0_{pp'} &= (\Phi^0_{pp'ij}), \quad \Phi^0_{pp'ij} = \varphi_{pp'}(T_{ij}) \in A[[T]]. \end{aligned} \tag{4.158}$$

By the way, if $\Phi_p^{0(\phi_{p'})}$ is the series obtained from Φ^0_p by applying $\phi_{p'}$ to the coefficients we have $\phi_p\phi_{p'}(T) = \Phi_{p'}^{0(\phi_p)}(\Phi^0_p(T))$ hence,

$$\Phi^0_{pp'} = \frac{1}{pp'}(\Phi_{p'}^{0(\phi_p)}(\Phi^0_p(T)) - (\Phi_p^{0(\phi_{p'})}(\Phi^0_{p'}(T)))). \tag{4.159}$$

Recall:

DEFINITION 4.75. The *trivial* adelic connection on G is $\delta_0 = (\delta_{0p})$ where

$$\delta_{0p}x = 0.$$

For this adelic connection

$$\phi_{0p}(T) = \Phi^0_{0p} = (1+T)^{(p)} - 1.$$

This adelic connection is global along the identity 1 of G and it is flat. Note that the family (δ_{0p}) above does not arise from a global connection on G but, rather, from a global connection on the affine space $\mathbb{A}^{n^2}$ in which G naturally embeds as an open set. We set

$$\overline{\delta} = (\overline{\delta}_p) = (\delta_{\overline{p}}) = \delta_0 = (\delta_{0p}), \quad \phi_{\overline{p}} = \phi_{0p} \tag{4.160}$$

on $A[[T]]$. So $\phi_{\overline{p}} = \phi_p$ on A but not on $A[[T]]$. The $(1,1)$-curvature $(\varphi_{p\overline{p}'})$ of δ with respect to $\overline{\delta} = \delta_0$ is defined and one can consider the corresponding matrices:

$$\Phi^0_{p\overline{p}'} = \frac{1}{p\overline{p}'}(\phi_p\phi_{\overline{p}'}(T) - \phi_{\overline{p}'}\phi_p(T)) \tag{4.161}$$

for $p \neq p'$ and

$$\Phi^0_{p\overline{p}} = \frac{1}{p}(\phi_p\phi_{\overline{p}}(T) - \phi_{\overline{p}}\phi_p(T)). \tag{4.162}$$

Formulas analogous to 4.159 hold in this $(1,1)$-case as well. The curvature $(\varphi_{\overline{p}\overline{p}'})$ of $\overline{\delta}_0$ is, of course, also defined but note that it vanishes because the $\phi_{\overline{p}}$'s commute.

Let us discuss actions of Frobenius lifts on forms. We have that

$$\Omega := \Omega^{for}_{A[[T]]/A} := \lim_{\leftarrow} \Omega_{\frac{A[[T]]}{(T)^n}/A} = \sum_{i,j=1}^{n} A[[T]] \cdot dT_{ij},$$

is free with basis (dT_{ij}). Recall that, since we assumed δ is global along 1 we have that all δ_p send $I := (T)$ into I, so we have induced $\mathbb{Z}$-module maps

$$\frac{\phi_p}{p}, \frac{\phi_{\bar{p}}}{p} : \Omega \to \Omega, \tag{4.163}$$

and the commutators

$$\varphi^{\Omega}_{pp'} := \left[\frac{\phi_p}{p}, \frac{\phi_{p'}}{p'}\right], \quad \varphi^{\Omega}_{p\bar{p}'} := \left[\frac{\phi_p}{p}, \frac{\phi_{\bar{p}'}}{p'}\right] : \Omega \to \Omega. \tag{4.164}$$

It is then clear that for all p, p',

$$\varphi^{\Omega}_{pp'}(dT_{ij}) = \sum_{k,l=1}^{n} \frac{\partial \Phi^0_{pp'ij}}{\partial T_{kl}} dT_{kl},$$

and, for $p \neq p'$,

$$\varphi^{\Omega}_{p\bar{p}'}(dT_{ij}) = \sum_{k,l=1}^{n} \frac{\partial \Phi^0_{p\bar{p}'ij}}{\partial T_{kl}} dT_{kl}, \quad \varphi^{\Omega}_{p\bar{p}}(dT_{ij}) = \frac{1}{p} \sum_{k,l=1}^{n} \frac{\partial \Phi^0_{p\bar{p}ij}}{\partial T_{kl}} dT_{kl}.$$

DEFINITION 4.76. Denote by $\mathcal{A}_p$, $\mathcal{F}_{pp'}$ the matrices, with entries in $A[[T]]$, of the endomorphisms

$$\frac{\phi_p}{p} : \Omega \to \Omega, \quad \varphi^{\Omega}_{pp'} = \left[\frac{\phi_p}{p}, \frac{\phi_{p'}}{p'}\right] : \Omega \to \Omega,$$

respectively, with respect to the $A[[T]]$-basis

$$dT^{col} = \{dT_{11}, dT_{21}, ..., dT_{nn}\}.$$

We define $\mathcal{A}_{\bar{p}}$ and $\mathcal{F}_{p\bar{p}'}$ similarly.

REMARK 4.77.

(1) The matrix $\mathcal{F}_{pp'}$ of $\varphi^{\Omega}_{pp'}$ with respect to dT^{col}, evaluated at $T = 0$, is the same as the matrix, with coefficients in A, of

$$\varphi^{(1)}_{pp'} : I/I^2 \to I/I^2,$$

$I = (T)$, with respect to (T^{col} mod I^2).

(2) For $p \neq p'$, the matrix $\mathcal{F}_{p\bar{p}'}$ of $\varphi^{\Omega}_{p\bar{p}'}$ with respect to dT^{col}, evaluated at $T = 0$, is the same as the matrix, with coefficients in A, of

$$\varphi^{(1)}_{p\bar{p}'} : I/I^2 \to I/I^2,$$

with respect to (T^{col} mod I^2).

(3) The matrix $\mathcal{F}_{p\bar{p}}$ of $\varphi^{\Omega}_{p\bar{p}}$ with respect to dT^{col}, evaluated at $T = 0$, is $1/p$ times the matrix, with coefficients in A, of

$$\varphi^{(1)}_{p\bar{p}} : I/I^2 \to I/I^2,$$

with respect to (T^{col} mod I^2).

REMARK 4.78. The first Chern form of δ (relative to the basis (dT_{ij})) is given by the matrix $\rho = \rho[\delta] = (\rho_{pp'})$,

$$\rho_{pp'} := \mathrm{tr}\left(\frac{\partial \Phi^0_{pp'}}{\partial T}\right) = \sum_{ij} \frac{\partial \Phi^0_{pp'ij}}{\partial T_{ij}} \in A[[T]]. \tag{4.165}$$

Similarly the *Chern* $(1,1)$-*form* is given by

$$\rho_{p\bar{p}'} := \mathrm{tr}\left(\frac{\partial \Phi^0_{p\bar{p}'}}{\partial T}\right) = \sum_{ij} \frac{\partial \Phi^0_{p\bar{p}'ij}}{\partial T_{ij}} \in A[[T]] \quad \text{for} \quad p \neq p' \quad \text{and} \tag{4.166}$$

$$\rho_{p\bar{p}} := \frac{1}{p}\mathrm{tr}\left(\frac{\partial \Phi^0_{p\bar{p}}}{\partial T}\right) = \frac{1}{p}\sum_{ij} \frac{\partial \Phi^0_{p\bar{p}ij}}{\partial T_{ij}} \in A[[T]]. \tag{4.167}$$

Here is some basic notation to be used later.

DEFINITION 4.79. We denote by $\rho^{(\nu)}_{pp'}$ and $\rho^{(\nu)}_{p\bar{p}'}$ the images of $\rho_{pp'}$ and $\rho_{p\bar{p}'}$ in $A[[T]]/(T)^{\nu+1}$ respectively.

REMARK 4.80. If

$$\mathcal{A}_p = (\mathcal{A}^{kl}_{pij}) = \begin{pmatrix} \mathcal{A}^{11}_{p11} & \mathcal{A}^{21}_{p11} & \dots & \mathcal{A}^{nn}_{p11} \\ \mathcal{A}^{11}_{p21} & \mathcal{A}^{21}_{p21} & \dots & \mathcal{A}^{nn}_{p21} \\ \dots & \dots & \dots & \dots \\ \mathcal{A}^{11}_{pnn} & \mathcal{A}^{21}_{pnn} & \dots & \mathcal{A}^{nn}_{pnn} \end{pmatrix} \tag{4.168}$$

and, similarly,

$$\mathcal{F}_{pp'} = (\mathcal{F}^{kl}_{pp'ij}) \tag{4.169}$$

then

$$\mathcal{A}^{kl}_{pij} = \frac{1}{p}\frac{\partial \Phi^0_{pij}}{\partial T_{kl}} \quad \text{and} \quad \mathcal{F}^{kl}_{pp'ij} = \frac{\partial \Phi^0_{pp'ij}}{\partial T_{kl}}. \tag{4.170}$$

We have the following equality:

$$\mathcal{F}_{pp'} := \phi_p(\mathcal{A}_{p'}) \cdot \mathcal{A}_p - \phi_{p'}(\mathcal{A}_p) \cdot \mathcal{A}_{p'}. \tag{4.171}$$

Similarly in the $(1,1)$ case we have

$$\mathcal{F}_{p\bar{p}'} := \phi_p(\mathcal{A}_{\bar{p}'}) \cdot \mathcal{A}_p - \phi_{\bar{p}'}(\mathcal{A}_p) \cdot \mathcal{A}_{\bar{p}'}. \tag{4.172}$$

By the way, for any primes p, p', p'', by the additivitity of $\phi_p, \phi_{p'}, \phi_{p''}$ we have

$$\sum_{(pp'p'')} \left[\frac{\phi_p}{p}, \left[\frac{\phi_{p'}}{p'}, \frac{\phi_{p''}}{p''}\right]\right] = 0 \tag{4.173}$$

in the ring of group endomorphisms of Ω.

Here is another basic notation that will be used later.

DEFINITION 4.81. We let $\mathcal{A}^{(0)}_p$ and $\mathcal{F}^{(0)}_{pp'}$ be obtained from $\mathcal{A}_p$ and $\mathcal{F}_{pp'}$ by setting $T = 0$. We define $\mathcal{A}^{(0)}_{\bar{p}}$ and $\mathcal{F}^{(0)}_{p\bar{p}'}$ similarly.

REMARK 4.82. Setting $T=0$ in 4.171 and 4.172 we get

$$\mathcal{F}^{(0)}_{pp'} := \phi_p(\mathcal{A}^{(0)}_{p'})\cdot\mathcal{A}^{(0)}_{p} - \phi_{p'}(\mathcal{A}^{(0)}_{p})\cdot\mathcal{A}^{(0)}_{p'} \tag{4.174}$$

and

$$\mathcal{F}^{(0)}_{p\bar{p}'} := \phi_p(\mathcal{A}^{(0)}_{\bar{p}'})\cdot\mathcal{A}^{(0)}_{p} - \phi_{\bar{p}'}(\mathcal{A}^{(0)}_{p})\cdot\mathcal{A}^{(0)}_{\bar{p}'}. \tag{4.175}$$

Also evaluating 4.173 on our basis dT and setting $T=0$ we get

$$\sum_{(pp'p'')} \{\phi_p(\mathcal{F}^{(0)}_{p'p''})\cdot\mathcal{A}^{(0)}_{p} - \phi_{p'}\phi_{p''}(\mathcal{A}^{(0)}_{p})\cdot\mathcal{F}^{(0)}_{p'p''}\} = 0; \tag{4.176}$$

this can be viewed as a "Bianchi identity." Alternatively 4.176 follows from the equality 4.171.

Note that in the $(1,1)$ case, since $\phi_{\bar{p}}(T) = (1+T)^{(p)} - 1$, we have

$$\mathcal{A}^{(0)}_{\bar{p}} = \text{diag}(E_{11}, ..., E_{nn}), \tag{4.177}$$

where E_{ii} is the $n\times n$ matrix with a 1 in position (i,i) and 0 in all the other positions. By 4.172, setting

$$\text{tr}^*(\mathcal{A}^{(0)}_{p}) := \left(\sum_i \mathcal{A}^{ii}_{pii}\right)_{|T=0}, \tag{4.178}$$

we have:

$$\rho^{(0)}_{p\bar{p}'} = (1-\phi_{\bar{p}'})\left(\text{tr}^*(\mathcal{A}^{(0)}_{p})\right). \tag{4.179}$$

This last equation can be viewed as an analogue of the "Dolbeault exactness" of the Chern $(1,1)$-form in the geometric case; cf. 2.56; it implies the following analogue of "Dolbeault closedness" (cf. 2.57):

$$(1-\phi_{\bar{p}''})\rho^{(0)}_{p\bar{p}'} = (1-\phi_{\bar{p}'})\rho^{(0)}_{p\bar{p}''}. \tag{4.180}$$

By the way if, for $a\in A$, we denote by $a : A\to A$ the multiplication by a, then 4.179 and 4.180 can be rewritten as equalities of additive maps $A\to A$,

$$\rho^{(0)}_{p\bar{p}'}\cdot\phi_{\bar{p}'} = [\text{tr}^*(\mathcal{A}^{(0)}_{p}), \phi_{\bar{p}'}], \tag{4.181}$$

$$[\rho^{(0)}_{p\bar{p}'}\cdot\phi_{\bar{p}'}, \phi_{\bar{p}''}] = [\rho^{(0)}_{p\bar{p}''}\cdot\phi_{\bar{p}''}, \phi_{\bar{p}'}]. \tag{4.182}$$

REMARK 4.83. In our superalgebra formalism, for

$$\bar{\phi} := \sum_p \phi_{\bar{p}}\cdot d\bar{p} \in \Lambda^{0,1}(\text{End}(A)),\quad \alpha := \sum_p \text{tr}^*(\mathcal{A}^{(0)}_{p})\cdot dp \in \Lambda^{1,0}(A), \tag{4.183}$$

$$\eta := \sum_{p\bar{p}'} \rho^{(0)}_{p\bar{p}'}\cdot\phi_{\bar{p}'}\cdot dp\wedge d\bar{p}' \in \Lambda^{1,1}(\text{End}(A)), \tag{4.184}$$

we have

$$[\bar{\phi}, \alpha] = \eta. \tag{4.185}$$

Note that, in our situation, the group scheme structure of $G = GL_n$ defines a formal group law

$$\mathcal{G} \in \mathfrak{g}(A[[T^{(1)}, T^{(2)}]])$$

given, in matrix form, by

$$\mathcal{G}(T^{(1)}, T^{(2)}) = T^{(1)} + T^{(2)} + T^{(1)}T^{(2)}. \tag{4.186}$$

Then, by our discussion on adelic connections on formal groups, we have at our disposal the $\mathcal{G}$-curvatures $(\varphi^{\mathcal{G}}_{pp'})$ and $(\varphi^{\mathcal{G}}_{p\bar{p}'})$ and the corresponding matrices $\Phi^{\mathcal{G}}_{pp'}$ and $\Phi^{\mathcal{G}}_{p\bar{p}'}$ with entries in $A[[T]]$, as in 3.25, 3.38, 3.39. Note that if a and b are $n \times n$ matrices with coefficients in an ideal J of a J-adically complete ring C we have

$$a +_{\mathcal{G}} b = a + b + ab, \quad a -_{\mathcal{G}} b = (a - b)(1 + b)^{-1}.$$

Hence

$$\Phi^{\mathcal{G}}_{pp'}(T) = \Phi^0_{pp'}(T) \cdot (1 + \Phi^0_p(\Phi^0_{p'}(T)))^{-1}. \tag{4.187}$$

So for any ideal J in $A[[T]]$ we have

$$\Phi^0_{pp'} \equiv 0 \mod J \text{ if and only if } \Phi^{\mathcal{G}}_{pp'} \equiv 0 \mod J. \tag{4.188}$$

Moreover, since $\Phi^0_p(0) = 0$, we have

$$\Phi^0_p(\Phi^0_{p'}(T)) \equiv 0 \mod (pp') + (T)^{\min\{p,p'\}},$$

and hence we have

$$\Phi^{\mathcal{G}}_{pp'}(T) \equiv \Phi^0_{pp'}(T) \mod (pp') + (T)^{\min\{p,p'\}}. \tag{4.189}$$

In the $(1,1)$ setting, we have

$$\Phi^0_{p\bar{p}'} \equiv 0 \mod J \text{ if and only if } \Phi^{\mathcal{G}}_{p\bar{p}'} \equiv 0 \mod J, \tag{4.190}$$

$$\Phi^0_{p\bar{p}} \equiv 0 \mod J \text{ if and only if } \Phi^{\mathcal{G}}_{p\bar{p}} \equiv 0 \mod J, \tag{4.191}$$

$$\Phi^{\mathcal{G}}_{p\bar{p}'}(T) \equiv \Phi^0_{p\bar{p}'}(T) \mod (pp') + (T)^{\min\{p,p'\}}, \text{ for } p \neq p', \tag{4.192}$$

$$\Phi^{\mathcal{G}}_{p\bar{p}}(T) \equiv \Phi^0_{p\bar{p}}(T) \mod (p) + (T)^{p^2}. \tag{4.193}$$

Now recall that by taking "usual" traces of the Jacobian matrices of $\Phi^0_{pp'}$ and $\Phi^0_{p\bar{p}'}$ we were led to our corresponding first Chern forms; here is a variant of that construction which uses a different kind of "trace" and is applied to the matrices $\Phi^{\mathcal{G}}_{pp'}$ and $\Phi^{\mathcal{G}}_{p\bar{p}'}$ themselves rather than to their Jacobians.

Let C be any ring, J any ideal in C, and let $\mathfrak{g}(J)$ be the set of $n \times n$ matrices with coefficients in J. Assume, for simplicity, that C is m-torsion free for some integer $m \in \mathbb{Z}$. Then, for any $a \in \mathfrak{g}(J)$, define

$$\text{tr}_m(a) := \frac{1}{m} \cdot (\det(1 + ma) - 1) \in J. \tag{4.194}$$

(This can be viewed as a variant of the "quantum derivative," in the sense of [**84**], of the determinant.) We have:

$$\text{tr}_m(a) \equiv \text{tr}(a) \mod m, \tag{4.195}$$

Then we can consider the following analogue of the curvature potential in 2.48:

DEFINITION 4.84. The *curvature potential* of δ is the family $\psi = (\psi_{pp'})$ where

$$\psi_{pp'} := \text{tr}_{pp'}(\Phi^{\mathcal{G}}_{pp'}) \in A[[T]].$$

The $(1,1)$-*curvature potential* of δ is the family $\psi = (\psi_{p\bar{p}'})$ where

$$\psi_{p\bar{p}'} := \text{tr}_{pp'}(\Phi^{\mathcal{G}}_{p\bar{p}'}) \in A[[T]] \quad \text{for} \quad p \neq p',$$

$$\psi_{p\bar{p}} := \text{tr}_p(\Phi^{\mathcal{G}}_{p\bar{p}}) \in A[[T]].$$

One immediately gets the following expression:

$$\psi_{pp'} = \frac{1}{pp'}\left(\frac{\det(1+\phi_p\phi_{p'}(T))}{\det(1+\phi_{p'}\phi_p(T))} - 1\right). \tag{4.196}$$

and similarly for $(\psi_{p\overline{p}'})$ if $p \neq p'$. Also

$$\psi_{p\overline{p}} = \frac{1}{p}\left(\frac{\det(1+\phi_p\phi_{\overline{p}}(T))}{\det(1+\phi_{\overline{p}}\phi_p(T))} - 1\right). \tag{4.197}$$

The above discussion has an analogue in the case of the 3-curvature. Indeed, for an adelic connection $\delta = (\delta_p)$ on G that is global along 1 one can define the 3-curvature of δ as the 3-curvature of the corresponding global connection on $A[[T]]$ i.e. as the family $(\varphi_{pp'p''})$ of additive maps

$$\varphi_{pp'p''} := [\phi_p, \varphi_{p'p''}] = \frac{1}{p'p''}[\phi_p,[\phi_{p'},\phi_{p''}]] : A[[T]] \to A[[T]]. \tag{4.198}$$

One can also consider the family of additive maps

$$\varphi^{\Omega}_{pp'p''} := \left[\frac{\phi_p}{p},\left[\frac{\phi_{p'}}{p'},\frac{\phi_{p''}}{p''}\right]\right] : \Omega \to \Omega. \tag{4.199}$$

If we set $\varphi_{pp'p''}(T) = \Phi^0_{pp'p''}$, $\Phi^0_{pp'p''} = (\Phi^0_{pp'p''ij})$ we have

$$\varphi^{\Omega}_{pp'p''}(dT_{ij}) = \frac{1}{p}\sum_{k,l=1}^{n}\frac{\partial\Phi^0_{pp'p''ij}}{\partial T_{kl}}dT_{kl}. \tag{4.200}$$

In what follows we give a definition of mean $(1,1)$-curvature and scalar $(1,1)$-curvature, analogous to 2.136 and 2.138; we proceed analytically, as follows. Recall that A is of the form $\mathcal{O}_K[1/M]$ where $\mathcal{O}_K$ is the ring of integers of an Abelian number field K.

DEFINITION 4.85. Let s denote a (complex) variable and denote by $A[[\mathbb{N}^{-s}]]$ the ring of formal Dirichlet series $\sum a_n n^{-s}$ with coefficients $a_n \in A$. Furthermore denote by $A\{\{\mathbb{N}^{-s}\}\}$ the subring of all Dirichlet series that are convergent on some half plane $Re(s) > s_0$ (with s_0 depending on the series).

Let δ be an adelic connection on GL_n that is global along 1 and let $(\varphi_{p\overline{p}'})$ and $(\rho_{p\overline{p}'})$ be the $(1,1)$-curvature and the Chern $(1,1)$-form of δ. Let $\varphi_{p\overline{p}}(T) = \Phi^0_{p\overline{p}} = (\Phi^0_{p\overline{p}ij})$ be as in 4.162.

DEFINITION 4.86. The *mean* $(1,1)$*-curvature* $K(s) = (K_{ij}(s))$ and the *scalar* $(1,1)$*-curvature* $\sigma(s)$ of δ are defined by the following formulas respectively:

$$K_{ij}(s) := \sum_{p\in\mathcal{V}} p^{-s}\cdot\Phi^0_{p\overline{p}ij} \in A[[\mathbb{N}^{-s}]][[T]], \tag{4.201}$$

$$\sigma(s) := \sum_{p\in\mathcal{V}} p^{-s}\cdot\rho_{p\overline{p}} \in A[[\mathbb{N}^{-s}]][[T]]. \tag{4.202}$$

REMARK 4.87. The above definition corresponds, intuitively, to considering the "metric" on *Spec* $\mathbb{Z}$ given by:

$$\text{“}\sum_p p^{-s}dpd\overline{p}.\text{''}$$

As we shall see, in significant examples of our theory, the coefficients of the above series actually belong to $A\{\{\mathbb{N}^{-s}\}\}$ rather than $A[[\mathbb{N}^{-s}]]$. One is then led to ask for complex analytic properties of these objects.

Similarly one can give the following:

DEFINITION 4.88. The *Yang-Mills form* attached to $\delta = (\delta_p)$ is the family (Y_p) of matrices with entries in $A[[\mathbb{N}^{-s}]][[T]]$ given by:

$$Y_p(s) := \sum_{p' \in \mathcal{V}} (p')^{-s} \Phi^0_{p'p'p}.$$

REMARK 4.89. The above definition corresponds, intuitively, to considering the "diagonal metric" on *Spec* $\mathbb{Z}$ given by

$$\text{``}\sum_p p^{-s} dp^2.\text{''}$$

4.8.3. Adelic connections on GL_n global along tori. So far, in this section, we considered adelic connection that are global along the identity $1 = 1_n \in GL_n(A)$. We would like to consider a generalization of this (that will play a role later) by considering adelic connections that are global along a torus in GL_n. Among the tori of GL_n a special role, in our discussion, will be played by what we will call *partition tori* which we now proceed to define. Note that adelic connections global along non-trivial tori will admit gauge actions by those tori; cf. Definition 3.62.

DEFINITION 4.90. By a *partition* of an integer $m \geq 0$ we understand an s-tuple of integers $\mu = (m_1, ..., m_s)$ with $s \geq 0$, such that

$$m = m_1 + ... + m_s, \quad m_i \in \mathbb{Z}, \quad m_i \geq 1. \tag{4.203}$$

Given an integer $n \geq 1$ there are two extreme cases of interest for partitions of integers $m \leq n$. One is when $s = 0$, which happens if and only if $m = 0$; in this case we write $\mu = \overline{0}$. At the opposite end we have the case $s = n$ which happens if and only if $m = n = 1 + ... + 1$; in this case we write $\mu = \overline{1}$. An intermediate case of interest is when $m = n$ and $s = 1$; in this case we write $\mu = \overline{n}$.

Going back to an arbitrary partition μ consider the embedding

$$\epsilon_\mu : GL_{m_1} \times ... \times GL_{m_s} \to GL_n, \quad (a_1, ..., a_s) \mapsto \text{diag}(a_1, ..., a_s, 1_{n-m}). \tag{4.204}$$

Next let C_{m_i} be the center of GL_{m_i}.

DEFINITION 4.91. The *partition torus* in GL_n attached to μ is the subgroup of GL_n defined by

$$D_\mu := \epsilon_\mu(C_{m_1} \times ... \times C_{m_s}) \subset G = GL_n.$$

REMARK 4.92. The following hold:

(1) for $\mu = \overline{0}$ we have that $D_\mu = D_{\overline{0}}$ is the trivial group.
(2) for $\mu = \overline{1}$ we have $D_\mu = D_{\overline{1}} = T$, the diagonal torus of G.
(3) for $\mu = \overline{n}$ we have $D_\mu = D_{\overline{n}} =: Z$, the center of G.

Fix now a partition μ defined by $m = m_1 + ... + m_s \leq n$. Set $k_i = m_1 + ... + m_i$. So $k_1 = m_1$ and $k_s = m$. Then, for any ring C, we may consider the map

$$\{a \in GL_n(C); a_{k_1k_1}, ..., a_{k_sk_s} \in C^\times\} \to D_\mu(C) \tag{4.205}$$

defined by

$$a \mapsto a^\mu := \text{diag}(a_{k_1k_1} \cdot 1_{m_1}, a_{k_2k_2} \cdot 1_{m_2}, ..., a_{k_sk_s} \cdot 1_{m_s}, 1, ..., 1). \tag{4.206}$$

REMARK 4.93. The following hold:
(1) for $\mu = \overline{0}$ we have $a^{\mu} = a^{\overline{0}} = 1$, the identity matrix;
(2) for $\mu = \overline{1}$, $a^{\mu} = \text{diag}(a_{11}, ..., a_{nn})$;
(3) for $\mu = \overline{n}$, $a^{\mu} = a_{nn} \cdot 1_n$.

In general, for a matrix of indeterminates $x = (x_{ij})$ we may consider the matrix

$$x^{\mu} \in D_{\mu}(A[x]). \tag{4.207}$$

Also let

$$t = \{t_1, ..., t_s\}, \ \ t_i = x_{k_i k_i}, \ \ \text{the distinct entries unequal to 1 of } x^{\mu}, \tag{4.208}$$

and let T be the set

$$T = \{\text{non-zero entries of the matrix } x - x^{\mu}\}. \tag{4.209}$$

Again the use the letter T to denote both the diagonal torus and the variables above should not lead to any confusion as the two meanings are always clear from context. We have a natural identification

$$D_{\mu} = Spec\ A[t, t^{-1}]. \tag{4.210}$$

If $(GL_n)_1$ is the open set of GL_n defined by

$$(GL_n)_1 := Spec\ A[x, \det(x)^{-1}, t^{-1}] \tag{4.211}$$

then $D_{\mu} \subset (GL_n)_1$ and 4.205 defines a morphism $(GL_n)_1 \to D_{\mu}$ whose composition with the inclusion $D_{\mu} \subset (GL_n)_1$ is the identity of D_{μ}. Moreover the kernel of the surjection $\mathcal{O}((GL_n)_1) \to \mathcal{O}(D_{\mu})$ is clearly generated by the members of the set T, which form a regular sequence. So the above plus Lemma 3.51 yield:

LEMMA 4.94. *D_{μ} is a retract in GL_n and*

$$\mathcal{O}((GL_n)^{\widehat{D_{\mu}}}) \simeq A[t, t^{-1}][[T]], \quad \mathcal{O}(((GL_n)^{\widehat{p}}))^{\widehat{D_{\mu}}}) \simeq A[t, t^{-1}]^{\widehat{p}}[[T]].$$

Note that, for $\mu = \overline{0}$, the set of variables t is empty, $x^{\mu} = x^{\overline{0}} = 1$, the identity matrix, and the set T consists of the entries of $x - 1$; so we may identify T with the matrix $x - 1$ as in 4.151; so for $\mu = \overline{0}$ our paradigm here reduces to the paradigm of subsection 4.8.2.

On the other hand, for $\mu \neq 0$, one may consider the set $\Lambda^1(G, D_{\mu})$ of all adelic connections on G that are global along D_{μ}; then we have a natural gauge action

$$D_{\mu}(A) \times \Lambda^1(G, D_{\mu}) \to \Lambda^1(G, D_{\mu}), \tag{4.212}$$

cf. Definition 3.62. Let us make this action explicit in case $\mu = \overline{1}$. So in this case $D_{\mu} = T$, the maximal diagonal torus. Indeed consider, in this case, an adelic connection $\delta = (\delta_p)$ and consider the induced p-derivations δ_p on the ring

$$C := A[t, t^{-1}][[T_{ij}; i \neq j]], \tag{4.213}$$

where $t = \{t_1, ..., t_n\}$. As usual we still denote by t the matrix $\text{diag}(t_1, ..., t_n)$. Set

$$\delta_p t_i = \Delta^0_{pi}(t, T) \in C, \ \ \delta_p T_{ij} = \Delta^0_{pij}(t, T) \in C, \ \ i \neq j. \tag{4.214}$$

On the other hand let

$$d = \text{diag}(d_1, ..., d_n) \in T(A), \ \ d_i \in A^{\times}. \tag{4.215}$$

Then a trivial computation shows that the adelic connection $\tilde{\delta} := d \star \delta$ satisfies

$$\tilde{\delta}_p t_i = \tilde{\Delta}^0_{pi}(t, T) \in C, \ \ \tilde{\delta}_p T_{ij} = \tilde{\Delta}^0_{pij}(t, T) \in C, \ \ i \neq j, \tag{4.216}$$

where

$$\tilde{\Delta}^0_{pi}(t,T) = \phi_p(d_i)\cdot\Delta^0_{pi}(d^{-1}t,T) + \frac{\delta_p d_i}{d_i^p}t_i^p, \tag{4.217}$$

$$\tilde{\Delta}^0_{pij}(t,T) = \Delta^0_{pij}(d^{-1}t,T). \tag{4.218}$$

These formulas should be viewed as arithmetic analogues of the formulas 2.64.

4.8.4. Adelic connections on GL_n that are global on $\mathfrak{gl}_n$. We may consider $G = GL_n$ as an open set of $\mathfrak{g} := \mathfrak{gl}_n$ and equip the latter with the action of GL_n by left translation. This makes $\mathfrak{g}$ a partial equivariant compactification of G in the sense of the discussion before Definition 3.63. In particular, by Definition 3.63, we have at our disposal a gauge action

$$G(A)\times\Lambda^1(\mathfrak{g})\to\Lambda^1(\mathfrak{g}),\quad (g,\delta)\mapsto g\star\delta. \tag{4.219}$$

Recall that $\Lambda^1(\mathfrak{g})$ consists of all global connections on $\mathfrak{g}$, i.e., on $A[x]$; such connections (δ_p) are in bijection with families of matrices (Δ_p) with entries in $A[x]$ the bijection being given by the formula $\delta_p x = \Delta_p$. The global connections on $\mathfrak{g}$ induce, of course, adelic connections on G which can be said to be *global on* $\mathfrak{g}$. The gauge action 4.219 is given explicitly as follows: if $g\in G(A)$ then $\tilde{\delta} := g\star\delta$ is given by $\tilde{\delta}_p x = \tilde{\Delta}_p$ where

$$\tilde{\Delta}_p(x) = \phi_p(g)\cdot\Delta_p(g^{-1}x) + \frac{1}{p}(\phi_p(g)\cdot(g^{-1}x)^{(p)} - x^{(p)}). \tag{4.220}$$

This latter formula should be viewed, again, as an arithmetic analogue of formula 2.64.

One can also introduce an analogue of translation action 2.66 as follows. Fix $\delta\in\Lambda^1(\mathfrak{g})$. Let $\phi_p : A[x]\to A[x]$ be the attached Frobenius lifts and set $\phi_p(x) = \Phi_p$, $\Phi_p = \Phi_p(x) = x^{(p)} + p\Delta_p(x)$. Assume, for simplicity, that $\Phi_p\in\mathbb{Z}[1/M][x]$ for all p, where $\mathbb{Z}[1/M]\subset A$. On the other hand one can consider a family $\alpha := (\alpha_p)$ of elements of $\mathfrak{g}(A)$, the family (ϵ_p) where $\epsilon_p := 1 + p\alpha_p$, and the global connection

$$\delta^\alpha\in\Lambda^1(\mathfrak{g})$$

attached to the Frobenius lifts $\phi^\alpha = (\phi_p^{\alpha_p})$,

$$\phi_p^{\alpha_p}(x) = \epsilon_p\Phi_p(x).$$

We may consider the curvature $(\varphi^\alpha_{pp'})$ of the global connection δ^α; this is an element in the set $\Lambda^2(\mathfrak{g})$ of all anti-symmetric $\mathcal{V}\times\mathcal{V}$-matrices whose entries are $\phi_p\phi_{p'}$-semilinear maps of A-modules $A[x]\to A[x]$. On the other hand we may consider the family of equations

$$\phi_p(u) = \epsilon_p\Phi_p(u), \tag{4.221}$$

with unknown $u\in G(A)$. The curvature $(\varphi^\alpha_{pp'})$ of δ^α is related to the *integrability* of the system 4.221 as follows. Assume 4.221 has a solution $u\in G(A)$. Since ϕ_p and $\phi_{p'}$ commute on A we get $\phi_p\phi_{p'}(u) = \phi_{p'}\phi_p(u)$ hence

$$\phi_{p'}(\epsilon_p)\Phi_p(\epsilon_{p'}\Phi_{p'}(u)) = \phi_p(\epsilon_{p'})\Phi_{p'}(\epsilon_p\Phi_p(u)),$$

which can be reinterpreted as saying that all the entries of the matrix $\varphi_{pp'}(x)$, which are polynomials in $A[x]$, vanish for $x = u$.

4.9. Non-existence of classical δ-cocycles on GL_n

Recall that, in classical differential geometry, the logarithmic derivative 2.107 is a cocycle in the sense that it satisfies the identity 2.109. In this section we show, by closely following [**44**], that, in arithmetic differential geometry, no such cocycles exist. This justifies a posteriori our introduction of arithmetic logarithmic derivatives; cf. Definition 3.137. In what follows we place ourselves in the case when our ground ring is the δ-ring of local type $A = R = W(\mathbb{F}_p^a)$; so GL_n is the general linear group over R. We freely use the terminology in section 3.6. As there, for any smooth R-scheme X, we write X instead of $X(R)$.

DEFINITION 4.95. A *classical δ-cocycle* of GL_n with values in $\mathfrak{gl}_n$ is a δ-map $f : GL_n \to \mathfrak{gl}_n$ with the property that for all $g_1, g_2 \in GL_n$ we have

$$f(g_1g_2) = f(g_1) + g_1 \cdot f(g_2) \cdot g_1^{-1}. \tag{4.222}$$

If in addition f is (induced by) a regular map, i.e., by a morphism of schemes over R, then we say f is a *regular cocycle.*

THEOREM 4.96. *Let $f : GL_n \to \mathfrak{gl}_n$ be a classical δ-cocycle. Then there exists a δ-homomorphism $\omega : \mathbb{G}_m \to \mathbb{G}_a$ and there exists $v \in \mathfrak{gl}_n$ such that for all $g \in GL_n$ we have:*

$$f(g) = \omega(\det(g))1_n + gvg^{-1} - v.$$

The corresponding statement in algebraic geometry is well known and follows easily from Whitehead's lemma; cf. Lemma 4.97 below, for which we are indebted to A. Minchenko. We will need this algebraic geometric statement as a step in the proof of Theorem 4.96.

LEMMA 4.97. *Let $f : GL_n(F) \to \mathfrak{gl}_n(F)$ be a regular map of algebraic varieties over an algebraically closed field F of characteristic zero and assume f is a cocycle i.e. for all $g_1, g_2 \in GL_n(F)$ equation 4.222 holds. Then f is a coboundary, i.e. there exists $v \in \mathfrak{gl}_n(F)$ such that for all $g \in GL_n(F)$, $f(g) = gvg^{-1} - v$.*

PROOF. If in 4.222 we set $g_1 = x \in GL_n(F)$, $g_2 = 1 + \epsilon\xi \in GL_n(F[\epsilon])$ where 1 is the identity, $\epsilon^2 = 0$, and $\xi \in \mathfrak{gl}_n(F)$ and if we take the coefficient of ϵ we get

$$d_x f(x\xi) = x d_1 f(\xi) x^{-1},$$

equivalently

$$d_x f(\xi) = x d_1 f(x^{-1}\xi) x^{-1}. \tag{4.223}$$

Here d_x is, of course, the differential at x. Similarly setting $g_1 = 1 + \epsilon\xi$, $g_2 = x$ in 4.222 we get

$$d_x f(\xi) = d_1 f(\xi x^{-1}) + [\xi x^{-1}, f(x)]. \tag{4.224}$$

From 4.223 and 4.224 we get

$$x d_1 f(x^{-1}\xi) x^{-1} = d_1 f(\xi x^{-1}) + [\xi x^{-1}, f(x)] \tag{4.225}$$

for all $x \in GL_n(F)$ and $\xi \in \mathfrak{gl}_n(F)$ and hence for all $x \in GL_n(E)$, $\xi \in \mathfrak{gl}_n(E)$, E an F-algebra. Setting in 4.225 $x = 1 + \epsilon\eta \in GL_n(F[\epsilon])$, with $\eta \in \mathfrak{gl}_n(F)$ and taking the coefficient of ϵ we get

$$d_1 f([\xi, \eta]) = [d_1 f(\xi), \eta] + [\xi, d_1 f(\eta)]$$

so $d_1 f$ is a derivation on $\mathfrak{gl}_n(F)$. Note that $d_1 f$ preserves $\mathfrak{sl}_n(F)$. Indeed by 4.222 we have

$$\text{tr } f(g_1 g_2) = \text{tr } f(g_1) + \text{tr } f(g_2).$$

Since GL_n has no regular non-zero homomorphism into $\mathbb{G}_a$ we get tr $f = 0$ hence f takes values in $\mathfrak{sl}_n(F)$. Now by the first Whitehead lemma [**79**], p. 77, all derivations of $\mathfrak{sl}_n(F)$ are inner hence there exists $v \in \mathfrak{sl}_n(F)$ such that

$$d_1 f(\xi) = [\xi, v] \tag{4.226}$$

for all $\xi \in \mathfrak{sl}_n(F)$. By 4.225, for all $\lambda \in F$, $f(\lambda 1)$ centralizes all of $\mathfrak{gl}_n(F)$ and, since $f(\lambda 1) \in \mathfrak{sl}_n(F)$, we must have $f(\lambda 1) = 0$. We conclude that 4.226 holds for all $\xi \in \mathfrak{gl}_n(F)$. Let $f^v : GL_n(F) \to \mathfrak{gl}_n(F)$ be the coboundary $f^v(g) = gvg^{-1} - v$ and note that $d_1 f = d_1 f^v$. Since, by 4.225, $d_x f$ is determined by $d_1 f$ it follows that $d_x f = d_x f^v$ for all $x \in GL_n(F)$. Since $f(1) = f^v(1) = 0$ it follows that $f = f^v$. □

As a consequence, going back to our situation over $R = W(\mathbb{F}^a)$, we get:

LEMMA 4.98. *Let $f : GL_n \to \mathfrak{gl}_n$ be a map of schemes over R and assume f is a cocycle i.e. for all $g_1, g_2 \in GL_n(R)$ equation 4.222 holds. Then f is a coboundary, i.e. there exists $v \in \mathfrak{gl}_n(R)$ such that for all $g \in GL_n(R)$, $f(g) = gvg^{-1} - v$.*

PROOF. Let K be the fraction field of R and F its algebraic closure. By Lemma 4.97 there exists $v_F \in \mathfrak{gl}_n(F)$ such that

$$f(g) = gv_F g^{-1} - v_F$$

for all $g \in GL_n(F)$. Denoting by v_K the sum of the conjugates of v_F over K we have

$$f(g) = gv_K g^{-1} - v_K$$

for all $g \in GL_n(K)$. Let $m \geq 1$ be an integer such that $p^m v_K \in \mathfrak{gl}_n(R)$. If

$$w = p^m v_K - \alpha 1 = (w_{ij})$$

then we have

$$gwg^{-1} - w = p^m f(g) \in p^m \mathfrak{gl}_n(R)$$

for all $g \in GL_n(R)$. Taking g to be an arbitrary diagonal matrix we get that $w_{ij} \equiv 0$ mod p^m for $i \neq j$. Taking $g = 1 + e_{ij}$, $i \neq j$, with e_{ij} having all its entries 0 except the (i, j) entry which is 1 we get $w_{ii} \equiv w_{jj}$ mod p^m; since $w_{11} \equiv 0$ mod p^m we get that all the entries of w are divisible by p^m in R. So $w = p^m v$, $v \in \mathfrak{gl}_n(R)$, and hence

$$f(g) = gvg^{-1} - v$$

for all $g \in GL_n(R)$, which proves our claim. □

LEMMA 4.99. *Any δ-homomorphism $f : \mathbb{G}_a \to \mathbb{G}_a$, viewed as a map $f : R \to R$, has the form*

$$f(a) = \sum_{i=0}^{r} \lambda_i \phi^i(a), \quad a \in R$$

for some $\lambda_0, \dots, \lambda_r \in R$.

PROOF. Note that by [**23**] there exists $\nu \geq 1$ and $\lambda_i \in R$ such that

$$p^\nu f(a) = \sum \lambda_i \phi^i(a), \quad a \in R.$$

We are left to check that if for $c_i \in R$ we have that for any $a \in R$,

$$\sum c_i \phi^i(a) \in pR$$

then we must have $c_i \in pR$ for all i. But this follows from the fact that

$$\sum c_i a^{p^i} \equiv 0 \mod p$$

for all $a \in R$ and from the fact that the residue field of R is algebraically closed. $\square$

Next recall from [**23**], p. 313, that there exists a remarkable δ-homomorphism $\psi : \mathbb{G}_m \to \mathbb{G}_a$ given on R-points by

$$\psi(a) = \sum_{n \geq 1} (-1)^{n-1} \frac{p^{n-1}}{n} \left(\frac{\delta a}{a^p} \right)^n.$$

With this notation we have:

LEMMA 4.100. *Any δ-homomorphism $f : \mathbb{G}_m \to \mathbb{G}_a$, viewed as a map $f : R^\times \to R$, has the form*

$$f(a) = \sum_{i=0}^{r} \lambda_i \phi^i(\psi(a)), \quad a \in R^\times$$

for some $\lambda_0, ..., \lambda_r \in R$.

PROOF. Note that by [**23**] there exist $\nu \geq 1$ such that $p^\nu f$ has the above form; we conclude exactly as in Lemma 4.99. $\square$

REMARK 4.101. Lemmas 4.99 and 4.100 are analogues of results in differential algebra, due to Cassidy [**51**].

If instead of looking at δ-homomorphisms $\mathbb{G}_m \to \mathbb{G}_a$ we are looking at cocycles of non-trivial actions the situation is very different:

LEMMA 4.102. *Assume $f : \mathbb{G}_m \to \mathbb{G}_a$ is a δ-map and $0 \neq s \in \mathbb{Z}$ is an integer such that for any $a_1, a_2 \in R^\times$ we have*

$$f(a_1 a_2) = f(a_1) + a_1^s f(a_2).$$

Then there exists $\mu \in R$ such that

$$f(a) = \mu(1 - a^s), \quad a \in R^\times;$$

in particular f is a regular cocyle.

PROOF. By induction on $r \geq 1$ one gets, for all $a \in R^\times$,

$$f(a^r) = (1 + a^s + a^{2s} + ... + a^{(r-1)s}) f(a).$$

In particular for any $\nu \geq 0$ one gets

$$(1 - a^s) f(a^{p^\nu}) = (1 - a^{sp^\nu}) f(a). \tag{4.227}$$

Now recall that

$$\mathcal{O}(J^n(\mathbb{G}_m)) = R[x, x^{-1}, x', ..., x^{(n)}]\hat{}.$$

Identify the point set map f with a restricted power series

$$f \in R[x, x^{-1}, x', ..., x^{(n)}]\hat{}$$

and write

$$f=\sum_{i_1,\dots,i_n\geq 0} f_{i_1\dots i_n}(x)(x')^{i_1}\dots(x^{(n)})^{i_n}$$

where $f_{i_1\dots i_n}\in R[x,x^{-1}]\hat{}$ tend to 0 p-adically for $|i_1|+\dots+|i_n|\to\infty$. By 4.227 we get

$$(4.228)\qquad (1-x^s)\sum_{i_1,\dots,i_n\geq 0} f_{i_1\dots i_n}(x^{p^\nu})\delta(x^{p^\nu})^{i_1}\dots\delta^n(x^{p^\nu})^{i_n}$$

$$=(1-x^{sp^\nu})\sum_{i_1,\dots,i_n\geq 0} f_{i_1\dots i_n}(x)(x')^{i_1}\dots(x^{(n)})^{i_n}$$

By induction on ν it is easy to see that for any $\nu\geq r\geq 1$ we have

$$(4.229)\qquad \delta^r(x^{p^\nu})\in p^{\nu-r+1}R[x,x',\dots,x^{(r)}].$$

Fix $(i_1,\dots,i_n)\neq(0,\dots,0)$ and let $\nu\geq n$ be arbitrary. Picking out the coefficient of $(x')^{i_1}\dots(x^{(n)})^{i_n}$ in the equation 4.228 we get

$$(1-x^{sp^\nu})f_{i_1\dots i_n}\in p^{\nu-n+1}R[x,x^{-1}]\hat{}.$$

Since ν is arbitrary we get

$$f_{i_1\dots i_n}=0 \quad\text{for}\quad (i_1,\dots,i_n)\neq(0,\dots,0),$$

hence $f=f_{0\dots 0}$. For $\nu=1$ we may rewrite equation 4.228 as

$$(1-x^s)f_{0\dots 0}(x^p)=(1-x^{sp})f_{0\dots 0}(x).$$

The latter is an equality in $R[x,x^{-1}]\hat{}$ so we can also view it as an equality in the ring

$$\mathcal{A}=\{\sum_{i=-\infty}^{\infty} c_ix^i; c_i\to 0 \ \text{ for } \ i\to-\infty\}.$$

If $s\geq 1$ we get the following equality in $\mathcal{A}$:

$$g(x^p)=g(x),\quad g(x)=\frac{f_{0\dots 0}(x)}{1-x^s}=f_{0\dots 0}(x)(1+x^s+x^{2s}+\dots);$$

if $s\leq -1$ we get the following equality in $\mathcal{A}$:

$$g(x^p)=g(x),\quad g(x)=\frac{-x^{-s}f_{0\dots 0}(x)}{1-x^{-s}}=-x^{-s}f_{0\dots 0}(x)(1+x^{-s}+x^{-2s}+\dots).$$

In either case we get $g=\mu\in R$ and we are done. □

LEMMA 4.103. *Assume we are given a δ-map $\epsilon:\mathbb{G}_m\times\mathbb{G}_a^{n-1}\to\mathfrak{gl}_{n-1}$ and a row vector $\lambda\in R^{n-1}$ such that for any $(a_1,b_1),(a_2,b_2)\in R^\times\times R^{n-1}$,*

$$(4.230)\qquad \epsilon(a_1a_2,b_1+a_1b_2)=\epsilon(a_1,b_1)+\epsilon(a_2,b_2)+a_1^{-1}(a_2^{-1}-1)\lambda^tb_1.$$

Then, for any $a_1,a_2,a\in R^\times$, $b\in R^{n-1}$,

$$(4.231)\qquad \epsilon(a_1a_2,0)=\epsilon(a_1,0)+\epsilon(a_2,0),$$

$$(4.232)\qquad \epsilon(a,b)=\epsilon(a,0)+a^{-1}\lambda^tb.$$

PROOF. Setting $b_1 = b_2 = 0$ in equation 4.230 we get 4.231. Setting $a_1 = a_2 = 1$ in equation 4.230 we get

$$\epsilon(1, b_1 + b_2) = \epsilon(1, b_1) + \epsilon(1, b_2).$$

So if $\epsilon = (\epsilon_{kl})$ then by Lemma 4.99 there exist $\lambda_{ikl} \in R^{n-1}$, $i \geq 0$, such that

$$\epsilon_{kl}(1, b) = \sum \phi^i(b)\lambda^t_{ikl}, \quad b \in R^{n-1}. \tag{4.233}$$

Now setting $a_1 = a \in R^\times$, $a_2 = 1$, $b_1 = 0$, $b_2 = a^{-1}b \in R^{n-1}$ in equation 4.230 we get

$$\epsilon_{kl}(a, b) = \epsilon_{kl}(a, 0) + \epsilon_{kl}(1, a^{-1}b) = \epsilon_{kl}(a, 0) + \sum \phi^i(a)^{-1}\phi^i(b)\lambda^t_{ikl}. \tag{4.234}$$

Setting $a_1 = 1$, $a_2 = a \in R^\times$, $b_1 = b$, $b_2 = 0$, in equation 4.230 we get

$$\epsilon_{kl}(a, b) = \epsilon_{kl}(1, b) + \epsilon_{kl}(a, 0) + (a^{-1} - 1)(\lambda^t b)_{kl} \tag{4.235}$$

hence, using equation 4.234 we get

$$\sum(\phi^i(a)^{-1} - 1)\phi^i(b)\lambda^t_{ikl} = (a^{-1} - 1)(\lambda^t b)_{kl}. \tag{4.236}$$

This can be viewed as an identity in $\phi^i(a), \phi^i(b)$ so we get $\lambda_{ilk} = 0$ for all $i \geq 1$ and all k, l and also $b\lambda^t_{0kl} = (\lambda^t b)_{kl}$. Then 4.232 follows from 4.234. □

LEMMA 4.104. *Assume we are given a δ-map $\alpha : \mathbb{G}_m \times \mathbb{G}_a^{n-1} \to \mathbb{A}^1$ and a row vector $\lambda \in R^{n-1}$ such that for any $(a_1, b_1), (a_2, b_2) \in R^\times \times R^{n-1}$,*

$$\alpha(a_1a_2, b_1 + a_1b_2) = \alpha(a_1, b_1) + \alpha(a_2, b_2) + a_1^{-1}(1 - a_2^{-1})b_1\lambda^t. \tag{4.237}$$

Then for any $a_1, a_2, a \in R^\times$, $b \in R^{n-1}$,

$$\alpha(a_1a_2, 0) = \alpha(a_1, 0) + \alpha(a_2, 0), \tag{4.238}$$

$$\alpha(a, b) = \alpha(a, 0) - a^{-1}b\lambda^t. \tag{4.239}$$

PROOF. Entirely similar to the proof of Lemma 4.103. □

LEMMA 4.105. *Assume we are given a δ-map $\beta : \mathbb{G}_m \times \mathbb{G}_a^{n-1} \to \mathbb{A}^{n-1}$, a δ-homomorphism $\psi : \mathbb{G}_m \to \mathfrak{gl}_{n-1}$, and a row vector $\lambda \in R^{n-1}$ such that for any $(a_1, b_1), (a_2, b_2) \in R^\times \times R^{n-1}$,*

$$\begin{aligned} \beta(a_1a_2, b_1 + a_1b_2) &= \beta(a_1, b_1) + a_1\beta(a_2, b_2) \\ &\quad + b_1\psi(a_2) \\ &\quad + a_2^{-1}b_2\lambda^t b_1 + a_2^{-1}b_1\lambda^t b_2 + a_1^{-1}(a_2^{-1} - 1)b_1\lambda^t b_1. \end{aligned} \tag{4.240}$$

Then $\psi = 0$ and there exist $\mu \in R^{n-1}$ and $\nu \in \mathfrak{gl}_{n-1}(R)$ such that for any $a \in R^\times$, $b \in R^{n-1}$ we have

$$\beta(a, b) = (1 - a)\mu + a^{-1}b\lambda^t b + b\nu. \tag{4.241}$$

PROOF. Setting $b_1 = b_2 = 0$ in equation 4.240 we get

$$\beta(a_1a_2, 0) = \beta(a_1, 0) + a_1\beta(a_2, 0).$$

So by Lemma 4.102 there exists $\mu \in R^{n-1}$ such that

$$\beta(a, 0) = (1 - a)\mu, \quad a \in R^\times.$$

Setting $a_1 = a_2 = 1$ in equation 4.240 we get

$$\beta(1, b_1 + b_2) = \beta(1, b_1) + \beta(1, b_2) + b_2\lambda^t b_1 + b_1\lambda^t b_2,$$

because $\psi(1) = 0$. Let $\beta^*(b) = \beta(1, b) - b\lambda^t b$. Then

$$\beta^*(b_1 + b_2) = \beta^*(b_1) + \beta^*(b_2).$$

So by Lemma 4.99 there exist $\lambda_i \in \mathfrak{gl}_{n-1}(R)$, $i \geq 0$, such that

$$\beta^*(b) = \sum \phi^i(b)\lambda_i, \quad b \in R^{n-1}$$

hence

$$\beta(1, b) = b\lambda^t b + \sum \phi^i(b)\lambda_i. \tag{4.242}$$

Now setting $a_1 = a \in R^\times$, $a_2 = 1$, $b_1 = 0$, $b_2 = a^{-1}b \in R^{n-1}$ in equation 4.240 we get

$$\begin{aligned} \beta(a, b) &= \beta(a, 0) + a\beta(1, a^{-1}b) \\ &= (1 - a)\mu + a\left(a^{-2}b\lambda^t b + \sum \phi^i(a)^{-1}\phi^i(b)\lambda_i\right). \end{aligned} \tag{4.243}$$

Setting $a_1 = 1$, $a_2 = a \in R^\times$, $b_1 = b$, $b_2 = 0$ in equation 4.240 we get

$$\beta(a, b) = \beta(1, b) + \beta(a, 0) + b\psi(a) + (a^{-1} - 1)b\lambda^t b, \tag{4.244}$$

hence, using equation 4.243 we get, after cancellation,

$$\sum (a\phi^i(a)^{-1} - 1)\phi^i(b)\lambda_i = b\psi(a). \tag{4.245}$$

Let $a = \zeta$ be an arbitrary root of unity. Since $\psi(\zeta) = 0$ and $\phi(\zeta) = \zeta^p$ we get

$$\sum (\zeta^{1-p^i} - 1)\phi^i(b)\lambda_i = 0.$$

Since, for each ζ this can be seen as an identity in $\phi^i(b)$ we get that for all ζ and all $i \geq 0$,

$$(\zeta^{1-p^i} - 1)\lambda_i = 0.$$

We get that $\lambda_i = 0$ for all $i \geq 1$. By 4.245 we get $\psi = 0$. By 4.243, if $\nu := \lambda_0$, we get 4.241. $\square$

LEMMA 4.106. *Let x be an $n \times n$ matrix of indeterminates over $\mathbb{Z}$ and let $\Delta_i(x)$ be the determinant of the matrix obtained from x by removing the first i rows and the first i columns. Let*

$$\Delta(x) = \prod_{i=1}^{n-1} \Delta_i(x).$$

Let W be the group of all $n \times n$ matrices obtained from the identity matrix 1_n by permuting its columns. Then there exist $w_0, ..., w_N \in W$ and there exist elements

$$a_1(x), ..., a_N(x) \in \mathbb{Z}[x, \Delta(x)^{-1}],$$

and $1 \times (n-1)$ vectors

$$b_1(x), ..., b_N(x) \in \mathbb{Z}[x, \Delta(x)^{-1}]^{n-1},$$

such that if

$$s_i(x) = \begin{pmatrix} a_i(x) & b_i(x) \\ 0 & 1_{n-1} \end{pmatrix}$$

then

$$x = w_0 \cdot s_1(x) \cdot w_1 \cdot s_2(x) \cdot w_2 \cdot ... \cdot s_N(x) \cdot w_N.$$

PROOF. We proceed by induction on n. Write

$$x = \begin{pmatrix} u & y \\ z^t & w \end{pmatrix}$$

where u is one variable, y, z are $1 \times (n-1)$ matrices of variables and w is an $(n-1) \times (n-1)$ matrix of variables. We may write

$$x = \begin{pmatrix} u - yw^{-1}z^t & yw^{-1} \\ 0 & 1_{n-1} \end{pmatrix} \begin{pmatrix} 1 & 0 \\ 0 & w \end{pmatrix} \begin{pmatrix} 1 & 0 \\ w^{-1}z^t & 1_{n-1} \end{pmatrix}. \tag{4.246}$$

Now for $n \times n$ matrices m, m' we say m and m' are W-equivalent if $WmW = Wm'W$. We then have that $\begin{pmatrix} 1 & 0 \\ 0 & w \end{pmatrix}$ is W-equivalent to $\begin{pmatrix} w & 0 \\ 0 & 1 \end{pmatrix}$. Also, if

$$w^{-1}z^t = (c_1(x), ..., c_{n-1}(x))^t$$

then

$$\begin{pmatrix} 1 & 0 \\ w^{-1}z^t & 1_{n-1} \end{pmatrix} = \begin{pmatrix} 1 & 0 & 0 \\ c_1(x) & 1 & 0 \\ 0 & 0 & 1_{n-2} \end{pmatrix} \begin{pmatrix} 1 & 0 & 0 & 0 \\ 0 & 1 & 0 & 0 \\ c_2(x) & 0 & 1 & 0 \\ 0 & 0 & 0 & 1_{n-3} \end{pmatrix} \cdots$$

and each of the matrices in the right hand side of the above equation is W-equivalent to its transpose. We may conclude from 4.246 by the induction hypothesis applied to w. □

PROOF OF THEOREM 4.96. Write matrices in $GL_n(R)$ in block from

$$\begin{pmatrix} a & b \\ c^t & d \end{pmatrix}$$

where $a \in R$, $b, c \in R^{n-1}$ are viewed as a $1 \times (n-1)$ matrices, c^t is the transpose of c, and d is an $(n-1) \times (n-1)$ matrix. Also let 1_{n-1} be the identity $(n-1) \times (n-1)$ matrix. We may write

$$f \begin{pmatrix} a & b \\ 0 & 1_{n-1} \end{pmatrix} = \begin{pmatrix} \alpha(a,b) & \beta(a,b) \\ \gamma(a,b)^t & \epsilon(a,b) \end{pmatrix}$$

where $\alpha, \beta, \gamma, \epsilon$ are δ-morphisms from $\mathbb{G}_m \times \mathbb{G}_a^{n-1}$ to

$$\mathbb{A}^1,\ \mathbb{A}^{n-1},\ \mathbb{A}^{n-1},\ \mathfrak{gl}_{n-1},$$

respectively. If we let $(a_i, b_i) \in R^\times \times R^{n-1}$, $i = 1, 2$, and if we set

$$\alpha_i = \alpha(a_i, b_i),\ \ \beta_i = \beta(a_i, b_i),\ \ \gamma_i = \gamma(a_i, b_i),\ \ \epsilon_i = \epsilon(a_i, b_i),\ \ i = 1, 2,$$

then the cocycle relation for f yields

$$\begin{array}{llcl} & (1)\ \ \alpha(a_1a_2, b_1 + a_1b_2) & = & \alpha_1 + \alpha_2 + a_1^{-1}b_1\gamma_2^t, \\ & (2)\ \ \beta(a_1a_2, b_1 + a_1b_2) & = & \beta_1 + a_1\beta_2 - \alpha_2 b_1 + b_1\epsilon_2 - a_1^{-1}b_1\gamma_2^t b_1, \\ (4.247) & & & \\ & (3)\ \ \gamma(a_1a_2, b_1 + a_1b_2) & = & \gamma_1 + a_1^{-1}\gamma_2, \\ & (4)\ \ \epsilon(a_1a_2, b_1 + a_1b_2) & = & \epsilon_1 + \epsilon_2 - a_1^{-1}\gamma_2^t b_1. \end{array}$$

By equation 4.247 (3) and by Lemma 4.102 there exists $\lambda \in R^{n-1}$ such that

$$\gamma(a, b) = (1 - a^{-1})\lambda, \ \ a \in R^\times, b \in R^{n-1}. \tag{4.248}$$

So equation 4.247 (4) reads

$$\epsilon(a_1a_2, b_1 + a_1b_2) = \epsilon(a_1, b_1) + \epsilon(a_2, b_2) + a_1^{-1}(a_2^{-1} - 1)\lambda^t b_1. \tag{4.249}$$

By Lemma 4.103 we get

$$\epsilon(a_1a_2, 0) = \epsilon(a_1, 0) + \epsilon(a_2, 0), \tag{4.250}$$

$$\epsilon(a, b) = \epsilon(a, 0) + a^{-1}\lambda^t b. \tag{4.251}$$

Similarly, equation 4.247 (1) reads

$$\alpha(a_1a_2, b_1 + a_1b_2) = \alpha(a_1, b_1) + \alpha(a_2, b_2) + a_1^{-1}(1 - a_2^{-1})b_1\lambda^t. \tag{4.252}$$

By Lemma 4.104 we get

$$\alpha(a_1a_2, 0) = \alpha(a_1, 0) + \alpha(a_2, 0), \tag{4.253}$$

$$\alpha(a, b) = \alpha(a, 0) - a^{-1}b\lambda^t. \tag{4.254}$$

Finally by equations 4.248, 4.251, 4.254, equation 4.247 (2) reads
(4.255)

$$\begin{aligned}\beta(a_1a_2, b_1 + a_1b_2) &= \beta(a_1, b_1) + a_1\beta(a_2, b_2)\\ &\quad +b_1(\epsilon(a_2, 0) - \alpha(a_2, 0)1_{n-1})\\ &\quad +a_2^{-1}b_2\lambda^t b_1 + a_2^{-1}b_1\lambda^t b_2 + a_1^{-1}(a_2^{-1} - 1)b_1\lambda^t b_1.\end{aligned}$$

By Lemma 4.105

$$\epsilon(a, 0) = \alpha(a, 0)1_{n-1},\ a \in R^\times \tag{4.256}$$

and there exist $\mu \in R^{n-1}$, $\nu \in \mathfrak{gl}_{n-1}(R)$ such that

$$\beta(a, b) = (1 - a)\mu + a^{-1}b\lambda^t b + b\nu. \tag{4.257}$$

Let $\omega : \mathbb{G}_m \to \mathbb{G}_a$ be the δ-homomorphism defined by $\omega(a) = \alpha(a, 0)$. Consider the classical δ-cocycle $f_0 : GL_n \to \mathfrak{gl}_n$ defined by

$$f_0(g) := f(g) - \omega(\det\ g)1_n.$$

By Lemma 4.98 it is sufficient to show that f_0 is induced on points by a regular map. Let $H \subset GL_n$ the natural closed subgroup scheme such that

$$H(R) = \{\begin{pmatrix} a & b \\ 0 & 1_{n-1} \end{pmatrix}; a \in R^\times, b \in R^{n-1}\}.$$

By equations 4.254, 4.251, 4.248, the restriction of f_0 to H is a regular map. On the other hand by Lemma 4.106 there exist finitely many matrices $w_0, w_1, ..., w_N \in W$ such that the map

$$H \times ... \times H \to GL_n$$

defined on points by

$$(h_1, ..., h_N) \mapsto w_0h_1w_1 \cdots h_Nw_N$$

has a rational section

$$s : U \to H \times ... \times H,\ s(g) = (s_1(g), ..., s_N(g))$$

defined on a Zariski open set U of GL_n which meets the special fiber at p. So for $g \in U(R)$ we have

$$f_0(g) = f_0(w_0 \cdot s_1(g) \cdot w_1 \cdot ... \cdot s_N(g) \cdot w_N).$$

By the cocycle condition we get that the restriction of f_0 to U is regular and hence, using translations and the cocycle condition again we get that f_0 is regular on the whole of GL_n. $\square$

4.10. Non-existence of δ-subgroups of simple groups

In this section we, again, place ourselves in the case when our ground ring is the δ-ring of local type $A = R = W(k)$, $k := \mathbb{F}_p^a$. We again freely use the terminology in section 3.6.

Our aim here is to prove an arithmetic analogue of Cassidy's theorem 2.174 by closely following [**26**].

Assume G/R is a smooth group scheme of finite type. We let

$$G^n = J^n(G)$$

be the p-jet spaces of G, cf. Definition 3.114, and we set

$$\mathcal{O}^n(G) = \mathcal{O}(G^n) = \mathcal{O}(J^n(G)). \tag{4.258}$$

We have natural homomorphisms

$$\mu_n^* : \mathcal{O}^n(G) \to \mathcal{O}^n(G \times G),\ s_n^* : \mathcal{O}^n(G) \to \mathcal{O}^n(G),\ \epsilon_n^* : \mathcal{O}^n(G) \to R$$

induced by multiplication, inverse and unit

$$\mu : G \times G \to G,\ s : G \to G,\ \epsilon : Spec\ R \to G.$$

We also have homomorphisms

$$pr_{1n}^*, pr_{2n}^* : \mathcal{O}^n(G) \to \mathcal{O}^n(G \times G)$$

induced by the projections

$$pr_1, pr_2 : G \times G \to G.$$

We denote by $J^*(G)$ the sequence $(J^n(G))$.

DEFINITION 4.107. A *δ-subgroup* of G is a sequence $S^* = (S^n)$ of p-formal schemes $S^n = Spf\ \mathcal{O}^n(G)/J_n$ where $J_n \subset \mathcal{O}^n(G)$, $n \geq 0$, are ideals satisfying the following conditions:

(1) $J_{n+1} \cap \mathcal{O}^n(G) = J_n$,
(2) $\mu_n^* J_n \subset (pr_{1n}^* J_n, pr_{2n}^* J_n)$,
(3) $s_n^* J_n = J_n$,
(4) $\epsilon_n^* J_n = 0$,
(5) $(J_n : p^\infty) = J_n$.

By condition (5) in Definition 4.107, the $R-$algebras $\mathcal{O}^n(G)/J_n$ are flat. These algebras are also p-adically separated by [**103**], p. 170, hence p-adically complete. Clearly S^n are closed p-formal subgroups of G^n. Set

$$G_0 := G \otimes_R k,\quad S_0^n = S^n \otimes_R k. \tag{4.259}$$

DEFINITION 4.108. We say S^* of $J^*(G)$ is *Zariski dense mod p* in G if the homomorphisms $S_0^n \to G_0$ are surjective for all n.

DEFINITION 4.109. By a *simple group over R* we understand a smooth affine group scheme of finite type G/R for which G_0/k is a simple algebraic group, i.e., G_0 is connected, non-commutative, and has no non-trivial closed connected normal subgroup other than G_0. We will say that G/R is *good at p* if either $p \neq 2, 3$ in case G_0 is not of type A_n (cf. [**76**]) or p does not divide $n + 1$ in case G_0 is of type A_n.

The condition "good at p" ensures that the adjoint representation of G_0 on its Lie algebra $L(G_0)$ is irreducible.

Our main result is the following:

THEOREM 4.110. *Let G/R be a simple group, good at p, and let S^* be a δ-subgroup of G that is Zariski dense mod p in G. Then $S^* = J^*(G)$.*

The proof of the Theorem will consist in a series of Lemmas.

To state the first Lemma we introduce some notation.

DEFINITION 4.111. Let $\mathcal{F} = \mathcal{F}(x_1, x_2) \in R[[x_1, x_2]]^g$ be a (not necessarily commutative) formal group law in g variables; here x_1 and x_2 are g-tuples of variables. For any integer $n \geq 1$ we define a group p-formal scheme $\mathcal{F}\{n\}$, called the *n-twist* of $\mathcal{F}$, by considering the p-adic completion $\widehat{\mathbb{A}^g}$ of the affine space of dimension g equipped with the multiplication given by the g-uple of restricted power series

$$p^{-n}\mathcal{F}(p^n x_1, p^n x_2).$$

Note that $\mathcal{F}\{n\}$ is non-commutative as soon as $\mathcal{F}$ is so; this will be the case in our applications. Tensoring $\mathcal{F}\{n\}$ with $R_m := R/p^{m+1}R$ over R yields a group scheme $\mathcal{F}\{n\}_m$ over R_m. With these notations we have the following "rigidity lemma" which says roughly speaking that "non reduced local finite subgroup schemes of $\mathcal{F}\{n\}_0$ cannot be deformed to (flat) subgroup schemes of $\mathcal{F}\{n\}_1$":

LEMMA 4.112. *Let $H_1 \subset \mathcal{F}\{n\}_1$ be a closed subgroup scheme flat over R_1. Assume $H_0 := H_1 \otimes k$ is connected, zero dimensional. Then H_0 is reduced (hence trivial as a group scheme over k).*

PROOF. We have $\mathcal{F}\{n\}_1 = Spec\ R_1[x]$ with multiplication given by

$$\mu : R_1[x] \to R_1[x_1, x_2], \quad \mu(x) = x_1 + x_2 + pQ(x_1, x_2)$$

where Q is the vector of quadratic forms in $\mathcal{F}$. In particular $\mathcal{F}\{n\}_0$ is an algebraic vector group (i.e. it is isomorphic to $\mathbb{G}_a^g$.) We denote by $x_{0i}, x_{1i}, x_{2i}, Q_i$ the components of x, x_1, x_2, Q. By the structure theorem of subgroup schemes of vector groups [**58**] p. 521, (6.6), we may assume, after making a linear change of the coordinates x, that H_0 is defined by the ideal

$$(x_{01}^{p^{e_1}}, ..., x_{0g}^{p^{e_g}}, p) \subset R_1[x].$$

Let $I \subset R_1[x]$ be the ideal defining H_1. We have

$$(I, p) = (x_{01}^{p^{e_1}}, ..., x_{0g}^{p^{e_g}}, p)$$

hence for any i, $1 \leq i \leq g$ we may write $x_{0i}^{p^{e_i}} = f_i + pg_i$ where $f_i \in I$ and $g_i \in R_1[x]$.

Claim. The ideal I is generated by $f_1, ..., f_g$.

Indeed, let $h \in I$; since $(I, p) = (f_1, ..., f_g, p)$ we may write $h = h_1 + ph_2$, $h_1 \in (f_1, ..., f_g)$, $h_2 \in R_1[x]$, so $ph_2 \in I$. Since $R_1[x]/I$ is flat over R_1 it follows that $I : p = (I, p)$; in particular $h_2 \in (I, p)$ so we may write $h_2 = h_3 + ph_4$, $h_3 \in I$, $h_4 \in R_1[x]$. Applying the above proceedure to h_3 in place of h we may write $h_3 = h_5 + ph_6$, $h_5 \in (f_1, ..., f_g)$, $h_6 \in R_1[x]$. We get $h = h_1 + ph_5 \in (f_1, ..., f_g)$ which proves our Claim.

Now for each i we have

$$\begin{aligned}\mu(f_i(x)) &= (x_{1i}+x_{2i}+pQ_i(x_1,x_2))^{p^{e_i}} - pg_i(x_1+x_2+pQ(x_1,x_2)) \\ &= (x_{1i}+x_{2i})^{p^{e_i}} - pg_i(x_1+x_2) \\ &= x_{1i}^{p^{e_i}} + x_{2i}^{p^{e_i}} + pC_{p^{e_i}}(x_{1i},x_{2i}) - pg_i(x_1+x_2) \\ &\equiv pg_i(x_1)+pg_i(x_2)+pC_{p^{e_i}}(x_{1i},x_{2i}) - pg_i(x_1+x_2) \ \ mod\ \ I_{12}\end{aligned}$$

in the ring $R_1[x_1,x_2]$ where

$$I_{12} := (f_1(x_1),...,f_g(x_1),f_1(x_2),...,f_g(x_2)) \subset R_1[x_1,x_2],$$

and $C_q(x,y)$ denotes as usual the polynomial with integer coefficients

$$((x+y)^q - x^q - y^q)/p,$$

q a power of p. By the Claim above we also have

$$\mu(f_i(x)) \equiv 0 \quad mod \quad I_{12}.$$

By R_1-flatness of $R_1[x_1,x_2]/I_{12}$ we get

$$C_{p^{e_i}}(x_{1i},x_{2i}) - g_i(x_1+x_2) + g_i(x_1) + g_i(x_2) \equiv 0 \quad mod \ \ (I_{12},p) \tag{4.260}$$

in the ring $R_1[x_1,x_2]$. Since

$$(I_{12},p) = (x_{11}^{p^{e_1}},...,x_{1g}^{p^{e_g}},x_{21}^{p^{e_1}},...,x_{2g}^{p^{e_g}},p),$$

by taking the image of 4.260 in $k[x_{1i},x_{2i}]$ (via the specialisation $x_{1j}\mapsto 0, x_{2j}\mapsto 0$ for $j\neq i$), we have

$$C_{p^{e_i}}(x_{1i},x_{2i}) - \gamma_i(x_{1i}+x_{2i}) + \gamma_i(x_{1i}) + \gamma_i(x_{2i}) = \lambda_{1i}(x_{1i},x_{2i})x_{1i}^{p^{e_i}} + \lambda_{2i}(x_{1i},x_{2i})x_{2i}^{p^{e_i}}$$

in this ring for some polynomials $\gamma_i \in k[x]$ and $\lambda_{1i},\lambda_{2i}\in k[x_{1i},x_{2i}]$. Taking the homogenous components of degree p^{e_i} we may assume that $\gamma_i \in k\cdot x^{p^{e_i}}$ and that $\lambda_{1i},\lambda_{2i}\in k$ which immediately gives a contradiction if $e_i\geq 1$ (look at the coefficient of $x_{1i}^{p^{e_i-1}} x_{2i}^{p^{e_i}-p^{e_i-1}}$). We conclude that $e_1 = ... = e_g = 0$ which proves Lemma 4.112. □

LEMMA 4.113. *Let G/R be a smooth affine group scheme of finite type, let $G^n = J^n(G)$ be the $p-$jet spaces of G and G_0^n their reductions modulo p. Assume that for some $n\geq 1$ there exists a closed $p-$formal subgroup $S\subset G^n$ flat over R such that the ring homomorphism $\mathcal{O}(G^{n-1})\to\mathcal{O}(S)$ is injective and the $k-$group scheme homomorphism $S_0 := S\otimes k\subset G_0^n \to G_0^{n-1}$ has a zero dimensional connected kernel. Then $n=1$ and the identity component of G_0 is a torus.*

PROOF. First we claim that $S_0\to G_0^{n-1}$ is surjective. Indeed, if we set $A = \mathcal{O}(G^{n-1})$ and $B = \mathcal{O}(S)$ then by hypothesis (and by Zariski's Main Theorem) B/pB is a finite $A/pA-$module generated, say, by the images of $b_1,...,b_m\in B$. Since B is $p-$adically separated and A is $p-$adically complete it follows easily that actually $b_1,...,b_m$ generate B as an $A-$module. So *Spec* $B\to$ *Spec* A is finite and dominant hence surjective, so *Spec* $B/pB\to$ *Spec* A/pA is surjective and our claim is proved.

Now let the lower index 1 denote, as usual, the result of tensoring with R_1. Since G_0^{n-1} is reduced it follows that S_0 is generically flat over G_0^{n-1} hence flat over G_0^{n-1}. This plus the fact that both S_1 and G_1^{n-1} are flat over R_1 implies that S_1 is flat over G_1^{n-1} (cf. [**103**], p. 152.) Hence $H_1 := Ker(S_1 \to G_1^{n-1})$ is flat over R_1. Note that H_1 is a closed subgroup scheme of M_1 where $M := Ker(G^n \to G^{n-1})$ and on the other hand, by [**23**], (2.2), M is a twist of the formal group law of G. But $H_0 = Ker(S_0 \to G_0^{n-1})$ hence H_0 is connected and zero dimensional. By Lemma 4.112 above H_0 is trivial hence $S_0 \to G_0^{n-1}$ is an isomorphism. This implies that

$$S \to G^{n-1}$$

itself is an isomorphism, due to the general fact that if $A \to B$ is an injective homomorphism of $p-$adically complete flat $R-$algebras such that $A/pA \to B/pB$ is an isomorphism then $A \to B$ is also an isomorphism.

Let us assume that either $n \geq 2$ or the identity component of G_0 is not a torus; we will derive a contradiction and this will end our proof.

Note that, in both cases, G_0^{n-1} contains an algebraic subgroup isomorphic to $\mathbb{G}_{a/k}$. (This is clear if $n \geq 2$. This is also clear if the unipotent radical of G_0 is non trivial. Finally if G_0 is reductive but not a torus, apply, say, [**76**], pp. 160-161.) Now since, by what was already proved, $G^n \to G^{n-1}$ has a section in the category of group p-formal schemes it follows from the universality property of $p-$jet spaces that there is a $p-$derivation of $\mathcal{O}_{G^{n-1}}$ into itself prolonging the natural $p-$derivation of $\mathcal{O}_{G^{n-1}}$ into $\mathcal{O}_{G^n}$. By the universality property of $p-$jet spaces again, one gets sections $G^{n-1} \to G^m$ (in the category of group p-formal schemes) of the projections $G^m \to G^{n-1}$ for $m \geq n$, compatible with the homomorphisms $G^{m+1} \to G^m$. Reducing modulo p we get a closed embedding

$$G_0^{n-1} \to G_0^{\infty} := \lim_{\leftarrow} G_0^m$$

hence an injective homomorphism

$$\mathbb{G}_a(k) \to G_0^{n-1}(k) \to G_0^{\infty}(k) \simeq G(R),$$

where the last isomorphism is as in 3.117. Let Γ be the image of this homomorphism; it is an infinite commutative group of exponent p. Hence so will be its Zariski closure Γ^* in $G(L)$ where L is some algebraically closed field containing R. But since *char* $L = 0$ the multiplication by p map on the identity component of Γ^* is surjective, hence this identity component is trivial, contradicting the infinity of Γ. This ends the proof of Lemma 4.113. □

The following Lemma is essentially trivial, but it is useful to have it stated separately:

LEMMA 4.114. *Assume G/R is an affine smooth group scheme of finite type and $S \subset G^n$ is closed $p-$formal subgroup flat over R. Assume $S_0 = G_0^n$. Then $S = G^n$.*

PROOF. Indeed this follows from the general fact that if A is a $p-$adically separated $R-$algebra and $I \subset A$ is an ideal such that A/I is flat over R and $I \subset pA$ then $I = 0$. □

To state the next Lemma, let us consider the following situation. Assume G/R is an affine smooth group scheme of finite type, let G^n be the $p-$jet spaces of G and G_0^n their reduction modulo p. Let N_0^n denote the scheme theoretic kernel of $G_0^n \to G_0^{n-1}$; it is a $k-$group scheme of finite type, actually an algebraic group.

Let $L(N_0^n)$ be the Lie algebra of N_0^n, viewed as an algebraic vector group over k; in other words we identify it with

$$Spec\ S(\epsilon^*\Omega_{N_0^n/k}),$$

where $\epsilon : Spec\ k \to N_0^n$ is the unit. For any $k-$scheme Y let Y^F denote the $k-$scheme obtained by taking the pull back of Y/k via the absolute Frobenius morphism

$$F : Spec\ k \to Spec\ k.$$

In particular we may consider the algebraic vector group $L(N_0^n)^F$. There is an action on it of the algebraic group $(G_0^n)^F$ coming from the adjoint action of G_0^n on $L(G_0^n)$. Composing with the relative Frobenius $k-$homomorphism

$$G_0^n \to (G_0^n)^F$$

we get an action of G_0^n on $L(N_0^n)^F$. On the other hand we have an action of G_0^n on N_0^{n+1} coming from the action of G_0^{n+1} by conjugation on N_0^{n+1} (since N_0^{n+1} is commutative, the $G_0^{n+1}-$action on N_0^{n+1} factors through a G_0^n-action).

LEMMA 4.115. *There is a natural isomorphism $N_0^{n+1} \simeq L(N_0^n)^F$ which is G_0^n-equivariant with respect to the two G_0^n-actions described above.*

PROOF. The isomorphism is defined as follows: for any affine $k-$scheme $Z = Spec\ C$ the set $\mathrm{Hom}_k(Z, N_0^{n+1})$ is identified, due to the universality property of $p-$jet spaces, with the set of $p-$derivations D of

$$\mathcal{O}(G^n) \stackrel{\epsilon}{\to} R \stackrel{c}{\to} k \stackrel{\eta}{\to} C$$

(ϵ the counit of $\mathcal{O}(G^n)$, c the reduction mod p) that prolong the $p-$derivation

$$\mathcal{O}(G^{n-1}) \stackrel{\delta}{\to} \mathcal{O}(G^n) \stackrel{\epsilon}{\to} R \stackrel{c}{\to} k \stackrel{\eta}{\to} C.$$

For each such D the map

$$D - \eta \circ c \circ \delta \circ \epsilon : \mathcal{O}(G^n) \stackrel{\epsilon}{\to} R \stackrel{\delta}{\to} R \stackrel{c}{\to} k \stackrel{\eta}{\to} C$$

is a (usual) derivation of the homomorphism

$$\mathcal{O}(G^n) \stackrel{\epsilon}{\to} R \stackrel{c}{\to} k \stackrel{\eta}{\to} C \stackrel{F}{\to} C$$

vanishing on $\mathcal{O}(G^{n-1})$ hence

$$D - \eta \circ c \circ \delta \circ \epsilon$$

defines an element of

$$\mathrm{Hom}_k(Z, L(N_0^n)^F).$$

The map

$$\mathrm{Hom}_k(Z, N_0^{n+1}) \to \mathrm{Hom}_k(Z, L(N_0^n)^F)$$

defined above is easily checked to be a group isomorphism (argument similar to the one in [**20**] pp. 1398-1399.) To prove the equivariance claimed in the Lemma it is enough to prove equivariance for $G = GL_m$. But in this case equivariance follows once we check the following formula

$$(x^{-1}yx)^{(n)} \equiv (x^{(p^n)})^{-1} y^{(n)} x^{(p^n)} \ mod\ (p, y-1, y', ..., y^{(n-1)}),$$

valid in the ring

$$A_n := R[x, y, x', y', ..., x^{(n)}, y^{(n)}, det(x)^{-1}, det(y)^{-1}]\hat{},$$

where x, y are $g \times g$ matrices of indeterminates, 1 is the identity matrix and $x^{(p^i)}$ stands for the matrix obtained from x by raising all entries to the power p^i while $y^{(i)}$ stands for $\delta^i y$. In order to check the formula above we will check by induction on n that

$$(x^{-1}yx)^{(n)} \equiv \phi^n(x)^{-1}y^{(n)}\phi^n(x) \ \ mod \ \ J_n$$

in the ring A_n where $J_n = (y-1, y', ..., y^{(n-1)})$ and $\phi : A_{n-1} \to A_n$ is the unique ring homomorphism extending $\phi : R \to R$ such that

$$\phi(x) = x^{(p)} + px', ..., \phi(x^{(n-1)}) = (x^{(n-1)})^{(p)} + px^{(n)}.$$

For $n = 0$ this is clear. Assuming the latter formula holds for n we have the following computation in the ring $A_{n+1} \otimes K$, where $\equiv$ means congruence modulo $J_{n+1} \otimes K$:

$$\begin{aligned}(x^{-1}yx)^{(n+1)} &= p^{-1}[\phi^{n+1}(x)^{-1}((y^{(n)})^{(p)}+py^{(n+1)})\phi^{n+1}(x)-(\phi^n(x)^{-1}y^{(n)}\phi^n(x))^{(p)}] \\ &\equiv \phi^{n+1}(x)^{-1}y^{(n+1)}\phi^{n+1}(x).\end{aligned}$$

Now the above congruence implies actually that the same congruence holds in the ring A_{n+1} modulo the ideal J_{n+1}, because A_{n+1}/J_{n+1} is flat over R. This completes the induction and hence the proof of the lemma. □

PROOF OF THEOREM 4.110. Recall our hypothesis. We have a simple group G/R good at p, we let G^n be the $p-$jet spaces of G, and we assume we are given, for each $n \geq 0$, a closed $p-$formal subgroup $S^n \subset G^n$ flat over R such that the projections $G^n \to G^{n-1}$ induce homomorphisms $S^n \to S^{n-1}$. We also assume the ring homomorphisms $\mathcal{O}(S^n) \to \mathcal{O}(S^{n+1})$ are injective and the $k-$group homomorphisms $S_0^n \to G_0$ are surjective. The conclusion we want to prove is that $S^n = G^n$ for all $n \geq 0$. For the proof let us borrow notations from the discussion before Lemma 4.115. Also for any algebraic group Γ over k denote by Γ° or by $\Gamma^\circ(k)$ or by $(\Gamma(k))^\circ$ the identity component of Γ. In particular we let $(S_0^n)^\circ$ denote the connected component of S_0^n.

Claim 1. $L(N_0^n)(k)$, viewed as a $k-$vector space, is a simple $(S_0^n(k))^\circ-$module with respect to the representation induced from the adjoint representation.

Indeed, let $T_0^m(k)$ be the image of

$$(S_0^n(k))^\circ \to G_0^m(k)$$

where $0 \leq m \leq n$ and let us prove by induction on m that $L(N_0^m)(k)$ is a simple $T_0^m(k)-$module (under the adjoint action). For $m = 0$ our hypothesis implies

$$T_0^0(k) = G_0(k)$$

hence $L(N_0^0)(k) = L(G_0)(k)$ is a simple $T_0^0(k)-$module. Assume now $L(N_0^m)(k)$ is a simple $T_0^m(k)-$module. Then $L(N_0^m)^F(k)$ is a simple $T_0^m(k)-$module with respect to the action induced via the relative Frobenius $G_0^m \to (G_0^m)^F$. By Lemma 4.115 there is a $G_0^m(k)-$equivariant vector space isomorphism

$$N_0^{m+1} \simeq L(N_0^m)^F(k).$$

Hence this isomorphism is in particular $T_0^m(k)-$equivariant with respect to the induced $T_0^m(k)-$actions. Since the homomorphism $T_0^{m+1}(k) \to T_0^m(k)$ is surjective it follows that $N_0^{m+1}(k)$ is a simple $T_0^{m+1}(k)-$module. But on the other hand, since N_0^{m+1} is an algebraic vector group, there is a $T_0^{m+1}(k)$-equivariant isomorphism

$$N_0^{m+1}(k) \simeq L(N_0^{m+1})(k).$$

So $L(N_0^{m+1})(k)$ is a simple $T_0^{m+1}(k)-$ module. This ends the induction and proves our Claim 1.

Next we shall prove by induction on $n \geq 0$ that $S^n = G^n$. This is clear for $n = 0$ by Lemma 4.114. Assume $S^n = G^n$ and $S^{n+1} \neq G^{n+1}$ for some $n \geq 0$ and seek a contradiction. Consider the closed $k-$subgroup schemes $S_0^{n+1}, N_0^{n+1} \subset G_0^{n+1}$.

Claim 2. The scheme theoretic intersection $S_0^{n+1} \cap N_0^{n+1}$ is connected and zero dimensional.

Assume for the moment this Claim is true and let's note that this ends our proof. For if Claim 2 holds then the $k-$group scheme homomorphism $S_0^{n+1} \to G_0^n$ has a connected zero dimensional kernel. On the other hand by the induction hypothesis the ring homomorphism $\mathcal{O}(G^n) = \mathcal{O}(S^n) \to \mathcal{O}(S^{n+1})$ is injective. So, by Lemma 4.113, G_0 must be a torus, a contradiction.

Let us then check Claim 2. Set

$$M := S_0^{n+1} \cap N_0^{n+1}.$$

We must prove that the group $M(k)$ is trivial. Let M° be the connected component of M. Since N_0^{n+1} is an algebraic vector group, $M^\circ(k)$ is a vector subspace of the $k-$vector space $N_0^{n+1}(k)$ [**58**] p. 523. On the other hand, by Claim 1, $N_0^{n+1}(k)$ is a simple $(S_0^{n+1}(k))^\circ-$module. Since $(S_0^{n+1})^\circ$ acts on M (and hence on M°) by conjugation it follows that $M^\circ(k)$ is a $(S_0^{n+1}(k))^\circ$-submodule of $N_0^{n+1}(k)$, hence either $M^\circ(k) = 0$ or

$$M^\circ(k) = N_0^{n+1}(k).$$

If the latter holds then

$$N_0^{n+1}(k) \subset S_0^{n+1}(k)$$

hence, since $S_0^n = G_0^n$ we get $S_0^{n+1}(k) = G_0^{n+1}(k)$, hence $S_0^{n+1} = G_0^{n+1}$ hence $S^{n+1} = G^{n+1}$ by Lemma 4.114, which we assumed is not the case. So we must have $M^\circ(k) = 0$, i.e. $M(k)$ is a finite group. Since $(S_0^{n+1}(k))^\circ$ is connected its action on the finite group $M(k)$ must be trivial. But then $M(k)$ must be trivial for otherwise if $0 \neq x \in M(k)$ then $k \cdot x \subset N_0^{n+1}(k)$ would be a non zero proper $(S_0^{n+1}(k))^\circ-$submodule, which is a contradiction. This concludes the proof of Claim 2 and we are done. □

4.11. Non-existence of invariant adelic connections on GL_n

In this section we place ourselves again in the case when our ground ring is the δ-ring of local type $A = R = W(k)$, $k := \mathbb{F}_p^a$.

A main corollary of Theorem 4.110 is the following non-existence result:

COROLLARY 4.116. *Let G be a simple group over R which is good at p. Then there is no p-adic connection on G that is right invariant with respect to G; i.e., there is no Frobenius lift ϕ on $\widehat{G}$ such that the following diagram is commutative:*

$$\begin{array}{ccc} \widehat{G} \times \widehat{G} & \to & \widehat{G} \\ \phi \times \phi \downarrow & & \downarrow \phi \\ \widehat{G} \times \widehat{G} & \to & \widehat{G} \end{array} \tag{4.261}$$

Here the horizontal arrows are given by multiplication.

PROOF. Assume such a ϕ exists. Then, by the universality property of p-jet spaces ϕ induces a compatible system of group homomorphisms $\sigma_n : \widehat{G} \to J^n(G)$

which are sections of the projections $J^n(G) \to \widehat{G}$. Then the sequence $S^n := \sigma_n(\widehat{G})$ defines a δ-subgroup of G that is Zariski dense mod p in G, contradicting Theorem 4.110. □

Using the above we will prove:

THEOREM 4.117. *Assume $p \nmid n$, $n \geq 2$. Then there is no p-adic connection on GL_n that is right invariant with respect to GL_n; i.e., there is no Frobenius lift ϕ on $\widehat{GL_n}$ making the following diagram commutative:*

$$\begin{array}{ccc} \widehat{GL_n} \times \widehat{GL_n} & \to & \widehat{GL_n} \\ \phi \times \phi \downarrow & & \downarrow \phi \\ \widehat{GL_n} \times \widehat{GL_n} & \to & \widehat{GL_n} \end{array} \tag{4.262}$$

PROOF. Assume such a ϕ exists and set $\phi(x) = \Phi$, $\Phi = (\Phi_{ij})$,

$$\Phi_{ij} = \Phi_{ij}(x) \in \mathcal{O}(\widehat{GL_n}).$$

Let u, v be two matrices with entries indeterminates. Then the commutativity of the diagram in the statement is equivalent to the equality

$$\Phi(uv) = \Phi(u)\Phi(v). \tag{4.263}$$

Consider the series

$$\gamma(x) = \left(\frac{\det(x)^p}{\det(\Phi(x))}\right)_p^{1/n} \in \mathcal{O}(\widehat{GL_n}) \tag{4.264}$$

and the matrix of series

$$\Phi_1(x) = \gamma(x) \cdot \Phi(x). \tag{4.265}$$

Clearly

$$\gamma(uv) = \gamma(u)\gamma(v)$$

and hence

$$\Phi_1(uv) = \Phi_1(u)\Phi_1(v). \tag{4.266}$$

Define the Frobenius lift ϕ_1 on $\widehat{GL_n}$ by the formula

$$\phi_1(x) = \Phi_1; \tag{4.267}$$

by 4.266 ϕ_1 is right invariant with respect to GL_n. Now note that

$$\phi_1(\det(x)) = \det(\phi_1(x)) = \det(\gamma(x)\Phi(x)) = \gamma(x)^n \det(\Phi(x)) = (\det(x))^p.$$

The latter implies that ϕ_1 induces a Frobenius lift on SL_n; this Frobenius lift is, of course, still right invariant with respect to SL_n. But now, since $p \nmid n$, we have that SL_n is good at p and we get a contradiction by Corollary 4.116. □

More generally we have:

COROLLARY 4.118. *Assume $p \nmid n$, $n \geq 2$. Then there are no Frobenius lifts ϕ, ϕ_1 on $\widehat{GL_n}$ making the following diagram commutative:*

$$\begin{array}{ccc} \widehat{GL_n} \times \widehat{GL_n} & \to & \widehat{GL_n} \\ \phi \times \phi_1 \downarrow & & \downarrow \phi \\ \widehat{GL_n} \times \widehat{GL_n} & \to & \widehat{GL_n} \end{array} \tag{4.268}$$

PROOF. Indeed assume such Frobenius lifts exist and set $\phi(x) = \Phi$, $\phi_1(x) = \Phi_1$. For two matrices of indeterminates u, v we get

$$\Phi(uv) = \Phi(u)\Phi_1(v). \tag{4.269}$$

Setting $u = 1$ we get

$$\Phi(v) = \Phi(1)\Phi_1(v). \tag{4.270}$$

Replacing v by uv in 4.270 we get

$$\Phi(uv) = \Phi(1)\Phi_1(uv). \tag{4.271}$$

Replacing v by u in 4.270 we get

$$\Phi(u) = \Phi(1)\Phi_1(u). \tag{4.272}$$

Combining 4.269, 4.271, and 4.272 we get

$$\Phi(1)\Phi_1(u)\Phi_1(v) = \Phi(u)\Phi_1(v) = \Phi(uv) = \Phi(1)\Phi_1(uv), \tag{4.273}$$

hence

$$\Phi_1(uv) = \Phi_1(u)\Phi_1(v);$$

the latter implies that ϕ_1 is right invariant with respect to GL_n, contradicting Theorem 4.117. $\square$

On the other hand we have the following elementary result mentioned in the Introduction; this result does not assume $p \nmid n$ and does not use Theorem 4.117:

PROPOSITION 4.119. *Assume $n \geq 2$. Then there is no Frobenius lift ϕ on $\widehat{GL_n}$ making the following diagram commutative:*

$$\begin{array}{ccc} \widehat{GL_n} \times \widehat{GL_n} & \rightarrow & \widehat{GL_n} \\ \phi \times \phi_0 \downarrow & & \downarrow \phi \\ \widehat{GL_n} \times \widehat{GL_n} & \rightarrow & \widehat{GL_n} \end{array} \tag{4.274}$$

where $\phi_0(x) = x^{(p)}$.

PROOF. Same argument as in the proof of Corollary 4.118 except that, instead of Theorem 4.117, one uses the trivial observation that ϕ_0 is not right invariant with respect to GL_n for $n \geq 2$. $\square$

CHAPTER 5

Curvature and Galois groups of Ehresmann connections

In this chapter we consider, in some detail, Ehresmann connections on GL_n. These connections are, by definition, global on $\mathfrak{gl}_n$, so they possess curvature. We will start by placing ourselves in the case when A is a δ-ring of global type, $A = \mathcal{O}_K[1/M]$, and comment on the explicit form of curvature. Then we return to the case when A is a δ-ring of local type, $A \subset R$, and examine solutions of the corresponding arithmetic differential equations, especially their existence, uniqueness, rationality, and symmetries (i.e., Galois groups). Finally we prove results about the structure of Galois groups of the equations defined by Ehresmann connections in the "generic case."

5.1. Gauge and curvature formulas

In this section we place ourselves in the case when A is a δ-ring of global type, $A = \mathcal{O}_K[1/M]$, cf. Definition 3.25. Say A is equipped with a global connection (δ_p) indexed by a set of primes $\mathcal{V}$. Also set $G = GL_n = Spec\ A[x, \det(x)^{-1}]$ and $\mathfrak{g} = Spec\ A[x]$.

DEFINITION 5.1. Let $\alpha = (\alpha_p)$ be a family, indexed by $\mathcal{V}$ with $\alpha_p \in \mathfrak{g}(A)$. The global connection $\delta^{\mathfrak{g},\alpha} = (\delta_p^{\mathfrak{g},\alpha})$ on $\mathfrak{g}$ defined by

$$\delta_p^{\mathfrak{g},\alpha} x = \alpha_p \cdot x^{(p)} \tag{5.1}$$

is called the *Ehresmann connection* attached to α. The induced adelic connection $\delta^{G,\alpha}$ on G will be called the *Ehresmann connection* on G attached to α. Denote by $\Lambda^1_{Ehr}(G)$ the set of Ehresmann connections on G.

We sometimes simply write δ_p^{α}, or even δ_p, instead of $\delta_p^{\mathfrak{g},\alpha}$ or $\delta_p^{G,\alpha}$. The Frobenius lifts (ϕ_p) attached to (δ_p) are given by

$$\phi_p(x) = \epsilon_p \cdot x^{(p)}, \quad \epsilon_p := 1 + p\alpha_p. \tag{5.2}$$

Note that δ^{α} is an adelic connection that is global on $\mathfrak{g}$ so the discussion in subsection 4.8.4 applies. In particular, if $N \subset G$ is, as usual, the normalizer of the diagonal torus then the gauge action in 4.219 induces an action

$$N(A) \times \Lambda^1_{Ehr}(G) \to \Lambda^1_{Ehr}(G), \quad g \star \delta^{\alpha} = \delta^{g\star\alpha}, \tag{5.3}$$

where, by formula 4.220, we have $g \star \alpha = (g \star_p \alpha_p)$,

$$g \star_p \alpha_p = \phi_p(g) \cdot \alpha_p \cdot (g^{(p)})^{-1} + \delta_p g \cdot (g^{(p)})^{-1}. \tag{5.4}$$

This should be viewed as an analogue of the formula 2.64 giving the gauge action in classical differential geometry.

One can, on the other hand, define the curvature $(\varphi_{pp'})$ as a family of $\phi_p\phi_{p'}$-semilinear endomorphisms of A-module,

$$\varphi_{pp'} : A[x] \to A[x]. \tag{5.5}$$

With notation as in 5.2 we have

$$\varphi_{pp'}(x) = \Phi_{pp'}(x) := \frac{1}{pp'}\left(\phi_p(\epsilon_{p'})(\epsilon_p x^{(p)})^{(p')} - \phi_{p'}(\epsilon_p)(\epsilon_{p'}x^{(p')})^{(p)}\right). \tag{5.6}$$

In particular, evaluating at $x = 1$ (the identity matrix) we have

$$\Phi_{pp'}(1) = \frac{1}{pp'}\left(p(\delta_p\epsilon_{p'})\epsilon_p^{(p')} - p'(\delta_{p'}\epsilon_p)\epsilon_{p'}^{(p)} + [\epsilon_{p'}^{(p)}, \epsilon_p^{(p')}]\right). \tag{5.7}$$

The latter formula can be viewed as an analogue of the formula 2.22 in classical differential geometry.

5.2. Existence, uniqueness, and rationality of solutions

5.2.1. Review and notation. For the remainder of this chapter we revert to the case when A is a δ-ring of local type i.e., $A \subset R = W(\mathbb{F}_p^a)$ is a valuation δ-subring of R (cf. Definition 3.27) and we set $k = \mathbb{F}_p^a$. We denote the unique p-derivation on R by $\delta = \delta_p$ and its unique Frobenius lift by $\phi = \phi_p$. We denote by R^δ the monoid of constants

$$R^\delta = \{\lambda \in R; \delta\lambda = 0\}.$$

So R^δ consists of 0 and all roots of unity in R. Recall that the reduction mod p map $R^\delta \to k$ is an isomorphism of monoids. Also we denote by K the fraction field of R. As usual, for a smooth scheme X over R, we will often write X instead of $X(R)$. As usual in the one prime case, $\widehat{\ }$ means p-adic completion.

In what follows we set $G = GL_n = Spec\ A[x, \det(x)^{-1}]$ and we consider a p-derivation $\delta = \delta^G$ on $\widehat{G}$, i.e., on $\mathcal{O}(\widehat{G})$. Set

$$\delta^G x = \Delta, \quad \Delta = \Delta(x) \in \mathfrak{gl}_n(\mathcal{O}(\widehat{G})), \quad \Phi(x) = x^{(p)} + p\Delta(x). \tag{5.8}$$

Moreover for $\alpha \in \mathfrak{gl}_n(A)$ set

$$\Delta^\alpha(x) = \alpha \cdot \Phi(x) + \Delta(x) = \alpha x^{(p)} + \epsilon\Delta(x), \quad \epsilon := 1 + p\alpha. \tag{5.9}$$

Recall from Proposition 4.2 that by a Δ-*linear equation* we understand an equation of the form

$$\delta u = \Delta^\alpha(u) \tag{5.10}$$

where $u \in G = GL_n = GL_n(R)$; u is a referred to as a *solution* to the equation 5.10. If $\Phi^\alpha(x) = \epsilon \cdot \Phi(x)$ then 5.10 is equivalent to

$$\phi(u) = \Phi^\alpha(u). \tag{5.11}$$

DEFINITION 5.2. The set G^α of all $u \in GL_n$ such that 5.10 (equivalently 5.11) holds is called the *solution set* of 5.10 (or of 5.11).

In the formalism above $\delta^{G,\alpha}x = \Delta^\alpha$ is an Ehresmann connection if $\delta^G x = \Delta = 0$ in which case we say δ^G is *trivial*. In what follows we are primarily interested in properties of solutions to Ehresmann connections; but some of our discussion is pertinent to other examples. Here are the main examples we have in mind.

EXAMPLE 5.3. (Ehresmann connections). This corresponds to the case δ^G is trivial, i.e., $\Delta = 0$. So in this case $\Phi(x) = x^{(p)}$ and 5.10 and 5.11 become

$$\delta u = \alpha \cdot u^{(p)} \tag{5.12}$$

and

$$\phi(u) = \epsilon \cdot u^{(p)} \tag{5.13}$$

respectively. It is worth noting that 5.13 is *not* an instance of a linear difference equation in the sense of [**114**]. Indeed a linear difference equation for ϕ has the form

$$\phi(u) = \epsilon \cdot u \tag{5.14}$$

rather than the form 5.13.

EXAMPLE 5.4. (Special linear connection). Recall δ^G is called *special linear* (cf. Definition 4.65) if

$$\Delta(x) = \frac{\lambda(x) - 1}{p} \cdot x^{(p)}$$

where $p \nmid n$ and

$$\lambda(x) := \left(\frac{\det(x^{(p)})}{\det(x)^p}\right)_p^{-1/n}. \tag{5.15}$$

Here the $(\ )_p^{-1/n}$ operation is, as usual, the principal root operation 4.21. In this case $\Phi(x) = \lambda(x) \cdot x^{(p)}$ and the equations 5.10 and 5.11 become

$$\delta u = \left(\lambda(u) \cdot \alpha + \frac{\lambda(u) - 1}{p}\right) \cdot u^{(p)} \tag{5.16}$$

and

$$\phi(u) = \lambda(u) \cdot \epsilon \cdot u^{(p)} \tag{5.17}$$

respectively. Note that, in this case, $\Phi(u) \in SL_n$ for any $u \in SL_n$. (This justifies the words *special linear*.) It is useful to recall, in this context, the δ-Lie algebra $\mathfrak{sl}_{n,\delta}$ of SL_n; cf. Remark 4.15. Its R-points form the set of all $\alpha \in \mathfrak{gl}_n$ such that $1 + p\alpha \in SL_n$, in other words

$$\mathfrak{sl}_{n,\delta} = \{\alpha \in \mathfrak{gl}_n; \mathrm{tr}(\alpha) + ... + p^{n-1}\det(\alpha) = 0\}.$$

Recall that this is not a subgroup of $(\mathfrak{gl}_n, +)$ where $+$ is the usual addition of matrices but rather a subgroup of $(\mathfrak{gl}_n, +_\delta)$ where

$$a +_\delta b := a + b + pab.$$

One should compare this with the Lie algebra $\mathfrak{sl}_n$ of SL_n which is given by

$$\mathfrak{sl}_n = \{\alpha \in \mathfrak{gl}_n;\ \mathrm{tr}(\alpha) = 0\}.$$

Note also that if $\alpha \in \mathfrak{sl}_{n,\delta}$ then $\det(x)$ is a prime integral for the connection $\delta^{G,\alpha}x = \Delta^{\alpha}(x)$. Indeed if $\epsilon = 1 + p\alpha$ then

$$\begin{aligned}\phi^{\alpha}(\det(x)) &= \det(\phi^{\alpha}(x)) \\ &= \det\left(\lambda(x)\cdot\epsilon\cdot x^{(p)}\right) \\ &= \lambda(x)^n\cdot\det(\epsilon)\cdot\det(x^{(p)}) \\ &= \det(x)^p,\end{aligned}$$

hence $\delta^{G,\alpha}(\det(x)) = 0$.

EXAMPLE 5.5. (real Chern connections). Let $q \in GL_n$, $q^t = \pm q$, be split; cf. 4.16. Let $SO(q) \subset SL_n$ be the identity component of the subgroup defined by the equations $x^tqx = q$. Recall that if $\delta^G x = \Delta$ is the *real Chern connection* attached to q then

$$\Delta(x) = x^{(p)}\cdot\frac{1}{p}(\Lambda(x)-1),\quad \Lambda(x) = \{((x^{(p)})^tqx^{(p)})^{-1}(x^tqx)^{(p)}\}_p^{1/2};$$

cf. Theorem 4.23. In this case we have $\Phi(x) = x^{(p)}\cdot\Lambda(x)$. Recall from 4.45 that, in this case,

$$\Phi(x)^tq\Phi(x) = (x^tqx)^{(p)}. \tag{5.18}$$

Note also that, in this case, $\Phi(u) \in SO(q)$ for any $u \in SO(q)$. Recall the δ-Lie algebra $\mathfrak{so}(q)_\delta$ of $SO(q)$; its points form the set of all $\alpha \in \mathfrak{gl}_n$ such that $1 + p\alpha \in SO(q)$, in other words

$$\mathfrak{so}(q)_\delta = \{\alpha \in \mathfrak{gl}_n; \alpha^tq + q\alpha + p\alpha^tq\alpha = 0\}.$$

This is, again, a subgroup of $(\mathfrak{gl}_n, +_\delta)$ and differs from the Lie algebra $\mathfrak{so}(q)$ of $SO(q)$ which is given by

$$\mathfrak{so}(q) = \{\alpha \in \mathfrak{gl}_n; \alpha^tq + q\alpha = 0\}.$$

Note also that if $\alpha \in \mathfrak{so}(q)_\delta$ then $\mathcal{H}(x) := x^tqx$ is a prime integral for δ^G.

5.2.2. Main result. In the next theorem an extension of integral domains is called *generically finite* if the induced extension between fraction fields is finite.

THEOREM 5.6. *Let $u_0 \in GL_n$ and $\alpha \in \mathfrak{gl}_n$ and let Δ be arbitrary. Then the following hold:*

(1) *There is a unique $u \in GL_n$ satisfying $\delta u = \Delta^{\alpha}(u)$ such that $u \equiv u_0$ mod p.*
(2) *If Δ, u_0, and α have entries in a complete valuation subring A of R then u also has entries in A.*
(3) *If $u_0 \in SL_n$, $\alpha \in \mathfrak{sl}_{n,\delta}$, and δ^G is special linear then $u \in SL_n$.*
(4) *If $u_0 \in SO(q)$, $\alpha \in \mathfrak{so}(q)_\delta$, and δ^G is real Chern, attached to q, then $u \in SO(q)$.*
(5) *Assume u_0 and α have entries in a valuation δ-subring A of R with finite residue field. Assume that either δ^G is trivial or δ^G is special linear and $u \in SL_n$. Then u has entries in a δ-subring of R which is generically finite over A.*

REMARK 5.7. It is worth comparing Equation 5.13 with the familiar linear equations in analysis in the case $n = 1$; in case $n = 1$ Equation 5.13 is, of course,

$$\phi(u) = \epsilon \cdot u^p \tag{5.19}$$

where $\epsilon = 1 + p\alpha$, $\alpha \in R$, $u \in R^\times$. This equation can be solved as follows. Write $\epsilon = \exp(p\beta)$, where $\exp : pR \to 1 + pR$ is the group isomorphism given by the p-dic exponential and $\beta \in R$. Then the set of solutions to 5.19 consists of all $u \in R^\times$ of the form

$$u = \zeta \cdot \exp\left(\sum_{n=1}^{\infty} p^n \phi^{-n}(\beta)\right) \tag{5.20}$$

where $\zeta \in R^\times$, $\delta\zeta = 0$. On the other hand consider the group homomorphism $\psi : R^\times \to R$ defined by

$$u \mapsto \psi(u) = \frac{1}{p} \log\left(\frac{\phi(u)}{u^p}\right) = \sum_{n=1}^{\infty} (-1)^{n-1} \frac{p^{n-1}}{n} \left(\frac{\delta u}{u^p}\right)^n \tag{5.21}$$

where log is the p-adic logarithm. Then Equation 5.19 is equivalent to the equation

$$\psi(u) = \beta \tag{5.22}$$

Now the homomorphism ψ above should be viewed as an analogue of the logarithmic derivative map $\text{Mer}(D)^\times \to \text{Mer}(D)$,

$$u \mapsto u'/u,$$

where $\text{Mer}(D)$ is the field of meromorphic functions on a disk $D \subset \mathbb{C}$, say, and $u' = \frac{du}{dz}$, where z is a complex variable. So the analogue, in analysis, of Equation 5.22 is the equation

$$\frac{u'}{u} = \beta, \tag{5.23}$$

where $\beta \in \text{Mer}(D)$. For β holomorphic in D the solutions to Equation 5.23 are of the form

$$u = c \cdot \exp\left(\int \beta dz\right) \tag{5.24}$$

where exp is the complex exponential and $c \in \mathbb{C}$. Hence the elements 5.20 in $R^\times$ should be viewed as arithmetic analogues of the functions 5.24 in $\text{Mer}(D)$.

5.2.3. Proofs. Theorem 5.6 follows from a series of propositions and remarks that we now state and prove. The following proposition settles the existence and uniqueness statement in assertion (1) of Theorem 5.6.

PROPOSITION 5.8. *Let $u_0 \in GL_n(R)$, and $\alpha \in \mathfrak{gl}_n(R)$. Then the Δ-linear equation $\delta u = \alpha \cdot \Phi(u) + \Delta(u)$ has a unique solution $u \in GL_n(R)$ such that $u \equiv u_0$ mod p.*

PROOF. Recall that the equation above is equivalent to $\phi(u) = \epsilon \cdot \Phi(u)$ where $\epsilon = 1 + p\alpha$. To check the uniqueness of the solution assume $\phi(u) = \epsilon \cdot \Phi(u)$ and $\phi(v) = \epsilon \cdot \Phi(v)$ with $u, v \in GL_n(R)$, $u \equiv v$ mod p. Then we prove by induction that $u \equiv v$ mod p^n. Indeed if the latter is the case then $u^{(p)} \equiv v^{(p)}$ mod p^{n+1} and $\Delta(u) \equiv \Delta(v)$ mod p^{n+1} hence $\Phi(u) \equiv \Phi(v)$ mod p^{n+1}. Hence $\phi(u) \equiv \phi(v)$ mod p^{n+1}. Hence $u \equiv v$ mod p^{n+1}.

To check the existence of a solution u such that $u \equiv u_0 \bmod p$ we define a sequence of matrices $u_n \in GL_n(R)$ by the formula

$$u_{n+1} = \phi^{-1}(\epsilon \cdot \Phi(u_n)), \quad n \geq 0.$$

We claim that for all $n \geq 0$ we have

$$\phi(u_n) \equiv \epsilon \cdot \Phi(u_n) \mod p^{n+1}.$$

Assuming the claim we get $u_{n+1} \equiv u_n \bmod p^{n+1}$ hence u_n converges p-adically to some $u \in GL_n(R)$. Also $\phi(u) = \epsilon \cdot \Phi(u)$ which ends our proof. We are left with checking the claim. We proceed by induction. The case $n = 0$ is clear. Assume now $\phi(u_n) \equiv \epsilon \cdot \Phi(u_n) \bmod p^{n+1}$. Hence

$$\phi^{-1}(\epsilon \cdot \Phi(u_n)) \equiv u_n \ \ mod \ \ p^{n+1},$$

hence

$$\Phi(\phi^{-1}(\epsilon \cdot \Phi(u_n))) \equiv \Phi(u_n) \ \ mod \ \ p^{n+2},$$

hence

$$\begin{aligned} \epsilon \cdot \Phi(u_{n+1}) &= \epsilon \cdot \Phi(\phi^{-1}(\phi(u_{n+1}))) \\ &= \epsilon \cdot \Phi(\phi^{-1}(\epsilon \cdot \Phi(u_n))) \\ &\equiv \epsilon \cdot \Phi(u_n) \mod p^{n+2} \\ &= \phi(u_{n+1}), \end{aligned}$$

and the induction step follows. □

REMARK 5.9. If, in Proposition 5.8, $\Delta = 0$, $n = 1$, and $u_0 \equiv \zeta \bmod p$ where $\zeta \in R$ is a root of unity, the solution u has a closed form:

$$u = \zeta \cdot \epsilon_{-1} \cdot \epsilon_{-2}^{p} \cdot \epsilon_{-3}^{p^2} \cdot \ldots \quad \text{(convergent product)}$$

where $\epsilon_i = \phi^i(\epsilon)$ for $i \in \mathbb{Z}$. This computation implies the formula in Remark 5.7.

REMARK 5.10. If in Proposition 5.8 δ^G is special linear, $u_0 \in SL_n(R)$, and $\alpha \in \mathfrak{sl}_{n,\delta}$ then $u \in SL_n(R)$. Indeed this follows because $\Phi(a) \in SL_n(R)$ and $\phi^{-1}(a) \in SL_n(R)$ for all $a \in SL_n(R)$; hence if u_n is as in the proof of that Proposition then $u_n \in SL_n(R)$. Similarly if δ^G is real Chern, attached to q, $u_0 \in SO(q)$, and $\alpha \in \mathfrak{so}(q)_\delta$ then $u \in SO(q)$. The above proves assertions (3) and (4) in Theorem 5.6. Alternatively these assertions can be deduced as follows. Let $\mathcal{I}(x)$ be $\det(x)$ or $x^t qx$ respectively and let u be such that $\delta u = \Delta^\alpha(u)$ with $u \equiv u_0 \bmod p$ with either $u_0 \in SL_n$, $\alpha \in \mathfrak{sl}_{n,\delta}$ or $u_0 \in SO(q)$, $\alpha \in \mathfrak{so}(q)_\delta$ respectively. By the discussion in Examples 5.4 and 5.5 we have $\delta(\mathcal{I}(u)) = 0$ hence $\mathcal{I}(u)$ is either 0 or a root of unity in R. On the other hand we have $\mathcal{I}(u) \equiv \mathcal{I}(u_0) \bmod p$ hence, since $\mathcal{I}(u_0) = 0$, we have $\mathcal{I}(u) \equiv 0 \bmod p$. Since $\mathcal{I}(u)$ is either 0 or a root of unity we conclude it must be 0, hence $u \in SL_n$ or $u \in SO(q)$ respectively.

REMARK 5.11. In notation of Propositon 5.8 the natural reduction map

$$G^\alpha \to GL_n(k)$$

is a bijection. So each solution set G^α has a natural structure of group; but of course with this structure G^α is not a subgroup of $GL_n(R)$.

Let us address the question of "rationality" of solutions of Δ-linear equations.

Recall that an extension $A \subset A'$ of subrings of R is called *generically finite* if the extension of their fraction fields is finite; if in addition A is a valuation subring then A' is a localization of a finite extension of A; if, in addition A is complete then any generically finite extension of A in R is finite.

PROPOSITION 5.12. *Assume A is a complete valuation subring of R (hence also a δ-subring) with finite residue field. If in Proposition 5.8 we have*

$$\Delta \in \mathfrak{gl}_n(A[x,\det(x)]\widehat{\ }), \quad u_0 \in GL_n(A), \quad \alpha \in \mathfrak{gl}_n(A)$$

then $u \in GL_n(A)$.

PROOF. Let $A = R^{\phi^\nu}$. Then $\phi^\nu(u_0) = u_0$ and $\phi^\nu(\alpha) = \alpha$ hence $\phi^\nu(\epsilon) = \epsilon$, where $\epsilon = 1 + p\alpha$. Also $\phi^\nu(\Delta(a)) = \Delta(\phi^\nu(a))$, and hence $\phi^\nu(\Phi(a)) = \Phi(\phi^\nu(a))$, for all $a \in GL_n(R)$. Since $\phi(u) = \epsilon \cdot \Phi(u)$ and $u \equiv u_0$ mod p it follows that

$$\phi^{\nu+1}(u) = \phi^\nu(\epsilon)(\phi^\nu(\Phi(u))) = \epsilon \cdot \Phi((\phi^\nu(u)))$$

and $\phi^\nu(u) \equiv \phi^\nu(u_0) \equiv u_0$ mod p. By the uniqueness in Proposition 5.8 it follows that $\phi^\nu(u) = u$ hence $u \in GL_n(A)$. □

PROPOSITION 5.13. *Assume A is a valuation δ-subring of R with finite residue field. Assume in Proposition 5.8 that one of the following holds:*

(1) *δ^G is trivial, $u_0 \in GL_n(A)$, and $\alpha \in \mathfrak{gl}_n(A)$.*
(2) *δ^G is special linear, $u_0 \in SL_n(A)$, and $\alpha \in \mathfrak{sl}_{n,\delta} \cap \mathfrak{gl}_n(A)$.*

Then there exists a generically finite extension of valuation δ-subrings $A \subset A'$ of R such that $u \in GL_n(A')$.

PROOF. Assume we are in case (2); case (1) is similar (and indeed slightly easier).

By Proposition 5.12 if $\widehat{A}$ is the completion of A then $u \in GL_n(\widehat{A})$ hence there exists $\nu \geq 0$ such that $\phi^{\nu+1}(u) = u$. Let $N = n^2$ and identify the points of $\mathbb{A}^N$ with $n \times n$ matrices. Let

$$\lambda_\nu(u) = \phi^\nu(\lambda(u)) \cdot \phi^{\nu-1}(\lambda(u))^p \cdot \ldots \cdot \lambda(u)^{p^\nu}.$$

Using $\phi(u) = \lambda(u) \cdot \epsilon \cdot u^{(p)}$, and setting $\epsilon_j = \phi^j(\epsilon)$, we get

$$u = \phi^{\nu+1}(u) = \lambda_\nu(u) \cdot \varphi(u), \tag{5.25}$$

where

$$\varphi : \mathbb{A}^N \to \mathbb{A}^N$$

is the morphism of schemes over A defined on points by

$$\varphi(v) = \epsilon_\nu(\epsilon_{\nu-1}(\epsilon_{\nu-2}(\ldots(\epsilon v^{(p)})^{(p)})^{(p)}\ldots)^{(p)}.$$

Let K^a be an algebraic closure of the fraction field K of R, let F be the fraction field of A, and let F^a be the algebraic closure of F in K^a. Note that

$$\varphi : \mathbb{A}^N(K^a) \to \mathbb{A}^N(K^a)$$

is obtained by composing maps $\eta \mapsto \epsilon_j \eta$ with copies of the map $\eta \mapsto \eta^{(p)}$; both these types of maps are given by homogeneous polynomials (of degree 1 and p respectively) and have the property that the pre-image of 0 is 0. Hence φ is given by

$$\varphi(\eta) = (\Phi_1(\eta), \ldots, \Phi_N(\eta))$$

where

$$\Phi_1, ..., \Phi_N \in F[x_1, ..., x_N]$$

are homogeneous polynomials of degree $p^{\nu+1} > 1$ and $\varphi^{-1}(0) = \{0\}$; hence $\Phi_1, ..., \Phi_N$ have no common zero in $\mathbb{A}^N(K^a)$ except at the origin. Consider an extra variable x_0 and consider the projective variety $V \subset \mathbb{P}^N$ defined by the equations

$$\Phi_j(x_1, ..., x_N) - x_0^{p^{\nu+1}-1} x_j = 0. \tag{5.26}$$

Clearly the intersection of V with the hyperplane $x_0 = 0$ is empty. So V has dimension zero hence $V(K^a)$ is finite. Since V is defined over F we have $V(K^a) = V(F^a)$. By equation 5.25 the point

$$(\lambda_\nu(u)_p^{-1/(p^{\nu+1}-1)} : u) \in \mathbb{P}^N(K)$$

belongs to $V(K)$ hence it belongs to $V(F^a)$. (Here the $1/(p^{\nu+1}-1)$-power is, again, the principal root 4.21). It follows that

$$\lambda_\nu(u)_p^{1/(p^{\nu+1}-1)} \cdot u \in \mathbb{A}^N(F^a) = \mathfrak{gl}_n(F^a) \tag{5.27}$$

hence

$$\det(\lambda_\nu(u)_p^{1/(p^{\nu+1}-1)} \cdot u) \in F^a.$$

Since, by Remark 5.10, $\det(u) = 1$ we get $(\lambda_\nu(u)_p^{1/(p^{\nu+1}-1)})^n \in F^a$ hence

$$\lambda_\nu(u)_p^{1/(p^{\nu+1}-1)} \in F^a.$$

By 5.27 again we get $u \in GL_n(F^a)$. Since $\phi(A) \subset A$ we have $\phi(F) \subset F$ hence $\phi(F^a) \subset F^a$. Set

$$F' := F(u, \phi(u), ..., \phi^\nu(u)) \subset F^a.$$

Then F' is a finite extension of F and $\phi(F') \subset F'$. Hence $A' := R \cap F'$ is a valuation δ-subring of R generically finite over A and u has coefficients in A' which ends our proof. □

Note that the Propositions in this subsection imply Theorem 5.6. The consideration of the variety cut out by equations 5.26 is a trick from [**65**] and is an indication of an interesting link between our paradigm and the arithmetic of dynamical systems on projective space.

5.3. Galois groups: generalities

We continue to use the notation in subsection 5.2.1. So we denote by G the group $GL_n(R)$ and by $\mathfrak{gl}_n$ the Lie algebra $\mathfrak{gl}_n(R)$. Let δ^G be a p-derivation on $\mathcal{O}(\widehat{G})$ where $G = Spec\ A[x, \det(x)^{-1}]$, $A \subset R$ a valuation δ-subring of R. Let $\delta^G x = \Delta(x) \in \mathfrak{gl}_n(\mathcal{O}(\widehat{G}))$, and let $\Phi(x) = x^{(p)} + p\Delta(x)$. Let $\alpha \in \mathfrak{gl}_n(A)$, $\Delta^\alpha(x) = \alpha \cdot \Phi(x) + \Delta(x)$, $\epsilon = 1 + p\alpha$, and consider the equivalent Δ-linear equations

$$\delta u = \Delta^\alpha(u), \tag{5.28}$$

$$\phi(u) = \Phi^\alpha(u). \tag{5.29}$$

Let G^α be the set of solutions to Equation 5.28 in G, let $u \in G^\alpha$ be a fixed solution, let

$$\Phi_u(x) = \Phi(u)^{-1}\Phi(ux), \quad \Delta_u(x) = \frac{1}{p}(\Phi_u(v) - v^{(p)}), \tag{5.30}$$

and let G_u be the set of solutions $v \in G$ to the Δ_u-linear equation

$$\delta v = \Delta_u(v), \tag{5.31}$$

equivalently to the equation

$$\phi(v) = \Phi_u(v). \tag{5.32}$$

Note that

$$uG_u \subset G^\alpha.$$

Indeed if $c \in G_u$ we have

$$\phi(uc) = \phi(u) \cdot \phi(c) = \epsilon \cdot \Phi(u) \cdot \phi(c) = \epsilon \cdot \Phi(uc)$$

so $uc \in G^\alpha$.

DEFINITION 5.14. Consider the valuation δ-subring $A \subset R$, let $\alpha \in \mathfrak{gl}_n(A)$ and let $u \in GL_n(R)$ be a solution of 5.28. Consider the subring $A\{u\}$ of R generated by A and the entries of $u, \delta u, \delta^2 u, ...$; so $\delta A \subset A$. Consider the group $Aut_\delta(A\{u\}/A)$ of all A-algebra automorphisms σ of $A\{u\}$ such that $\sigma \circ \delta = \delta \circ \sigma$ on $A\{u\}$. Consider furthermore the subgroup $\tilde{G}_{u/A}$ of $Aut_\delta(A\{u\}/A)$ consisting of all $\sigma \in Aut_\delta(A\{u\}/A)$ such that $u^{-1} \cdot \sigma(u) \in GL_n(A)$. Consider the injective group homomorphism,

$$\tilde{G}_{u/A} \to GL_n(A) \tag{5.33}$$

sending any σ into $c_\sigma := u^{-1} \cdot \sigma(u)$. Then define the *$\delta$-Galois group* of u/A as the image $G_{u/A}$ of 5.33. In particular $G_{u/A} \simeq \tilde{G}_{u/A}$.

THEOREM 5.15.

(1) *Assume δ^G is special linear and let*

$$\alpha \in \mathfrak{gl}_n(A), \quad u \in G^\alpha.$$

Then for any $c \in G_{u/A}$ we have $\delta(\det(c)) = 0$.

(2) *Assume δ^G is real Chern , attached to q, and let*

$$\alpha \in \mathfrak{gl}_n(A), \quad u \in SO(q) \cap G^\alpha.$$

Then for any $c \in G_{u/A}$ we have $\delta(c^t q c) = 0$.

Cf. Propositions 5.16 plus 5.20 below.

Theorem 5.15 shows that if δ^G is either special linear or real Chern the δ-Galois group $G_{u/A}$ is "close to being contained" in SL_n and $SO(q)$ respectively, provided u is in these groups respectively. Indeed $G_{u/A}$ being contained in SL_n (respectively $SO(q)$) means $\det(c) = 1$ (respectively $\det(c) = 1$ and $c^t q c = q$) for $c \in G_{u/A}$. Theorem 5.15, however, merely guarantees that $\delta(\det(c)) = 0$ or $\delta(c^t q c) = 0$, which is a "slightly" weaker property.

In the special cases of interest to us the δ-Galois group has a "δ-theoretic description/bound" which we now discuss. Let $x', x'', ...$ be new matrices of indeterminates and consider, as usual, the polynomial ring

$$A\{x\} := A[x, x', x'', ...].$$

Recall that there is a unique ring endomorphism ϕ of $A\{x\}$ whose restriction to A is ϕ and such that $\phi(x) = x^{(p)} + px'$, $\phi(x') = (x')^{(p)} + px''$, etc. Define, as usual, the map

$$\delta : A\{x\} \to A\{x\}$$

by the formula

$$\delta f = \frac{\phi(f) - f^p}{p}.$$

We let $I_{u/A}$ be the kernel of the unique A-algebra map $A\{x\} \to R$, sending

$$x \mapsto u, \quad x' \to \delta u, \quad x'' \mapsto \delta^2 u, \quad \text{etc.}$$

So $I_{u/A}$ is the *ideal of δ-algebraic relations among the entries of u*; note that $A\{u\}$ is then the image of the map $A\{x\} \to R$ above. We let $\Sigma_{u/A}$ be the subgroup of $GL_n(A)$ consisting of all matrices c such that the A-automorphism

$$\sigma_c : A\{x\} \to A\{x\}$$

defined by

$$\sigma_c(x) = xc, \quad \sigma(x') = \delta(xc), \quad \sigma(x'') = \delta^2(xc), \quad \text{etc.},$$

satisfies

$$\sigma_c(I_{u/A}) = I_{u/A}.$$

Similarly let $I^0_{u/A}$ be the kernel of the map

$$A[x] \to A[u], \quad x \mapsto u,$$

and let $\Sigma^0_{u/A}$ be the subgroup of $GL_n(A)$ consisting of all matrices c such that the A-automorphism

$$\sigma^0_c : A[x] \to A[x],$$

defined by

$$\sigma_c(x) = xc,$$

satisfies

$$\sigma^0_c(I^0_{u/A}) = I^0_{u/A}.$$

Here is the "δ-theoretic description/bound" of the δ-Galois group in our cases of interest:

PROPOSITION 5.16.

(1) $G_{u/A} = \Sigma_{u/A}$.
(2) *If δ^G is trivial, special linear, or real Chern we have $G_{u/A} \subset G_u$.*
(3) *If δ^G is trivial we have $G_{u/A} = \Sigma^0_{u/A} \cap G_u$.*

PROOF. To check assertion (1) let, first, $c \in G_{u/A}$, so there exists $\sigma \in \tilde{G}_{u/A}$ with $\sigma u = cu$. Then we claim that $c \in \Sigma_{u/A}$. Indeed this follows from the commutativity of the diagram

$$\begin{array}{ccc} A\{x\} & \stackrel{\sigma_c}{\to} & A\{x\} \\ \downarrow & & \downarrow \\ A\{u\} & \stackrel{\sigma}{\to} & A\{u\} \end{array} \tag{5.34}$$

Conversely, if $c \in \Sigma_{u/A}$ then $\sigma_c : A\{x\} \to A\{x\}$ obviously induces an automorphism $\sigma : A\{u\} \to A\{u\}$ commuting with δ and sending u into uc so $c \in G_{u/A}$.

To check assertions (2) and (3) we need a preliminary discussion in which we assume that we are in the hypothesis of assertion (2).

Let us start with an A-algebra automorphism σ of $A\{u\}$ such that

$$c := u^{-1} \cdot \sigma(u) \in GL_n(A)$$

and let $\epsilon = 1 + p\alpha$. We may (uniquely) extend σ to an automorphism of $S^{-1}A\{u\}$ where S is the multiplicative system consisting of all elements of $A\{u\}$ of the form $\det(u)^m + pf$ where $m \in \mathbb{Z}_{\geq 0}$, $f \in A\{u\}$. We claim that

$$\sigma(\Phi(u)) = \Phi(\sigma(u)). \tag{5.35}$$

(The left hand side makes sense because $\Phi(u) = \epsilon^{-1} \cdot \phi(u)$ has entries in $S^{-1}A\{u\}$.) This is clear if δ^G is trivial because in this case $\Phi(x) = x^{(p)}$ has polynomial entries. Let us check 5.35 in case δ^G is real Chern ; the case when it is special linear is similar. Indeed, since

$$\Lambda(u) = (u^{(p)})^{-1}\Phi(u) = (u^{(p)})^{-1}\epsilon^{-1}\phi(u) = (u^{(p)})^{-1}(1 + p\alpha)^{-1}(u^{(p)} + p\delta u) \tag{5.36}$$

it follows that $\Lambda(u)$ has entries in $S^{-1}A\{u\}$ so

$$\sigma(\Phi(u)) = \sigma(u^{(p)} \cdot \Lambda(u)) = \sigma(u^{(p)}) \cdot \sigma(\Lambda(u)) = (uc)^{(p)} \cdot \sigma(\Lambda(u)), \tag{5.37}$$

$$\Phi(\sigma u) = (uc)^{(p)} \cdot \Lambda(uc). \tag{5.38}$$

So it is enough to check that $\sigma(\Lambda(u)) = \Lambda(uc)$. Since both matrices in the latter equality are $\equiv 1 \bmod p$ in $GL_n(R)$ (for the first one use 5.36) it is enough to check that their squares are equal. But, since $M(x) := \Lambda(x)^2$ has entries rational functions of x, we get:

$$\sigma(\Lambda(u))^2 = \sigma(\Lambda(u)^2) = \sigma(M(u)) = M(\sigma u) = M(uc) = \Lambda(uc)^2,$$

which concludes the proof of 5.35. Using 5.35 we get

$$\phi(\sigma(u)) = \phi(uc) = \phi(u) \cdot \phi(c) = \epsilon \cdot \Phi(u) \cdot \phi(c), \tag{5.39}$$

$$\sigma(\phi(u)) = \sigma(\epsilon \cdot \Phi(u)) = \epsilon \cdot \sigma(\Phi(u)) = \epsilon \cdot \Phi(\sigma(u)) = \epsilon \cdot \Phi(uc). \tag{5.40}$$

To check assertion (2) let $c \in G_{u/A}$ and let us prove that $c \in G_u$. Let $\sigma \in \tilde{G}_{u/A}$, $\sigma u = uc$. Since $\sigma \circ \delta = \delta \circ \sigma$ on $A\{u\}$ it follows that $\sigma \circ \phi = \phi \circ \sigma$ on $A\{u\}$ so, by 5.39 and 5.40, $\Phi(uc) = \Phi(u) \cdot \phi(c)$ hence $c \in G_u$.

To check assertion (3), assume δ^G is trivial (hence $A\{u\} = A[u]$). Let, first, $c \in G_{u/A}$ and let us prove that

$$c \in \Sigma^0_{u/A} \cap G_u.$$

By assertion (2) we already know that $c \in G_u$. Also $c \in \Sigma^0_{u/A}$ by the commutativity of the diagram

$$\begin{array}{ccc} A[x] & \overset{\sigma^0_c}{\to} & A[x] \\ \downarrow & & \downarrow \\ A[u] & \overset{\sigma}{\to} & A[u] \end{array} \tag{5.41}$$

Conversely let $c \in \Sigma^0_{u/A} \cap G_u$ and let us prove that $c \in G_{u/A}$. Indeed since $\sigma^0_c(I^0_{u/A}) = I^0_{u/A}$, it follows that $\sigma^0_c : A[x] \to A[x]$ induces an automorphism $\sigma : A[u] \to A[u]$ with $\sigma(u) = uc$. On the other hand since $c \in G_u$ we have $\Phi(uc) = \Phi(u) \cdot \phi(c)$ hence, by 5.39 and 5.40, $\phi(\sigma(u)) = \sigma(\phi(u))$. It follows that $\sigma \circ \phi = \phi \circ \sigma$ on $A[u]$ hence $\sigma \circ \delta = \delta \circ \sigma$ on $A[u]$. So $\sigma \in \tilde{G}_{u/A}$, hence $c \in G_{u/A}$ and we are done. $\square$

For our discussion below we recall that we denote by T, W, N the torus of diagonal matrices in G, the Weyl group of permutation matrices in G and the normalizer of T in G respectively; so

$$N = TW = WT.$$

Also if

$$G^\delta = \{a \in G; \delta a = 0\}$$

we set

$$T^\delta = T \cap G^\delta, \quad N^\delta = N \cap G^\delta = T^\delta W = WT^\delta;$$

G^δ is a subset of G while T^δ and N^δ are subgroups of G. Note that, in our notation, $\delta^G x = \Delta$ on G is right invariant with respect to N if and only if

$$\Phi(ac) = \Phi(a) \cdot c^{(p)}$$

for all $a \in G$ and all $c \in N$.

EXAMPLE 5.17. If δ^G is trivial, special linear, or real Chern then it is right invariant with respect to N. By the way in case δ^G is trivial, right invariance of δ^G with respect to N simply means that $(ac)^{(p)} = a^{(p)}c^{(p)}$ for $a \in G$ and $c \in N$.

LEMMA 5.18. *If δ^G is right invariant with respect to N then $N^\delta \subset G_u$.*

PROOF. Trivial. □

LEMMA 5.19. *Assume δ^G is trivial and set $N_{u/A} = N^\delta \cap \Sigma^0_{u/A}$.*

(1) *Assume the entries of one of the rows of u are algebraically independent over A. Then $G_{u/A} \subset N^\delta$ hence*
$$G_{u/A} = N_{u/A}.$$
(2) *Assume the entries of u are algebraically independent over A; then*
$$G_{u/A} = N^\delta \cap GL_n(A).$$
(3) *Assume σ is an A-automorphism of $A[u]$ such that $\sigma(u) = uc$ with $c \in GL_n(A) \cap G_u$. Then $c \in G_{u/A}$.*
(4) *Assume $n = 1$. Then $G_{u/A} \subset N^\delta = G^\delta$.*
(5) *We have an equality*
$$\bigcap_{u \in G} G_u = N^\delta.$$

PROOF. To prove assertion (1) let $c \in G_{u/A}$, hence $c \in G_u$, i.e. $(uc)^{(p)} = u^{(p)}\phi(c)$. If $c = (c_{ij})$ then for all m and j

$$\sum_{i=1}^n u_{mi}^p \phi(c_{ij}) = (\sum_{i=1}^n u_{mi}c_{ij})^p.$$

Let m be such that $u_{m1}, ..., u_{mn}$ are algebraically independent over A. Identifying the coefficients of the monomials in $u_{m1}, ..., u_{mn}$ in the latter equality we get that for each j there exists an index $\tau(j)$ such that $c_{ij} = 0$ for all $i \neq \tau(j)$ and such that $c^p_{\tau(j)j} = \phi(c_{\tau(j)j})$. Since c is non-singular we must have that τ is a permutation and $c \in N^\delta$.

To prove assertion (2) note that $G_{u/A} \subset N^\delta \cap GL_n(A)$ by assertion (1). Also $N^\delta \subset G_u$ by Lemma 5.18 and, since $A[x] \to A[u]$ is an isomorphism, we also have $\Sigma^0_{u/A} = GL_n(A)$; hence, using Proposition 5.16,

$$N^\delta \cap GL_n(A) \subset G_u \cap \Sigma^0_{u/A} = G_{u/A}.$$

To prove assertion (3) let $\sigma^0_c : A[x] \to A[x]$ be the unique A-algebra homomorphism such that $\sigma^0_c(x) = xc$. Then $\sigma^0_c(I^0_{u/A}) = I^0_{u/A}$ by the commutativity of the diagram 5.41; hence $c \in \Sigma^0_{u/A}$, hence $c \in G_{u/A}$.

To prove assertion (4) let $c \in G_{u/A}$; then $uc \in G^\alpha$ hence $\phi(u)\phi(c) = \epsilon u^p c^p$ where $\epsilon = 1 + p\alpha$. Since $\phi(u) = \epsilon u^p$ we get $\phi(c) = c^p$ hence $c \in G^\delta = N^\delta$.

To prove assertion (5) note that the inclusion $\supset$ follows from Lemma 5.18. To prove the inclusion $\subset$ let c be in the intersection. Since R is uncountable one can find u with entries algebraically independent over the ring generated by the entries of $c, \delta c, \delta^2 c, \ldots$. Then one concludes that $c \in N^\delta$ by using the same argument as in the proof of assertion (1). □

PROPOSITION 5.20.

(1) *Assume δ^G is special linear and let $u \in GL_n$. Then $\mathcal{I}(x) = \det(x)$ is a prime integral for the Δ_u-linear equation $\delta v = \Delta_u(v)$; equivalently for any $v \in G_u$ we have $\delta(\det(v)) = 0$.*

(2) *Assume δ^G is real Chern and let $u \in SO(q)$. Then $\mathcal{I}(x) = x^t q x$ is a prime integral for the Δ_u-linear equation $\delta v = \Delta_u(v)$; equivalently for any $v \in G_u$ we have $\delta(v^t q v) = 0$.*

PROOF. To check assertion (1) note that since $v \in G_u$ we have

$$\lambda(uv) \cdot (uv)^{(p)} = \lambda(u) \cdot u^{(p)} \cdot \phi(v).$$

Taking determinants we get

$$\lambda(uv)^n \cdot \det((uv)^{(p)}) = \lambda(u)^n \cdot \det(u^{(p)}) \cdot \det(\phi(v)).$$

Taking into account the definition of $\lambda(x)$ we get

$$(\det(uv))^p = \det(u)^p \cdot \det(\phi(v))$$

hence $\det(v)^p = \det(\phi(v)) = \phi(\det(v))$ which implies $\delta(\det(v)) = 0$. The fact that this is equivalent to $\det(x)$ being a prime integral follows from the fact that the solutions to the corresponding equation are Zariski dense in G; cf. Proposition 5.8.

To check assertion (2) note that by Equation 5.32 we have

$$\phi(v) = \Phi(u)^{-1}\Phi(uv).$$

On the other hand, using 5.18, we get

$$\Phi(u)^t q \Phi(u) = (u^t q u)^{(p)} = q^{(p)} = q,$$

hence

$$(\Phi(u)^t)^{-1} q \Phi(u)^{-1} = q,$$

hence

$$\begin{aligned}\phi(v^t q v) &= \phi(v)^t q \phi(v)\\ &= \Phi(uv)^t(\Phi(u)^t)^{-1} q \Phi(u)^{-1}\Phi(uv)\\ &= \Phi(uv)^t q \Phi(uv)\\ &= (v^t u^t q u v)^{(p)}\\ &= (v^t q v)^{(p)},\end{aligned}$$

which implies that $\delta(v^t q v) = 0$. We conclude as in the proof of assertion (1). □

Our next task will be to compute/bound the δ-Galois group $G_{u/A}$ in case δ^G is trivial (hence $\delta^{G,\alpha}$ is Ehresmann). One of the morals that will emerge in the next section will be that this group tends to be contained in N; but this is not always the case as shown in the following:

EXAMPLE 5.21. Let $A = \mathbb{Z}_{(p)}$, $n = 2$, and assume $p \equiv 1 \bmod 3$. Consider the matrices

$$u = \begin{pmatrix} 1 & \zeta \\ 1 & \zeta^2 \end{pmatrix}, \quad c = \begin{pmatrix} 1 & -1 \\ 0 & -1 \end{pmatrix}, \quad uc = \begin{pmatrix} 1 & \zeta^2 \\ 1 & \zeta \end{pmatrix},$$

where $\zeta \in \mathbb{Z}_p \subset R$ is a cubic root of unity. Note that $\det u = \zeta^2 - \zeta \not\equiv 0 \bmod p$ so $u, c, uc \in GL_2(R)$. Then u is a solution to the Δ-linear equation equation

$$\delta u = 0,$$

where $\Delta = 0$. We will show that $G_{u/A} \not\subset N$. Indeed $u, c, uc \in G^{\delta} \backslash N$ and $u^{(p)} = u$, $c^{(p)} = c$, $(uc)^{(p)} = uc$ so $c \in G_u$. Also we have $A[u] = \mathbb{Z}_{(p)}[\zeta]$ and the unique non-trivial automorphism σ of $\mathbb{Z}_{(p)}[\zeta]$ sending $\sigma(\zeta) = \zeta^2$ satisfies $\sigma(u) = uc$. By assertion (3) in Lemma 5.19 we have $c \in G_{u/A}$; so in particular $G_{u/A} \not\subset N$, and our claim is proved. By the way in this case $G_{u/A} = \langle c \rangle$ is cyclic of order 2.

We end our discussion here by recording the following:

PROPOSITION 5.22. *Assume δ^G is trivial and A is a valuation δ-subring of R with finite residue field. Then $G_{u/A}$ is a finite group.*

PROOF. Let $u_0 \in G^{\delta}$ be the unique element such that $u \equiv u_0 \bmod p$. Let F be the field of fractions of A, let F' be the field generated by F and the roots of unity appearing as entries in u_0, and let $A' = R \cap F'$. Then A' is a valuation δ-subring of R generically finite over A and $u_0 \in GL_n(A')$. In particular A' has a finite residue field. Since $\alpha \in GL_n(A')$, by Proposition 5.13, we get $u \in GL_n(A'')$ for some generically finite extension A'' of A'. Then, by the equality $A\{u\} = A[u]$, $G_{u/A}$ is finite. □

5.4. Galois groups: the generic case

Our aim here is to show that the δ-Galois group $G_{u/A}$ attached to a solution u of an equation

$$\delta u = \alpha \cdot u^{(p)}$$

is "close to being a subgroup of N" if α is "sufficiently general." We will consider two types of genericity: in the sense of Baire and in the sense of Zariski. The above equation is, of course, attached to an Ehresmann connection

$$\delta^{G,\alpha} x = \Delta^{\alpha}(x) = \alpha \cdot x^{(p)}$$

so, everywhere here δ^G is trivial, i.e., $\Delta = 0$.

We first address genericity in the sense of Baire. So recall that $R = W(\mathbb{F}_p^a)$ and K is the fraction field of R. In what follows we view R as a complete metric space with respect to the p-adic metric. So we can talk about open balls in R. Any open ball has the form $X = b + p^N R$ for some $b \in R$ and $N \in \mathbb{Z}_{\geq 0}$; any such X is also closed and is, again, a complete metric space with respect to the induced metric. Now recall that a subset of a metric space is called *of the first category* if it is a countable union of subsets each of which has the property that its closure has an empty interior. By the Baire-Hausdorff theorem [**123**], p. 11, any subset of the first category in a non-empty complete metric space X is different from X. This applies then to any open ball X in R.

THEOREM 5.23. *There exists a subset Ω of the first category in the metric space*

$$X = \{u \in GL_n(R); u \equiv 1 \quad mod \quad p\}$$

such that for any $u \in X \backslash \Omega$ the following holds. Let $\alpha = \delta u \cdot (u^{(p)})^{-1}$. Then there exists a valuation δ-subring A of R containing R^{δ} such that $\alpha \in \mathfrak{gl}_n(A)$ and such that the δ-Galois group $G_{u/A}$ equals N^{δ}.

For the proof we need a preparation.

LEMMA 5.24. *Let $x, x', ..., x^{(r)}$ are a m-tuples of indeterminates and let $f \in R[x, x', ..., x^{(r)}]\hat{}$. Assume the map $f_* : R^m \to R$ defined by*

$$f_*(a) = f(a, \delta a, ..., \delta^m a)$$

vanishes on a product of open balls. Then f vanishes on the whole of R^m.

PROOF. Recall that $f = 0$ if and only if $f_* = 0$. So it is enough to show that for any $b_j \in R$, $1 \leq j \leq m$, the R-algebra homomorphism

$$R[x, x', ..., x^{(r)}]\hat{} \to R[x, x', ..., x^{(r)}]\hat{}, \quad x_j^{(i)} \mapsto \delta^i(b_j + p^N x_j),$$

is injective. To check this we may assume $b_j = 0$ for all j. But then the assertion follows from the fact that

$$R[x, x', ..., x^{(r)}]\hat{} \subset K[[x, x', ..., x^{(r)}]] = K[[x, \phi(x), ..., \phi^r(x)]]$$

and from the fact that the endomorphism of $K[[x, \phi(x), ..., \phi^r(x)]]$ defined by $\phi^i(x) \mapsto p^N \phi^i(x)$ is injective. $\square$

LEMMA 5.25. *Let E be a countable subfield of K and let $X_1, ..., X_m \subset R$ be open balls. Then one can find a subset Ω of the first category in the metric space $X = X_1 \times ... \times X_m$ such that for all $u = (u_1, ..., u_m) \in X \backslash \Omega$ the family*

$$(\delta^i u_j)_{i \geq 0, 1 \leq j \leq m}$$

is algebraically independent over E.

PROOF. Let $\mathcal{F} = E[x, x', x'', ...]$ be the polynomial ring where each of x, x', x'', ... is an m-tuple of indeterminates . Hence $\mathcal{F}$ is countable. Then for each $f \in \mathcal{F}$ with $f \neq 0$ set

$$X_f := \{u \in X; f(u, \delta u, \delta^2 u, ...) = 0\}.$$

Now we claim that each X_f is closed in the metric space X and has empty interior; indeed X_f is the zero locus in X of $f_* : R^m \to R$ and our claim follows from Lemma 5.24. The present Lemma follows now by taking

$$\Omega = \bigcup_{0 \neq f \in \mathcal{F}} X_f.$$

□

PROOF OF THEOREM 5.23. Let E be the subfield of K generated over $\mathbb{Q}$ by all the roots of unity in K; i.e. $E = \mathbb{Q}(R^\delta)$. Now X in the Theorem is a product of balls so by Lemma 5.25 there exists a subset of the first category $\Omega \subset X$ such that for all $u \in X \backslash \Omega$ the family

$$(\delta^r u_{ij})_{r \geq 0, 1 \leq i,j \leq n}$$

is algebraically independent over E. Let

$$\epsilon = \phi(u) \cdot (u^{(p)})^{-1}, \quad \alpha = \frac{\epsilon - 1}{p}$$

and consider the fields

$$F_s = E(\delta^r \alpha_{ij}; 0 \leq r \leq s,\ 1 \leq i,j \leq n) = E(\phi^r(\epsilon_{ij}); 0 \leq r \leq s,\ 1 \leq i,j \leq n)$$

and $F = \cup_s F_s$. Let A be a valuation δ-subring of $R \cap F$ containing R^δ and the entries of α (e.g. one can take the "maximal" choice" $A = R \cap F$). Note that for $s \geq 1$ we have equalities of fields

$$E(\delta^r u_{ij}; 0 \leq r \leq s,\ 1 \leq i,j \leq n) = F_{s-1}(u_{ij}; 1 \leq i,j \leq n). \tag{5.42}$$

Now the field in the left hand side of 5.42 has transcendence degree $(s+1)n^2$ over E. Since F_{s-1} has transcendence degree at most sn^2 over E it follows from 5.42 that $(u_{ij})_{ij}$ are algebraically independent over F_{s-1}. Since this is true for all s it follows that $(u_{ij})_{ij}$ are algebraically independent over F. By assertion (2) in Lemma 5.19, $G_{u/A} = N^\delta$. □

Next we consider genericity in the sense of Zariski. The next Theorem shows that the δ-Galois group cannot be "too large" at least if we take our data in a Zariski open set of the set of all data. In the statement below by a Zariski K-closed set in $GL_n(R)$ we understand the intersection of $GL_n(R)$ with a Zariski K-closed set of $GL_n(K^a)$; in other words a K-closed set of $GL_n(R)$ is the zero set in $GL_n(R)$ of a collection of polynomials with coefficients in K in n^2 variables. A subgroup Γ of $GL_n(R)$ is called diagonalizable if there exists $g \in GL_n(K^a)$ such that $g^{-1}\Gamma g$ consists of diagonal matrices.

THEOREM 5.26. *There exists a Zariski K-closed set Ω in $G = GL_n(R)$ not containing* 1 *such that for any $u \in G \backslash \Omega$ the following holds. Let $\alpha = \delta u \cdot (u^{(p)})^{-1}$ and let A be a valuation δ-subring of R containing the entries of α. Then the δ-Galois group $G_{u/A}$ contains a normal subgroup of finite index which is diagonalizable.*

In order to prove Theorem 5.26 we need a series of Lemmas: 5.27, 5.31, 5.32. In the discussion below (pertaining to these Lemmas only!) it is convenient to temporarily change some of the notation used so far. Indeed we let $\mathcal{C}$ be an uncountable algebraically closed field of characteristic zero (such as K^a or $\mathbb{C}$) and all schemes will be schemes over $\mathcal{C}$. By a variety we will understand a reduced (not necessarily irreducible) scheme of finite type over $\mathcal{C}$. We use the same letter X to denote a variety X over $\mathcal{C}$ and its set $X(\mathcal{C})$ of $\mathcal{C}$-points. In particular we denote by G the group scheme GL_n over $\mathcal{C}$ and also the "abstract" group $GL_n(\mathcal{C})$; we denote by T the group scheme of diagonal matrices over $\mathcal{C}$ and also the "abstract" group $T(\mathcal{C})$ of diagonal matrices with entries in $\mathcal{C}$. If X is a variety and $x \in X$ is a point we always understand x is a $\mathcal{C}$-point and we denote by $\dim_x X$ the maximum of the dimensions of the irreducible components of X passing through x. Also, in what follows, we let p be any integer ≥ 2 (not necessarily prime).

LEMMA 5.27. *Let $X \subset G$ be the Zariski closed subset consisting of all $v \in G$ satisfying the following properties:*

(1) $(v^m)^{(p)} = (v^{(p)})^m$ *for all* $m \geq 0$,

(2) $(v^m)^{(p)}(v^{-m})^{(p)} = 1$ *for all* $m \geq 0$.

Then X has exactly one irreducible component passing through 1 *and that component is T.*

REMARK 5.28. The equalities (1) and (2) are viewed as equalities in $\mathfrak{g} = \mathfrak{gl}_n(\mathcal{C})$; note however that, by (1) and (2), for any $v \in X$ we have that $(v^m)^{(p)} \in G$ for all $m \in \mathbb{Z}$ and hence (1) holds for all $m \in \mathbb{Z}$ as an equality in G.

REMARK 5.29. The set X contains the group $N = WT = TW$ generated by the Weyl group W and the group T of diagonal matrices with entries in $\mathcal{C}$. It is not clear whether X actually coincides with the group N.

REMARK 5.30. Let $\mathbb{X}$ be the closed subscheme of G defined by the equations (1) and (2) in the statement of Lemma 5.27; hence the variety $\mathbb{X}_{red}$ coincides with X. It is interesting to note that tangent space of $\mathbb{X}$ at 1 is the whole of the tangent space of G i.e. the Lie algebra $L(G)$ of G; indeed, equations (1) and (2) are easily seen to hold when v is replaced by $1+\epsilon\xi$, where $\epsilon^2 = 0$ and ξ is an arbitrary element of $\mathfrak{gl}_n(\mathcal{C})$. In particular $\mathbb{X}$ is not reduced.

PROOF OF LEMMA 5.27. Let $v \in X$, let $\langle v \rangle \subset G$ be the group generated by v, let $H_v \subset G$ be the Zariski closure of $\langle v \rangle$ in G (which is an algebraic subgroup of G, cf. [**76**], p. 54), and let H_v° be the identity component of H_v. Clearly H_v is commutative.

Claim. For all $v \in X$ we have $H_v^\circ \subset T$.

To check the claim note first that $\langle v \rangle \subset X$ hence $H_v \subset X$. Denote by

$$\Phi : G \to \mathfrak{g}$$

the map $\Phi(v) = v^{(p)}$. Clearly we have

$$\Phi(v^r v^s) = \Phi(v^r)\Phi(v^s)$$

for all $r, s \in \mathbb{Z}$ hence we have

$$\Phi(gh) = \Phi(g)\Phi(h)$$

for all $g, h \in H_v$. Let

$$\varphi : H_v \to \mathfrak{g}$$

be the restriction of Φ; then the regular map φ takes values in G and is a group homomorphism hence its image

$$H_v' := \varphi(H_v) \subset G$$

is a subgroup which is constructible. Hence H_v' is a closed subgroup of G (cf. [**76**], p. 54) and hence $\varphi : H_v \to H_v'$ is an algebraic group homomorphism. Consider the commutative diagram of (possibly reducible) varieties

$$\begin{array}{ccc} H_v & \subset & G \\ \varphi \downarrow & & \downarrow \Phi \\ H_v' & \subset & \mathfrak{g} \end{array}$$

and the induced tangent maps between the corresponding tangent spaces at the identity

$$\begin{array}{ccc} L(H_v) & \subset & L(G) \\ d_1\varphi \downarrow & & \downarrow d_1\Phi \\ L(H_v') & \subset & L(G) \end{array}$$

(Here $L(\ )$ denotes the Lie algebra functor. The linear map $d_1\Phi$ is not a Lie algebra map. The map $d_1\varphi$, on the other hand, is, of course, a Lie algebra map because its source and target are abelian.) One can compute $d_1\Phi$ explicitly: letting

$$v = 1 + \epsilon\xi \in GL_n(\mathcal{C}[\epsilon]), \quad \epsilon^2 = 0,$$

we have

$$\Phi(v) = (1 + \epsilon\xi)^{(p)} = \text{diag}(1 + \epsilon p\xi_{11}, ..., 1 + \epsilon p\xi_{nn}).$$

Hence the image of $d_1\Phi$ is contained in the Lie algebra $L(T)$ of the torus T. Since $d_1\varphi$ is surjective (because we are in characteristic zero) it follows that $L(H_v') \subset L(T)$. Hence the identity component $(H_v')^\circ$ of H_v' is contained in T. Now, clearly, $\Phi^{-1}(T) = T$. Hence

$$H_v^\circ \subset \Phi^{-1}((H_v')^\circ) \subset \Phi^{-1}(T) = T$$

and our claim is proved.

For any subtorus $S \subset T$ let us denote by $C(S)$ the centralizer of S in G; moreover, for any integer $e \geq 1$ denote by $S^{1/e}$ the set of all $v \in G$ such that $v^e \in S$. By the above Claim and by the commutativity of H_v it follows that for any $v \in X$ we have that H_v° is a subtorus of T and there exists $e \geq 1$ such that

$$v \in C(H_v^\circ) \cap H_v^{1/e}.$$

In particular we have

$$X = \bigcup_{S,e} (C(S) \cap S^{1/e} \cap X)$$

where S runs through the (countable!) set of subtori of T and e runs through the set of positive integers. Since $\mathcal{C}$ is uncountable no irreducible variety over $\mathcal{C}$ is a countable union of proper closed subvarieties; in particular, applying this to the irreducible components of X it follows that there exists $e \geq 1$ and finitely many subtori $S_1, ..., S_q \subset T$ such that

$$X = \bigcup_{i=1}^{q} (C(S_i) \cap S_i^{1/e} \cap X). \tag{5.43}$$

To conclude the proof of the Lemma we assume (as we always can) that $\mathcal{C} = \mathbb{C}$. Let V be an irreducible component of X passing through 1. We will prove that $V = T$ and this will end the proof. Assume $V \neq T$ and seek a contradiction. Since $V \neq T$ it follows that $V \not\subset T$ hence $V \backslash T$ is Zariski open in V hence dense in V in the complex topology. So there exists a sequence $x_n \to 1$ (in the complex topology) with $x_n \in X \backslash T$. By 5.43 and by replacing x_n with a subsequence we may assume

$$x_n \in C(S_i) \cap S_i^{1/e} \cap X$$

for some i. Let $[x_n] \in C(S_i)/S_i$ be the class of x_n and choose an embedding

$$\rho : C(S_i)/S_i \to GL_\nu(\mathcal{C})$$

for some ν. Then $\rho([x_n]) \to 1$ hence the eigenvalues of $\rho([x_n])$ tend to 1. But $[x_n]^e = 1$, hence $\rho([x_n])^e = 1$, for all n. So the eigenvalues of $\rho([x_n])$ are e-th roots of unity so they form a discrete set. We get that for n sufficiently big the eigenvalues of $\rho([x_n])$ are equal to 1. But a matrix of finite order with all eigenvalues equal to 1 must be the identity. Hence $\rho([x_n]) = 1$ hence $[x_n] = 1$ hence $x_n \in S_i \subset T$ for some n, a contradiction. This ends the proof of the Lemma. □

The next lemma is completely standard; we just include it for convenience.

LEMMA 5.31. *Let $\pi : Z \to Y$ be a morphism of varieties over $\mathcal{C}$ and assume $\sigma : Y \to Z$ is a section of π. Assume Y is irreducible and for $y \in Y$ consider the variety $\pi^{-1}(y)$. Let $y_0 \in Y$ and assume the point $\sigma(y_0)$ is a connected component of $\pi^{-1}(y_0)$. Then there exists a Zariski open set $U \subset Y$ containing y_0 such that for all $y \in U$ the point $\sigma(y)$ is a connected component of $\pi^{-1}(y)$.*

PROOF. This is a standard consequence of the semicontuinty theorem for the local dimension of fibers. Indeed let $Z^1, ..., Z^m$ be the irreducible components of Z, let $S = \sigma(Y)$ and assume $\sigma(y_0) \in Z^i$ for $1 \leq i \leq r$ and $\sigma(y_0) \notin Z^j$ for $r < j \leq m$. Let

$$U_0 = \pi(S \backslash \bigcup_{j>r} Z^j).$$

Also let $Y^i \subset Y$ be the closure of $\pi(Z^i)$ and let $\pi_i : Z^i \to Y^i$ for $i \leq r$ be induced by π. By the semicontinuity theorem in [**76**], p.33, for $i \leq r$, there exist closed sets $T^i \subset Z^i$ not containing $\sigma(y_0)$ such that

$$\dim_x \pi_i^{-1}(\pi(x)) \leq \dim_{\sigma(y_0)} \pi_i^{-1}(y_0) \text{ for all } x \in Z^i \backslash T^i. \tag{5.44}$$

Consider the closed set

$$T := T^1 \cup ... \cup T^r \cup Z^{r+1} \cup ... \cup Z^m$$

in Z and the open subset

$$U = \pi(S \backslash T) = Y \backslash \pi(S \cap T)$$

of Y. Then $y_0 \in U$. Let $y \in U$ and let F be an irreducible component of $\pi^{-1}(y)$ passing through $\sigma(y)$. Then $F \not\subset Z^j$ for $j > r$ because if one assumes the contrary then

$$\sigma(y) \in S \cap Z^j \subset S \cap T$$

hence $y \in \pi(S \cap T)$, a contradiction. So $F \subset Z^i$ for some $i \leq r$ and hence $F \subset \pi_i^{-1}(y)$. Since $y \notin \pi(S \cap T)$ we have $\sigma(y) \notin T$ hence $\sigma(y) \notin T^i$; on the other hand $\sigma(y) \in F \subset Z^i$, hence $\sigma(y) \in Z^i \backslash T^i$. So by 5.44 we get

$$\dim_{\sigma(y)} F \leq \dim_{\sigma(y)} \pi_i^{-1}(y) \leq \dim_{\sigma(y_0)} \pi_i^{-1}(y_0) \leq \dim_{\sigma(y_0)} \pi^{-1}(y_0) = 0.$$

So $\dim_{\sigma(y)} F = 0$ hence $F = \{\sigma(y)\}$ and we are done. □

LEMMA 5.32. *Let Y be the Zariski open set of $G = GL_n(\mathcal{C})$ consisting of all $u \in G$ such that $u^{(p)}$ is invertible. Let $\Psi : Y \times G \to \mathfrak{g}$ be the morphism defined by*

$$\Psi(u, v) = (u^{(p)})^{-1}(uv)^{(p)}.$$

For each $u \in Y$ let $X_u \subset G$ be the Zariski closed set consisting of all $v \in G$ such that

(1) $\Psi(u, v^m) = \Psi(u, v)^m$ *for all* $m \geq 0$,
(2) $\Psi(u, v^m)\Psi(u, v^{-m}) = 1$ *for all* $m \geq 0$.

Then there exists a Zariski open set $U \subset Y$ containing 1 with the property that for any $u \in U$ and for any connected closed subgroup $S \subset G$ contained in X_u we have that S is a torus.

PROOF. Let $Z \subset Y \times G$ be the closed set defined by the equations (1) and (2) together with the equation $(v-1)^n = 0$. Note that this latter equation is equivalent to asking that v be unipotent. Let

$$\pi : Z \to Y, \quad \pi(u, v) = u,$$

and let

$$pr_G : Y \times G \to G$$

be the second projection. Then $pr_G(\pi^{-1}(u))$ coincides with the set of unipotent matrices in X_u. Also note that X_1 coincides with X in Lemma 5.27. Now, by Lemma 5.27, there is exactly one irreducible component of X_1 passing through 1 and that component is a torus so it does not contain unipotent matrices with the exception of 1 itself. In particular 1 is a connected component of $pr_G(\pi^{-1}(1))$. Now π has a section

$$\sigma : Y \to Z, \quad \sigma(u) = (u, 1).$$

By Lemma 5.31 there exists a Zariski open set U of Y containing 1 such that for all $u \in U$ we have that $(u, 1)$ is a connected component of $\pi^{-1}(u)$. So 1 is a connected component of the set of unipotent matrices in X_u. Now let $S \subset G$ be a closed connected subgroup contained in X_u. Then 1 is a connected component of the set of unipotent matrices in S. This implies that S contains no unipotent matrix except 1 (because any unipotent matrix $\neq 1$ is contained in a subgroup isomorphic to the additive group). So the unipotent radical of S is trivial, hence a torus by [**76**], p. 161. □

REMARK 5.33. Exactly as in Remark 5.30, if $\mathbb{X}_u$ is the subscheme of G defined by equations (1) and (2) in Lemma 5.32 then $(\mathbb{X}_u)_{red} = X_u$ and the tangent space to $\mathbb{X}_u$ at 1 is, again, the whole of the Lie algebra $L(G) = \mathfrak{gl}_n(\mathcal{C})$.

PROOF OF THEOREM 5.26. Consider the situation and notation in Lemma 5.32 with $\mathcal{C} = K^a$. Choose a polynomial $F \in K^a[x]$ such that

$$1 \in D(F) := \{v \in GL_n(K^a); F(v) \neq 0\} \subset U.$$

Replacing F by the product of its conjugates over K we may assume $F \in K[x]$ and hence that $F \in R[x]$. Now let

$$u \in D(F) \cap GL_n(R), \quad \alpha = \delta u \cdot (u^{(p)})^{-1},$$

and let $A \subset R$ be a valuation δ-subring containing the entries of α. Let $\overline{G_{u/A}}$ be the Zariski closure of $G_{u/A}$ in $GL_n(K^a)$. We want to show that the connected component

$$(\overline{G_{u/A}})^\circ$$

of $\overline{G_{u/A}}$ is a torus in $GL_n(K^a)$. Note that $u^{(p)}$ is invertible so $u \in Y$. Let $c \in G_{u/A}$ hence $c^m \in G_{u/A} \subset G_u$ for all $m \in \mathbb{Z}$. Hence

$$(uc^m)^{(p)} = u^{(p)}\phi(c^m),$$

hence

$$\Psi(u, c^m) = \phi(c^m).$$

We claim that $c \in X_u$; indeed for $m \geq 0$ we have

$$\Psi(u, c^m) = \phi(c^m) = \phi(c)^m = \Psi(u, c)^m$$

and also

$$\Psi(u, c^m)\Psi(u, c^{-m}) = \phi(c^m)\phi(c^{-m}) = \phi(1) = 1.$$

Since c was arbitrary in $G_{u/A}$ we conclude that $G_{u/A} \subset X_u$ hence $\overline{G_{u/A}} \subset X_u$. By Lemma 5.32, $(\overline{G_{u/A}})^\circ$ is a torus. Then clearly

$$G_{u/A} \cap (\overline{G_{u/A}})^\circ$$

is a normal subgroup of finite index in $G_{u/A}$ which is diagonalizable. $\square$

CHAPTER 6

Curvature of Chern connections

In this chapter we study, in some detail, Chern connection on GL_n attached to symmetric/anti-symmetric matrices. We will concentrate on real Chern connections (which are more interesting) and leave the case of complex Chern connections to the reader.

We will start by showing that, for matrices q with entries 0 or roots of unity, real Chern connections are global along the identity; more generally, under certain restrictions on q, real Chern connections are global along certain tori. In particular, for such q's, one can attach curvature via the method of analytic continuation between primes. Then the problem arises as to whether this curvature vanishes; we will prove a number of results about the vanishing/non-vanishing of this curvature. We will also complement our discussion by proving some convergence estimates showing that mean and scalar curvatures have coefficients in the ring of convergent Dirichlet series. Finally we will prove that real Chern connections admit algebraization by correspondences; as we saw, this leads to a different type of curvature and, again, one is led to the problem of whether this different type of curvature vanishes. We will prove vanishing/non-vanishing results for this new type of curvature.

6.1. Analytic continuation along tori

We begin by defining real Chern connections in the global setting. So we place ourselves in the case when A is a δ-ring of global type, $A = \mathcal{O}_K[1/M]$, cf. Definition 3.25; so A is equipped with a global connection $(\delta_p) = (\delta_p^A)$ indexed by a set of primes $\mathcal{V}$. Let $G = GL_n$ be the general linear group scheme over A. Assume $q^t = \pm q \in G(A)$ and consider the involution $x^\tau = q^{-1}(x^t)^{-1}q$ on GL_n. Consider the associated maps

$$\mathcal{H}_q : G \to G, \quad \mathcal{B}_q : G \times G \to G, \quad \mathcal{H}_q(x) = x^t q x, \quad \mathcal{B}_q(x, y) = x^t q y.$$

Let $(\phi_{0p}) = (\phi_{0p}^G)$ be the Frobenius lifts attached to the trivial adelic connection on G, $\phi_{0p}(x) = x^{(p)}$.

DEFINITION 6.1. A Frobenius lift ϕ_p on $G^{\widehat{p}}$ is $\mathcal{H}_q$-*horizontal*, respectively $\mathcal{B}_q$-*symmetric*, with respect to ϕ_{0p} if the left, respectively the right, diagram below is commutative.

$$(6.1)\qquad \begin{array}{ccc} G^{\widehat{p}} & \xrightarrow{\phi_p} & G^{\widehat{p}} \\ \mathcal{H}_q \downarrow & & \downarrow \mathcal{H}_q \\ G^{\widehat{p}} & \xrightarrow{\phi_{0p}} & G^{\widehat{p}} \end{array} \qquad \begin{array}{ccc} G^{\widehat{p}} & \xrightarrow{\phi_{0p} \times \phi_p} & G^{\widehat{p}} \times G^{\widehat{p}} \\ \phi_p \times \phi_{0p} \downarrow & & \downarrow \mathcal{B}_q \\ G^{\widehat{p}} \times G^{\widehat{p}} & \xrightarrow{\mathcal{B}_q} & G^{\widehat{p}} \end{array}$$

Then the "local" Theorem 4.23 immediately implies the following "global" result:

THEOREM 6.2. *Let $q \in GL_n(A)$, $q^t = \pm q$. There exists a unique adelic connection (δ_p) on G, with attached Frobenius lifts (ϕ_p), such that, for all $p \in \mathcal{V}$, ϕ_p is $\mathcal{H}_q$-horizontal and $\mathcal{B}_q$-symmetric with respect to ϕ_{0p}. It is given by $\phi_p(x) = \Phi_p(x)$, where $\Phi_p(x)$ is the matrix*

$$\Phi_p(x) = x^{(p)} \cdot \{(x^{(p)t}\phi_p(q)x^{(p)})^{-1}(x^t qx)^{(p)}\}_p^{1/2}. \tag{6.2}$$

DEFINITION 6.3. The adelic connection (δ_p) on G in Theorem 6.2 is called the *real Chern connection* attached to q.

REMARK 6.4. More generally one can consider families $q = (q_p)$ with $q_p \in GL_n(A^{\widehat{p}})$, $q_p^t = \pm q_p$, and one can attach to them real Chern connections. This is in line with our principle that adelic objects are arithmetic analogues of C^∞ geometric objects. For simplicity we will not pursue this here but the reader should be aware that some of the theory extends to this more general setting.

For special q's the real Chern connection $\delta = (\delta_p)$ has additional properties that we now address.

DEFINITION 6.5. Let us say that q is *δ-constant* if $\delta_p q = 0$ for all $p \in \mathcal{V}$.

Recall that q is *split* if q is one of the matrices 0.73. Any split q is, of course, δ-constant. The identity component $S = SO(q)$ of the subgroup scheme of G defined by $x^t qx = x$, with q split as above is referred to as the *split classical group* defined by q; these groups are denoted by $Sp_{2r}, SO_{2r}, SO_{2r+1}$ respectively. The real Chern connection attached to a split q is sometimes referred to as being attached to $Sp_{2r}, SO_{2r}, SO_{2r+1}$ respectively. Recall that we denoted by N the subgroup scheme of GL_n generated by the diagonal subgroup T and the Weyl group W of permutation matrices.

DEFINITION 6.6. Let $H \subset G = GL_n$ be a smooth closed subgroup scheme. Consider an adelic connection $\delta = (\delta_p)$ on G with Frobenius lifts (ϕ_p). One says H is *δ-horizontal* if, for all $p \in \mathcal{V}$, $H^{\widehat{p}}$ is δ_p-horizontal. One says δ is *left invariant* (respectively *right invariant*) with respect to H if H is δ-horizontal and, for all $p \in \mathcal{V}$, the left (respectively right) diagram below is commutative

$$\begin{array}{ccccccc} H^{\widehat{p}} \times G^{\widehat{p}} & \longrightarrow & G^{\widehat{p}} & \quad & G^{\widehat{p}} \times H^{\widehat{p}} & \longrightarrow & G^{\widehat{p}} \\ \phi_p \times \phi_p \downarrow & & \downarrow \phi_p & & \phi_p \times \phi_p \downarrow & & \downarrow \phi_p \\ H^{\widehat{p}} \times G^{\widehat{p}} & \longrightarrow & G^{\widehat{p}} & & G^{\widehat{p}} \times H^{\widehat{p}} & \longrightarrow & G^{\widehat{p}} \end{array} \tag{6.3}$$

where the horizontal arrows are given by multiplication. *Bi-invariant* means left and right invariant.

The local Theorem 4.30 immediately implies the global theorem below:

THEOREM 6.7. *Let $\delta = (\delta_p)$ be the real Chern connection on GL_n attached to a split q and let $S = SO(q)$ be the split classical group defined by q. Then S is δ-horizontal and δ is left invariant with respect to $N \cap S$ and right invariant with respect to N.*

REMARK 6.8. Note the following compatibility of real Chern connection with the action of the permutation group W. Consider a permutation matrix $w \in W$, let $\tilde{q} = wqw^{-1}$, and let

$$\tilde{\Phi}(x) = x^{(p)}\{(x^{(p)t}\phi_p(\tilde{q})x^{(p)})^{-1}(x^t\tilde{q}x)^{(p)}\}_p^{1/2}.$$

Then, using $w^t = w^{-1}$, one checks that

$$w\Phi_p(x)w^{-1} = \tilde{\Phi}_p(wxw^{-1}). \tag{6.4}$$

The latter translates into the commutativity of the following diagram:

$$\begin{array}{ccc} G^{\widehat{p}} & \xrightarrow{\phi_p} & G^{\widehat{p}} \\ c_w \downarrow & & \downarrow c_w \\ G^{\widehat{p}} & \xrightarrow{\tilde{\phi}_p} & G^{\widehat{p}} \end{array} \tag{6.5}$$

where c_w is conjugation, $c_w(x) = wxw^{-1}$, and $(\tilde{\phi}_p)$ is the family of Frobenius lifts corresponding to the real Chern connection attached to $\tilde{q}$. The above applies, in particular, to the case when q is split and n is even; in that case there exists $w \in W$ such that

$$wqw^{-1} = \text{diag}(q_2, ..., q_2),$$

where q_2 is 2×2 and split. Similarly, if q is split and n is odd there exists $w \in W$ such that

$$wqw^{-1} = \text{diag}(1, q_2, ..., q_2),$$

where q_2 is 2×2 and split and 1 is the 1×1 identity matrix.

We also have the following compatibility with the embedding $\iota : GL_n \to GL_{n'}$ given by

$$x \mapsto \begin{pmatrix} x & 0 \\ 0 & 1 \end{pmatrix},$$

where $1 = 1_{n'-n}$ is the identity matrix. Indeed let

$$\Phi'(x') = (x')^{(p)}\{(x')^{(p)t}\phi_p(q')(x')^{(p)})^{-1}((x')^tq'x')^{(p)}\}_p^{1/2},$$

where x' is an $n' \times n'$-matrix of indeterminates and

$$q' = \begin{pmatrix} q & 0 \\ 0 & q_0 \end{pmatrix},$$

$(q')^t = \pm q'$, with q_0 a δ-constant matrix. Then we have

$$\Phi'_p \begin{pmatrix} x & 0 \\ 0 & 1 \end{pmatrix} = \begin{pmatrix} \Phi_p(x) & 0 \\ 0 & 1 \end{pmatrix}. \tag{6.6}$$

The latter translates into the commutativity of the following diagram:

$$\begin{array}{ccc} GL_n^{\widehat{p}} & \xrightarrow{\phi_p} & GL_n^{\widehat{p}} \\ \iota \downarrow & & \downarrow \iota \\ GL_{n'}^{\widehat{p}} & \xrightarrow{\phi'_p} & GL_{n'}^{\widehat{p}}, \end{array} \tag{6.7}$$

where (ϕ'_p) is the family of Frobenius lifts corresponding to the real Chern connection attached to q'. In particular assume one is given a sequence

$$q := (q^{(1)}, q^{(2)}, q^{(3)}, ...) \in \prod_{k=1}^{\infty} GL_{r_k}(A) \tag{6.8}$$

with $q^{(k)}$ symmetric (respectively anti-symmetric), δ-constant matrices. Set $n_k = r_1 + ... + r_k$. Then the Frobenius lifts on GL_{n_k} corresponding to the real Chern connections attached to the $n_k \times n_k$ matrices $\text{diag}(q^{(1)}, ..., q^{(k)})$ are compatible (i.e. commute) with the embeddings defined by the ι's:

$$GL_{n_1} \xrightarrow{\iota} GL_{n_2} \xrightarrow{\iota} GL_{n_3} \xrightarrow{\iota} ...$$

The simplest cases of such a construction are when all $n_r = 1$ and $q^{(k)} = 1$ or when all $n_r = 2$ and $q^{(k)}$ are split; in the second case (although not in the first) the matrices $\text{diag}(q^{(1)}, ..., q^{(k)})$ are W-conjugate to the corresponding split matrix of dimension n_k. The above considerations lead to an obvious notion of *adelic connection* on

$$GL_\infty = \lim_{\rightarrow} GL_n$$

and to a way of attaching an adelic connection on the latter to any sequence q as above. The group GL_∞ could be viewed as the "frame bundle of *Spec* $\mathbb{Z}$," once a bijection is fixed between the set of primes $\mathcal{V}$ and the index set for the matrices $q^{(k)}$; the covariant dependence on such bijections is controlled by the above mentioned behavior of real Chern connections with respect to the W-action by conjugation. Fixing a bijection as above can be viewed as a rough analogue of Cartan's idea of *soldering* [**116**], p. 351.

Next we address the issue of globality of real Chern connection along various retracts of $G = GL_n$; this is crucial if one wants to attach curvature to real Chern connections using the method of analytic continuation between primes.

THEOREM 6.9. *Assume q is δ-constant. Then the real Chern connection (δ_p) on GL_n attached to q is global along* 1.

REMARK 6.10. In particular, if q is δ-constant then $\delta := (\delta_p) \in \Lambda^1(G, 1)$ and δ has a well defined curvature $(\varphi_{pp'}) \in \Lambda^2(G, 1)$ and a well defined $(1, 1)$-curvature $(\varphi_{p\overline{p}'}) \in \Lambda^2(G, 1)$ with respect to the trivial adelic connection $\overline{\delta} = \delta_0$, $\delta_{\overline{p}} x = \delta_{0p} x = 0$. Of course the curvature $(\varphi_{\overline{pp}'})$ of $\overline{\delta}$ vanishes. Also the truncations $\varphi^{(\nu)}_{pp'}$ and $\varphi^{(\nu)}_{p\overline{p}'}$ are defined.

We would like to prove a more general result about real Chern connections being global along partition tori in $G = GL_n$. Recall that for μ a partition of an integer m with $0 \leq m \leq n$, defined by

$$m = m_1 + ... + m_s, \quad m_i \in \mathbb{Z}, \quad m_i \geq 1, \quad s \geq 0,$$

we attached the *partition torus*

$$D_\mu := \epsilon_\mu(C_{m_1} \times ... \times C_{m_s}) \subset G = GL_n$$

where

$$\epsilon_\mu : GL_{m_1} \times ... \times GL_{m_s} \to GL_n, \quad (a_1, ..., a_s) \mapsto \text{diag}(a_1, ..., a_s, 1_{n-m})$$

and C_{m_i} is the center of GL_{m_i}. Cf Definition 4.91. Recall from Lemma 4.94 that D_μ is a retract in G.

DEFINITION 6.11. Denote by $G_\mu \subset G$ the centralizer of D_μ in G. Hence

$$G_\mu = \epsilon_{\mu'}(GL_{m_1} \times ... \times GL_{m_s} \times GL_{n-m}) \subset G,$$

where $\mu' = \mu$ if $m = n$ and μ' is the partition of n given, for $m < n$, by

$$n = m_1 + ... + m_s + (n - m).$$

REMARK 6.12. Recall that:

(1) $s = 0$ if and only if $m = 0$ in which case we write $\mu = \overline{0}$ and we have that $D_\mu = \{1\}$ is the trivial group and $G_\mu = G$.
(2) $s = n$ if and only if $m = n = 1 + ... + 1$ in which case we write $\mu = \overline{1}$ and we have $D_\mu = G_\mu = T$, the diagonal maximal torus of G.
(3) for $n = m$ and $s = 1$, in which case we write $\mu = \overline{n}$, we have $D_\mu = Z$, the center of G and $G_\mu = G$.

Next consider any ring C in which 2 is invertible and consider the associative ring $\mathfrak{g}(C)$ of $n \times n$ matrices with entries in C.

DEFINITION 6.13. For any matrix $a \in \mathfrak{g}(C)$ set

$$\text{jor}(a) = \det\left(\mathfrak{g}(C) \to \mathfrak{g}(C),\ b \mapsto \frac{ab + ba}{2}\right) \in C.$$

REMARK 6.14. Here "jor" stands for "Jordan;" indeed the expression

$$\frac{ab + ba}{2}$$

is the *Jordan product* of a and b [**80**]. For instance, for $\lambda \in C^\times$ we have

$$\text{jor}(\lambda \cdot 1_n) = \lambda^{n^2}. \tag{6.9}$$

In particular, for $n = 1$,

$$\text{jor}(a) = a. \tag{6.10}$$

On the other hand, for $n = 2$,

$$\text{jor}(a) = \frac{1}{4}(\text{tr}(a))^2 \cdot \det(a). \tag{6.11}$$

For $n \geq 2$ arbitrary and $a = \text{diag}(a_1, ..., a_n)$ a diagonal matrix we have

$$\text{jor}(a) = 2^{n-n^2} a_1 ... a_n \cdot \prod_{i<j} (a_i + a_j)^2. \tag{6.12}$$

It is also convenient to make the following:

DEFINITION 6.15. Let $q_1, q_2 \in \mathbb{Z}$ be two non-zero integers and $p \in \mathbb{Z}$ an odd prime; then set

$$M_p(q_1, q_2) := \left(\frac{q_1}{p}\right) q_1^{\frac{p-1}{2}} + \left(\frac{q_2}{p}\right) q_2^{\frac{p-1}{2}} \in \mathbb{Z}. \tag{6.13}$$

Here $\left(\frac{*}{p}\right)$ denotes, as usual, the Legendre symbol.

REMARK 6.16. Note that for p odd not dividing $q_1 q_2$ we have

$$M_p(q_1, q_2) \equiv 2 \mod p \tag{6.14}$$

hence

$$p \nmid M_p(q_1, q_2). \tag{6.15}$$

DEFINITION 6.17. Let

$$\mathcal{V} = \{p_1, p_2, ...\}$$

be a (finite or infinite) set of odd primes and

$$\mathcal{Q} = \{q_1, ..., q_n\}$$

be a finite set of non-zero integers. We say that the pair $(\mathcal{V}, \mathcal{Q})$ is *global* if the following conditions hold:

(1) for any k and i, $p_k \nmid q_i$;
(2) for any $k \neq l$ and $i \neq j$, $p_k \nmid M_{p_l}(q_i, q_j)$.

Then Theorem 6.9 is the special case $\mu = \bar{0}$ of assertion (2) in the following:

THEOREM 6.18. *Let D_μ be a partition torus and let $q \in G_\mu(A)$, $q^t = \pm q$. Let $\delta = (\delta_p)$ be the real Chern connection on G attached to q and let (ϕ_p) be the attached Frobenius lifts. Furthermore for each $p \in \mathcal{V}$, set*

$$\gamma_p := (\phi_p(q)^{-1} q^{(p)})^{1/2} \in GL_n(A^{\widehat{p}}).$$

Then:

(1) *Assume that, for all $p \in \mathcal{V}$ we have $\gamma_p \in D_\mu(A^{\widehat{p}})$. Then δ fixes D_μ.*
(2) *Assume that, for all $p \in \mathcal{V}$ we have $\gamma_p \in D_\mu(A)$ and $\text{jor}(\gamma_p) \in A^\times$. Then δ is global along D_μ.*
(3) *Assume $\mu = \bar{1}$; in particular $D_\mu = G_\mu = T$. Let $\mathcal{Q} = \{q_1, ..., q_n\}$ be a set of integers and assume the pair $(\mathcal{V}, \mathcal{Q})$ is global. Let M be the supernatural number equal to the product of all primes in $\mathbb{Z}$ not belonging to $\mathcal{V}$. Assume $A = \mathbb{Z}[1/M]$ and let $q := diag(q_1, ..., q_n)$; in particular $q \in T(A)$. Then δ on G is global along T and the global connection on T induced by δ is flat.*
(4) *Assume $\mu = \bar{n}$; in particular $D_\mu = Z$, $G_\mu = G$. Assume $q = \lambda \cdot q_0$, with $\lambda \in \mathbb{Z}[1/M]^\times$ and $q_0 \in G(A)$, q_0 δ-constant. Then δ on G is global along Z and the global connection on Z induced by δ is flat.*

REMARK 6.19. In particular, under the assumptions of either assertion (2), (3) or (4) above, the curvature $(\varphi_{pp'})$ and $(1,1)$-curvature $(\varphi_{p\overline{p}'})$ of δ are defined, with $\varphi_{pp'}, \varphi_{p\overline{p}'} : \mathcal{O}(G^{\widehat{D_\mu}}) \to \mathcal{O}(G^{\widehat{D_\mu}})$.

REMARK 6.20. The q in assertion (4) should be "morally" viewed as a "metric" that is "conformally equivalent" to the δ-constant "metric" q_0.

For the proof we need the following:

LEMMA 6.21. *Let C be a ring, J an ideal, and assume C is J-adically complete. Let $\gamma, \beta \in \mathfrak{g}(C)$ with $\beta \equiv 0 \mod J$ and $\text{jor}(\gamma) \in C^\times$. Then there exists $\eta \in \mathfrak{g}(C)$ such that $\eta^2 = \gamma^2 + \beta$ and $\eta \equiv \gamma \mod J$.*

PROOF. A Hensel type argument: one constructs by induction a sequence $\eta_n \in \mathfrak{g}(C)$ such that $\eta_1 = \gamma$, $\eta_{n+1} \equiv \eta_n \mod J^n$ and $\eta_n^2 \equiv \gamma^2 + \beta \mod J^n$. Indeed, assuming $\eta_1, .., \eta_n$ were found, it is enough to set $\eta_{n+1} = \eta_n + z$ where z is a solution to

$$z\gamma + \gamma z = \gamma^2 + \beta - \eta_n^2.$$

Such a z exists because $\text{jor}(\gamma) \in C^\times$. □

PROOF OF THEOREM 6.18. We adopt the notation in the discussion before Lemma 4.94; in particular we use the notation x^μ, t, T introduced in 4.207, 4.208, 4.209.

To prove assertion (1) let (δ_p) be the real Chern connection attached to q, with attached Frobenius lifts (ϕ_p). Set $\Phi_p = \phi_p(x)$. Hence $\phi_p(x^\mu) = \Phi_p^\mu$ and

$$\phi_p(x - x^\mu) = \Phi_p - \Phi_p^\mu.$$

By Lemmas 3.54 and 3.55, in order to prove that δ_p fixes D_μ it is enough to show that the image of $\phi_p(x-x^\mu) = \Phi_p - \Phi_p^\mu$ in $A[t,t^{-1}]^{\widehat{p}}[[T]]$ is in the ideal (T) generated by T. But we have the following congruences mod (T):

$$\Phi_p - \Phi_p^\mu := \Phi_p(x) - \Phi_p(x)^\mu \equiv \Phi_p(x^\mu) - (\Phi_p(x^\mu))^\mu \mod (T).$$

Claim. $\Phi_p(x^\mu) \in D_\mu(A[t,t^{-1}]^{\widehat{p}})$.

This implies that $(\Phi_p(x^\mu))^\mu = \Phi_p(x^\mu)$ which ends the proof of assertion (1). The claim follows from the following computation:

(6.16)

$$\begin{aligned}
\Phi_p(x^\mu) &= (x^\mu)^{(p)}\{((x^\mu)^{(p)t}\phi_p(q)(x^\mu)^{(p)})^{-1}((x^\mu)^t q x^\mu)^{(p)}\}_p^{1/2}\\
&= (x^\mu)^p\{((x^\mu)^p\phi_p(q)(x^\mu)^p)^{-1}((x^\mu)^p q^{(p)}(x^\mu)^p\}_p^{1/2} \quad (x^\mu \text{ is diagonal})\\
&= (x^\mu)^p\{\phi_p(q)^{-1}q^{(p)}\}_p^{1/2} \quad (x^\mu \text{ commutes with } q^{(p)} \text{ and } \phi_p(q))\\
&= \gamma_p \cdot (x^\mu)^p \in D_\mu(A[t,t^{-1}]^{\widehat{p}}) \quad (\text{because } \gamma_p \in D_\mu(A^{\widehat{p}})).
\end{aligned}$$

This ends the proof of assertion (1).

To prove assertion (2), by Lemma 3.56, it is enough to show that $\Phi_p(x)$ has entries in $A[t,t^{-1}][[T]]$. Setting

$$v(x) = (x^{(p)t}\phi_p(q)x^{(p)})^{-1}(x^t q x)^{(p)}$$

it is enough to show that $v(x)_p^{1/2}$ (which is the unique square root of $v(x)$ with entries in $A[t,t^{-1}]^{\widehat{p}}[[T]]$ that is $\equiv 1 \bmod p$) has entries in $A[t,t^{-1}][[T]]$. Set

$$\beta = v - \phi_p(q)^{-1}q^{(p)} = v - \gamma_p^2.$$

Claim 1. β has entries in $p(T)A[t,t^{-1}][[T]]$. Assuming Claim 1 and using the fact that $A[t,t^{-1}][[T]]$ is T-adically complete, and hence $p(T)A[t,t^{-1}][[T]]$-adically complete, it follows from Lemma 6.21 that there exists a matrix η with entries in $A[t,t^{-1}][[T]]$ such that $\eta^2 = v$ and $\eta \equiv \gamma_p \bmod p(T)A[t,t^{-1}][[T]]$. In particular

$$\eta \equiv \gamma_p \equiv 1 \mod p.$$

So $v(x)_p^{1/2} = \eta$, hence $v(x)_p^{1/2}$ has entries in $A[t,t^{-1}][[T]]$ and assertion (2) is proved if we prove our Claim 1 above.

To prove our Claim 1 it is enough to check that:

Claim 2. β has entries in $pA[t,t^{-1}]^{\widehat{p}}[[T]]$;

Claim 3. β has entries in $(T)A[t,t^{-1}]^{\widehat{p}}[[T]]$;

Claim 4. β, equivalently v, has entries in $A[t,t^{-1}][[T]]$.

Claim 2 is clear because $v \equiv \gamma_p \equiv 1 \bmod p$. To check Claim 3 we use again the fact that $x \equiv x^\mu \bmod (T)$ plus the fact that x^μ is diagonal and commutes with $\phi_p(q)$ and $q^{(p)}$. Let us prove Claim 4. Note that

$$\det(x^{(p)}) = \det((x^\mu + (x - x^\mu))^{(p)}) \in \det(x^\mu) + (T)A[t,T],$$

hence

$$\det(x^{(p)})^{-1} \in A[t,t^{-1}][[T]].$$

But $v(x)$ is $\det(x^{(p)})^{-2}$ times a matrix with coefficients in $A[t,T]$ and Claim 4 is proved.

Let us prove assertion (3). First note that

$$\gamma_p := (\phi(q)^{-1}q^{(p)})^{1/2} = (q^{p-1})^{1/2} = \left(\frac{q}{p}\right) \cdot q^{(p-1)/2} \tag{6.17}$$

where

$$\left(\frac{q}{p}\right) := \operatorname{diag}\left(\left(\frac{q_1}{p}\right), ..., \left(\frac{q_n}{p}\right)\right) \tag{6.18}$$

and $\left(\frac{q_i}{p}\right)$ are the Legendre symbols. So $\gamma_p \in T(A)$ and, by 6.12 and Definition 6.17, $\operatorname{jor}(\gamma_p) \in A^\times$ for all p. Hence, by assertion (2), δ is global along T. We are left to prove the flatness of the global connection on T induced by δ on G. This follows from the following computation, for any two primes $p, p' \in \mathcal{V}$; in this computation we are using the formulas 6.16 and 6.17:

$$\begin{aligned}
\Phi_{p'}\Phi_p(x^\mu) &= \Phi_{p'}\left(q^{(p-1)/2}\left(\tfrac{q}{p}\right)(x^\mu)^p\right)\\
&= q^{(p'-1)/2}\left(\tfrac{q}{p'}\right)\left(q^{(p-1)/2}\left(\tfrac{q}{p}\right)(x^\mu)^p\right)^{p'}\\
&= q^{(p'-1+p'(p-1))/2}\left(\tfrac{q}{p'}\right)\left(\tfrac{q}{p}\right)^{p'}(x^\mu)^{pp'}\\
&= q^{(pp'-1)/2}\left(\tfrac{q}{p}\right)\left(\tfrac{q}{p'}\right)(x^\mu)^{pp'}\\
&= \Phi_p\Phi_{p'}(x^\mu).
\end{aligned}$$

Finally to check assertion (4) note that globality follows directly from assertion (2) while flatness follows as in the proof of assertion (3). □

We next give examples when assertion (3) in the above Theorem applies, i.e., we give examples of global pairs $(\mathcal{V}, \mathcal{Q})$.

EXAMPLE 6.22. Let $s \geq 1$ be an integer and set

$$\mathcal{V} = \{p \mid p \equiv 1 \bmod 2^s, \quad p \not\equiv 1 \bmod 2^{s+1}\}, \tag{6.19}$$

In particular:

(1) for $s = 1$, $\mathcal{V} = \{p \mid p \equiv 3 \bmod 4\}$,
(2) for $s = 2$, $\mathcal{V} = \{p \mid p \equiv 1 \bmod 4, \quad p \not\equiv 1 \bmod 8\}$, etc.

Let $a_1, ..., a_n \in \mathbb{Z}$ be integers not divisible by any of the primes in $\mathcal{V}$ and set

$$\mathcal{Q} := \{q_1, ..., q_n\}, \quad q_i = a_i^2. \tag{6.20}$$

Then we claim that the pair $(\mathcal{V}, \mathcal{Q})$ is global. Indeed, by 6.13, we have

$$M_p(q_i, q_j) = a_i^{p-1} + a_j^{p-1} \tag{6.21}$$

which is a sum of two 2^s-powers of integers. Now any odd prime dividing a sum of two 2^s-powers of integers either divides both integers or is congruent to 1 mod 2^{s+1}. Hence none of the prime divisors of 6.21 is in $\mathcal{V}$. Also, by construction no p in $\mathcal{V}$ divides any of the q_i's. So our claim is proved.

Here is yet another example of global pairs.

EXAMPLE 6.23. Let

$$\mathcal{Q} = \{q_1, ..., q_m\}$$

be a set of non-zero integers whose squares are distinct. We claim that there exists an infinite set of primes

$$\mathcal{V} = \{p_1, p_2, p_3, ...\}$$

such that the pair

$$(\mathcal{Q}, \mathcal{V})$$

is global. To prove this let

$$Q := \prod_{i=1}^{m} q_i \cdot \prod_{1 \leq i < j \leq m} (q_i^2 - q_j^2);$$

so $Q \neq 0$. Also let us say that a set of odd primes $\{p_1, ..., p_n\}$ has property P if

(1) $p_k \nmid Q$ for $k = 1, ..., n$,
(2) $p_k \equiv 2 \bmod 3$ for $k = 1, ..., n$,
(3) $p_k \nmid M_{p_l}(q_i, q_j)$ for $k, l = 1, ..., n$, $k \neq l$, and $i, j = 1, ..., m$, $i \neq j$.

Then it is enough to prove that any set of odd primes, of cardinality n, and having property P, is contained in a set of odd primes, of cardinality $n+1$, and having property P; for then one can construct an increasing sequence of sets $(\mathcal{V}_n)$ with property P, $|\mathcal{V}_n| = n$, and one can take $\mathcal{V} = \cup_{n=1}^{\infty} \mathcal{V}_n$. So let $\{p_1, ..., p_n\}$ be a set of odd primes having property P. It is enough to find a prime $p_{n+1} = p$, larger than $p_1, ..., p_n$, such that

(1') $p \nmid Q$,
(2') $p \equiv 2 \bmod 3$,
(3') $p_k \nmid M_p(q_i, q_j)$ for $k = 1, ..., n$, $i \neq j$,
(4') $p \nmid M_{p_k}(q_i, q_j)$ for $k = 1, ..., n$, $i \neq j$.

To prove the existence of such a p recall first the general fact that if a, b, n, m are integers such that $(n, m) = 1$ then the system of congruences

$$\begin{array}{lcccc} x & \equiv & a & \bmod & n \\ x & \equiv & b & \bmod & m \end{array} \tag{6.22}$$

has solution

$$x \equiv amm' + bnn' \mod mn,$$

where m', n' are integers such that $mm' \equiv 1 \bmod n$ and $nn' \equiv 1 \bmod m$. Assume now, in addition, that $(a, n) = 1$ and $(b, m) = 1$. Then one trivially checks that

$$(amm' + bnn', mn) = 1$$

so it follows, by Dirichlet, that the system 6.22 has infinitely many primes among its solutions. In particular the system

$$\begin{array}{lcccc} p & \equiv & 3 & \bmod & (p_1 - 1)...(p_n - 1) \\ p & \equiv & 2 & \bmod & 3 \end{array} \tag{6.23}$$

has infinitely many primes p among its solutions. Take such a prime solution p, larger than $p_1, ..., p_n$, and satisfying conditions (1') and (4'). Then by 6.23 p satisfies (2'). We claim that p also satisfies (3') which will end our proof. Indeed, if $p_k | M_p(q_i, q_j)$ for some $k \in \{1, ..., m\}$ and some $i \neq j$ then

$$q_1^{p-1} \equiv q_2^{p-1} \mod p_k.$$

Now, by property (1), $p_k \nmid q_i$ and $p_k \nmid q_j$. Also, by 6.23,

$$p - 1 \equiv 3 - 1 = 2 \mod p_k - 1.$$

By Fermat's little theorem, we get

$$q_j^2 \equiv q_j^{p-1} \equiv q_i^{p-1} \equiv q_i^2 \mod p_k$$

so $p_k | q_i^2 - q_j^2$, which contradicts property (1). This ends our proof. □

REMARK 6.24. For the proof of the next Theorem note that the commutativity of the two diagrams 6.1 is equivalent to the following two equalities respectively:

$$\begin{aligned} \Phi_p(x)^t \phi_p(q) \Phi_p(x) &= (x^t q x)^{(p)}, \\ \Phi_p(x)^t \phi_p(q) x^{(p)} &= x^{(p)t} \phi_p(q) \Phi_p(x), \end{aligned} \tag{6.24}$$

and hence (in case δ is global along 1) equivalent to the following two equalities respectively:

$$\begin{aligned} (1 + \Phi_p^0(T)^t) \phi_p(q)(1 + \Phi_p^0(T)) &= ((1 + T^t) q (1 + T))^{(p)} \\ (1 + \Phi_p^0(T)^t) \phi_p(q)(1 + T)^{(p)} &= (1 + T^t)^{(p)} \phi_p(q)(1 + \Phi_p^0(T)) \end{aligned} \tag{6.25}$$

THEOREM 6.25. *Let q be split and let $S = SO(q)$ be the classical group defined by q. Let $\delta = (\delta_p)$ be an adelic connection on $G = GL_n$ that is $\mathcal{H}_q$-horizontal and global along* 1. *Let $\varphi_{pp'}^{\mathcal{G}}$ be the $\mathcal{G}$-curvature of (δ_p), and set $\Phi_{pp'}^{\mathcal{G}} = \varphi_{pp'}^{\mathcal{G}}(T)$. Then*

$$1 + pp' \Phi_{pp'}^{\mathcal{G}} \in S(A[[T]]). \tag{6.26}$$

PROOF. Let $\varphi_{pp'}$ be the curvature of (δ_p) and, as usual, $\Phi_{pp'}^0 = \varphi_{pp'}(T)$. Then it follows from 6.25 that

$$\begin{aligned} &(1 + \Phi_p^0(\Phi_{p'}^0(T))^t) q (1 + \Phi_p^0(\Phi_{p'}^0(T))) \\ &= ((1 + \Phi_{p'}^0(T)^t) q (1 + \Phi_{p'}^0(T)))^{(p)} \\ &= (((1 + T^t) q (1 + T))^{(p')})^{(p)}, \\ &= (((1 + T^t) q (1 + T))^{(p)})^{(p')} \\ &= (1 + \Phi_{p'}^0(\Phi_p^0(T))^t) q (1 + \Phi_{p'}^0(\Phi_p^0(T))); \end{aligned}$$

this, together with 4.187, immediately implies 6.26. □

REMARK 6.26. Note that 6.26 should be viewed as an analogue of the fact that if a classical connection takes values in the Lie algebra of an algebraic subgroup H of GL_n defined by polynomials with constant coefficients then the curvature of that connection takes values in the Lie algebra $\mathfrak{h}$ of H; cf. 2.32. Indeed, for any A-algebra C, the set

$$\begin{aligned} \mathfrak{so}_{\delta_p \delta_{p'}}(q)(C) &= \{M \in \mathfrak{g}(C); 1 + pp'M \in S(C)\} \\ &= \{M \in \mathfrak{g}(C); M^t q + qM + pp'M^t q M = 0\} \end{aligned} \tag{6.27}$$

should be viewed as the set of C-points of an arithmetic analogue (at pp') of the set of C-points of the Lie algebra of $SO(q)$,

$$\mathfrak{so}(q)(C) = \{M \in \mathfrak{g}(C); M^t q + qM = 0\}. \tag{6.28}$$

So 6.26 says that

$$\Phi^{\mathcal{G}}_{pp'} \in \mathfrak{so}_{\delta_p \delta_{p'}}(q)(A[[T]])$$

for the $\mathcal{G}$-curvature of any $\mathcal{H}_q$-horizontal adelic connection. No such statement holds for our usual curvature $\varphi_{pp'}$ (or rather for its corresponding matrix $\Phi^0_{pp'}$); this suggests that $\mathcal{G}$-curvature $\varphi^{\mathcal{G}}_{pp'}$ is the "correct" concept while our usual curvature $\varphi_{pp'}$ should be viewed as an "approximation" (or "linearization") of the "correct" concept. On the other hand the linearized concept of curvature, $\varphi_{pp'}$, is useful enough to deserve study: in particular it recaptures the matrix of $\varphi^{\Omega}_{pp'}$ acting on the Kähler differentials Ω.

REMARK 6.27. If q is δ-constant and $\tilde{q} = wqw^{-1}$ for some permutation matrix $w \in W$ then $\tilde{q}$ is also δ-constant and the curvatures $\varphi_{pp'}$ and $\tilde{\varphi}_{pp'}$ of the real Chern connections attached to q and $\tilde{q}$ respectively are related as follows:

$$\tilde{\varphi}_{pp'} = c_w \circ \varphi_{pp'} \circ c_w^{-1},$$

where $c_w : A[[T]] \to A[[T]]$ is given by $T \mapsto wTw^{-1}$. Similarly the corresponding $(1,1)$-curvatures $\varphi_{p\overline{p}'}$ and $\tilde{\varphi}_{p\overline{p}'}$ are related by

$$\tilde{\varphi}_{p\overline{p}'} = c_w \circ \varphi_{p\overline{p}'} \circ c_w^{-1}$$

In particular the curvature, respectively the $(1,1)$-curvature of the real Chern connection attached to q vanishes if and only if the same holds for $\tilde{q}$; and a similar statement holds for the corresponding truncations. Later we will prove vanishing/non-vanishing for various types of curvatures of real Chern connections attached to various q's. By what was just said, the same vanishing/non-vanishing results will hold for q replaced by any $\tilde{q}$ of the form $\tilde{q} = wqw^{-1}$, with $w \in W$.

REMARK 6.28. The compatibility of real Chern connections with the embeddings $GL_n \to GL_{n'}$ gives the following picture for curvature. Assume q is a sequence as in 6.8. Since the matrices in the sequence are δ-constant the real Chern connections on GL_{n_k} have well defined curvatures

$$\varphi_{pp',k} : A[[T_{n_k \times n_k}]] \to A[[T_{n_k \times n_k}]]$$

and $(1,1)$-curvatures

$$\varphi_{p\overline{p}',k} : A[[T_{n_k \times n_k}]] \to A[[T_{n_k \times n_k}]],$$

where $T_{n_k \times n_k}$ is the corresponding $n_k \times n_k$-matrix of indeterminates. Moreover these curvatures and $(1,1)$-curvatures are compatible with the surjections

$$A[[T_{n_{k+1} \times n_{k+1}}]] \to A[[T_{n_k \times n_k}]]$$

sending the upper $n_k \times n_k$ corner of $T_{n_{k+1} \times n_{k+1}}$ into $T_{n_k \times n_k}$ and all the other entries of $T_{n_{k+1} \times n_{k+1}}$ into 0. We get then a corresponding compatibility between first Chern forms, mean and scalar curvatures, and $(1,1)$-curvatures. Let us also remark that, if we pursue the analogy between GL_∞ and the "frame bundle of *Spec* $\mathbb{Z}$" then one is tempted to consider, as a "fundamental case" the case when the two "metrics," (p^{-s}) and q, in the background of the theory coincide; this corresponds to the case when $q = (q_2, q_2, q_2, ...)$ with $q_2 = \begin{pmatrix} 0 & 1 \\ 1 & 0 \end{pmatrix}$.

The above discussion of the real Chern connection has a complex analogue which we only briefly discuss here. First the "local" Theorem 4.31 immediately implies the following "global" result:

THEOREM 6.29. *Let $q \in GL_n(A)$, $q^t = \pm q$. There exists a unique adelic connection (δ_p) on GL_n, with attached Frobenius lifts (ϕ_p), such that, for all $p \in \mathcal{V}$, ϕ_p is $(\mathcal{H}_q, \mathcal{B}_q)$-horizontal with respect to ϕ_{0p}. It is given by $\phi_p(x) = \Phi_p(x)$, where $\Phi_p(x)$ is the matrix*

$$\Phi_p(x) = \phi_p(q)^{-1}(x^{(p)t})^{-1}(x^t q x)^{(p)}. \tag{6.29}$$

DEFINITION 6.30. The adelic connection (ϕ_p) on GL_n in Theorem 6.29 is called the *complex Chern connection* attached to q.

We then have the following easy analogue of Theorem 6.18 whose proof, analogous to that of Theorem 6.18, but much simpler, is left to the reader.

THEOREM 6.31. *Let D_μ be a partition torus and let $q \in G_\mu(A)$, $q^t = \pm q$. Let $\delta = (\delta_p)$ be the complex Chern connection on G attached to q and let (ϕ_p) be the attached Frobenius lifts. Assume that for all $p \in \mathcal{V}$ we have*

$$\phi_p(q)^{-1}q^{(p)} \in D_\mu(A).$$

Then δ is global along D_μ. Moreover if $t = x^\mu$ and $T = x - x^\mu$ then $\phi_p(x) = \Phi_p(x)$ where $\Phi_p(x)$ is the matrix with coefficients in $A[t, t^{-1}][[T]]$ defined by

$$\Phi_p(x) = \phi_p(q)^{-1}(t^{-p} - t^{-p}T^{(p)t}t^{-p} + t^{-p}T^{(p)t}t^{-p}T^{(p)t}t^{-p} - \ldots)((t+T^t)q(t+T))^{(p)}.$$

We end by discussing the globality of *special linear adelic connection.*

DEFINITION 6.32. Assume no prime in $\mathcal{V}$ divides n and assume n is invertible in A. The *special linear connection* on $G = GL_n$ (or the *adelic connection attached to* SL_n) is the adelic connection (δ_p) whose Frobenius lifts are given by $\phi_p : \mathcal{O}(G)^{\widehat{p}} \to \mathcal{O}(G)^{\widehat{p}}$, $x \mapsto \phi_p(x) = \Phi_p$, where Φ_p is the $n \times n$ matrix with coefficients in $\mathcal{O}(G)^{\widehat{p}}$ defined by

$$\Phi_p(x) = \lambda_p(x) \cdot x^{(p)}, \quad \lambda_p(x) = \left(\frac{\det(x^{(p)})}{(\det(x))^p}\right)_p^{-1/n}.$$

THEOREM 6.33. *The special linear connection (δ_p) on GL_n is global along* 1.

PROOF. It is enough to check that $\lambda_p(1+T)$ belongs to $\mathbb{Z}[1/n][[T]]$. For this it is enough to check that

$$\frac{\det((1+T)^{(p)})}{(\det(1+T))^p} - 1 \in p(T)\mathbb{Z}[[T]]$$

which is clear. □

REMARK 6.34. By assertion (1) in Remark 4.36 and Theorem 6.33 it follows that the adelic connection attached to SL_2 coincides with the real Chern connection attached to Sp_2. This does not follows a priori from the identification of SL_2 and Sp_2 but rather from the direct comparison of the formulas of the two adelic connections.

6.2. Non-vanishing/vanishing of curvature via analytic continuation

In this section we again place ourselves in the case when A is a δ-ring of global type, $A = \mathcal{O}_K[1/M]$, cf. Definition 3.25; so A is equipped with a global connection $(\delta_p) = (\delta_p^A)$ indexed by a set of primes $\mathcal{V}$. Recall that for $q \in GL_n(A)$, $q^t = \pm q$ we defined the real Chern connection (δ_p) on $G = GL_n$ attached to q (Definition 6.3). Also for q δ-constant we showed that the real Chern connection attached to q is global along 1 (Theorem 6.9) and hence this real Chern connection has a well defined curvature $(\varphi_{pp'})$ and $(1,1)$-curvature $(\varphi_{p\overline{p}'})$ (Remark 6.10),

$$\varphi_{pp'}, \varphi_{p\overline{p}'} : A[[T]] \to A[[T]]. \tag{6.30}$$

For $I = (T)$ we also have the truncations of the above

$$\varphi_{pp'}^{(\nu)}, \varphi_{p\overline{p}'}^{(\nu)} : I/I^{\nu+1} \to I/I^{\nu+1}. \tag{6.31}$$

We also have a well defined 3-curvature $(\varphi_{pp'p''})$,

$$\varphi_{pp'p''} : A[[T]] \to A[[T]]. \tag{6.32}$$

We will use some of the basic notation introduced in subsection 4.8.2 in relation to these objects; as a rule we will tend to recall, each time, the relevant notation. In particular we set

$$\Phi_p^0 = \phi_p(T), \quad \Phi_{pp'}^0 = \varphi_{pp'}(T), \quad \Phi_{p\overline{p}'}^0 = \varphi_{p\overline{p}'}(T). \tag{6.33}$$

Finally we refer to 4.165 and 4.166 for formulas of the first Chern form and first Chern $(1,1)$-form in terms of $\Phi_{pp'}^0$ and $\Phi_{p\overline{p}'}^0$ above.

In this section we would like to compute (or prove vanishing/non-vanishing) for these curvatures. A similar (but more complicated) analysis can be made for real Chern connections that are global along general partition tori; we leave such an analysis to the reader. Finally the whole analysis of curvature of real Chern connections can be redone (and is actually much simpler) for complex Chern connections; we also leave this to the reader.

Here is our first main non-vanishing result.

THEOREM 6.35. *Let $\delta = (\delta_p)$ be the real Chern connection on GL_n attached to a split q, where $n \geq 4$. Then the curvature of δ satisfies $\varphi_{pp'}^{(pp')} \neq 0$ for all $p \neq p'$. In particular, $\varphi_{pp'} \neq 0$ for $p \neq p'$.*

It would be interesting to find the minimum ν such that $\varphi_{pp'}^{(\nu)} \neq 0$. In regards to this we will prove:

THEOREM 6.36. *Let $\delta = (\delta_p)$ be the real Chern connection on GL_n attached to a split q, where $n = 2r \geq 2$. Then the curvature of δ satisfies $\varphi_{pp'}^{(2)} = 0$ for all p, p'.*

Now the real Chern connection on GL_n attached to a split q is global along 1 and hence, by Definition 3.59, one can attach to this situation the completed holonomy rings $\widehat{\mathfrak{hol}}$ and $\widehat{\mathfrak{hol}}^{\Omega}$. Then in particular, by Corollary 3.14, we have:

COROLLARY 6.37. *Let $\delta = (\delta_p)$ be the real Chern connection on GL_n attached to a split q, where $n = 2r \geq 2$. Then the completed holonomy rings $\widehat{\mathfrak{hol}}$ and $\widehat{\mathfrak{hol}}^{\Omega}$ are pro-nilpotent.*

REMARK 6.38. Note that by Theorem 6.35 if $\delta = (\delta_p)$ is the real Chern connection on GL_n attached to a split q, where $n = 2r \geq 4$ then $\widehat{\mathfrak{hol}} \neq 0$ while by Corollary 6.37 $\widehat{\mathfrak{hol}}$ is pro-nilpotent. It is interesting to contrast this situation with that in Corollary 3.47 which says that for Galois connections indexed by cofinite sets of primes holonomy cannot be nilpotent unless it vanishes. So real Chern connection attached to split q's behave quite differently from Galois connections.

On the other hand, in the case of SL_n, we will prove:

THEOREM 6.39. *Let $\delta = (\delta_p)$ be the special linear connection on GL_n. Then the curvature of δ satisfies $\varphi_{pp'} = 0$ for all p, p'.*

REMARK 6.40. It follows from Remark 6.34 and Theorem 6.39 that the real Chern connection on GL_2 attached to Sp_2 is flat i.e. its curvature satisfies $\varphi_{pp'} = 0$ for all p, p'. So Theorem 6.35 fails for the split q defining Sp_2. Whether Theorem 6.35 holds for the split q defining SO_2 or SO_3 is not clear at this point.

One can complement Theorem 6.35 as follows; recall from Definition 3.19 the definition of $\mathcal{G}$-curvature.

THEOREM 6.41. *Let $\delta = (\delta_p)$ be the real Chern connection on GL_n attached to a split q, where $n \geq 4$ and let $\varphi_{pp'}$ and $\varphi^{\mathcal{G}}_{pp'}$ be the curvature and the $\mathcal{G}$-curvature of δ respectively. Set $\varphi_{pp'}(T) = \Phi^0_{pp'}$, $\varphi^{\mathcal{G}}_{pp'}(T) = \Phi^{\mathcal{G}}_{pp'}$. Then the following hold:*

1. *For any prime ℓ with $p \not\equiv p'$ mod ℓ we have $\Phi^0_{pp'} \not\equiv 0$ mod ℓ.*
2. *If $p \neq p'$ then $\Phi^{\mathcal{G}}_{pp'} \not\equiv 0$ mod p, $\Phi^{\mathcal{G}}_{pp'} \not\equiv 0$ mod p'.*

On the other hand for the first Chern form we have:

THEOREM 6.42. *Let $\rho_{pp'}$ be the first Chern form of the real Chern connection on GL_n attached to a split q, where $n = 2r \geq 4$. Then for $p \neq p'$ we have $\rho_{pp'} \neq 0$.*

Results more refined than Theorem 6.35 can actually be proved as follows.

Let S be SO_n or Sp_n and let $V_0 \subset G := GL_n$ be the subgroup scheme over A whose points are the matrices

$$\begin{pmatrix} 1_r & b \\ 0 & 1_r \end{pmatrix}, \begin{pmatrix} 1_r & b \\ 0 & 1_r \end{pmatrix}, \begin{pmatrix} 1 & 0 & 0 \\ 0 & 1_r & b \\ 0 & 0 & 1_r \end{pmatrix},$$

in case S is $Sp_{2r}, SO_{2r}, SO_{2r+1}$, respectively; in the above b is an $r \times r$ matrix. So V_0 is isomorphic to the r^2 fold product of the additive group scheme. Also let V_0^+ be the subgroup scheme of V_0 with points as above with $b^t = b$ and V_0^- the subgroup scheme of V_0 with points as above with $b^t = -b$. Finally let $V \subset G$ be the subgroup scheme of upper triangular matrices with 1s on the diagonal. Then $V \cap SO_n = V_0^-$ and $V \cap Sp_n = V_0^+$. Theorem 6.35 follows then trivially from assertion (1) of the following:

THEOREM 6.43. *The real Chern connection $\delta = (\delta_p)$ on $G = GL_n$ attached to a split q induces adelic connections $\delta = (\delta_p)$ on each of the 3 groups V_0, V_0^+, V_0^-, that are global along 1. Moreover:*

1. *For δ on V_0 and $n \geq 4$ the curvature satisfies $\varphi^{(pp'-1)}_{pp'} = 0$ and $\varphi^{(pp')}_{pp'} \neq 0$.*
2. *For δ on V_0^+ the curvature satisfies $\varphi_{pp'} = 0$ for all p, p'.*
3. *For δ on V_0^- the curvature satisfies $\varphi_{pp'} = 0$ for all p, p'.*

Theorem 6.35 also follows trivially from the Theorem 6.44 below. Let $H \subset G = GL_n$ be the subgroup scheme whose points are the matrices

$$\begin{pmatrix} a & 0 \\ 0 & d \end{pmatrix}, \quad \begin{pmatrix} a & 0 \\ 0 & d \end{pmatrix}, \quad \begin{pmatrix} 1 & 0 & 0 \\ 0 & a & 0 \\ 0 & 0 & d \end{pmatrix},$$

in case S is $Sp_{2r}, SO_{2r}, SO_{2r+1}$, respectively, where a and d are $r \times r$ matrices. By the way, $H \cap S$ is the subgroup of H of all matrices in H such that $d = (a^t)^{-1}$. We have:

THEOREM 6.44. *Let q be split and let $S = SO(q)$. Then the real Chern connection $\delta = (\delta_p)$ on $G = GL_n$ attached to q induces adelic connections $\delta = (\delta_p)$ on H, S, and $H \cap S$, which are global along 1. Moreover,*

1. *For δ on H and $n \geq 4$, the curvature satisfies $\varphi_{pp'}^{(pp')} \neq 0$ for $p \neq p'$.*
2. *For δ on $H \cap S$ and $n \geq 6$, the curvature satisfies $\varphi_{pp'}^{(pp')} \neq 0$ for $p \neq p'$.*
3. *For δ on S and $n \geq 6$, the curvature satisfies $\varphi_{pp'} \neq 0$ for $p \neq p'$.*
4. *For δ on S and $n = 2$, the curvature satisfies $\varphi_{pp'} = 0$ for all p, p'.*

For the next Theorem recall the notion of curvature potential; cf. Definition 4.84.

THEOREM 6.45. *Let q be split, let $n = 2r \geq 2$, and let $(\varphi_{p\overline{p}'})$, $(\rho_{p\overline{p}'})$, $(\psi_{p\overline{p}'})$ be the $(1,1)$-curvature, the Chern $(1,1)$-form, and the $(1,1)$-curvature potential of the real Chern connection on G attached to q. Then the following hold.*

1. *Assume $q^t = q$. Then, for all p, p' we have $\rho_{p\overline{p}'} \neq 0$.*
2. *For all p, p' we have $\varphi_{p\overline{p}'}^{(1)} = 0$, $\varphi_{p\overline{p}'}^{(2)} \neq 0$ and $\rho_{p\overline{p}'}^{(2)} = 0$. In particular $\varphi_{p\overline{p}'} \neq 0$.*
3. *$\psi_{p\overline{p}'} \not\equiv 0$ mod $(T)^3$ for $p \neq p'$. In particular $\psi_{p\overline{p}'} \neq 0$ for $p \neq p'$.*

Note that the above theorem says nothing about the vanishing/non-vanishing of the curvature potential $\psi_{pp'}$.

We also have the following result on the 3-curvature:

THEOREM 6.46. *Let $q = q^t$ be split, let $n = 2r \geq 4$, and let $(\varphi_{pp'p''})$ be the 3-curvature of the real Chern connection δ attached to q. Then for all $p' \neq p''$, all p, and all non-zero integers a, we have that $a \cdot \varphi_{pp'p''}$ cannot be written as a $\mathbb{Z}$-linear combination of the components of the curvature $(\varphi_{pp'})$.*

COROLLARY 6.47. *Under the hypotheses of Theorem 6.46 we have:*

1. *$\varphi_{pp'p''} \neq 0$ for $p' \neq p''$,*
2. *the Yang-Mills form satisfies $Y_p \neq 0$ for all p,*
3. *δ does not have a $\mathbb{Q}$-horizontal curvature.*

On the other hand we have, of course, $\varphi_{pp'p''} = 0$ for $p' = p''$.

So far we discussed split q's. Next we discuss non-split q's.

We start by recording general formulas for Frobenius lifts and curvature mod $(T)^2$.

THEOREM 6.48. *Assume* $q^t = \pm q \in GL_n(A)$ *is* δ*-constant, let* $\delta = (\delta_p)$ *be the real Chern connection on* G *attached to* q*, and let* $\phi_p : A[[T]] \to A[[T]]$ *be the induced Frobenius lifts. Let* (q_p^{jk}) *be the inverse of the matrix* $q^{(p)} = (q_{jk}^p)$ *and set*

$$L_{jk} := \sum_{s \neq j} T_{sj} q_{sk} + \sum_{s \neq k} T_{sk} q_{sj}, \tag{6.34}$$

$$L_{pjk} = p \cdot \{\delta_{jk} T_{jj} + \frac{1}{2} \sum_m q_p^{jm} q_{mk}^{p-1} L_{mk}\}, \tag{6.35}$$

$$L_p := (L_{pjk}). \tag{6.36}$$

Then we have the following congruences

$$\phi_p(T) \equiv L_p \quad mod \quad (T)^2. \tag{6.37}$$

In the statement above δ_{jk} is the Kronecker symbol.

In particular, in the notation above, we have:

COROLLARY 6.49. *If* $\varphi_{pp'} : A[[T]] \to A[[T]]$ *is the curvature of the real Chern connection then:*

$$\varphi_{pp'}(T) \equiv \frac{1}{pp'} \left(L_{p'}^{(\phi_p)} \circ L_p - L_p^{(\phi_{p'})} \circ L_{p'} \right) \quad mod \quad (T)^2, \quad for \quad p \neq p', \tag{6.38}$$

$$\varphi_{pp}(T) \equiv \frac{1}{p} \left(L_p^{(\phi_p)} \circ L_p - L_p^{(\phi_p)} \circ L_p \right) \quad mod \quad (T)^2. \tag{6.39}$$

In the above statement $L_p^{(\phi_{p'})}$ is obtained, as usual, from L_p by applying $\phi_{p'}$ to its coefficients; and similarly for $L_{p'}^{(\phi_p)}$, etc.

We will then deduce:

THEOREM 6.50. *Let* $n = 2$, $A = \mathbb{Z}[1/M, \zeta_N]$, $2N|M$*, and*

$$q = \begin{pmatrix} \zeta_{11} & \zeta_{12} \\ \zeta_{21} & \zeta_{22} \end{pmatrix},$$

where ζ_{ij} *are powers of* ζ_N *such that* $\zeta_{11} = \zeta_{22}$, $\zeta_{12} = \zeta_{21}$*, and* $\zeta := \zeta_{12}/\zeta_{22} \neq \pm 1$*. Then we have the following formulas:*

(1) $\rho_{pp'}^{(0)} = \frac{\zeta^{pp'+1}(\zeta^{2p} - \zeta^{2p'})}{(1-\zeta^{2pp'})(1-\zeta^{2p})(1-\zeta^{2p'})}$;

(2) $\varphi_{p\overline{p}}^{(1)}(T) = \frac{p}{2} \cdot \begin{pmatrix} -\frac{\zeta^{2p-1}}{1-\zeta^{2p}} T_{12} + \frac{2\zeta - \zeta^{2p-1}}{1-\zeta^{2p}} T_{21} & 0 \\ 0 & -\frac{\zeta^{2p-1}}{1-\zeta^{2p}} T_{21} + \frac{2\zeta - \zeta^{2p-1}}{1-\zeta^{2p}} T_{12} \end{pmatrix}$;

(3) $\rho_{p\overline{p}'}^{(0)} = 0$.

Let us also record the following formulae for the mean $(1,1)$-curvature $K_{ij}(s)$ and the scalar $(1,1)$-curvature $\sigma(s)$; cf. Definition 4.86:

COROLLARY 6.51. *Under the assumptions of Theorem* 6.50 *we have:*

$$K_{11}(s) \equiv -\frac{1}{2} \left(\sum_p p^{1-s} \frac{\zeta^{2p-1}}{1-\zeta^{2p}} \right) T_{12} + \frac{1}{2} \left(\sum_p p^{1-s} \cdot \frac{2\zeta - \zeta^{2p-1}}{1-\zeta^{2p}} \right) T_{21} \quad mod \quad (T)^2,$$

$$K_{22}(s) \equiv -\frac{1}{2} \left(\sum_p p^{1-s} \frac{\zeta^{2p-1}}{1-\zeta^{2p}} \right) T_{21} + \frac{1}{2} \left(\sum_p p^{1-s} \cdot \frac{2\zeta - \zeta^{2p-1}}{1-\zeta^{2p}} \right) T_{12} \quad mod \quad (T)^2,$$

$$K_{12}(s) \equiv K_{21}(s) \equiv \sigma(s) \equiv 0 \quad mod \quad (T)^2.$$

Finally we have the following theorem in the "unitary case;" recall from Theorem 4.28 that the real Chern connection on GL_{2r} attached to the identity matrix $q = 1_{2r}$ induces an adelic connection on the complexified unitary group U_r^c; let us refer to this adelic connection as the real Chern connection on U_r^c attached to $q = 1_{2r}$.

THEOREM 6.52. *Let $n = 2r \geq 2$ and let $\varphi_{p\overline{p}'}$ be the $(1,1)$-curvature of the real Chern connection on U_r^c attached to $q = 1_{2r}$. Then $\varphi_{p\overline{p}'} \neq 0$ for all p, p'.*

In what follows we prove our Theorems.

PROOF OF THEOREM 6.39. To show that ϕ_p and $\phi_{p'}$ commute on $A[[T]]$ recall that $T = x - 1$ and note that $x^{(pp')} = x^{(p)(p')} = x^{(p')(p)}$. Then

$$\phi_p(T) = \lambda_p(1+T)\cdot(1+T)^{(p)} - 1$$

and hence

$$\begin{aligned}
\phi_p(\phi_{p'}(T)) &= \lambda_{p'}(\lambda_p(x)\cdot x^{(p)})\cdot(\lambda_p(x)\cdot x^{(p)})^{(p')} - 1 \\
&= \lambda_{p'}(x^{(p)})\cdot(\lambda_p(x))^{p'}\cdot x^{(pp')} - 1 \\
&= \left(\frac{\det(x^{(pp')})}{\det(x^{(p)})^{p'}}\right)_p^{-1/n}\cdot\left(\frac{\det(x^{(p)})^{p'}}{(\det(x))^{pp'}}\right)_p^{-1/n}\cdot x^{(pp')} - 1 \\
&= \left(\frac{\det(x^{(pp')})}{(\det(x))^{pp'}}\right)_p^{-1/n}\cdot x^{(pp')} - 1 \\
&= \phi_{p'}(\phi_p(T)).
\end{aligned}$$

□

In some of the next arguments certain polynomials will play a recurring role. Let v, w be two variables and let p, p', p'' be primes. Then consider the following two polynomials (one for the upper sign and one for the lower sign):

$$f_p(v,w) := \frac{1}{2}(\pm(\pm v + w)^p + v^p \mp w^p) \in \mathbb{Z}[1/2][v,w]. \tag{6.40}$$

Also consider the corresponding polynomials in $\mathbb{Z}[1/2][u,v]$:

$$f_{pp'}(v,w) := f_{p'}(f_p(v,w), f_p(w,v)), \tag{6.41}$$

$$g_{pp'}(v,w) := f_{pp'}(v,w) - f_{p'p}(v,w), \tag{6.42}$$

$$f_{pp'p''}(v,w) := f_p(f_{p'}(f_{p''}(v,w), f_{p''}(w,v)), f_{p'}(f_{p''}(w,v), f_{p''}(v,w))), \tag{6.43}$$

$$g_{pp'p''}(v,w) := f_{pp'p''}(v,w) - f_{p'pp''}(v,w) - f_{p''pp'}(v,w) + f_{p''p'p}(v,w). \tag{6.44}$$

LEMMA 6.53. *For $p \neq p'$ we have $g_{pp'}(v,w) \neq 0$.*

PROOF. We have:

$$\begin{aligned} f_{pp'}(v,w) &:= \tfrac{1}{2}(v \pm w)^{pp'} \\ &+\tfrac{1}{2}\left(\tfrac{\pm(\pm v+w)^p+v^p\mp w^p}{2}\right)^{p'} \\ &\mp\tfrac{1}{2}\left(\tfrac{\pm(v\pm w)^p\mp v^p+w^p}{2}\right)^{p'}. \end{aligned} \tag{6.45}$$

Consider the polynomial $f_{pp'}(v,1) \in \mathbb{Z}[1/2][v]$. We have the following congruences mod (v^3) in $\mathbb{Z}[1/2][v]$:

$$\begin{aligned} f_{pp'}(v,1) &\equiv \tfrac{1}{2}\left(\pm 1 + pp'v \pm \tfrac{pp'(pp'-1)}{2}v^2\right) \\ &\mp\tfrac{1}{2}\left(1 \pm \tfrac{p}{2}v + \tfrac{p(p-1)}{4}v^2\right)^{p'} \\ &\equiv \tfrac{1}{2}\left(\pm 1 + pp'v \pm \tfrac{pp'(pp'-1)}{2}v^2\right) \\ &\mp\tfrac{1}{2}\left(1 + p'(\pm\tfrac{p}{2}v + \tfrac{p(p-1)}{4}v^2) + \tfrac{p'(p'-1)}{2}(\pm\tfrac{p}{2}v)^2\right) \\ &\equiv \tfrac{pp'}{4}v \pm \tfrac{pp'}{16}(3pp'-p-2)v^2 \mod (v^3). \end{aligned}$$

Similarly

$$f_{p'p}(v,1) \equiv \frac{p'p}{4}v \pm \frac{p'p}{16}(3p'p-p'-2)v^2 \mod (v^3)$$

hence we get

$$g_{pp'}(v,1) \equiv \pm\frac{pp'}{16}(p'-p)v^2 \mod (v^3). \tag{6.46}$$

This ends our proof. □

LEMMA 6.54. *For $p \neq p'$ we have*

$$\frac{\partial}{\partial v}(g_{pp'}(v,w)) + \frac{\partial}{\partial w}(g_{pp'}(w,v)) \neq 0.$$

PROOF. For the choice of the upper sign one immediately checks that we have $g_{pp'}(w,v) = -g_{pp'}(v,w)$ while for the choice of the lower sign we have $g_{pp'}(w,v) = g_{pp'}(v,w)$. So our Lemma follows from Lemma 6.53 above and Lemma 6.55 below. □

LEMMA 6.55. *Let $g \in K[v,w]$ be a polynomial in 2 variables with coefficients in a field K of characteristic zero. Assume one of the following holds:*

(1) *$g(w,v) = -g(v,w)$ and $\frac{\partial g}{\partial v} = \frac{\partial g}{\partial w}$.*
(2) *$g(w,v) = g(v,w)$, $\frac{\partial g}{\partial v} = -\frac{\partial g}{\partial w}$, and g is homogeneous of odd degree.*

Then $g = 0$.

Note that the hypothesis of "odd degree" in (2) above is necessary as shown by the example of $g(v,w) = (v-w)^2$, say.

PROOF. Write $g(v,w)=\sum a_{ij}v^iw^j$. In case (1) we get

$$a_{ji}=-a_{ij},\quad ia_{ij}=(j+1)a_{i-1,j+1}.$$

So if $a_{ij}\neq 0$ for some i,j and $d:=i+j$ then

$$a_{d0}=\frac{1}{d}\cdot a_{d-1,1}=\frac{1}{d}\cdot\frac{2}{d-1}\cdot a_{d-2,2}=\ldots=\frac{1}{d}\cdot\frac{2}{d-1}\cdot\ldots\cdot\frac{d}{1}\cdot a_{0,d}=a_{0,d},$$

a contradiction. In case (2) we get

$$a_{ji}=a_{ij},\quad ia_{ij}=-(j+1)a_{i-1,j+1},$$

so if $a_{ij}\neq 0$ for some i,j with $d:=i+j$ odd then

$$a_{d0}=\left(-\frac{1}{d}\right)\cdot\left(-\frac{2}{d-1}\right)\cdot\ldots\cdot\left(-\frac{d}{1}\right)\cdot a_{0,d}=(-1)^d\cdot a_{0,d}=-a_{0,d},$$

a contradiction. □

In the next lemma we assume that we choose the upper sign in the definition of $f_p(v,w)$.

LEMMA 6.56. *$g_{pp'p''}(v,w)\neq 0$ for $p\neq p'$.*

PROOF. An immediate computation gives:

$$f_{pp'p''}(v,w)=\frac{1}{2}(v+w)^{pp'p''}+\frac{1}{2}h_{pp'p''}(v,w),$$

where

$$\begin{aligned} h_{pp'p''}(v,w) &= \left(\tfrac{(v+w)^{p'p''}+f_{p''}(v,w)^{p'}-f_{p''}(w,v)^{p'}}{2}\right)^p \\ &\quad-\left(\tfrac{(v+w)^{p'p''}+f_{p''}(w,v)^{p'}-f_{p''}(v,w)^{p'}}{2}\right)^p. \end{aligned}$$

Next, an easy computation as in the proof of Lemma 6.53 gives the following congruences mod v^3 in $\mathbb{Z}[1/2][v]$:

$$h_{pp'p''}(v,w)\equiv-\left(1+\frac{3}{4}pp'p''v+\frac{pp'p''}{32}(p'p''+2p''-12+9pp'p'')v^2\right)\mod (v^3).$$

Hence

$$\begin{aligned} 2g_{pp'p''}(v,1) &\equiv h_{pp'p''}(v,1)-h_{p'pp''}(v,1)-h_{p''pp'}(v,1)+h_{p''p'p}(v,1)\\ &\equiv -\tfrac{pp'p''}{32}\left((p'p''+2p'')-(pp''+2p'')-(pp'+2p')+(p'p+2p)\right)v^2\\ &\equiv -\tfrac{pp'p''}{32}(p''-2)(p'-p)v^2\mod (v^3). \end{aligned}$$

which ends our proof. □

PROOF OF THEOREM 6.43. We may assume our ring $A=\mathbb{Z}[1/2]$. In what follows we assume that q is split, defining SO_{2r} or Sp_{2r} respectively. The case SO_{2r+1} can be treated similarly. Let us make the convention that the upper sign in $\pm,\mp$ refers to the SO_{2r} case and the lower sign refers to the Sp_{2r} case. Write T as a block matrix $\begin{pmatrix} A & B\\ C & D\end{pmatrix}$, where A,B,C,D are $r\times r$ matrices. (The use of the letter

A for this matrix and also for the ring $\mathbb{Z}[1/2]$ should not lead to any confusion.) A direct computation gives:

$$\Phi_p^0\begin{pmatrix} 0 & B \\ 0 & 0 \end{pmatrix} = \begin{pmatrix} 0 & F_p(B) \\ 0 & 0 \end{pmatrix}, \tag{6.47}$$

where

$$F_p(B) := \frac{1}{2}(\pm(\pm B + B^t)^{(p)} + B^{(p)} \mp B^{(p)t}). \tag{6.48}$$

For the convenience of the reader let us check 6.47 in case $q^t = -q$; the case $q^t = q$ is similar. So we set

$$x = \begin{pmatrix} 1 & B \\ 0 & 1 \end{pmatrix}, \quad q = \begin{pmatrix} 0 & 1 \\ -1 & 0 \end{pmatrix}$$

in the formula 4.36. We get

$$x^{(p)t}\phi(q)x^{(p)} = \begin{pmatrix} 1 & 0 \\ B^{(p)t} & 1 \end{pmatrix}\begin{pmatrix} 0 & 1 \\ -1 & 0 \end{pmatrix}\begin{pmatrix} 1 & B^{(p)} \\ 0 & 1 \end{pmatrix} = \begin{pmatrix} 0 & 1 \\ -1 & B^{(p)t} - B^{(p)} \end{pmatrix}.$$

Hence

$$(x^{(p)t}\phi(q)x^{(p)})^{-1} = \begin{pmatrix} B^{(p)t} - B^{(p)} & -1 \\ 1 & 0 \end{pmatrix}.$$

On the other hand

$$(x^t q x)^{(p)} = \begin{pmatrix} 0 & 1 \\ -1 & (B^t - B)^{(p)} \end{pmatrix}.$$

One gets

$$(x^{(p)t}\phi(q)x^{(p)})^{-1}(x^t q x)^{(p)} = \begin{pmatrix} 1 & B^{(p)t} - B^{(p)} - (B^t - B)^{(p)} \\ 0 & 1 \end{pmatrix},$$

hence

$$\{(x^{(p)t}\phi(q)x^{(p)})^{-1}(x^t q x)^{(p)}\}_p^{1/2} = \begin{pmatrix} 1 & \frac{1}{2}(B^{(p)t} - B^{(p)} - (B^t - B)^{(p)}) \\ 0 & 1 \end{pmatrix} =: M,$$

hence

$$\Phi_p\begin{pmatrix} 1 & B \\ 0 & 1 \end{pmatrix} = \begin{pmatrix} 1 & B^{(p)} \\ 0 & 1 \end{pmatrix} \cdot M = \begin{pmatrix} 1 & \frac{1}{2}(B^{(p)t} + B^{(p)} - (B^t - B)^{(p)}) \\ 0 & 1 \end{pmatrix}$$

which ends the proof of 6.47.

Now by 6.48 $F_p(B)$ has coefficients in $\mathbb{Z}[1/2][B]$, hence the ring endomorphism

$$\phi_p^B : \mathbb{Z}[1/2][[B]] \to \mathbb{Z}[1/2][[B]], \quad B \mapsto \phi_p^B(B) = F_p(B)$$

sends $\mathbb{Z}[1/2][B]$ into itself. We have, of course, a commutative diagram

$$\begin{array}{ccc} \mathbb{Z}[1/2][[B]] & \stackrel{\phi_p^B}{\longrightarrow} & \mathbb{Z}[1/2][[B]] \\ u\uparrow & & \uparrow u \\ \mathbb{Z}[1/2][[T]] & \stackrel{\phi_p}{\longrightarrow} & \mathbb{Z}[1/2][[T]] \end{array} \tag{6.49}$$

where u is the surjective homomorphism sending $A, C, D \mapsto 0$, $B \mapsto B$. In order to prove assertion (1) assume ϕ_p^B and $\phi_{p'}^B$ commute on $\mathbb{Z}[1/2][[B]]$ for some $p \neq p'$ and let us derive a contradiction. Clearly we have

$$F_p(B_{ij}) = F_p(B)_{ij} = \frac{1}{2}(\pm(\pm B_{ij} + B_{ji})^p + B_{ij}^p \mp B_{ji}^p) = f_p(B_{ij}, B_{ji}).$$

Since $\phi_p^B\phi_{p'}^B(B) = F_{p'}(F_p(B))$ we have

$$\begin{aligned}\phi_p^B\phi_{p'}^B(B_{ij}) &= \phi_p^B\phi_{p'}^B(B)_{ij} \\ &= F_{p'}(F_p(B))_{ij} \\ &= f_{p'}(F_p(B)_{ij}, F_p(B)_{ji}) \\ &= f_{p'}(f_p(B_{ij}, B_{ji}), f_p(B_{ji}, B_{ij})) \\ &= f_{pp'}(B_{ij}, B_{ji}),\end{aligned}$$

where $f_{pp'}$ is an in 6.41. In particular if

$$pp'\varphi_{pp'}^B = \phi_p^B \circ \phi_{p'}^B - \phi_{p'}^B \circ \phi_p^B : \mathbb{Z}[1/2][[B]] \to \mathbb{Z}[1/2][[B]]$$

denotes the map induced by the map $pp'\varphi_{pp'}$ on $\mathbb{Z}[1/2][[T]]$ then we have

$$pp'\varphi_{pp'}^B(B_{ij}) = g_{pp'}(B_{ij}, B_{ji}). \tag{6.50}$$

Choose now two indices $i \neq j$ (this is possible because $n \geq 4$). Then we get $g_{pp'}(v, w) = 0$ which contradicts Lemma 6.53.

To prove assertions (2) and (3) assume that either $S = SO_n$ and $b^t = -b$ or $S = Sp_n$ and $b^t = b$. Then, clearly, $F_p(b) = b^{(p)}$. In particular for any such b and any two primes p, p' we have

$$F_{p'}(F_p(b)) = F_p(F_{p'}(b)).$$

On the other hand let us assume that either $S = SO_n$ and $b^t = b$ or $S = Sp_n$ and $b^t = -b$. Then, clearly, $F_p(b) = 2^{p-1}b^{(p)}$. In particular for any such b and any two primes p, p' we have

$$F_{p'}(F_p(b)) = 2^{p'-1}(2^{p-1}b^{(p)})^{(p')} = 2^{pp'-1}b^{(pp')} = F_p(F_{p'}(b)).$$

Assertions (2) and (3) follow. □

PROOF OF THEOREM 6.41. We adopt the notation in the proof of Theorem 6.43. Note that assertion (2) in Theorem 6.41 follows from assertion (1) and from the congruence 4.188. To check assertion (1) note that, if one assumes $\Phi_{pp'}^0 \equiv 0 \bmod \ell$ then, by the diagram 6.49 we get that $\frac{1}{pp'}g_{pp'}(v, w) \equiv 0 \bmod \ell$, which, together with 6.46, leads to a contradiction. □

PROOF OF THEOREM 6.46. In the notation of the proof of Theorem 6.43 let us still denote by $\varphi_{pp'p''}^B : \mathbb{Z}[1/2][[B]] \to \mathbb{Z}[1/2][[B]]$ the map induced by the map $\varphi_{pp'p''}$ on $\mathbb{Z}[1/2][[T]]$; then we have

$$\begin{aligned}p'p''\varphi_{pp'p''}^B(B) &= F_{p''}(F_{p'}(F_p(B))) - F_{p'}(F_{p''}(F_p(B))) \\ &\quad -F_p(F_{p''}(F_{p'}(B))) + F_p(F_{p'}(F_{p''}(B))).\end{aligned}$$

Exactly as in loc. cit. we have

$$F_{p''}(F_{p'}(F_p(B_{ij}))) = f_{p''p'p}(B_{ij}, B_{ji})$$

hence

$$p'p''\varphi_{pp'p''}^B(B_{ij}) = g_{p''p'p}(B_{ij}, B_{ji}) \tag{6.51}$$

and, taking any two indices i, j with $i \neq j$ (which is possible because $n \geq 4$) we conclude, by Lemma 6.56, that $\varphi^B_{pp'p''} \neq 0$ for $p' \neq p''$. Assume now that

$$a \cdot \varphi_{pp'p''} = \sum_{ll'} a_{ll'} \varphi_{ll'}$$

for some non-zero $a \in \mathbb{Z}$ and some $a_{ll'} \in \mathbb{Z}$, and seek a contradiction; this will conclude our proof. Now by the latter equality we get an equality of endomorphisms of $\mathbb{Z}[1/2][[B]]$,

$$a \cdot \varphi^B_{pp'p''} = \sum_{ll'} a_{ll'} \varphi^B_{ll'},$$

hence, by 6.50 and 6.51, an equality of polynomials in $\mathbb{Q}[u, v]$,

$$\frac{a}{p'p''} \cdot g_{p''p'p}(u, v) = \sum_{ll'} \frac{a_{ll'}}{ll'} \cdot g_{ll'}(u, v).$$

But note that $g_{p''p'p}(u, v)$ is homogeneous of degree $pp'p''$ while $g_{ll'}(u, v)$ are homogeneous of degree ll'; this leads to a contradiction. $\square$

PROOF OF THEOREM 6.52. Again, we may assume $A = \mathbb{Z}[1/2]$ and write $T = \begin{pmatrix} A & B \\ C & D \end{pmatrix}$. Then A, B are retract coordinates for GL^c_r: if $a = 1 + A$, $b = B$, $c = C$, $d = 1 + D$, then GL^c_r is defined in GL_{2r} by the equations

$$a = d, \quad b = -c.$$

Also U^c_r is defined in GL^c_r by the equations

$$aa^t + bb^t = 1, \quad ba^t = ab^t,$$

equivalently by the equations:

$$F := A + A^t + AA^t + BB^t, \; G := (B + BA^t) - (B^t + AB^t).$$

Similarly U^c_1 is defined in

$$GL^c_1 = Spec\ \mathbb{Z}[1/2][\alpha, \beta, \frac{1}{(1+\alpha)^2 + \beta^2}]$$

by the equation

$$f := 2\alpha + \alpha^2 + \beta^2.$$

Consider the embedding of GL^c_1 into GL^c_r given by the homomorphism between their coordinate rings

$$u : \begin{pmatrix} A & B \\ -B & A \end{pmatrix} \mapsto \begin{pmatrix} \alpha \cdot 1_r & \beta \cdot 1_r \\ -\beta \cdot 1_r & \alpha \cdot 1_r \end{pmatrix}.$$

The ideal of the image of GL^c_1 via this embedding is easily seen to be sent into itself by the Frobenius lifts ϕ_p on $(GL^c_r)^{\widehat{p}}$; hence the ideal of U^c_1 is sent into itself by the Frobenius lifts ϕ_p on $(U^c_r)^{\widehat{p}}$. So we have a commutative diagram

$$(6.52) \qquad \begin{array}{ccc} \mathbb{Z}[1/2][[\alpha, \beta]]/(f) & \xrightarrow{\phi_p} & \mathbb{Z}[1/2][[\alpha, \beta]]/(f) \\ u \uparrow & & \uparrow u \\ \mathbb{Z}[1/2][[A, B]]/(F, G) & \xrightarrow{\phi_p} & \mathbb{Z}[1/2][[A, B]]/(F, G) \\ v \uparrow & & \uparrow v \\ \mathbb{Z}[1/2][[T]] & \xrightarrow{\phi_p} & \mathbb{Z}[1/2][[T]] \end{array}$$

where v sends A into A, B into B, C into $-B$ and D into A. An easy computation yields

$$\phi_p\begin{pmatrix} \alpha & \beta \\ -\beta & \alpha \end{pmatrix} = \begin{pmatrix} \alpha_p & \beta_p \\ -\beta_p & \alpha_p \end{pmatrix}$$

where

$$\alpha_p = (1+\alpha)^p K_p(\alpha,\beta) - 1, \quad \beta_p = \beta^p K_p(\alpha,\beta),$$
$$K_p(\alpha,\beta) := \left(\frac{((1+\alpha)^2+\beta^2)^p}{(1+\alpha)^{2p}+\beta^{2p}}\right)^{1/2}.$$

On the other hand we have

$$\phi_{\overline{p}'}\begin{pmatrix} \alpha & \beta \\ -\beta & \alpha \end{pmatrix} = \begin{pmatrix} (1+\alpha)^{p'}-1 & \beta^{p'} \\ -\beta^{p'} & (1+\alpha)^{p'} \end{pmatrix}.$$

The equality $\varphi_{p\overline{p}'} = 0$ implies that ϕ_p and $\phi_{\overline{p}'}$ commute on $\mathbb{Z}[1/2][[\alpha,\beta]]/(f)$ which immediately implies

$$K_p(\alpha,\beta)^{p'} = K_p((1+\alpha)^{p'}-1, \beta^{p'}).$$

So if we assume the latter equation, we square it, and we use the relation

$$\beta^2 = -2\alpha - \alpha^2,$$

we get the following equality of polynomials in α:

$$(1+\alpha)^{2pp'} - (2\alpha+\alpha^2)^{pp'} = ((1+\alpha)^{2p} - (2\alpha+\alpha^2)^p)^{p'} \cdot ((1+\alpha)^{2p'} - (2\alpha+\alpha^2)^{p'})^p.$$

Picking out the coefficients of α we get $2pp' = 4pp'$, a contradiction; this ends the proof. $\square$

PROOF OF THEOREM 6.44. For simplicity we only treat the case $n = 2r$; the case $n = 2r+1$ can be reduced to the case $n = 2r$. Write again $T = \begin{pmatrix} A & B \\ C & D \end{pmatrix}$, where A, B, C, D are $r \times r$ matrices and set $a = 1 + A$, $d = 1 + D$. The embedding $H \subset G$ corresponds to the homomorphism sending A, D into themselves and B, C into 0. We have

$$\Phi_p\begin{pmatrix} a & 0 \\ 0 & d \end{pmatrix} = \begin{pmatrix} S_p(a,d) & 0 \\ 0 & S_p(d,a) \end{pmatrix},$$

where $\Phi_p(1+T) = \Phi_p^0(T) + 1$ and

$$S_p(a,d) := a^{(p)} \cdot \{(a^{(p)})^{-1}(d^{(p)t})^{-1}(d^t a)^{(p)}\}_p^{1/2}.$$

This immediately implies that H is preserved by all ϕ_ps and the induced adelic connection on H is global along 1. We have an induced adelic connection on $H \cap S$ given by

(6.53)
$$\Phi_p\begin{pmatrix} a & 0 \\ 0 & (a^t)^{-1} \end{pmatrix} = \begin{pmatrix} \Sigma_p(a) & 0 \\ 0 & \Sigma_p((a^t)^{-1}) \end{pmatrix} = \begin{pmatrix} \Sigma_p(a) & 0 \\ 0 & (\Sigma_p(a)^t)^{-1} \end{pmatrix},$$

where

(6.54)
$$\Sigma_p(a) := a^{(p)} \cdot \{(a^{-1})^{(p)} a^{(p)}\}_p^{-1/2}.$$

In particular this adelic connection on $H \cap S$ is global along 1. Finally the ideal of S in $G^{\widehat{p}}$ is preserved by all ϕ_ps by Theorem 6.7 and the induced adelic connection on S is global along 1 because this is true for GL_n and because, by 6.24, Φ_p^0 sends $x^t q x - q$ into $(x^t q x)^{(p)} - q$ which has entries in the ideal $(x^t q x - q)\mathbb{Z}[1/2][[T]]$.

To prove assertions (1), (2), (3), (4) of the theorem note first that assertion (3) follows from assertion (2); also assertion (4) is trivial in case $q^t = q$ and follows from Theorem 6.39 in case $q^t = -q$.

To prove assertion (1) let $E(v)$ be the $r \times r$ matrix with v in its upper right corner, 1's on the diagonal, and zeros everywhere else and let H_0 be the subgroup of H consisting of all matrices for which $a = E(v)$ and $d = E(w)^t$ for some v and w. It is enough to check that H_0 is preserved by ϕ_p and the curvature on H_0 satisfies $\varphi_{pp'} \neq 0$ for $p \neq p'$. For v and w two variables we have

$$\Phi_p \begin{pmatrix} E(v) & 0 \\ 0 & E(w)^t \end{pmatrix} = \begin{pmatrix} E(f_p(v,w)) & 0 \\ 0 & E(f_p(w,v))^t \end{pmatrix},$$

where

$$f_p(v,w) := \frac{1}{2}((v+w)^p + v^p - w^p).$$

Hence

$$\Phi_{p'}\Phi_p \begin{pmatrix} E(v) & 0 \\ 0 & E(w)^t \end{pmatrix} = \begin{pmatrix} E(f_{pp'}(v,w)) & 0 \\ 0 & E(f_{pp'}(w,v))^t \end{pmatrix},$$

where $f_{pp'}$ is the polynomial in 6.45 (with the choice of the upper sign). So if we assume $\varphi_{pp'} = 0$ for δ on H_0, for some $p \neq p'$, then $f_{pp'} = f_{p'p}$ which contradicts Lemma 6.53. Assertion (1) follows.

To prove assertion (2) assume $n = 2r \geq 6$ and $\varphi_{pp'} = 0$ for some $p \neq p'$ and δ on $H \cap S$ and seek a contradiction. We may assume $r = 3$. In view of equation 6.53 we have

$$\Sigma_{p'}\Sigma_p(a) = \Sigma_p\Sigma_{p'}(a). \tag{6.55}$$

A trivial computation using 6.54 yields

$$\Sigma_p\Sigma_{p'} \begin{pmatrix} 1 & u & v \\ 0 & 1 & w \\ 0 & 0 & 1 \end{pmatrix} = \begin{pmatrix} 1 & u^{pp'} & f_{pp'}(v, uw - v) \\ 0 & 1 & w^{pp'} \\ 0 & 0 & 1 \end{pmatrix},$$

where $f_{pp'}$ is the polynomial 6.45 (with the choice of the upper sign). From 6.55 we get $f_{pp'} = f_{p'p}$ which contradicts Lemma 6.53; this ends our proof. □

In the next Lemma we again let $A = \mathbb{Z}[1/2]$ and we let $T = \begin{pmatrix} A & B \\ C & D \end{pmatrix}$, where A, B, C, D are $r \times r$ matrices. We denote by (T) the ideal in $\mathbb{Z}[1/2][[T]]$ generated by the entries of T. We further assume $q^t = \pm q$ is split. The upper sign in the proof below corresponds to the case $q^t = q$ and the lower sign corresponds to the case $q^t = -q$. Note that Lemma 6.57 below immediately implies Theorem 6.36.

LEMMA 6.57. *$\Phi^0_{pp'} \equiv 0$ mod $(T)^3$.*

PROOF. We proceed by computing $\Phi_p(1+T)$ modulo $(T)^3$. Let $1 \leq i, j, k \leq r$ and set $\underline{i} = i + r$. We have

$$(1+T)^{(p)} = \begin{pmatrix} (1+T_{11})^p & T_{12}^p & \cdots \\ T_{21}^p & (1+T_{22})^p & \cdots \\ \vdots & \vdots & \ddots \end{pmatrix} \equiv \begin{pmatrix} E_1 & 0 \\ 0 & E_2 \end{pmatrix} \operatorname{mod}(T)^3,$$

where E_1 and E_2 are diagonal matrices with

$$(E_1)_{ii} = 1 + pT_{ii} + \binom{p}{2} T_{ii}^2, \quad (E_2)_{ii} = 1 + pT_{\underline{ii}} + \binom{p}{2} T_{\underline{ii}}^2.$$

Therefore,

$$\begin{aligned}(1+T^t)^{(p)}q(1+T)^{(p)} &\equiv \begin{pmatrix} E_1 & 0 \\ 0 & E_2 \end{pmatrix} q \begin{pmatrix} E_1 & 0 \\ 0 & E_2 \end{pmatrix} \bmod (T)^3 \\ &= \begin{pmatrix} 0 & E_1E_2 \\ \pm E_2E_1 & 0 \end{pmatrix} \\ &= G,\end{aligned}$$

where the last equality defines G. Next, we have

$$((1+T^t)q(1+T))^{(p)} = \left(\begin{pmatrix} 1+A^t & C^t \\ B^t & 1+D^t \end{pmatrix} q \begin{pmatrix} 1+A & B \\ C & 1+D \end{pmatrix} \right)^{(p)}$$

$$= \begin{pmatrix} (C \pm C^t \pm C^tA + A^tC)^{(p)} & (1 + D + A^t + A^tD \pm C^tB)^{(p)} \\ (\pm 1 \pm A \pm D^t \pm D^tA + B^tC)^{(p)} & (\pm B + B^t + B^tD \pm D^tB)^{(p)} \end{pmatrix}.$$

Write

(6.56) $$A_{ij} = T_{ij}, \quad B_{ij} = T_{i\underline{j}}, \quad C_{ij} = T_{\underline{i}j}, \quad D_{ij} = T_{\underline{ij}},$$

hence

$$(A^tD)_{ij} = \sum_k T_{ki}T_{\underline{kj}}, \qquad (D^tA)_{ij} = \sum_k T_{\underline{ki}}T_{kj},$$

$$(B^tC)_{ij} = \sum_k T_{k\underline{i}}T_{\underline{k}j}, \qquad (C^tB)_{ij} = \sum_k T_{\underline{k}i}T_{k\underline{j}},$$

and define

(6.57) $$Q_i = \sum_k (T_{ki}T_{\underline{ki}} \pm T_{\underline{k}i}T_{k\underline{i}}).$$

Let H be the diagonal matrix with

$$H_{ii} = 1 + p(T_{ii} + T_{\underline{ii}} + Q_i) + \binom{p}{2}(T_{ii}^2 + T_{\underline{ii}}^2 + 2T_{ii}T_{\underline{ii}}).$$

Then we have

$$((1+T^t)q(1+T))^{(p)} \equiv \begin{pmatrix} 0 & H \\ \pm H & 0 \end{pmatrix} = F \mod (T)^3,$$

where the last equality is the definition of F. One trivially checks that

$$G^{-1}F = \begin{pmatrix} E_1^{-1}E_2^{-1}H & 0 \\ 0 & E_2^{-1}E_1^{-1}H \end{pmatrix} \equiv 1 \mod (T)^2.$$

By the binomial expansion

$$(1 + (G^{-1}F - 1))^{1/2} \equiv 1 + \frac{1}{2}(G^{-1}F - 1) \mod (T)^3.$$

So

$$\Phi_p(1+T) \equiv (1+T)^{(p)} + \frac{1}{2}(1+T)^{(p)}\frac{1}{2}(G^{-1}F-1)$$

$$\equiv \begin{pmatrix} E_1 & 0 \\ 0 & E_2 \end{pmatrix} + \frac{1}{2}\begin{pmatrix} E_1 & 0 \\ 0 & E_2 \end{pmatrix}\begin{pmatrix} E_1^{-1}E_2^{-1}H - 1 & 0 \\ 0 & E_2^{-1}E_1^{-1}H - 1 \end{pmatrix}$$

$$\equiv \frac{1}{2}\begin{pmatrix} E_2^{-1}H + E_1 & 0 \\ 0 & E_1^{-1}H + E_2 \end{pmatrix} \quad \mathrm{mod}\ (T)^3.$$

Set $X = E_2^{-1}H + E_1$ and $Y = E_1^{-1}H + E_2$ and note that X and Y are diagonal matrices. Then

$$\Phi_p(1+T) \equiv \frac{1}{2}\begin{pmatrix} X & 0 \\ 0 & Y \end{pmatrix} \quad \mathrm{mod}\ (T)^3. \tag{6.58}$$

On the other hand

$$\begin{aligned} X_{ii} \ &\equiv \ \left(1 - pT_{\underline{ii}} - \tbinom{p}{2}T_{\underline{ii}}^2 + p^2T_{\underline{ii}}^2\right)\times \\ &\quad \times\left(1 + p(T_{ii} + T_{\underline{ii}} + Q_i) + \tbinom{p}{2}(T_{ii}^2 + T_{\underline{ii}}^2 + 2T_{ii}T_{\underline{ii}})\right) \\ &\quad +\left(1 + pT_{ii} + \tbinom{p}{2}T_{ii}^2\right) \ \mathrm{mod}\ (T)^3 \\ &\equiv \ 2 + 2pT_{ii} - pT_{ii}T_{\underline{ii}} + 2\tbinom{p}{2}T_{ii}^2 + pQ_i \ \mathrm{mod}\ (T)^3, \end{aligned} \tag{6.59}$$

and similarly

$$Y_{ii} \equiv 2 + 2pT_{\underline{ii}} - pT_{ii}T_{\underline{ii}} + 2\binom{p}{2}T_{\underline{ii}}^2 + pQ_i \ \mathrm{mod}\ (T)^3. \tag{6.60}$$

By the way equations 6.58, 6.59, and 6.60 provide an explicit expression for

$$\Phi_p(1+T) \quad \mathrm{mod}\ (T)^3$$

and, in particular, one gets that $\Phi_p(1+T)$ is congruent modulo $(T)^3$ to a diagonal matrix.

Now, in order to compute $\Phi^0_{pp'}$ mod $(T)^3$ set

$$Z_p = \Phi_p(1+T) = \Phi_p^0(T) + 1.$$

Since Z_p is congruent mod $(T)^3$ to a diagonal matrix and since for any $n\times n$ diagonal matrix Δ and any $n\times n$ matrix Γ we have $(\Gamma\Delta)^{(p')} = \Gamma^{(p')}\Delta^{(p')}$, the following holds:

$$\begin{aligned} \Phi^0_{p'}(\Phi^0_p(T)) &= Z_p^{(p')}\left[\left(Z_p^{t(p')}qZ_p^{(p')}\right)^{-1}(Z_p^tqZ_p)^{(p')}\right]^{\frac{1}{2}} - 1 \\ &\equiv Z_p^{(p')}\left[\left((Z_p^tqZ_p)^{(p')}\right)^{-1}(Z_p^tqZ_p)^{(p')}\right]^{\frac{1}{2}} - 1 \ \mathrm{mod}\ (T)^3 \\ &= Z_p^{(p')} - 1. \end{aligned}$$

By 6.59 we get

$$(6.61)\quad (\frac{1}{2}X_{ii})^{p'} \equiv 1 + pp'T_{ii} - \frac{pp'}{2}T_{ii}T_{\underline{ii}} + \frac{pp'}{2}Q_i + \frac{pp'(pp'-1)}{2}T_{ii}^2 \mod (T)^3,$$

and similarly, by 6.60, we get

$$(6.62)\quad (\frac{1}{2}Y_{ii})^{p'} \equiv 1 + pp'T_{\underline{ii}} - \frac{pp'}{2}T_{ii}T_{\underline{ii}} + \frac{pp'}{2}Q_i + \frac{pp'(pp'-1)}{2}T_{\underline{ii}}^2 \mod (T)^3.$$

Consequently

$$Z_p^{(p')} \equiv Z_{p'}^{(p)} \mod (T)^3,$$

hence $\Phi^0_{pp'} \equiv 0 \bmod (T)^3$. □

REMARK 6.58. As already noticed in the proof above, Φ^0_p is congruent mod $(T)^3$ to a diagonal matrix. Note, on the other hand, that Φ^0_p is *not* congruent mod $(T)^4$ to a diagonal matrix (e.g. for $p = 3$). By the way, modulo $(T)^2$, the situation is simpler; we have

$$\Phi^0_p \equiv \mathrm{diag}(pT_{11}, ..., pT_{nn}) \mod (T)^2.$$

Note that, in particular, the matrix $\mathcal{A}_p^{(0)}$ with entries $\left(\frac{1}{p}\frac{\partial \Phi^0_{pij}}{\partial T_{kl}}\right)_{|T=0}$ (cf. 4.168 and the discussion after that) satisfies

$$(6.63)\qquad \mathcal{A}_p^{(0)} = \mathrm{diag}(E_{11}, ..., E_{nn})$$

where E_{ii} is the $n \times n$ matrix with a 1 in position (i, i) and 0 in all the other positions. So the $n^2 \times n^2$ matrix $\mathcal{A}_p^{(0)}$ has zero determinant (actually it has rank n); compare this with Remark 6.63 below where, as we shall see, for general 2×2 symmetric q with entries roots of unity the 4×4 matrices $\mathcal{A}_p^{(0)}$ are invertible.

In the next Lemma we assume, again, that $A = \mathbb{Z}[1/2]$, $n = 2r \geq 2$ and $q^t = \pm q$ is split. We let $\Phi^0_{p\overline{p}'ij}$ be the entries of the matrix $\Phi^0_{p\overline{p}'}$ where $i, j = 1, ..., n = 2r$. As before, we set $\underline{i} = i + r$ for $1 \leq i \leq r$. Also we use the notation in 6.56 and 6.57. Finally the recall the curvature potential $\psi_{p\overline{p}'}$; cf. Definition 4.84.

LEMMA 6.59.

(1) $\Phi^0_{p\overline{p}'ii} \equiv \Phi^0_{p\overline{p}'\underline{ii}} \equiv \frac{1}{2}(Q_i - T_{ii}T_{\underline{ii}})$ *mod* $(T)^3$ *for* $1 \leq i \leq r$ *and* $p \neq p'$.
(2) $\Phi^0_{p\overline{p}ii} \equiv \Phi^0_{p\overline{p}\underline{ii}} \equiv \frac{p}{2}(Q_i - T_{ii}T_{\underline{ii}})$ *mod* $(T)^3$ *for* $1 \leq i \leq r$.
(3) $\Phi^0_{p\overline{p}'ij} \equiv 0$ *mod* $(T)^3$ *for* $1 \leq i, j \leq n = 2r$, $i \neq j$.
(4) $\sum_{i,j=1}^n \frac{\partial \Phi^0_{p\overline{p}'ij}}{\partial T_{ij}} \equiv 0$ *mod* $(T)^3$ *for all* p, p'.
(5) $tr(\Phi^0_{p\overline{p}'}) \equiv (\sum_{k\neq i} T_{ki}T_{\underline{ki}}) \pm (\sum_{ki} T_{\underline{k}i}T_{k\underline{i}})$ *mod* $(T)^3$ *for* $p \neq p'$. *In particular, the curvature potential satisfies* $\psi_{p\overline{p}'} \not\equiv 0$ *mod* $(T)^3$ *for* $p \neq p'$.
(6) $tr(\Phi^0_{p\overline{p}}) \equiv p(\sum_{k\neq i} T_{ki}T_{\underline{ki}}) \pm p(\sum_{ki} T_{\underline{k}i}T_{k\underline{i}})$ *mod* $(T)^3$.

PROOF. Assertions (4), (5), (6) follow from assertions (1), (2), (3). To check assertions (1), (2), (3) we use the notation in the proof of Lemma 6.57. Set

$$Z_{\overline{p}'} := \Phi_{\overline{p}'}(1 + T) = \Phi^0_{\overline{p}'}(T) + 1 = (1 + T)^{(p')}.$$

Then exactly as in the proof of Lemma 6.57 we have

$$\Phi^0_{\overline{p}'}(\Phi^0_p(T)) = Z_p^{(p')} - 1,$$

$$\Phi^0_p(\Phi^0_{\overline{p}'}(T)) = Z_{\overline{p}'}^{(p)} - 1.$$

Now $Z_p^{(p')}$ mod $(T)^3$ was computed in 6.61 and 6.62 while $Z_{\overline{p'}}^{(p)}$ mod $(T)^3$ is, of course, trivial to compute; then assertions (1), (2), (3) in our Lemma follow. □

For the next Lemma we again assume $A = \mathbb{Z}[1/2]$, $n = 2r$, $q^t = \pm q$ is split, and write

$$(6.64)\qquad T = \begin{pmatrix} A & B \\ C & D \end{pmatrix}, \quad \Phi_p^0 = \begin{pmatrix} A_p & B_p \\ C_p & D_p \end{pmatrix}, \quad \Phi_{pp'}^0 = \begin{pmatrix} A_{pp'} & B_{pp'} \\ C_{pp'} & D_{pp'} \end{pmatrix}.$$

We also denote by (A, C, D) the ideal generated in $\mathbb{Z}[1/2][[T]]$ by the entries of A, C, D. The upper sign corresponds to the case $q^t = q$ and the lower sign corresponds to the case $q^t = -q$.

LEMMA 6.60. *We have the following congruences:*

(1) $A_p \equiv pA \pm \frac{p}{2}B^tC$, $D_p \equiv pD \pm \frac{p}{2}C^tB$, $C_p \equiv 0$ *mod* $(A, C, D)^2$.

(2) $A_{pp'} \equiv C_{pp'} \equiv D_{pp'} \equiv 0$ *mod* $(A, C, D)^2$.

PROOF. One has the following congruences mod $(A, C, D)^2$:

$$(1+T)^{(p)t}q(1+T)^{(p)} \equiv \begin{pmatrix} 1+pA^t & 0 \\ B^{(p)t} & 1+pD^t \end{pmatrix} \begin{pmatrix} 0 & 1 \\ \pm 1 & 0 \end{pmatrix} \begin{pmatrix} 1+pA & B^{(p)} \\ 0 & 1+pD \end{pmatrix}$$

$$\equiv \begin{pmatrix} 0 & 1+pA^t+pD \\ \pm 1 \pm pD^t \pm pA & \star \end{pmatrix}.$$

The inverse of the latter is congruent mod $(A, C, D)^2$ to

$$(6.65)\qquad \begin{pmatrix} \star & \pm 1 \mp pA \mp pD^t \\ 1 - pD - pA^t & 0 \end{pmatrix}.$$

On the other hand we have the following congruences mod $(A, C, D)^2$:

$$(6.66)\quad ((1+T)^tq(1+T))^{(p)} \equiv \begin{pmatrix} C \pm C^t & 1 + A^t + D \pm C^tB \\ \pm 1 \pm A \pm D^t + B^tC & \star \end{pmatrix}^{(p)}$$

$$\equiv \begin{pmatrix} 0 & 1 + pA^t + pD \pm pC^tB \\ \pm 1 \pm pA \pm pD^t + pB^tC & \star \end{pmatrix}.$$

The product of 6.65 with 6.66 is congruent mod $(A, C, D)^2$ to

$$(6.67)\qquad \begin{pmatrix} 1 \pm pB^tC & \star \\ 0 & 1 \pm pC^tB \end{pmatrix}.$$

So $1 + \Phi_p^0$ is congruent to the product between

$$(1+T)^{(p)} \equiv \begin{pmatrix} 1+pA & B^{(p)} \\ 0 & 1+pD \end{pmatrix}$$

the square root of 6.67; this product is congruent to

$$\begin{pmatrix} 1+pA\pm\frac{p}{2}B^tC & \star \\ 0 & 1+pD\pm\frac{p}{2}C^tB \end{pmatrix}.$$

Assertion (1) of the Lemma follows. In particular $A_p, C_p, D_p \in (A, C, D)$. In view of the latter we get the following congruence mod $(A, C, D)^2$:

$$\Phi^0_{p'}(\Phi^0_p(T)) \equiv \begin{pmatrix} pp'A\pm\frac{pp'}{2}B^tC & \star \\ 0 & pp'D\pm\frac{pp'}{2}C^tB \end{pmatrix};$$

since the diagonal elements of the latter are "symmetric in p and p'", this ends the proof of assertion (2). □

PROOF OF THEOREM 6.42. We will use the notation in the proof of Theorem 6.43. Let u be as in the diagram 6.49; then, clearly,

$$\frac{\partial}{\partial B_{ij}} \circ u = u \circ \frac{\partial}{\partial T_{i\underline{j}}}.$$

We will first prove the following:

$$\sum_{i,j=1}^{r} \frac{\partial \Phi^0_{pp'i\underline{j}}}{\partial T_{i\underline{j}}} \neq 0. \tag{6.68}$$

In order to prove 6.68, it is enough to show that

$$\sum_{i,j=1}^{r} \frac{\partial}{\partial B_{ij}} \left(u(\Phi_{pp'i\underline{j}})\right) \neq 0. \tag{6.69}$$

Now

$$u(\Phi_{pp'ij}) = \frac{1}{pp'}\left(f_{pp'}(B_{ij}, B_{ji}) - f_{p'p}(B_{ij}, B_{ji})\right) = \frac{1}{pp'} g_{pp'}(B_{ij}, B_{ji}).$$

On the other hand one trivially checks that $g_{pp'}(v, v) = 0$ so we get

$$\sum_{i,j=1}^{r} \frac{\partial}{\partial B_{ij}}\left(u(\Phi_{pp'ij})\right) = \frac{1}{pp'}\sum_{i<j}\left(\frac{\partial}{\partial B_{ij}}(g_{pp'}(B_{ij}, B_{ji})) + \frac{\partial}{\partial B_{ji}}(g_{pp'}(B_{ji}, B_{ij}))\right).$$

But the latter is non-zero by Lemma 6.54 and 6.68 is proved. To prove the non-vanishing in our theorem note that, by Lemma 6.60, we have that for all $i, j = 1, ..., r$,

$$\frac{\partial \Phi^0_{pp'ij}}{\partial T_{ij}} \in (A, C, D), \quad \frac{\partial \Phi^0_{pp'\underline{i}\underline{j}}}{\partial T_{\underline{i}\underline{j}}} \in (A, C, D), \quad \frac{\partial \Phi^0_{pp'i\underline{j}}}{\partial T_{i\underline{j}}} \in (A, C, D),$$

so, since $u(A, C, D) = 0$,

$$u\left(\sum_{i,j=1}^{n} \frac{\partial \Phi^0_{pp'ij}}{\partial T_{ij}}\right) = u\left(\sum_{i,j=1}^{r} \frac{\partial \Phi^0_{pp'i\underline{j}}}{\partial T_{i\underline{j}}}\right) = \sum_{i,j=1}^{r} \frac{\partial}{\partial B_{ij}}\left(u(\Phi_{pp'i\underline{j}})\right).$$

But the latter is non-zero by 6.69 and our theorem follows. □

In order to prove Theorem 6.45 we need an analogue of Lemma 6.60. We assume $A = \mathbb{Z}[1/2]$, $n = 2r$, $q^t = q$ is split, and, in addition to the notation in 6.64 write

$$\Phi^0_{p\overline{p}'} = \begin{pmatrix} A_{p\overline{p}'} & B_{p\overline{p}'} \\ C_{p\overline{p}'} & D_{p\overline{p}'} \end{pmatrix}. \tag{6.70}$$

LEMMA 6.61. *We have the following congruences:*

$$A_{p\overline{p}'} \equiv \frac{pp'}{2} B^t C, \ D_{p\overline{p}'} \equiv \frac{pp'}{2} C^t B, \quad C_{p\overline{p}'} \equiv 0 \quad mod(A, C, D)^2.$$

PROOF. A computation as in Lemma 6.60. □

PROOF OF THEOREM 6.45. Assertion (2) of the theorem follows directly from Lemma 6.59. Let us prove assertion (3) of the theorem. By 4.195 we have, for $p \neq p'$,

$$\mathrm{tr}_{p\overline{p}'}(\Phi^{\mathcal{G}}_{p\overline{p}'}) \equiv \mathrm{tr}(\Phi^{\mathcal{G}}_{p\overline{p}'}) \quad \mathrm{mod} \quad pp', \quad \text{hence mod} \quad (p) + (T)^3.$$

On the other hand, by 4.192,

$$\mathrm{tr}(\Phi^{\mathcal{G}}_{p\overline{p}'}) \equiv \mathrm{tr}(\Phi^{0}_{p\overline{p}'}) \quad \mathrm{mod} \quad (pp') + (T)^3, \quad \text{hence mod} \quad (p) + (T)^3.$$

Finally, by Lemma 6.59,

$$\mathrm{tr}(\Phi^{0}_{p\overline{p}'}) \not\equiv 0 \quad \mathrm{mod} \quad (p) + (T)^3.$$

Hence

$$\psi_{p\overline{p}'} = \mathrm{tr}_{p\overline{p}'}(\Phi^{\mathcal{G}}_{p\overline{p}'}) \not\equiv 0 \quad \mathrm{mod} \quad (p) + (T)^3, \quad \text{hence mod} \quad (T)^3,$$

and assertion (3) is proved.

Let us now prove assertion (1) of the theorem. We use, again, the notation in the proof of Theorem 6.43. By Lemma 6.61 we have, for $i, j = 1, ..., r$:

$$\frac{\partial \Phi_{p\overline{p}'ij}}{\partial A_{ij}} \equiv \frac{\partial \Phi_{p\overline{p}'i\underline{j}}}{\partial C_{ij}} \equiv \frac{\partial \Phi_{p\overline{p}'\underline{i}\underline{j}}}{\partial D_{ij}} \equiv 0 \quad \mathrm{mod} \quad (A, C, D).$$

So, as in the proof of Theorem 6.42, it is enough to show that

$$\sum_{i,j=1}^{r} \frac{\partial}{\partial B_{ij}} (u(\Phi^0_{p\overline{p}'i\underline{j}})) \neq 0. \tag{6.71}$$

Now, a computation as in the proof of Theorem 6.43 yields

$$\begin{aligned} u(\Phi^0_{p\overline{p}'}) &= \Phi^0_{p\overline{p}'} \begin{pmatrix} 0 & B \\ 0 & 0 \end{pmatrix} \\ &= \tfrac{1}{2}\left((B + B^t)^{(p)} + B^{(p)} - B^{(p)t}\right)^{(p')} \\ &\quad -\tfrac{1}{2}\left((B^{(p')} + B^{(p')t})^{(p)} + B^{(pp')} - B^{(pp')t}\right). \end{aligned}$$

Therefore, for each $i \neq j$, $u(\Phi^0_{p\overline{p}'i\underline{j}})$ belongs to the square of the ideal (B_{ij}, B_{ji}) generated by the 2 variables B_{ij} and B_{ji}. We get that, for each pair $i \neq j$,

$$\frac{\partial}{\partial B_{ij}} (u(\Phi^0_{p\overline{p}'i\underline{j}})) \in (B_{ij}, B_{ji}). \tag{6.72}$$

On the other hand a direct computation yields:

$$\frac{\partial}{\partial B_{ii}} (u(\Phi^0_{p\overline{p}'i\underline{i}})) = (2^{p'(p-1)} - 2^{p-1}) pp' B_{i\underline{i}}^{pp'-1}. \tag{6.73}$$

Equations 6.72 and 6.73 clearly imply

$$\sum_{i,j=1}^{r} \frac{\partial}{\partial B_{ij}}(u(\Phi^0_{p\overline{p}'i\underline{j}})) \in (2^{p'(p-1)} - 2^{p-1})pp' \left(\sum_{i=1}^{r} B_{i\underline{i}}^{pp'-1} \right) + \sum_{i\neq j}(B_{ij}, B_{ji})$$

and 6.71 follows. □

PROOF OF THEOREM 6.48. This follows by a direct computation using formula 4.36. Indeed, if we set

$$A = x^{(p)t}\phi(q)x^{(p)}, \quad B = (x^t qx)^{(p)},$$

then we have

$$\begin{aligned} A &= (1+T^t)^{(p)}q^{(p)}(1+T)^{(p)} \\ &\equiv ((1+pT_{ii}+pT_{jj})q_{ij}^p) \mod (T)^2, \end{aligned}$$

and

$$\begin{aligned} B &= ((1+T^t)q(1+T))^{(p)} \\ &\equiv (q_{ij}^p + pq_{ij}^{p-1}\{\textstyle\sum_m q_{im}T_{mj} + \sum_m q_{jm}T_{mi}\}) \mod (T)^2. \end{aligned}$$

Then, by formula 4.36,

$$\phi_p(T) = (1+T)^{(p)} \cdot \{A^{-1}B\}_p^{1/2} - 1$$

can be readily computed mod $(T)^2$ and leads to the desired result. □

We next prove Theorem 6.50. We may assume, in that theorem, that $A = \mathbb{Z}[1/2N, \zeta_N]$, $N \geq 1$. Consider, more generally, a matrix

$$q = \begin{pmatrix} \zeta_{11} & \zeta_{12} \\ \zeta_{21} & \zeta_{22} \end{pmatrix} \tag{6.74}$$

where ζ_{ij} are powers of ζ_N, $\zeta_{12} = \zeta_{21}$, and $\det(q) \neq 0$ (so $q \in GL_2(A)$). Let (ϕ_p) be the Frobenius lifts corresponding to the real Chern connection attached to q, let $T = \begin{pmatrix} A & B \\ C & D \end{pmatrix}$ be a 2×2 matrix of indeterminates, write

$$\phi_p(T) = \begin{pmatrix} A_p & B_p \\ C_p & D_p \end{pmatrix},$$

and write

$$\begin{pmatrix} A_p \\ C_p \\ B_p \\ D_p \end{pmatrix} \equiv p\mathcal{A}_p^{(0)} \cdot \begin{pmatrix} A \\ C \\ B \\ D \end{pmatrix} \mod (T)^2,$$

with $\mathcal{A}_p^{(0)}$ a 4×4 matrix with coefficients in $\mathbb{Z}[1/2N, \zeta_N]$. Cf. 4.168 and the discussion after that. It is convenient, in what follows, to set

$$\Delta_p := \zeta_{11}^p \zeta_{22}^p - \zeta_{12}^{2p}$$

and to let

$$w_{(24)} = [e_1, e_4, e_3, e_2]$$

be the permutation matrix obtained from the identity matrix by applying the transposition (24).

LEMMA 6.62. *We have:*

$$w_{(24)}\mathcal{A}_p^{(0)}w_{(24)} = \begin{pmatrix} 1_2 & \mathcal{N}_p \\ 0 & \mathcal{M}_p \end{pmatrix},$$

where 1_2 *is the identity* 2×2 *matrix and* $\mathcal{M}_p, \mathcal{N}_p$ *are the* 2×2 *matrices*

$$\mathcal{M}_p = \frac{1}{2\Delta_p}\begin{pmatrix} \zeta_{12}^{p-1}\zeta_{22}^{p}\zeta_{11} - 2\zeta_{12}^{p+1}\zeta_{22}^{p-1} & \zeta_{12}^{p-1}\zeta_{22}^{p+1} \\ \zeta_{12}^{p-1}\zeta_{11}^{p+1} & \zeta_{12}^{p-1}\zeta_{11}^{p}\zeta_{22} - 2\zeta_{12}^{p+1}\zeta_{11}^{p-1} \end{pmatrix},$$

$$\mathcal{N}_p = \frac{1}{2\Delta_p}\begin{pmatrix} -\zeta_{12}^{2p-1}\zeta_{11} & 2\zeta_{11}^{p-1}\zeta_{22}^{p}\zeta_{12} - \zeta_{12}^{2p-1}\zeta_{22} \\ 2\zeta_{11}^{p}\zeta_{22}^{p-1}\zeta_{12} - \zeta_{12}^{2p-1}\zeta_{11} & -\zeta_{12}^{2p-1}\zeta_{22} \end{pmatrix}.$$

PROOF. Follows from the formula in Theorem 6.48. □

REMARK 6.63. In particular, in the notation of the Lemma above,

$$\det(\mathcal{A}_p^{(0)}) = \det(\mathcal{M}_p) = -\zeta_{12}^{2p}\zeta_{11}^{p-1}\zeta_{22}^{p-1}\frac{\zeta_{11}\zeta_{22} - \zeta_{12}^2}{(\zeta_{11}^p\zeta_{22}^p - \zeta_{12}^{2p})^2} \tag{6.75}$$

is invertible in $\mathbb{Z}[1/2N, \zeta_N]$. Compare this with the fact that for q split and $n \geq 4$ even the corresponding $n^2 \times n^2$ matrix $\mathcal{A}_p^{(0)}$ has zero determinant (actually has rank n); cf. Remark 6.58.

REMARK 6.64. Note that the curvature $(\varphi_{pp'})$ of the real Chern connection attached to q in 6.74 satisfies

$$\varphi_{pp'}\begin{pmatrix} A \\ C \\ B \\ D \end{pmatrix} \equiv \mathcal{F}_{pp'}^{(0)} \cdot \begin{pmatrix} A \\ C \\ B \\ D \end{pmatrix} \quad \text{mod } (T)^2,$$

where, by 4.174,

$$\begin{aligned} \mathcal{F}_{pp'}^{(0)} &:= \phi_p(\mathcal{A}_{p'}^{(0)}) \cdot \mathcal{A}_p^{(0)} - \phi_{p'}(\mathcal{A}_p^{(0)}) \cdot \mathcal{A}_{p'}^{(0)} \\ &= w_{(24)} \cdot \begin{pmatrix} 0 & \mathcal{N}_{pp'} \\ 0 & \mathcal{M}_{pp'} \end{pmatrix} \cdot w_{(24)}, \\ \mathcal{M}_{pp'} &= \phi_p(\mathcal{M}_{p'}) \cdot \mathcal{M}_p - \phi_{p'}(\mathcal{M}_p) \cdot \mathcal{M}_{p'}, \\ \mathcal{N}_{pp'} &= \mathcal{N}_p - \mathcal{N}_{p'} + \phi_p(\mathcal{N}_{p'}) \cdot \mathcal{M}_p - \phi_{p'}(\mathcal{N}_p) \cdot \mathcal{M}_{p'}. \end{aligned} \tag{6.76}$$

Similarly, since

$$w_{(24)}\mathcal{A}_{\overline{p}}^{(0)}w_{(24)} = \begin{pmatrix} 1_2 & 0 \\ 0 & 0 \end{pmatrix},$$

we get that the $(1,1)$-curvature $(\varphi_{p\overline{p}'})$ of the real Chern connection attached to q in 6.74 satisfies

$$\varphi_{p\overline{p}'}\begin{pmatrix} A \\ C \\ B \\ D \end{pmatrix} \equiv \mathcal{F}_{p\overline{p}'}^{(0)} \cdot \begin{pmatrix} A \\ C \\ B \\ D \end{pmatrix} \quad \text{mod } (T)^2, \tag{6.77}$$

for $p \neq p'$ and

$$\varphi_{p\bar{p}} \begin{pmatrix} A \\ C \\ B \\ D \end{pmatrix} \equiv p \cdot \mathcal{F}^{(0)}_{p\bar{p}} \cdot \begin{pmatrix} A \\ C \\ B \\ D \end{pmatrix} \mod (T)^2, \tag{6.78}$$

where, by 4.175,

$$\begin{aligned} \mathcal{F}^{(0)}_{p\bar{p}'} &:= \phi_p(\mathcal{A}^{(0)}_{\bar{p}'}) \cdot \mathcal{A}^{(0)}_p - \phi_{p'}(\mathcal{A}^{(0)}_p) \cdot \mathcal{A}^{(0)}_{\bar{p}'} \\ &= w_{(24)} \cdot \begin{pmatrix} 0 & \mathcal{N}_p \\ 0 & 0 \end{pmatrix} \cdot w_{(24)}, \end{aligned} \tag{6.79}$$

Proof of Theorem 6.50. A direct computation using Lemma 6.62 and formula 6.76 shows that, for $\zeta_{11} = \zeta_{22} = 1$ and $\zeta_{12} = \zeta$ we have that

$$\operatorname{tr}(\mathcal{F}^{(0)}_{pp'}) = \operatorname{tr}(\mathcal{M}_{pp'}) = \frac{\zeta^{pp'+1}(\zeta^{2p} - \zeta^{2p'})}{(1-\zeta^{2pp'})(1-\zeta^{2p})(1-\zeta^{2p'})},$$

which ends the proof of assertion (1) in Theorem 6.50 in case $\zeta_{11} = 1$. The general case of assertion (1) follows from this case because the real Chern connection attached to q does not change if we replace q by $\zeta_{11}^{-1} q$. Assertions (2) and (3) follow similarly from 6.77, 6.78, and 6.79. □

6.3. Convergence estimates

Assume, in this section that $A = \mathbb{Z}[1/M, \zeta_N]$, M an even supernatural number, and view A as embedded into the complex field $\mathbb{C}$ equipped with the natural complex absolute value $|\ |$. Our aim in what follows is to address the problem of convergence for the mean and scalar curvatures. We need a preparation.

Let T be a d-tuple of indeterminates T_i, $1 \leq i \leq d$, and consider the power series ring $\mathbb{C}[[T]]$ in these d variables. If g is a series in this ring we write $g = \sum c_\alpha(g) T^\alpha$ where $\alpha = (\alpha_1, ..., \alpha_d)$ are multi-indices, $T^\alpha := T_1^{\alpha_1} ... T_d^{\alpha_d}$. We write $|\alpha| = \alpha_1 + ... + \alpha_d$. Let $\mathcal{V}$ be a set of prime numbers not dividing MN.

Definition 6.65. Let $f = (f_p)$ be a family of elements of $\mathbb{C}[[T]]$ indexed by $\mathcal{V}$ and let e be a natural number. We say that f has *slope* e if for any α there exists a positive real number $m_{f,\alpha}$ such that for all $p \in \mathcal{V}$ we have

$$|c_\alpha(f_p)| \leq m_{f,\alpha} \cdot p^{e|\alpha|}.$$

Lemma 6.66.

(1) *Let* $f = (f_p)$ *and* $g = (g_p)$ *have slope* e. *Then* $f + g := (f_p + g_p)$ *and* $fg := (f_p g_g)$ *have slope* e.
(2) *Let* $f = (f_p)$ *have slope* e *and let* $g_1 = (g_{1p}), ..., g_d = (g_{dp})$ *have slope* e'. *Assume* $g_{ip}(0) = 0$ *for all* i *and all* p. *Then* $f(g_1, ..., g_d) := (f_p(g_{1p}, ..., g_{dp}))$ *has slope* $e + e'$.
(3) *Let* $g = (g_p)$ *have slope* e *and assume* $|c_0(g_p)| \leq 1$ *for all* p. *Then* $g^p := (g_p^p)$ *has slope* $e + 1$.

PROOF. Assertion (1) is a trivial exercise. Assertion (2) follows from assertion (1) via the following computation:

$$\begin{aligned} |c_\alpha(f_p(g_{1p},...,g_{dp}))| &= |c_\alpha(\sum_\beta c_\beta(f_p) g_{1p}^{\beta_1}...g_{dp}^{\beta_d})| \\ &\leq \sum_{|\beta|\leq|\alpha|} |c_\beta(f_p)||c_\alpha(g_{1p}^{\beta_1}...g_{dp}^{\beta_d})| \\ &\leq \sum_{|\beta|\leq|\alpha|} m_{f,\beta}\cdot p^{e|\beta|}\cdot m_{g_{1p}^{\beta_1}...g_{dp}^{\beta_d},\alpha}\cdot p^{e'|\alpha|}, \end{aligned}$$

and from the fact that for each α there are only finitely many β's with $|\beta| \leq |\alpha|$.

Assertion (3) follows from assertion (2) and from the fact that if (λ_p) is a family of complex numbers with $|\lambda_p| \leq 1$ for all p then the family $((\lambda_p + T_1)^p)$ has slope 1. □

Let us assume now that $d = n^2$ and assume the variables $T = (T_{ij})$ are arranged as an $n \times n$ matrix. Assume $F = (F_p)$ is a family, indexed by $\mathcal{V}$, of $n \times n$ matrices $F_p = (F_{pij})$ with entries in $\mathbb{C}[[T]]$. We say that F has *slope* e if for any i and j the family (F_{pij}), indexed by p, has slope e.

LEMMA 6.67.

(1) *If $F = (F_p)$ and $G = (G_p)$ have slope e then $F + G := (F_p + G_p)$ and $FG := (F_pG_p)$ have slope e.*
(2) *If $F = (F_p)$ has slope e, $G = (G_p)$ has slope e', and $G_{pij}(0) = 0$ for all p, i, j then $F(G) := (F_p(G_p))$ has slope $e + e'$.*
(3) *If $F = (F_p)$ has slope e and $|c_0(F_{pij})| \leq 1$ for all p, i, j then $F^{(p)} := (F_p^{(p)})$ has slope $e + 1$.*
(4) *If $F = (F_p)$ has slope e and $F_p(0) = 1$ (the identity) for all p then, for any rational number θ, $F^\theta := (F_p^\theta)$ has slope e. Here*

$$F_p^\theta := \sum_k \binom{\theta}{k} (F_p - 1)^k.$$

PROOF. Assertions (1), (2), (3) follows directly from assertions (1), (2), (3) in Lemma 6.66. Assertion (4) follows directly from assertion (1). □

For the next Theorem recall the definition of mean and scalar $(1,1)$-curvature; cf. Definition 4.86.

THEOREM 6.68. *Let $A = \mathbb{Z}[1/M, \zeta_N] \subset \mathbb{C}$, let $\mathcal{V}$ be a set of primes not dividing MN, and let $q \in GL_n(A)$, $q^t = \pm q$, be such that each entry of q is either 0 or a power of ζ_N. Let $\delta = (\delta_p)$ be the real Chern connection on GL_n attached to q, with attached Frobenius lifts (ϕ_p). (Hence, by Theorem 6.9, δ is global along 1, so, in particular, it has a $(1,1)$-curvature $(\varphi_{p\overline{p'}})$). Consider the matrices $\Phi_p^0 = \phi_p(T)$ and $\Phi_{p\overline{p}}^0 = \varphi_{p\overline{p}}(T)$. Then:*

(1) *The family (Φ_p^0) has slope 1.*
(2) *The family $(p\Phi_{p\overline{p}}^0)$ has slope 2.*
(3) *For each multi-index α, the coefficient of T^α in the mean curvature $K_{ij}(s)$ is a Dirichlet series converging for $Re(s) > 2|\alpha|$.*
(4) *For each multi-index α, the coefficient of T^α in the scalar $(1,1)$-curvature $\sigma(s)$ is a Dirichlet series converging for $Re(s) > 2|\alpha| + 1$.*

PROOF. Clearly assertions (3) and (4) follow from assertion (2). On the other hand $\Phi^0_{\overline{p}}(T) = (1+T)^p - 1$ clearly has slope 1 and we have

$$p\Phi^0_{p\overline{p}} = \Phi^{0(\phi_p)}_{\overline{p}}(\Phi^0_p) - \Phi^{0(\phi_{\overline{p}})}_p(\Phi^0_{\overline{p}});$$

so, by assertion (3) in Lemma 6.67, assertion (2) follows from assertion (1). Let us prove assertion (1). The "diagonal" family of matrices $((1+T^t)q(1+T))$ has slope 0; since q has entries with absolute value ≤ 1, by assertion (1) in Lemma 6.67, the family of matrices

$$(G_p) := (((1+T^t)q(1+T))^{(p)})$$

has slope 1. Now note that the families $(\phi_p(q))$ and $(\phi_p(q)^{-1})$ contain only finitely many matrices so these families have slope 0. So, by assertions (1) and (3) in Lemma 6.67, the family

$$(F_p) := ((1+T^t)^{(p)}\phi_p(q)(1+T)^{(p)})$$

has slope 1. Hence the family $(\phi_p(q)^{-1}F_p)$ has slope 1. Since each member of the latter family evaluated at $T = 0$ is the identity, by assertion (4) of Lemma 6.67, the family $((\phi(q)^{-1}F_p)^{-1})$ has slope 1. Hence, by assertion (1) of the same Lemma the family $(F_p^{-1}) = ((\phi(q)^{-1}F_p)^{-1}\phi_p(q)^{-1})$ has slope 1. So the family $(F_p^{-1}G_p)$ has slope 1. Since the members of the latter family evaluated at $T = 0$ equal the identity, by assertions (1) and (4) of the same Lemma again, the family $\Phi^0_p = (1+T)^{(p)}\{F_p^{-1}G_p\}_p^{1/2} - 1$ has slope 1 and we are done. □

6.4. The cases $n = 1$ and $m = 1$

There are two extreme cases of our theory which we would like to take a closer look at in this section: one is that of GL_n for $n = 1$; the other is that of one prime, $m := |\mathcal{V}| = 1$. Throughout this section we let A be a δ-ring of global type, $A = \mathcal{O}_K[1/M]$.

Start with some $q \in GL_1(A)$, and let $\delta = (\delta_p)$ be the real Chern connection on GL_1 attached to q. Then, by 6.29,

$$\Phi_p(x) = \gamma_p \cdot x^p, \quad \gamma_p = (\phi_p(q)^{-1}q^p)_p^{1/2} \in A^{\widehat{p}}.$$

Note that if γ_p is a square in A for all p in $\mathcal{V}$ then the real Chern connection δ attached to q is global on $\mathfrak{gl}_1$ because $\Phi_p(x) \in A[x]$. Cf. section 4.8.4. In particular δ has a well defined curvature and $(1,1)$-curvature,

$$\varphi_{pp'}, \varphi_{p\overline{p}'} : A[x] \to A[x].$$

An example of this situation is when $A = \mathbb{Z}[1/M]$ for some M, cf. 6.17; in this case we have the following expression for γ_p involving the Legendre symbol:

$$\gamma_p = \left(\frac{q}{p}\right) \cdot q^{(p-1)/2}.$$

THEOREM 6.69. *Let $q \in GL_1(A) = A^\times$, assume $\gamma_p := (\phi_p(q)^{-1}q^p)_p^{1/2} \in A$ for all p in $\mathcal{V}$. Let δ be the real Chern connection on GL_1 attached to q. Then the following hold:*

(1) *The curvature $(\varphi_{pp'})$ of δ is given by*

$$\varphi_{pp'}(x) = \frac{1}{pp'}\{p(\delta_p\gamma_{p'})\gamma_p^{p'} - p'(\delta_{p'}\gamma_p)\gamma_{p'}^p\}x^{pp'}.$$

(2) *Assume* $A = \mathbb{Z}[1/M]$. *Then* δ *is flat, so the curvature* $\varphi_{pp'} = 0$ *for all* p, p'.

(3) *Let* $\overline{\delta}$ *be the adelic connection on* GL_1 *(global on* $\mathfrak{gl}_1$*) with* $\delta_{\overline{p}}x = 0$ *and let* $(\varphi_{p\overline{p}'})$ *be the* $(1,1)$*-curvature of* δ *with respect to* $\overline{\delta}$. *Then*

$$\varphi_{p\overline{p}}(x) = -\delta_{\overline{p}}(\gamma_p) \cdot x^{p^2}.$$

PROOF. Assertion (2) follows from assertion (4) of Theorem 6.18; indeed, since $n = 1$, we have $\text{jor}(\gamma_p) = \gamma_p \in A^\times$. Assertions (1) and (3) follow by direct computations. □

REMARK 6.70. Under the assumptions of Theorem 6.69 one can introduce the *scalar* $(1,1)$*-curvature* of δ by the following formula analogous to 4.202:

$$\sigma(s) := -\sum_{p\in\mathcal{V}} p^{-s} \cdot \delta_{\overline{p}}(\gamma_p)(1+T)^{p^2} \in A[[\mathbb{N}^{-s}]][[T]].$$

In the special case when $A = \mathbb{Z}[1/M]$ and $q \in A^\times$ satisfies $|q| < 1$ (real absolute value) the value of $\sigma(s)$ at $T = 0$ is a Dirichlet series

$$\sum_{p\in\mathcal{V}} \left(\frac{q}{p}\right) \cdot (q^{p(p-1)/2} - q^{(p-1)/2}) \cdot p^{-s-1} \in A\{\{\mathbb{N}^{-s}\}\}$$

that is convergent on $Re(s) > 0$.

Assume now that $\mathcal{V}$ consists of one odd prime p only and let δ be an adelic (i.e., p-adic) connection on $G = GL_n$. Then δ defines a global connection on $\mathcal{O}(G)^{\widehat{p}}$. Clearly the curvature $\varphi_{pp} = 0$. Now let $\overline{\delta}$ be another adelic connection on G. Then one can consider the $(1,1)$-curvature $\varphi_{p\overline{p}} : \mathcal{O}(G)^{\widehat{p}} \to \mathcal{O}(G)^{\widehat{p}}$. Natural gauge and translation actions can be then defined even though, in this case, δ and $\overline{\delta}$ are not assumed to fix the identity; this is left to the reader. Finally one can consider the Chern $(1,1)$-form; the latter is also non-trivial in general, even in the case $n = 1$ as shown in Theorem 6.71 below.

THEOREM 6.71. *Let* $q \in GL_n(A)$, $q^t = \pm q$. *Let* $\delta = (\delta_p)$ *be the real Chern connection attached to* q *and let* $\overline{\delta} = (\delta_{\overline{p}})$ *be the adelic connection with* $\delta_{\overline{p}}x = 0$*; here both adelic connections are indexed by the set* $\{p\}$ *with one prime* p. *Let* $\varphi_{p\overline{p}}$ *and* $\rho_{p\overline{p}}$ *be the* $(1,1)$*-curvature and Chern* $(1,1)$*-form of* δ *with respect to* $\overline{\delta}$ *and let* $\Phi_{p\overline{p}} := \varphi_{p\overline{p}}(x)$. *Then:*

(1) $\Phi_{p\overline{p}}(1) = -\delta_{\overline{p}}\left(\left(1 + p(q^{(p)})^{-1}\delta_p q\right)^{-1/2}\right)$.

(2) *If* $n = 1$ *then* $\Phi_{p\overline{p}} = \Phi_{p\overline{p}}(1) \cdot x^{p^2}$.

(3) $\Phi_{p\overline{p}}(1) = 0$ *if and only if each entry of* q *is either* 0 *or a root of unity. Moreover, if* $n = 1$, *then* $\varphi_{p\overline{p}} = 0$ *if and only* $\rho_{p\overline{p}} = 0$ *if and only if* q *is a root of unity.*

PROOF. The formulae in assertions (1) and (2) follows from a direct computation. Assertion (3) follows from assertions (1) and (2) plus the following two facts: (i) if $\delta_p a = 0$ or $\delta_{\overline{p}} a = 0$ for some $a \in A$ then a is a root of unity or 0; and (ii) if $1 + pa$ is a root of unity or 0 for some $a \in A$ then $a = 0$. □

6.5. Non-vanishing/vanishing of curvature via correspondences

As mentioned before there are two general approaches, in this book, towards curvature: one is via analytic continuation between primes while the other is via algebraization by correspondences. The first approach can only work for adelic connections that are either global along a retract (in application, a point, or even a torus D_μ) or are global on an equivariant compactification (in applications, $\mathfrak{gl}_n$). The second approach has the advantage of not requiring our adelic connections to be global along a subscheme or on a bigger scheme; the price to pay is that instead of endomorphisms we have to consider correspondences. We examine, in this section, the curvature of real Chern connections that can be derived from the correspondence method. A similar (and much simpler) analysis can be made for the complex Chern connection; we leave this to the reader.

For the basic concepts of *correspondence structures* and their *curvatures* we refer to Definitions 3.99 and 3.100. Also, in what follows we assume A is a δ-ring of global type: $A = \mathcal{O}_K[1/M]$ cf. Definition 3.25. We let $G = GL_n = Spec\ B$, $B = A[x, \det(x)^{-1}]$, and we denote by $E = K(x)$ the fraction field of G. So correspondence structures for adelic connections on G are correspondences on E. Also we let E^+ be the fixed field of $x \mapsto -x$. We also recall that for $q \in GL_n(A)$ with $q^t = \pm q$ we have an attached *real Chern connection* $\delta = (\delta_p)$; if (ϕ_p) are the attached Frobenius lifts on $\mathcal{O}(G)^{\widehat{p}} = A[x, \det(x)^{-1}]^{\widehat{p}}$ then ϕ_p is defined by $x \mapsto \phi_p(x) = \Phi_p$,

$$\Phi_p(x) = x^{(p)}\{(x^{(p)t}\phi_p(q)x^{(p)})^{-1}(x^t q x)^{(p)}\}_p^{1/2}. \tag{6.80}$$

Recall that a matrix $q \in GL_{2r}(\mathbb{Z})$ is *split* if it is one of the following:

$$\begin{pmatrix} 0 & 1_r \\ -1_r & 0 \end{pmatrix}, \quad \begin{pmatrix} 0 & 1_r \\ 1_r & 0 \end{pmatrix}, \tag{6.81}$$

where 1_r is the $r \times r$ identity matrix.

Here are our main results on correspondence structures.

THEOREM 6.72. *For any $q \in GL_n(A)$ with $q^t = \pm q$ the real Chern connection on GL_n attached to q admits a correspondence structure.*

THEOREM 6.73. *If $q \in GL_2(A)$ is split, with $q^t = -q$, then the real Chern connection on GL_2 attached to q admits a correspondence structure (Γ_p) with the following properties:*

1. *Each Γ_p is irreducible, is induced from E^+, and has left degree 2 and right degree $2p^4$;*
2. *(Γ_p) has upper $*$-curvature satisfying $\varphi^*_{pp'} = 0$ and upper $*$-$(1,1)$-curvature satisfying $\varphi^*_{p\overline{p}'} \neq 0$, hence the lower $*$-$(1,1)$-curvature $\varphi_{p\overline{p}'*} \neq 0$, for all p, p';*
3. *For all p, p', Γ_p and $\Gamma_{p'}$ partially commute, but do not commute.*

THEOREM 6.74. *If n is even and $q \in GL_n(A)$ is split, with $q^t = q$ then the real Chern connection on GL_n attached to q admits a correspondence structure (Γ_p) with the following properties:*

1. *Each Γ_p has left degree 2^n;*
2. *For $n = 2$, Γ_p is irreducible, it is partially induced from E^+, and it is not induced from E^+;*

(3) *For* $n = 2$, (Γ_p) *has upper* $*$*-*$(1,1)$*-curvature satisfying* $\varphi^*_{p\overline{p}'} \neq 0$, *hence the lower* $*$*-*$(1,1)$*-curvature* $\varphi_{p\overline{p}'*} \neq 0$, *for all* p, p'.

To prove the above theorems we need a preparation. Let us consider an $n \times n$ matrix y of indeterminates. Let B be a ring where 2 is invertible and let $\mathfrak{g}(B)$ be the B-module of all $n \times n$ matrices with entries in B, viewed as an associative ring. Consider the B-linear map

$$\mathfrak{g}(B) \to \mathfrak{g}(B), \quad z \mapsto \frac{zb + bz}{2}$$

(which is the Jordan multiplication by b). Then recall from Definition 6.13 that we denoted by $\text{jor}(b) \in B$ the determinant of this map.

In what follows, for a matrix M with entries in a ring we denote by (M) the ideal in that ring generated by the entries of M. Also, if B is a field, then dim_B below means dimension of a B-vector space. Finally let y be a $n \times n$ matrix of indeterminates. In particular we have the polynomial $\text{jor}(y) \in \mathbb{Z}[1/2][y]$.

LEMMA 6.75. *Let* B *be a ring in which* 2 *is invertible, let* $b \in GL_n(B)$, *and let*

$$C = B[y, jor(y)^{-1}]/(y^2 - b).$$

Then the following hold:

(1) C *is étale over* B.
(2) *If* B *is a field and* b *is a scalar matrix then* $dim_B C = 2$.
(3) *If* B *is a field and* b *has distinct eigenvalues then* $dim_B C = 2^n$.
(4) *If* B *is a field and* 1 *is not an eigenvalue of* b *then* $\det(y + 1) \in C^\times$.

PROOF. Assertion (4) is trivial. To check assertion (1) consider a commutative diagram of rings

$$\begin{array}{ccc} B & \xrightarrow{\pi} & C \\ v \downarrow & & \downarrow u \\ D & \xrightarrow{\rho} & D/I \end{array}$$

where π is the natural map and

$$I^2 = 0, \quad v(b) = \beta, \quad u(y) = \rho(\gamma), \quad \gamma \in \mathfrak{g}(D).$$

So $\gamma^2 = \beta + \epsilon$ with ϵ a matrix with entries in I. Since $\rho(\text{jor}(\gamma))$ is invertible in D/I and I is nilpotent it follows that $\text{jor}(\gamma)$ is invertible in D. Now $\text{jor}(\gamma)$ is a power of 2 times the determinant of the linear system

$$z\gamma + \gamma z = -\epsilon$$

with unknowns the entries of a matrix z hence by Cramer's rule this system has a unique solution $\gamma_1 \in \mathfrak{g}(D)$ and this solution is in $\mathfrak{g}(I)$. Note that $(\gamma + \gamma_1)^2 = \beta$; so there is a unique homomorphism $U : C \to D$ such that $U \circ \pi = v$ and $\rho \circ U = u$; it is given by $U(y) = \gamma + \gamma_1$.

To check assertion (2) say $b = \lambda \cdot 1$ with $\lambda \in B^\times$. Since, by assertion (1), C is a product of finite field extensions of B it is enough to show that there are exactly 2 B-algebra homomorphisms $C \to B^a$, where B^a is a fixed algebraic closure of B. Consider such a homomorphism, $y \mapsto \gamma$, where γ is a matrix with coefficients in B^a. So

$$\gamma^2 = \lambda \cdot 1, \quad \text{jor}(\gamma) \neq 0.$$

So γ is diagonalizable with eigenvalues $\pm\sqrt{\lambda} \in B^a$. It is enough to show that γ is a scalar matrix. Let us assume the contrary and derive a contradiction. Indeed, in this case, $\gamma = u\eta u^{-1}$, $u \in GL_n(B^a)$,

$$\eta = \sqrt{\lambda} \cdot \begin{pmatrix} 1 & 0 & 0 \\ 0 & -1 & 0 \\ 0 & 0 & \eta_{n-2} \end{pmatrix},$$

where η_{n-2} is a diagonal $(n-2) \times (n-2)$ matrix (with ± 1 on the diagonal). Since $\text{jor}(\gamma) \neq 0$ the map

$$\mathfrak{g}(B^a) \to \mathfrak{g}(B^a), \quad z \mapsto z\gamma + \gamma z = zu\eta u^{-1} + u\eta u^{-1}z,$$

is bijective; hence the map

$$z \mapsto u^{-1}(z\gamma + \gamma z)u = u^{-1}zu\eta + \eta u^{-1}zu$$

is bijective; hence the map

$$z \mapsto z\eta + \eta z$$

is bijective. But if one takes

$$z_0 = \begin{pmatrix} 0 & 1 & 0 \\ 1 & 0 & 0 \\ 0 & 0 & 0_{n-2} \end{pmatrix},$$

one gets $z_0\eta + \eta z_0 = 0$, a contradiction, which ends the proof of assertion (2).

To check assertion (3) let $\lambda_1, ..., \lambda_n \in (B^a)^\times$ be the distinct eigenvalues of b, let λ be the diagonal matrix with these eigenvalues on the diagonal, and write $b = u\lambda u^{-1}$, with $u \in GL_n(B^a)$. Again, it is enough to show that there are exactly 2^n B-algebra homomorphisms $C \to B^a$. Consider such a homomorphism, $y \mapsto \gamma$, where γ is a matrix with coefficients in B^a, and $\gamma^2 = b^2$. Hence γ has distinct eigenvalues, so is diagonalizable, so $\gamma = u\eta u^{-1}$ where

$$\eta = \text{diag}(\epsilon_1\sqrt{\lambda_1}, ..., \epsilon_n\sqrt{\lambda_n}), \quad \epsilon_i \in \{\pm 1\}$$

and $\sqrt{\lambda_i}$ is a fixed square root to λ_i. So there are at most 2^n homomorphisms $C \to B^a$. To conclude it is enough to show that for any choice of $\epsilon_1, ..., \epsilon_n \in \{\pm 1\}$, and for

$$\gamma = u\eta u^{-1}, \quad \eta = \text{diag}(\epsilon_1\sqrt{\lambda_1}, ..., \epsilon_n\sqrt{\lambda_n}),$$

we have $\text{jor}(\gamma) \neq 0$. Assume $\text{jor}(\gamma) = 0$ for some choice of the ϵ_is and seek a contradiction. From $\text{jor}(\gamma) = 0$ we know there exists $0 \neq z \in \mathfrak{g}(B^a)$ such that

$$\gamma z + z\gamma = 0.$$

Exactly as in the proof of assertion (2) one can find $0 \neq w \in \mathfrak{g}(B^a)$ such that

$$\eta w + w\eta = 0,$$

hence

$$\eta w\eta^{-1} = -w,$$

hence

$$\frac{\epsilon_i\sqrt{\lambda_i}}{\epsilon_j\sqrt{\lambda_j}} w_{ij} = -w_{ij}$$

for all i, j. Choose i, j such that $w_{ij} \neq 0$. Then $i \neq j$ and $\sqrt{\lambda_i} = \pm\sqrt{\lambda_j}$, hence $\lambda_i = \lambda_j$, a contradiction. Assertion (3) is proved. □

6.5.1. The case $q \in GL_n(A)$, $q^t = \pm q$. Let us fix such a q. We start by constructing a correspondence structure (Γ_p) for the real Chern connection attached to q. Then we will show, in the present subsection, that (Γ_p) possesses the required property in Theorem 6.72. In later subsections we will show that the same (Γ_p) satisfies the required properties in Theorem 6.73 and 6.74.

We begin by constructing (Γ_p).

Recall that we set $G = GL_n = Spec\ B$, $B = A[x, \det(x)^{-1}]$. Let y be a new $n \times n$ matrix of indeterminates. Define

$$\begin{aligned} g_p(x) &= \det(x) \cdot \det(x^{(p)}) \cdot \det((x^t qx)^{(p)}) \in A[x], \\ f_p(x) &= (x^{(p)t}\phi_p(q)x^{(p)})^{-1}(x^t qx)^{(p)} \in A[x, g_p(x)^{-1}], \end{aligned}$$

and consider the ring

$$C_p = A[x, g_p(x)^{-1}, y, \det(y+1)^{-1}, \mathrm{jor}(y)^{-1}]/(y^2 - f_p(x)),$$

and its spectrum,

$$Y_{p/G} = Spec\ C_p.$$

Note that for the identity matrix 1 we have $\mathrm{jor}(1) = 1$. Consider the correspondence

$$\Gamma_{p/G} = (Y_{p/G}, \pi_{p/G}, \varphi_{p/G})$$

on G where $\pi_{p/G}$ is defined by the natural map (still denoted by)

$$\pi_{p/G} : B \to C_p, \quad \pi_{p/G}(x) = x,$$

and $\varphi_{p/G}$ is defined by the map (still denoted by)

$$\varphi_{p/G} : B \to C_p, \quad \varphi_{p/G}(x) = x^{(p)}y.$$

Finally define the correspondence $\Gamma_p := \Gamma_{p/G} \otimes E$ on E.

LEMMA 6.76. *$\pi_{p/G}$ is étale.*

PROOF. This follows from Lemma 6.75. □

LEMMA 6.77. *$\pi_{p/G} \otimes \mathbb{F}_p : B \otimes \mathbb{F}_p \to C_p \otimes \mathbb{F}_p$ is an isomorphism whose inverse sends y into the identity matrix, 1.*

PROOF. We have

$$C_p \otimes \mathbb{F}_p = (A/pA)[x, \det(x)^{-1}, y, \det(y+1)^{-1}, \mathrm{jor}(y)^{-1}]/(y^2 - 1).$$

By Lemma 6.75, $\mathbb{F}_p[y, \mathrm{jor}(y)^{-1}]/(y^2 - 1)$ is étale over $\mathbb{F}_p$ so it is a finite product of finite fields. In order to prove our Lemma it is enough to prove that

$$Z := Spec\ \mathbb{F}_p[y, \det(y+1)^{-1}, \mathrm{jor}(y)^{-1}]/(y^2 - 1) \simeq Spec\ \mathbb{F}_p$$

under $y \mapsto 1$ so it is enough to prove that the scheme Z has exactly one $\mathbb{F}_p^a$-point (which must then be the point given by the identity matrix $y = 1$). This follows from Lemma 6.75. Alternatively, let

$$y \mapsto \gamma \in \mathfrak{g}(\mathbb{F}_p^a)$$

be an $\mathbb{F}_p^a$-point of Z; hence $\gamma^2 = 1$ and $\det(\gamma+1) \neq 0$. Since the minimal polynomial of γ divides $t^2 - 1$ and $p \neq 2$ the minimal polynomial has distinct roots so γ is diagonalizable, with eigenvalues ± 1. Since -1 is not an eigenvalue of γ, we must have $\gamma = 1$. □

LEMMA 6.78. *$\pi^{\widehat{p}}_{p/G} : B^{\widehat{p}} \to C^{\widehat{p}}_p$ is an isomorphism.*

PROOF. The Lemma follows from Lemma 6.77 and from the fact that p is a non-zero divisor in B and C_p (which, in its turn follows from the fact that C_p is étale over B, cf. Lemma 6.76). $\square$

LEMMA 6.79. *We have an equality $\varphi^{\widehat{p}}_{p/G} = \pi^{\widehat{p}}_{p/G} \circ \phi_p : B^{\widehat{p}} \to C^{\widehat{p}}_p$.*

PROOF. By 6.29, we have

$$\pi^{\widehat{p}}_{p/G}(\phi_p(x)) = x^{(p)} f_p(x)^{1/2}_p$$

(where the 1/2 power is the principal root) while

$$\varphi^{\widehat{p}}_{p/G}(x) = x^{(p)} y.$$

So it is enough to show that $f_p(x)^{1/2}_p$ and y are equal in $C^{\widehat{p}}_p$. Now this follows from the fact that their squares are equal in $C^{\widehat{p}}_p$ and they are both congruent to 1 in $C^{\widehat{p}}_p$; the fact that $f_p(x)^{1/2}_p \equiv 1 \bmod p$ is part of the definition of the principal root while the fact that $y \equiv 1 \bmod p$ in $C^{\widehat{p}}_p$ follows from Lemma 6.77 $\square$

PROOF OF THEOREM 6.72. The theorem follows from Lemmas 6.76, 6.78, and 6.79. $\square$

6.5.2. The case $q \in GL_2(A)$ split, $q^t = -q$. Let us fix such a q and concentrate on proving Theorem 6.73. Set $\Gamma_p = (Y_p, \pi_p, \varphi_p)$.

LEMMA 6.80. *Y_p is irreducible and π_p and φ_p have degrees 2 and $2p^4$ respectively.*

PROOF. If we set

$$x = \begin{pmatrix} a & b \\ c & d \end{pmatrix}, \quad \beta_p = \frac{\det(x)^p}{\det(x^{(p)})} = \frac{(ad-bc)^p}{a^p d^p - b^p c^p}$$

then

$$Y_p = Y_{p/G} \otimes E = Spec\ \frac{E[y, \det(y+1)^{-1}, \mathrm{jor}(y)^{-1}]}{(y^2 - \beta_p \cdot 1)}.$$

Let $t_p = t$ be an indeterminate and set

$$F_p = \frac{E[t]}{(t^2 - \beta_p)}. \tag{6.82}$$

Since β_p^{-1} is the inverse of a square times a product of coprime linear forms in $\mathbb{Q}[a, b, c, d]$ it follows that $t^2 - \beta_p$ is irreducible in $E[t]$, hence F_p is a field of degree 2 over E. We claim that

$$Y_p \simeq Spec\ F_p. \tag{6.83}$$

Indeed the map

$$E[y] \to F_p, \quad y \mapsto \begin{pmatrix} t & 0 \\ 0 & t \end{pmatrix}$$

sends $(y^2 - \beta_p \cdot 1)$ into 0 and induces a map

$$\mathcal{O}(Y_p) \to F_p$$

because jor $\begin{pmatrix} t & 0 \\ 0 & t \end{pmatrix} = t^4 \in F_p^\times$ and det $\begin{pmatrix} t+1 & 0 \\ 0 & t+1 \end{pmatrix} \in F_p^\times$. Since Y_p is the disjoint union of spectra of fields it follows that $Spec\ F_p$ is a connected component of Y_p. In order to show there are no other connected components it is enough to show that $\mathcal{O}(Y_p)$ has dimension at most two over E; this follows from Lemma 6.75. This ends the proof of the fact that Y_p is irreducible and of the fact that π_p has degree 2.

To prove that φ_p has degree $2p^4$ we need to show that, for $\theta := \theta_p \in F_p$ the class of $t = t_p$, the degree of the extension

$$K(a^p\theta, b^p\theta, c^p\theta, d^p\theta) \subset K(a, b, c, d, \theta) \tag{6.84}$$

is $2p^4$. Set

$$u = a/d, \quad v = b/d, \quad w = c/d, \quad M = K(u, v, w),$$

and note that

$$\theta^2 = \beta_p \in M.$$

Then the extension 6.84 is the composition of the following extensions:

$$\begin{array}{rcll} K(a^p\theta, b^p\theta, c^p\theta, d^p\theta) & = & K(u^p, v^p, w^p, d^p\theta) & \\ & \subset & M(d^p\theta) & (\star) \\ & \subset & M(\theta, d^p) & (\star\star) \\ & \subset & M(\theta, d) & (\star\star\star) \\ & = & K(a, b, c, d, \theta). & \end{array}$$

Now the extensions $(\star)$ and $(\star\star\star)$ obviously have degrees p^3 and p respectively. We claim that the extension $(\star\star)$ has degree 2. It has degree 1 or 2 so we need to show that

$$\theta \notin M(d^p\theta).$$

Assume the contrary. Then, writing z for the indeterminate d^p over M we have

$$\theta \cdot F(\theta z) = G(\theta z)$$

for two polynomials F, G with coefficients in M. Set $z = \theta^{2i-1}$ for $i = 1, 2, 3, \ldots$. We get

$$\theta \cdot F(\beta_p^i) = G(\beta_p^i)$$

for all i. Since β_p is not a root of unity we have $F(\beta_p^i) \neq 0$ for some i; hence we get $\theta \in M$, a contradiction. This ends the proof that φ_p has degree $2p^4$. $\square$

LEMMA 6.81. *For all p, p', Γ_p^* and $\Gamma_{p'}^*$ commute (i.e., $\varphi_{pp'}^* = 0$) and, for all p, Γ_p is induced from E^+.*

PROOF. View $Y_p = Spec\ F_p$ and view

$$\pi_p : E \to F_p = E + \theta_p E$$

as the natural inclusion, $\theta_p^2 = \beta_p$. So

$$\mathrm{tr}_{\pi_p}(\theta_p) = 0.$$

Also

$$\varphi_p(x) = \theta_p x^{(p)},$$

i.e. $\varphi_p(a) = \theta_p a^p$, $\varphi_p(b) = \theta_p b^p$, etc. Recall the spaces E^+ and E^- in 3.77. Note that $\beta_p \in E^+$ and $\varphi_p(E^+) \subset E^+$, hence

$$\Gamma_p^*(E^+) \subset E^+, \quad \Gamma^*_{|E^+} = 2\varphi_p. \tag{6.85}$$

Now we claim that $\varphi_p \circ \varphi_{p'}$ and $\varphi_{p'} \circ \varphi_p$ have the same value on any monomial of degree 2 in $\{a, b, c, d\}$. Indeed let $a^i b^j c^k d^l$ be such a monomial, $i + j + k + l = 2$. Then

$$\begin{aligned}
\Gamma_{p'}^* \Gamma_p^* (a^i b^j c^k d^l) &= \Gamma_{p'}^* \left(\mathrm{tr}_{\pi_p}\left(a^{pi} b^{pj} c^{pk} d^{pl} \tfrac{\det(x)^p}{\det(x^{(p)})}\right)\right) \\
&= 2\Gamma_{p'}^* \left(a^{pi} b^{pj} c^{pk} d^{pl} \tfrac{\det(x)^p}{\det(x^{(p)})}\right) \\
&= 4a^{pp'i} b^{pp'j} c^{pp'k} d^{pp'l} \cdot \left(\frac{\det(x)^{p'}}{\det(x^{(p')})}\right)^p \cdot \frac{\det(x^{(p')})^p}{\det(x^{(pp')})} \\
&= 4a^{pp'i} b^{pp'j} c^{pp'k} d^{pp'l} \cdot \frac{\det(x)^{pp'}}{\det(x^{(pp')})} \\
&= \Gamma_p^* \Gamma_{p'}^* (a^i b^j c^k d^l).
\end{aligned}$$

By the claim we just proved we get that $\varphi_p \circ \varphi_{p'}$ and $\varphi_{p'} \circ \varphi_p$ are equal as maps $E^+ \to E^+$. Hence Γ_p^* and $\Gamma_{p'}^*$ commute on E^+. On the other hand

$$\Gamma_p^*(E^-) \in \mathrm{tr}_{\pi_p}(\varphi_p(a)\varphi_p(E^+)) \subset \mathrm{tr}_{\pi_p}(\theta_p E) = 0. \tag{6.86}$$

Similarly $\Gamma_{p'}^*$ vanishes on E^-. So Γ_p^* and $\Gamma_{p'}^*$ commute on E^- as well, so Γ_p^* and $\Gamma_{p'}^*$ commute on E. On the other hand 6.85 and 6.86 show that Γ_p is induced from E^+. □

LEMMA 6.82. *For all p, p', $\varphi_{p\overline{p}'} \neq 0$.*

PROOF. Using the fact that $\Gamma_{\overline{p}'}^*(x) = x^{(p')}$, we have:

$$\begin{aligned}
\Gamma_{\overline{p}'}^* \Gamma_p^* (a^2) &= \Gamma_{\overline{p}'}^* \left(\mathrm{tr}_{\pi_p}\left(a^{2p} \tfrac{\det(x)^p}{\det(x^{(p)})}\right)\right) \\
&= 2a^{2pp'} \frac{\det(x^{(p')})^p}{\det(x^{(pp')})}, \\
\Gamma_p^* \Gamma_{\overline{p}'}^* (a^2) &= \Gamma_p^*(a^{2p'}) \\
&= 2a^{2pp'} \left(\tfrac{\det(x)^p}{\det(x^{(p)})}\right)^{p'}.
\end{aligned}$$

So we have

$$\Gamma_{\overline{p}'}^* \Gamma_p^* \neq \Gamma_p^* \Gamma_{\overline{p}'}^*$$

because

$$\frac{\det(x^{(p')})^p}{\det(x^{(pp')})} \neq \left(\frac{\det(x)^p}{\det(x^{(p)})}\right)^{p'};$$

the latter can be checked by taking $a = c = d = 1$, clearing out the denominators and picking out the coefficient of b. □

LEMMA 6.83. *For all p, p', Γ_p and $\Gamma_{p'}$ partially commute, but do not commute.*

PROOF. Recall from 6.82 and 6.83 that we established an identification

$$Y_p = Y_{\Gamma_p} = Spec\ F_p, \quad Y_{p'} = Y_{\Gamma_{p'}} = Spec\ F_{p'}$$

where

$$F_p = \frac{E[t_p]}{(t_p^2 - \beta_p)}, \quad F_{p'} = \frac{E[t_{p'}]}{(t_{p'}^2 - \beta_{p'})},$$

where $\pi_p, \pi_{p'}$ identify with the inclusions of E into F_p and $F_{p'}$ respectively and

$$\varphi_p : E \to F_p, \quad \varphi_{p'} : E \to F_{p'}$$

are given by

$$\varphi_p(x) = t_p x^{(p)}, \quad \varphi_{p'}(x) = t_{p'} x^{(p')}.$$

Then we have

$$Y_{\Gamma_p \circ \Gamma_{p'}} = Spec\ F_{pp'}, \quad Y_{\Gamma_{p'} \circ \Gamma_p} = Spec\ F_{p'p},$$

where

$$F_{pp'} = \frac{E[t_p, t_{p'}]}{(t_{p'}^2 - \beta_{p'}, t_p^2 - \varphi_{p'}(\beta_p))}, \quad F_{p'p} = \frac{E[t_p, t_{p'}]}{(t_p^2 - \beta_p, t_{p'}^2 - \varphi_p(\beta_{p'}))}.$$

We need to prove that the two maps

$$Spec(F_{pp'} \otimes_E F_{p'p}) \to Spec\ E \tag{6.87}$$

defined by the ring maps

$$E \to F_{pp'} \otimes_E F_{p'p}, \quad x \mapsto 1 \otimes t_{p'} t_p^{p'} x^{(pp')} \quad \text{and} \quad x \mapsto t_p t_{p'}^p x^{(pp')} \otimes 1,$$

do not coincide on

$$Spec(F_{pp'} \otimes_E F_{p'p})$$

but *do* coincide on a connected component of the latter. Note that $x^{(pp')}$ has entries in E so may be balanced left and right of the $\otimes$ sign. Set

$$\psi^{\pm} := t_p t_{p'}^p \otimes 1 \pm 1 \otimes t_{p'} t_p^{p'} \in F_{pp'} \otimes_E F_{p'p}.$$

We have that $\psi^+ \neq 0$ and $\psi^- \neq 0$ (otherwise we get a contradiction with the fact that $F_{pp'} \otimes_E F_{p'p}$ has dimension 16 over E). On the other hand we have

$$\psi^+ \psi^- = t_p^2 t_{p'}^{2p} \otimes 1 - 1 \otimes t_{p'}^2 t_p^{2p'} = \varphi_{p'}(\beta_p)\beta_{p'}^p - \varphi_p(\beta_{p'})\beta_p^{p'}.$$

Now an easy computation gives

$$\varphi_{p'}(\beta_p)\beta_{p'}^p = \frac{(\det(x))^{pp'}}{\det(x^{(pp')})} = \varphi_p(\beta_{p'})\beta_p^{p'},$$

hence $\psi^+\psi^- = 0$. So neither ψ^+ nor ψ^- is invertible in $F_{pp'} \otimes_E F_{p'p}$. Let M^+ and M^- be maximal ideals in $F_{pp'} \otimes_E F_{p'p}$ containing ψ^+ and ψ^- respectively. Then the two maps in 6.87 coincide on

$$Spec((F_{pp'} \otimes_E F_{p'p})/M^-)$$

and *do not* coincide on

$$Spec((F_{pp'} \otimes_E F_{p'p})/M^+).$$

□

PROOF OF THEOREM 6.73. The theorem follows from Lemmas 6.80, 6.81, 6.82, 6.83. □

6.5.3. The case $q \in GL_n(A)$, n even, q split, $q^t = q$. Fix such a q and let us concentrate on proving Theorem 6.74.

LEMMA 6.84. *Γ_p has left degree 2^n.*

PROOF. To check this it is enough to check that $f_p(x)$ has distinct eigenvalues in an algebraic closure E^a of E and none of these eigenvalues is 1; cf. Lemma 6.75. Let us introduce some notation. For any square matrix M with entries in a ring we denote by $Char(M)$ the characteristic polynomial of M; furthermore, for any monic polynomial f with coefficients in our ring we denote by $Dis(f)$ the discriminant of f; so we may consider the element $Dis(Char(M))$ in our ring. In particular we may consider the element

$$\mathcal{D}_p = Dis(Char(f_p(x))) \in A[x, g_p(x)^{-1}].$$

Let $n = 2r$ and write $q = q_n$. Then $q_n = w^{-1}\tilde{q}_n w$ where w is a permutation matrix and

$$\tilde{q}_n = \text{diag}(q_2, ..., q_2), \qquad q_2 = \begin{pmatrix} 0 & 1 \\ 1 & 0 \end{pmatrix}$$

Since w is a permutation matrix we have $w^{-1} = w^t$ and $(wx)^{(p)} = wx^{(p)}$ so

$$x^{(p)t} q_n x^{(p)} = x^{(p)t} w^t \tilde{q}_n w x^{(p)} = (wx)^{(p)t} \tilde{q}_n (wx)^{(p)},$$

$$x^t q_n x = x^t w^t \tilde{q}_n w x = (wx)^t \tilde{q}_n (wx),$$

and hence, for $\tilde{x} = wx$, we have

$$f_p(x) = (\tilde{x}^{(p)t} \tilde{q}_n \tilde{x}^{(p)})^{-1} (\tilde{x}^t \tilde{q}_n \tilde{x})^{(p)}. \tag{6.88}$$

For $1 \leq l, m \leq r$ consider the matrices

$$z_{lm} = \begin{pmatrix} \tilde{x}_{2l-1,2m-1} & \tilde{x}_{2l-1,2m} \\ \tilde{x}_{2l,2m-1} & \tilde{x}_{2l,2m} \end{pmatrix}.$$

Let $\mathcal{C}_p := \mathcal{C}(\tilde{x})$ be the characteristic polynomial of 6.88. Then if one sets $z_{lm} = 0$ in $\mathcal{C}_p(\tilde{x})$ for all $l \neq m$ one obtains the product $\mathcal{C}_p^1(z_{11})...\mathcal{C}_p^r(z_{rr})$ where $\mathcal{C}_p^l(z_{ll})$ is the characteristic polynomial of

$$f_p^l := (z_{ll}^{(p)t} q_2 z_{ll}^{(p)})^{-1} (z_{ll}^t q_2 z_{ll})^{(p)} =: \begin{pmatrix} u_l & v_l \\ w_l & u_l \end{pmatrix}.$$

Set

$$z_{ll} = \begin{pmatrix} a & b \\ c & d \end{pmatrix}.$$

Then, by a direct computation, using 6.29, we have $u_l = u$, $v_l = v$, $w_l = w$, where

$$\begin{aligned} u &= \frac{(a^p d^p + b^p c^p)(ad+bc)^p - 2^{p+1} a^p b^p c^p d^p}{(a^p d^p - b^p c^p)^2}, \\ v &= b^p d^p \cdot \frac{2^p (a^p d^p + b^p c^p) - 2(ad+bc)^p}{(a^p d^p - b^p c^p)^2}, \\ w &= a^p c^p \cdot \frac{2^p (a^p d^p + b^p c^p) - 2(ad+bc)^p}{(a^p d^p - b^p c^p)^2}. \end{aligned} \tag{6.89}$$

Note that $v_l, w_l \neq 0$; hence the discriminant of the characteristic polynomial of f_p^l, which equals $4v_l w_l$, is non-zero. One then easily gets that the discriminant of

$$\mathcal{C}_p^1(z_{11})...\mathcal{C}_p^r(z_{rr})$$

is non-zero in

$$A[z, g_p(z)^{-1}] := A[z_{11}, ..., z_{rr}, g_p(z_{11})^{-1}, ..., g_p(z_{rr})^{-1}],$$

where the latter $g_p(z_{ii})$ are the corresponding quantities attached to q_2. Let $\tilde{g}_p$ be the polynomial analogous to g_p constructed for $\tilde{q}_n$ rather than q_n. We conclude that $\mathcal{D}_p \neq 0$ in $A[x, g_p(x)^{-1}]$ because the image of $\mathcal{D}_p$ under the map

$$A[x, g_p(x)^{-1}] = A[\tilde{x}, \tilde{g}_p(\tilde{x})^{-1}] \to A[z, g_p(z)^{-1}]$$

$$z_{lm} \mapsto 0 \quad \text{for} \quad l \neq m,$$

$$z_{ll} \mapsto z_{ll}$$

is $\neq 0$. Finally to show that 1 is not an eigenvalue for $f_p(x)$ it is enough to show that 1 is not an eigenvalue for any of the matrices f_p^l. Assume 1 is an eigenvalue of f_p^l and seek a contradiction. Our assumption implies that

$$(u_l - 1)^2 - v_l w_l = 0,$$

hence $v_l w_l$ is a square in $K(a, b, c, d)$, where K is the fraction field of A. But $v_l w_l$ is a square in $K(a, b, c, d)$ times $a^p b^p c^p d^p$, hence $a^p b^p c^p d^p$ is a square in $K(a, b, c, d)$, a contradiction. □

Next we assume $n = 2$, hence $q = \begin{pmatrix} 0 & 1 \\ 1 & 0 \end{pmatrix}$. Write $x = \begin{pmatrix} a & b \\ c & d \end{pmatrix}$. Then recall from the proof of Lemma 6.84 that

$$f_p = \begin{pmatrix} u & v \\ w & u \end{pmatrix}$$

where u, v, w are given by 6.89. Recall also that, by the proof of Lemma 6.84, 1 is not an eigenvalue of f_p and $vw \neq 0$. So by Lemma 6.75 we have $Y_p = Spec\ F_p$ where

$$F_p = \frac{E[y, \text{jor}(y)^{-1}]}{(y^2 - f_p)}.$$

If $\begin{pmatrix} \alpha & \beta \\ \gamma & \delta \end{pmatrix}$ is the image of y in the 2×2 matrices with entries in F_p then π_p : $E \to F_p$ identifies with the natural inclusion, and

(6.90)
$$\varphi_p(a) = a^p\alpha + b^p\gamma, \quad \varphi_p(b) = a^p\beta + b^p\delta,$$
$$\varphi_p(c) = c^p\alpha + d^p\gamma, \quad \varphi_p(d) = c^p\beta + d^p\delta.$$

LEMMA 6.85. *For t a variable the map*

$$h : E[t] \to \frac{E[y]}{(y^2 - f_p)}, \quad t \mapsto \tau := \alpha + \delta,$$

sends $t^4 - 4ut^2 + 4vw$ into 0.

PROOF. We have

(6.91) $$\alpha^2 + \beta\gamma = \beta\gamma + \delta^2 = u, \quad (\alpha + \delta)\beta = v, \quad (\alpha + \delta)\gamma = w.$$

Set $\eta = \alpha - \delta$. Then

(6.92) $$\tau\eta = \alpha^2 - \delta^2 = 0, \quad \tau^2 + \eta^2 = 2(\alpha^2 + \delta^2).$$

By 6.91

$$\alpha^2 + \delta^2 + 2\beta\gamma = 2u,$$

hence, by 6.92,

$$\tau^2 + \eta^2 + 4\beta\gamma = 4u,$$

hence

$$\tau^4 + \tau^2\eta^2 + 4\beta\gamma\tau^2 = 4u\tau^2,$$

hence by 6.92 and 6.91,

$$\tau^4 - 4u\tau^2 + 4vw = 0. \tag{6.93}$$

□

LEMMA 6.86. *τ is invertible in $\frac{E[y]}{(y^2-f_p)}$.*

PROOF. Note that by Lemma 6.85 we have an induced homomorphism

$$\overline{h} : \frac{E[t]}{(t^4 - 4ut^2 + 4vw)} \to \frac{E[y]}{(y^2 - f_p)}.$$

Our Lemma follows now from the fact that the class of t in $\frac{E[t]}{(t^4-4ut^2+4vw)}$ is invertible (because $vw \neq 0$) and τ is the image of this class via the homomorphism $\overline{h}$. □

LEMMA 6.87. *The map h is surjective.*

PROOF. By 6.91 and Lemma 6.86,

$$\beta = \tau^{-1}v, \quad \gamma = \tau^{-1}w \tag{6.94}$$

belong to the image of the map h. Also, since by 6.92, $\tau\eta = 0$, we get $\eta = 0$ hence

$$\alpha = \delta = \tau/2, \tag{6.95}$$

so α and β belong to the image of the map h. □

LEMMA 6.88. *We have natural isomorphisms*

$$\frac{E[t]}{(t^4 - 4ut^2 + 4vw)} \simeq \frac{E[y]}{(y^2 - f_p)} \simeq F_p, \quad t \mapsto \tau := \alpha + \delta. \tag{6.96}$$

PROOF. By Lemma 6.86 and by 6.11 the image of $\text{jor}(y)$ in $\frac{E[y]}{(y^2-f_p)}$ is invertible. So $F_p \simeq \frac{E[y]}{(y^2-f_p)}$. Since F_p has dimension 4 over E and $\overline{h}$ is surjective (cf. Lemma 6.87), our Lemma follows. □

LEMMA 6.89. *Γ_p is partially induced from E^+.*

PROOF. Note that by 6.96 we have

$$\text{tr}_{\pi_p}(\tau) = \text{tr}_{\pi_p}(\tau^{-1}) = 0. \tag{6.97}$$

Recall the automorphism $\iota : E \to E = K(x)$, $\iota(x) = -x$ and the spaces E^+, E^- in 3.77. Note that

$$u, v, w \in E^+, \tag{6.98}$$

hence

$$\text{tr}_{\pi_p}(\tau^2) = 8u \in E^+, \quad \text{tr}_{\pi_p}(\tau^{-2}) = \frac{2u}{vw} \in E^+. \tag{6.99}$$

Using 6.90, 6.94, 6.95, 6.98, one gets that

$$\varphi_p(E^+) \subset E^+ + E^+\tau^2, \quad \varphi_p(E^-) \subset E^+\tau + E^+\tau^{-1}.$$

Hence by 6.97, 6.99 one gets that the map $\Gamma_p^* = \mathrm{tr}_{\pi_p} \circ \varphi_p : E \to F_p \to E$ satisfies

$$\Gamma_p^*(E^+) \subset E^+, \quad \Gamma_p^*(E^-) = 0, \tag{6.100}$$

and partial induction from E^+ is proved. □

LEMMA 6.90. *$\varphi_{p\bar{p}'}^* \neq 0$ and Γ_p^* is not induced from E^+.*

PROOF. Note that

$$\begin{aligned} \varphi_p(ab) &= a^{2p}\alpha\beta + a^p b^p(\alpha\delta + \beta\gamma) + b^{2p}\gamma\delta && \text{(by 6.90)} \\ &= a^{2p}\tfrac{v}{2} + b^{2p}\tfrac{w}{2} + a^p b^p(\tfrac{\tau^2}{4} + vw\tau^{-2}) && \text{(by 6.94, 6.95)} \\ &= a^{2p}\tfrac{v}{2} + b^{2p}\tfrac{w}{2} + ua^p b^p && \text{(by 6.93)} \\ &= 2^{p-1}a^p b^p && \text{(by 6.89).} \end{aligned} \tag{6.101}$$

One gets

$$\begin{aligned} \Gamma_{\bar{p}'}^*\Gamma_p^*(ab) &= \Gamma_{\bar{p}'}^*(\mathrm{tr}_{\pi_p}(\varphi_p(ab))) \\ &= \Gamma_{\bar{p}'}^*(\mathrm{tr}_{\pi_p}(2^{p-1}a^p b^p)) && \text{(by 6.101)} \\ &= \Gamma_{\bar{p}'}^*(2^{p+1}a^p b^p) && \text{(since } deg(\pi_p) = 4) \\ &= 2^{p+1}a^{pp'}b^{pp'}. \end{aligned}$$

On the other hand, similarly, we have

$$\begin{aligned} \Gamma_p^*\Gamma_{\bar{p}'}^*(ab) &= \Gamma_p^*(a^{p'}b^{p'}) \\ &= \mathrm{tr}_{\pi_p}((\varphi_p(ab))^{p'}) \\ &= \mathrm{tr}_{\pi_p}((2^{p-1}a^p b^p)^{p'}) \\ &= 4 \cdot 2^{(p-1)p'}a^{pp'}b^{pp'}. \end{aligned}$$

So we get

$$\varphi_{p\bar{p}'}^*(ab) = \frac{2^{p+1}(1 - 2^{(p-1)(p'-1)})}{pp'}a^{p^2}b^{p^2} \neq 0, \quad \text{for } p \neq p', \text{ and}$$

$$\varphi_{p\bar{p}}^*(ab) = \frac{2^{p+1}(1 - 2^{(p-1)^2})}{p}a^{p^2}b^{p^2} \neq 0,$$

so $\varphi_{p\bar{p}'}^* \neq 0$ for all p, p'. Finally a direct computation gives

$$\varphi_p(ad) = (a^p c^p \frac{v}{2} + b^p d^p \frac{w}{2} + ub^p c^p) + (a^p d^p - b^p c^p)\frac{\tau^2}{4} \notin E^+,$$

hence Γ_p^* is not induced from E^+. □

LEMMA 6.91. *The E-subalgebra $E[\tau^2]$ of F_p generated by τ^2 is a field.*

PROOF. Note that the discriminant of the polynomial $y^2 - 4uy + 4vw$, satisfied by τ^2, has the following simple form:

$$16u^2 - 16vw = 16 \cdot \frac{(ad+bc)^{2p} - 4^p a^p b^p c^p d^p}{(a^p d^p - b^p c^p)^2}. \tag{6.102}$$

We claim that this discriminant is not a square in the field $E = K(a, b, c, d)$. We need to show that

$$Q := (ad+bc)^{2p} - 4^p a^p b^p c^p d^p$$

is not a square in $K(a, b, c, d)$. Assume the contrary. Then $Q = P^2$ with $P \in K[a, b, c, d]$ because $K[a, b, c, d]$ is factorial. Consider the $\mathbb{G}_m \times \mathbb{G}_m$-action $\star$ on $K[a, b, c, d]$, where $\mathbb{G}_m = Spec\ K[t, t^{-1}]$, with points $g := (\lambda, \mu)$ of $\mathbb{G}_m \times \mathbb{G}_m$ acting via

$$g \star a = \lambda a, \quad g \star b = \mu b, \quad g \star c = \mu^{-1} c, \quad g \star d = \lambda^{-1} d.$$

The ring of invariants of this action is $K[ad, bc]$. Since Q is invariant we have

$$g \star P = \epsilon(g) \cdot P, \qquad \epsilon(g) \in \{\pm 1\},$$

for all g. Since $\mathbb{G}_m \times \mathbb{G}_m$ is connected we must have $\epsilon(g) = 1$ for all g, hence

$$P(a, b, c, d) = G(ad, bc)$$

for some polynomial G in 2 variables with K-coefficients. Hence, for s a variable, we have

$$f(s) := Q(s, 1, 1, 1) = (s+1)^{2p} - 4^p s^p = G(s, 1)^2$$

in $K[s]$, so all the complex roots of f are multiple. Now, if $f'(s) = df/ds$, we get

$$(s+1)f'(s) - 2pf(s) = 4^p p s^{p-1}(s-1).$$

So the only root of f is 1, hence $f(s) = (s-1)^{2p}$, a contradiction, and our Lemma is proved. □

LEMMA 6.92. *τ^2 is not a square in the field $E[\tau^2]$.*

PROOF. We have

$$\tau^2 = 2u \pm \sqrt{u^2 - vw}.$$

Assume τ^2 is a square in $E[\tau^2]$. Using 6.102, and setting

$$D := (ad+bc)^{2p} - 4^p a^p b^p c^p d^p,$$

we get that

$$2(a^p d^p + b^p c^p)(ad+bc)^p - 2^{p+2} a^p b^p c^p d^p \pm 2(a^p d^p - b^p c^p)\sqrt{D} = (X + Y\sqrt{D})^2$$

for some $X, Y \in E$. Set

$$A = ad, \quad B = bc.$$

We get

$$\begin{aligned} 2(A^p + B^p)(A+B)^p - 2^{p+2} A^p B^p &= X^2 + DY^2 \\ 2(A^p - B^p) &= 2XY, \end{aligned}$$

for some $X, Y \in E$. Again, consider the $\mathbb{G}_m \times \mathbb{G}_m$-action $\star$ on $E = K(a, b, c, d)$, $\mathbb{G}_m = Spec\ K[t, t^{-1}]$, with points $g := (\lambda, \mu)$ of the torus $\mathbb{G}_m \times \mathbb{G}_m$ acting via

$$g \star a = \lambda a, \quad g \star b = \mu b, \quad g \star c = \mu^{-1} c, \quad g \star d = \lambda^{-1} d,$$

respectively. The invariant field of this action is $K(A,B)$. By connectivity of the torus we get that X and Y are invariant and hence belong to $K(A,B)$. Set

$$s = B/A, \qquad Z = X/A.$$

We get

$$\begin{aligned} M := 2(1+s^p)(1+s)^p - 2^{p+2}s^p &= Z^2 + ((1+s)^{2p} - 4^p s^p)Y^2 = Z^2 + PY^2 \\ N := 2(1-s^p) &= 2ZY. \end{aligned}$$

Consider the action of $\mathbb{G}_m$ on $K(A,B)$ with λ acting on A, B by sending them into $\lambda A, \lambda B$ respectively. Since $\mathbb{G}_m$ is connected and s is invariant it follows that Z, Y are invariant hence belong to $K(s)$. Eliminating Z from the above system we get

$$PY^4 - MY^2 + N^2/4 = 0.$$

Since $Y \in K(s)$ we get that the discriminant,

$$M^2 - PN^2,$$

of the latter equation is a square in $K(s)$. In particular $M^2 - PN^2$ has even s-adic valuation. On the other hand we have

$$\begin{aligned} 4^{-1}(M^2 - PN^2) &= ((1+s^p)(1+s)^p - 2^{p+1}s^p)^2 - (1-s^p)^2((1+s)^{2p} - 4^p s^p) \\ &\equiv (4^p - 2^{p+2} + 4)s^p \mod s^{p+1}. \end{aligned}$$

Since

$$4^p - 2^{p+2} + 4 \neq 0$$

the s-adic valuation of $M^2 - PN^2$ is p which is odd, a contradiction. This ends the proof of our Lemma. □

LEMMA 6.93. *F_p is a field.*

PROOF. This follows directly from Lemmas 6.88, 6.91, and 6.92. □

PROOF OF THEOREM 6.74. The theorem follows from Lemmas 6.84, 6.89, 6.90, and 6.93. □

CHAPTER 7

Curvature of Levi-Cività connections

In this chapter we study Levi-Cività connections $(\delta_1, ..., \delta_n)$, $\delta_i = (\delta_{ip})$ on GL_n attached to symmetric matrices q. We will start by introducing and examining curvature (φ^{ij}) in the case of δ-rings of local type (hence in the case of one prime). Here we will prove a non-vanishing result for curvature. (A similar conclusion can be obtained for $n = 2$ in the case of Fedosov connections.) We will continue by considering an arbitrary set of primes, hence δ-rings of global type. We will show that, for q δ-constant, the connections δ_i are global along the identity. This will allow us to attach to them a *mixed curvature* $(\varphi^{ij}_{pp'})$ via the method of analytic continuation between primes.

7.1. The case $m = 1$: non-vanishing of curvature mod p

In this section we assume $A \subset R$ is a δ-ring of local type.

One can introduce the *curvature* of the Levi-Cività connection as in Definition 3.37. Indeed we have:

DEFINITION 7.1. Let $q \in GL_n(A)$, $q^t = q$, and let $(\delta_1^{GL_n}, ..., \delta_n^{GL_n})$ be the Levi-Cività connection attached to q; cf. Definition 4.39. The *curvature* of the Levi-Cività connection is the family (φ^{ij}) indexed by $i, j = 1, ..., n$ where φ^{ij} are the additive maps

$$\varphi^{ij} := \frac{1}{p}[\phi_i^{GL_n}, \phi_j^{GL_n}] : \mathcal{O}(\widehat{GL_n}) \to \mathcal{O}(\widehat{GL_n}), \tag{7.1}$$

and $\phi_i^{GL_n}$ are the Frobenius lifts attached to $\delta_i^{GL_n}$. Similarly the 3-*curvature* of the Levi-Cività connection is the family (φ^{ijk}) indexed by $i, j, k = 1, ..., n$ where φ^{ijk} are the additive maps

$$\varphi^{ijk} := \frac{1}{p}[\phi_i^{GL_n}, [\phi_j^{GL_n}, \phi_k^{GL_n}]] : \mathcal{O}(\widehat{GL_n}) \to \mathcal{O}(\widehat{GL_n}).$$

THEOREM 7.2. *Let $q \in GL_n(A)$, $q^t = q$, let $\Gamma = (\Gamma_{jk})$ be the Christoffel symbol of the 1st kind of the real Chern connection attached to q and let $\Gamma_i = (\Gamma_{ijk})$ be the Christoffel symbols of the 1st kind of the Levi-Cività connection attached to q; cf. Definition 4.40. Also let (φ^{ij}) be the curvature of the Levi-Cività connection attached to q. Then the following congruences hold:*

$$\begin{aligned}(x^{(p^2)t}q^{(p^2)}\varphi^{ij}(x))_{kl} &\equiv (\Gamma_{jlk} - \Gamma_{ilk})^p \quad mod \quad p \\ &\equiv (\Gamma_{jl} + \Gamma_{ik} - \Gamma_{jk} - \Gamma_{il})^p \quad mod \quad p \\ &\equiv \tfrac{1}{2}(\delta q_{jk} + \delta q_{il} - \delta q_{ik} - \delta q_{jl})^p \quad mod \quad (p, x-1).\end{aligned} \tag{7.2}$$

In particular the following are equivalent:

(1) $\varphi^{ij}(x) \equiv 0$ *mod* p;
(2) $\Gamma_{jk} + \Gamma_{il} \equiv \Gamma_{ik} + \Gamma_{jl}$ *mod* p *for all* $k, l = 1, ..., n$, *and they imply*
(3) $\delta q_{jk} + \delta q_{il} \equiv \delta q_{ik} + \delta q_{jl}$ *mod* p *for all* $k, l = 1, ..., n$.

If we denote by R_{ijkl} the left hand side of 7.2 then one immediately gets, by the second congruence in 7.2, the following congruences mod p:

$$R_{ijkl} \equiv -R_{ijlk}, \quad R_{ijkl} \equiv -R_{jikl}, \quad R_{lijk} + R_{ljki} + R_{lkij} \equiv 0, \quad R_{ijkl} \equiv R_{klij}.$$

By the way, one can view R_{ijkl} as an analogue of the classical covariant Riemann tensor. Also the last expression on the right hand side of 7.2 can be viewed as an "infinitesimal" analogue of the expression

$$\kappa(q_{jk}q_{il} - q_{ik}q_{jl})$$

giving the classical covariant Riemann tensor of a metric with constant sectional curvature. So, roughly speaking, all metrics, in the arithmetic setting, behave "mod p" like metrics with constant sectional curvature.

PROOF. Write, as usual, $\phi_i^{GL_n}(x) = x^{(p)} + p\Delta_i$. By 4.82 we have

$$\Delta_i = \phi(q^t)^{-1} \cdot (x^{(p)t})^{-1} \cdot \Gamma_i^t. \tag{7.3}$$

On the other hand we have

$$\begin{aligned} \phi_i^{GL_n}\phi_j^{GL_n}(x) &= \phi_i^{GL_n}(x^{(p)} + p\Delta_j) \\ &= (x^{(p)} + p\Delta_i)^{(p)} + p\phi_i^{GL_n}(\Delta_j) \\ &\equiv x^{(p^2)} + p\Delta_j^{(p)} \mod p^2. \end{aligned}$$

Hence, using 7.3,

$$\begin{aligned} \varphi^{ij}(x) &\equiv \Delta_j^{(p)} - \Delta_i^{(p)} \mod p \\ &\equiv (\Delta_j - \Delta_i)^{(p)} \mod p \\ &\equiv (q^{(p^2)t})^{-1}(x^{(p^2)t})^{-1}(\Gamma_j - \Gamma_i)^{(p)t} \mod p, \end{aligned}$$

which ends our proof by 4.87 and 4.88. □

The implication $1 \Rightarrow 3$ in the above theorem shows that the curvature (φ^{ij}) does not vanish mod p for a "general" symmetric $q \in GL_n(A)$. The next computation shows the non-vanishing of curvature mod p for q split.

COROLLARY 7.3. *Let* $n = 2r$, $q \in GL_n(A)$, $q^t = q$, q *split, and let* (φ^{ij}) *be the curvature of the Levi-Cività connection attached to* q. *Then for all* $i, j = 1, ..., n$ *with* $i \neq j$ *we have*

$$\varphi^{ij}(x) \not\equiv 0 \mod p.$$

PROOF. Assume first $n = 2$ and write $x = \begin{pmatrix} a & b \\ c & d \end{pmatrix}$. Using the congruence 4.85 we obtain

$$\begin{aligned} \Gamma_{11} &\equiv \tfrac{2^{p-1}-1}{p} a^p c^p \mod p \\ \Gamma_{12} = \Gamma_{21} &\equiv \tfrac{1}{2p}((ad+bc)^p - (a^p d^p + b^p c^p)) \mod p \\ \Gamma_{22} &\equiv \tfrac{2^{p-1}-1}{p} b^p d^p \mod p. \end{aligned}$$

One has

$$\Gamma_{11} + \Gamma_{22} \not\equiv \Gamma_{21} + \Gamma_{12} \mod p,$$

as one can see by looking at the coefficient of $a^{p-1}d^{p-1}bc$. Hence by Theorem 7.2, $\varphi^{12}(x) \not\equiv 0 \mod p$ and the case $r = 1$ of the corollary is proved.

Assume now $r \geq 2$ and write $x = \begin{pmatrix} A & B \\ C & D \end{pmatrix}$. Let

$$J = (p, A-1, C, D-1)$$

be the ideal generated in $A[x, \det(x)^{-1}]\hat{}$ by p and the entries of $A-1, C, D-1$. Using the congruence 4.85 we obtain

$$\Gamma \equiv \begin{pmatrix} 0 & 0 \\ 0 & E \end{pmatrix} \mod J,$$

where

$$E = \frac{1}{2p}\{(B+B^t)^{(p)} - B^{(p)} - B^{(p)t}\}.$$

So we have, for all $j, k = 1, ..., r$,

$$E_{jk} = \frac{1}{2p}\{(B_{jk} + B_{kj})^p - B_{jk}^p - B_{kj}^p\}.$$

Let $i, j \in \{1, ..., r\}$ and set $\underline{i} = i + r$, $\underline{j} = j + r$.

Claim 1. $\varphi^{i\underline{j}}(x) \not\equiv 0 \mod p$.

Indeed choose an index $1 \leq l \leq r$ with $l \neq j$ (which is possible because $r \geq 2$) and choose any index $1 \leq k \leq r$. Note that

$$E_{jl} \not\equiv 0 \mod J \tag{7.4}$$

hence

$$\Gamma_{ik} + \Gamma_{\underline{j}\underline{l}} \equiv E_{jl} \not\equiv 0 \equiv \Gamma_{i\underline{l}} + \Gamma_{\underline{j}k} \mod J, \text{ and hence mod } p.$$

By Theorem 7.2, $\varphi^{i\underline{j}}(x) \not\equiv 0 \mod p$ and Claim 1 is proved.

Claim 2. $\varphi^{\underline{i}\underline{j}}(x) \not\equiv 0 \mod p$.

Indeed choose $1 \leq l \leq r$ with $l \neq i$ and $1 \leq k \leq r$ arbitrary. Note that

$$\{B_{il}, B_{li}\} \cap \{B_{jl}, B_{lj}\} = \emptyset;$$

hence, by 7.4, we have

$$E_{il} \not\equiv E_{jl} \mod J.$$

Hence

$$\Gamma_{\underline{i}k} + \Gamma_{\underline{j}\underline{l}} \equiv E_{jl} \not\equiv E_{il} \equiv \Gamma_{\underline{i}\underline{l}} + \Gamma_{\underline{j}k} \mod J, \text{ and hence mod } p.$$

By Theorem 7.2, $\varphi^{\underline{i}\underline{j}}(x) \not\equiv 0 \mod p$ and Claim 2 is proved.

Claim 3. $\varphi^{ij}(x) \not\equiv 0 \mod p$.

Indeed one repeats the argument in the proof of Claim 2 by taking

$$J := (p, A-1, B, D-1).$$

Clearly Claims 1, 2, 3 finish the proof of the corollary in case $r \geq 2$ and we are done. □

A similar picture holds for Fedosov connections in case $n = 2$. Indeed we have:

DEFINITION 7.4. Let $q \in GL_2(A)$, $q^t = -q$, q split, and let $(\delta_1^{GL_2}, \delta_2^{GL_2})$ be a Fedosov connection attached to q; cf. Definition 4.44. The *curvature* of this connection is the family (φ^{ij}) indexed by $i, j = 1, 2$ where

$$\varphi^{ij} := \frac{1}{p}[\phi_i^{GL_2}, \phi_j^{GL_2}] : \mathcal{O}(\widehat{GL_2}) \to \mathcal{O}(\widehat{GL_2}), \tag{7.5}$$

and $\phi_i^{GL_2}$ are the Frobenius lifts attached to $\delta_i^{GL_2}$.

Then exactly as in the case of Levi-Cività connection we get

THEOREM 7.5. *Let $q \in GL_2(A)$, $q^t = -q$. Let (φ^{ij}) be the curvature of a Fedosov connection attached to q. Then the following congruence holds:*

$$\varphi^{ij}(x) \equiv (\Delta_j - \Delta_i)^{(p)} \quad mod \quad p. \tag{7.6}$$

EXAMPLE 7.6. In particular we have

$$\varphi^{12} \not\equiv 0 \mod p$$

for the Fedosov connection in Example 4.47.

7.2. Analytic continuation along the identity

We begin by defining Levi-Cività connections in the global setting. So we place ourselves here in the case when A is a δ-ring of global type, $A = \mathcal{O}_K[1/M]$, cf. Definition 3.25; A is equipped with a global connection $(\delta_p) = (\delta_p^A)$ indexed by a set of primes $\mathcal{V}$. Let $G = GL_n$ be the general linear group scheme over A. Assume $q^t = q \in G(A)$ and consider the involution $x^\tau = q^{-1}(x^t)^{-1}q$ on GL_n. Consider the associated map

$$\mathcal{H}_q : G \to G, \quad \mathcal{H}_q(x) = x^t q x.$$

Let $(\phi_{0p}) = (\phi_{0p}^G)$ be the Frobenius lifts attached to the trivial adelic connection on G, $\phi_{0p}(x) = x^{(p)}$. We have a well defined concept of a Frobenius lift ϕ_p on $G^{\widehat{p}}$ being $\mathcal{H}_q$-horizontal with respect to ϕ_{0p}; cf. Definition 6.1.

Then the "local" Theorem 4.38 immediately implies the following "global" result:

THEOREM 7.7. *Let $q \in G(A)$ with $q^t = q$. There exists a unique n-tuple of adelic connections $(\delta_1^G, ..., \delta_n^G)$ on $G = GL_n$, with $\delta_i^G = (\delta_{ip}^G)$, such that if (ϕ_{ip}) are the attached Frobenius lifts, then:*

(1) *ϕ_{ip} is $\mathcal{H}_q$-horizontal with respect to ϕ_{0p} for all $i = 1, ..., n$ and $p \in \mathcal{V}$;*
(2) *$\delta_{ip}^G x_{kj} = \delta_{jp}^G x_{ki}$ for all $i, j, k = 1, ..., n$ and $p \in \mathcal{V}$.*

DEFINITION 7.8. The n-tuple of adelic connections $(\delta_1^G, ..., \delta_n^G)$ on G in Theorem 7.7 is called the *Levi-Cività connection* attached to q.

Recall that q is called *δ-constant* if $\delta_p q = 0$ for all $p \in \mathcal{V}$. We have the following:

THEOREM 7.9. *Assume $q^t = q \in G(A)$ is δ-constant and let $(\delta_1^G, ..., \delta_n^G)$ be the Levi-Cività connection on G attached to q. Then for each $i = 1, ..., n$ the adelic connection δ_i^G is global along the identity, 1.*

PROOF. We already know from assertion (3) of Theorem 4.38 that δ_i^G fixes 1. Let ϕ_{ip} be the Frobenius lifts attached to $\delta_{ip} := \delta_{ip}^G$. Write $\phi_{ip}(x) = x^{(p)}\Lambda_{ip}(x)$, where $\Lambda_{ip}(x)$ is a matrix $\equiv 1 \mod p$ with coefficients in

$$A[x, \det(x)^{-1}]^{\widehat{p}}. \tag{7.7}$$

The condition that δ_{ip} on G be $\mathcal{H}_q$-horizontal with respect to δ_{0p} is equivalent to the equalities

$$\Lambda_{ip}^t A \Lambda_{ip} = B, \tag{7.8}$$

for all $i = 1, ..., n$, where A, B are as in the proof of Theorem 4.38. So there is a unique $\Lambda_p(x)$ with coefficients in 7.7 satisfying 7.8 and

$$\Lambda_{ip} \equiv 1 \mod p, \tag{7.9}$$

such that

$$\Delta_{ipkj} = \Delta_{jpki}, \tag{7.10}$$

where $\Delta_{ip} = (\Delta_{ipjk}) = p^{-1}x^{(p)}(\Lambda_{ip} - 1)$. Since δ_{ip} fixes 1 we have $\Lambda_{ip}(1) = 1$. Then 7.7 embeds into $A^{\widehat{p}}[[T]]$ (where $T := x - 1$) and 7.8, 7.9, 7.10 hold in $A^{\widehat{p}}[[T]]$. One checks the following stronger uniqueness property:

$$\Lambda_{ip}(x) = \Lambda_{ip}(1 + T) =: \Lambda_{ip}^0(T)$$

is the unique matrix with entries in $A^{\widehat{p}}[[T]]$ satisfying 7.8, 7.9, 7.10; this is done by taking any two such matrices and proving by induction on ν that they coincide mod p^ν. So it is enough to find matrices $\Lambda_{ip}^*(T)$ with entries in $A[[T]]$, such that

$$\Lambda_{ip}^{t*} A \Lambda_{ip}^* = B, \tag{7.11}$$

$$\Lambda_{ip}^* \equiv 1 \mod p, \tag{7.12}$$

$$\Delta_{ipkj}^* = \Delta_{jpki}^*, \tag{7.13}$$

where $\Delta_{ip}^* = p^{-1}x^{(p)}(\Lambda_{ip}^* - 1)$, because then, by the mentioned uniqueness, we will have that $\phi_{ip}(x)$ has entries in $A[[T]]$. The proof of the existence of Λ_{ip}^* given below is a (T)-adic version of the p-adic proof for the existence of Λ_p in Theorem 4.38. For completeness we redo here the argument; the objects below, with a $*$ superscript, roughly correspond to the corresponding objects without a $*$ in the proof of Theorem 4.38.

Fix p and drop p from the indices. We will construct by induction a sequence of n-tuples

$$(\Lambda_1^{*(\nu)}, ..., \Lambda_n^{*(\nu)}), \quad \nu \geq 1$$

of $n \times n$ matrices in entries in $A[[T]]$ with the following properties:

(i) $\Lambda_i^{*(1)} = 1$ for all i,
(ii) $\Lambda_i^{*(\nu+1)} \equiv \Lambda_i^{*(\nu)} \mod p(T)^\nu$ for all i and ν,
(iii) $\Lambda_i^{*(\nu)t} A \Lambda_i^{*(\nu)} \equiv B \mod p(T)^\nu$ for all i and ν,
(iv) If $\Delta_i^{*(\nu)} = p^{-1}x^{(p)}(\Lambda_i^{*(\nu)} - 1)$ then $\Delta_{ikj}^{*(\nu)} = \Delta_{jki}^{*(\nu)}$.

Then we are done by setting

$$\Lambda_i^* = \lim_{\nu\to\infty} \Lambda_i^{*(\nu)}.$$

To construct of our sequence of n-tuples define the n-tuple for $\nu = 1$ by condition (i); then (iii) is satisfied because $\phi(q) = q^{(p)}$ while (iv) is trivially satisfied. Now assume the n-tuple was constructed for some ν and seek the n-tuple for $\nu+1$ in the form

$$\Lambda_i^{*(\nu+1)} = \Lambda_i^{*(\nu)} + pZ_i^*, \tag{7.14}$$

with

$$Z_i^* \equiv 0 \mod (T)^\nu. \tag{7.15}$$

Write

$$\Lambda_i^{*(\nu)t} A \Lambda_i^{*(\nu)} = B - pC_i^*, \tag{7.16}$$

with

$$C_i^* \equiv 0 \mod (T)^\nu. \tag{7.17}$$

Also note that by (i) and (ii) we have

$$\Lambda_i^{*(\nu)} \equiv 1 \mod (T). \tag{7.18}$$

Using 7.18 we get

$$\begin{aligned} \Lambda_i^{*(\nu+1)t} A \Lambda_i^{*(\nu+1)} &\equiv \Lambda_i^{*(\nu)t} A \Lambda_i^{*(\nu)} \\ &+ p(\Lambda_i^{*(\nu)t} A Z_i^* + Z_i^{*t} A \Lambda_i^{*(\nu)}) \mod p(T)^{\nu+1} \\ &\equiv B + p(-C_i^* + AZ_i^* + Z_i^{*t} A) \mod p(T)^{\nu+1}. \end{aligned} \tag{7.19}$$

Now $A^t = A$ and $B^t = B$ so $C_i^{*t} = C_i^*$ so if $C_i^* = (C_{ijk}^*)$ then $C_{ijk}^* = C_{ikj}^*$. So there exists a (unique) family (D_{ijk}^*) of elements in $A[[T]]$ such that

$$D_{ijk}^* = D_{jik}^* \tag{7.20}$$

and

$$D_{ijk}^* + D_{ikj}^* = C_{ijk}^*; \tag{7.21}$$

explicitly,

$$D_{ijk}^* = \frac{1}{2}(C_{ijk}^* + C_{jik}^* - C_{kij}^*). \tag{7.22}$$

Cf. Remark 2.113. Now define the matrices $D_i^* = (D_{ijk}^*)$ hence

$$D_i^* + D_i^{*t} = C_i^*.$$

Setting

$$Z_i^* := A^{-1} D_i^{*t} \tag{7.23}$$

we get $D_i^{*t} = AZ_i^*$ and $D_i^* = Z_i^{*t} A$ so, by 7.19,

$$\Lambda_i^{*(\nu+1)t} A \Lambda_i^{*(\nu+1)} \equiv B \mod p(T)^{\nu+1}.$$

So condition (iii) holds for $\Lambda_i^{(\nu+1)}$ mod $p^{\nu+1}$. Condition (iv) for $\Delta_{ikj}^{*(\nu+1)}$ can be checked exactly as in the proof of Theorem 4.38. This ends our construction of the sequence $\Lambda_i^{*(\nu)}$ and hence the proof of our theorem. □

7.3. Mixed curvature

Let A be a δ-ring of global type and $q \in GL_n(A)$ with $q^t = q$, q δ-constant.

By our Theorems 7.7 and 7.9 we may introduce the Levi-Cività connection attached to q (in the sense of Definition 7.8) and its mixed curvature

$$(\varphi^{ij}_{pp'})$$

(in the sense of Definition 3.64).

REMARK 7.10. This mixed curvature has a varying "vertical" component, indexed by i, j, and a "horizontal" component, indexed by p, p'. This seems to not have a direct analogue in classical Riemannian geometry and deserves further understanding.

A direct computation, using the proof of Theorem 7.9 yields the following:

COROLLARY 7.11. *Assume $q^t = q \in GL_n(A)$ is δ-constant, let $(\delta_1^G, ..., \delta_n^G)$ be the Levi-Cività connection on G attached to q, $\delta_i^G = (\delta_{ip}^G)$, and let $\phi_{ip} : A[[T]] \to A[[T]]$ be the induced Frobenius lifts. Let (q_p^{jk}) be the inverse of the matrix $q^{(p)} = (q^p_{jk})$ and set*

$$L_{jk} := \sum_{s \neq j} T_{sj} q_{sk} + \sum_{s \neq k} T_{sk} q_{sj}, \tag{7.24}$$

$$L_{ipjk} = p \cdot \{\delta_{jk} T_{jj} + \frac{1}{2} \sum_m q_p^{jm} (q_{mk}^{p-1} L_{mk} + q_{mi}^{p-1} L_{mi} - q_{ki}^{p-1} L_{ki})\} \tag{7.25}$$

$$L_{ip} = (L_{ipjk}). \tag{7.26}$$

Then we have the following congruences

$$\phi_{ip}(T) \equiv L_{ip} \quad mod \quad (T)^2. \tag{7.27}$$

In the statement above δ_{jk} is the Kronecker symbol.

In particular we have:

COROLLARY 7.12. *If $\varphi^{ij}_{pp'} : A[[T]] \to A[[T]]$ is the mixed curvature of the Levi-Cività connection then:*

$$\varphi^{ij}_{pp'}(T) \equiv \frac{1}{pp'} \left(L_{jp'}^{(\phi_p)} \circ L_{ip} - L_{ip}^{(\phi_{p'})} \circ L_{jp'} \right) \quad mod \quad (T)^2, \quad for \quad p \neq p', \tag{7.28}$$

$$\varphi^{ij}_{pp}(T) \equiv \frac{1}{p} \left(L_{jp}^{(\phi_p)} \circ L_{ip} - L_{ip}^{(\phi_p)} \circ L_{jp} \right) \quad mod \quad (T)^2. \tag{7.29}$$

In the above statement $L_{ip}^{(\phi_{p'})}$ is obtained, as usual, from L_{ip} by applying $\phi_{p'}$ to its coefficients; and similarly for $L_{jp'}^{(\phi_p)}$, etc.

CHAPTER 8

Curvature of Lax connections

In this chapter we study, in some detail, Lax connections. Recall that these are adelic connections defined on certain open sets of GL_n. However these open sets do not contain the identity so analytic continuation along the identity is not possible. Instead, we will perform analytic continuation along certain torsion points (other than the identity) which will allow us to again define curvature. We end by proving vanishing/non-vanishing results for this curvature.

8.1. Analytic continuation along torsion points

We examine here in some detail the isocharacteristic and isospectral Lax connections. We will work over a δ-ring of global type, $A = \mathcal{O}_K[1/M]$, cf. Definition 3.25. We assume the notation and terminology introduced in section 4.5. In particular the have at our disposal the objects

$$G = GL_n,\ \mathfrak{g} = \mathfrak{gl}_n,\ T,\ D^*,\ D^{**},\ T^*,\ G^*,\ G^{**},\ \mathcal{P},\ \mathcal{C},\ \text{etc.}$$

introduced in loc.cit. These objects were defined over any δ-ring of local type but they are obviously defined over our δ-ring of global type as well. By a *regular n-torsion point* of $T(A)$ we mean, as usual, an n-torsion point of $T(A)$ belonging to $T^*(A)$. The local Theorem 4.56 immediately implies the following global statement:

THEOREM 8.1. *There exists a unique adelic connection $(\delta_p) = (\delta_p^{G^{**}})$ on $G^{**} \subset G = GL_n$, with Frobenius lifts $(\phi_p^{G^{**}})$, such that, for all p, $\phi_p^{G^{**}}$ is $\mathcal{P}$-horizontal with respect to the trivial $\phi_{0p}^{\mathbb{A}^n}$ on $\mathbb{A}^n = Spec\ A[z]$, $\phi_{0p}^{\mathbb{A}^n} z = z^{(p)}$, and $\phi_p^{G^{**}}$ is a T-deformation of the trivial p-adic connection δ_{0p}^G on G. Moreover, for each $p \equiv 1$ mod n, $\phi_p^{G^{**}}$ fixes individually all the regular n-torsion points of $T(A)$.*

As in the local case we make the following:

DEFINITION 8.2. The adelic connection $(\delta_p^{G^{**}})$ in Theorem 8.1 is called the *canonical isocharacteristic Lax connection* on G^{**}. Let (ϕ_p) be the family of Frobenius lifts attached to the canonical isocharacteristic Lax connection. If, in addition, a family $(\alpha_p(x))$ is given, where, for each p, $\alpha_p \in \mathfrak{g}(\mathcal{O}(G^{**})^{\widehat{p}})$ then the adelic connection $(\delta_p^{(\alpha_p)})$ defined by the Frobenius lifts

$$\phi_p^{(\epsilon_p)}(x) = \Phi_p^{(\epsilon_p)}(x) := \epsilon_p(x) \cdot \phi_p(x) \cdot \epsilon_p(x)^{-1}, \quad \epsilon_p(x) = 1 + p\alpha_p(x)$$

will be referred to as the *isocharacteristic Lax connection* on G^{**} attached to (α_p).

The family $(\phi_p^{(\epsilon_p)})$ is trivially seen to be still $\mathcal{P}$-horizontal with respect to $(\phi_{0p}) = (\phi_{0p}^{\mathbb{A}^n})$. In other words we have a commutative diagram

$$\begin{array}{ccc} (G^{**})^{\widehat{p}} & \stackrel{\phi_p^{(\epsilon_p)}}{\longrightarrow} & (G^{**})^{\widehat{p}} \\ \mathcal{P}\downarrow & & \downarrow\mathcal{P} \\ (\mathbb{A}^n)^{\widehat{p}} & \stackrel{\phi_{0p}}{\longrightarrow} & (\mathbb{A}^n)^{\widehat{p}} \end{array} \tag{8.1}$$

Note that if the canonical isocharacteristic Lax connection (δ_p) is given by

$$\phi_p(x) = x^{(p)} \cdot \Lambda_p(x)$$

then $\Lambda_p(x)$ is diagonal and the isocharacteristic Lax connection

$$(\delta_p^{((\Lambda_p-1)/p)})$$

is given by the family $(\phi_p^{(\epsilon_p)})$,

$$\phi_p^{(\epsilon_p)}(x) = \Lambda_p(x) \cdot \phi_p(x) \cdot \Lambda_p(x)^{-1} = \Lambda_p(x) \cdot x^{(p)}. \tag{8.2}$$

DEFINITION 8.3. We will say that a family (α_p) with $\alpha_p \in \mathfrak{g}(\mathcal{O}(G^{**})^{\widehat{p}})$ is global along a point $P \in G(A)$ (with retract coordinates T) if α_p vanish at P and the images of α_p in $\mathfrak{g}(A^{\widehat{p}}[[T]])$ belong to $\mathfrak{g}(A[[T]])$.

We will prove the following:

THEOREM 8.4. *Assume $\mathbb{Z}[1/n, \zeta_n] \subset A$ and let $\mathcal{V}$ be such that $p \equiv 1$ mod n for all $p \in \mathcal{V}$. Let P be a regular n-torsion point of $T(A)$, T the diagonal torus of $G = GL_n$. Then the canonical isocharacteristic Lax connection on G^{**} is global along P. Consequently if, in addition, a family (α_p) global along P is given then the isocharacteristic Lax connection $(\delta_p^{(\alpha_p)})$ on G^{**} is global along P. In particular the isocharacteristic Lax connection* 8.2 *is global along P.*

In particular, by the theorem above, one can consider the curvature of isocharacteristic Lax connections and also the $(1,1)$-curvature of isocharacteristic Lax connections (δ_p) (with respect to (δ_{0p})); since, in the case of classical differential geometry, the curvature of the isocharacteristic Lax connections are closely connected to the traceless curvature of Ehresmann connections one could consider the curvature of isocharacteristic Lax connections, in the arithmetic case, as an alternative path towards curvature in arithmetic. In particular one can ask for the vanishing of curvature and $(1,1)$-curvature of isocharacteristic Lax connections; we will address this question in the next section.

PROOF OF THEOREM 8.4. Let (ϕ_p) be the Frobenius lifts corresponding to the canonical isocharacteristic Lax connection. The ϕ_p's fix P by Theorem 8.1. Write $\phi_p(x) = x^{(p)}\Lambda_p(x)$, where $\Lambda_p(x)$ is a matrix $\equiv 1$ mod p with coefficients in

$$A[x, \det(x)^{-1}, D^{**}(x)^{-1}]^{\widehat{p}}. \tag{8.3}$$

The condition that $\phi_p^{G^{**}}$ on G^{**} be $\mathcal{P}$-horizontal with respect to δ_{0p} on $\mathbb{A}^n$ is equivalent to the equalities

$$\mathcal{P}_i(x^{(p)}\Lambda_p(x)) = \mathcal{P}_i(x)^p \tag{8.4}$$

for all $i = 1, ..., n$. Also the condition that $\phi_p^{G^{**}}$ on G^{**} be a T-deformation of δ_{0p} on G is equivalent to $\Lambda_p(x)$ being diagonal. So there is a unique diagonal $\Lambda_p(x)$

with coefficients in 8.3 satisfying 8.4 and $\equiv 1 \mod p$. Let now $P \in T^*(A)$ be given by

$$P = \mathrm{diag}(d_1, ..., d_n) =: d,$$

and consider the matrix $T = x - d$. (Again we are using the letter T to denote the entries of $x - d$ and also to denote the diagonal torus; this should lead to no confusion.) By our hypotheses $\{d_1, ..., d_n\} = \{1, \zeta_n, ..., \zeta_n^{n-1}\}$; hence $\det(s \cdot 1_n - d) = s^n - 1$. Since ϕ_p fixes P we have $\Lambda_p(d) = 1$. Then 8.3 embeds into $A^{\widehat{p}}[[T]]$ and 8.4 holds therefore in $A^{\widehat{p}}[[T]]$. One checks the following stronger uniqueness property: $\Lambda_p(x) = \Lambda_p(d+T) =: \Lambda_p^0(T)$ is the unique diagonal matrix with entries in $A^{\widehat{p}}[[T]]$ satisfying 8.4 and $\equiv 1 \mod p$; this is done by taking any two such matrices and proving by induction on ν that they coincide mod p^ν. So it is enough to find a diagonal matrix $\Lambda_p^*(T)$ with entries in $A[[T]]$, which is $\equiv 1 \mod p$, such that

$$\mathcal{P}_i(x^{(p)}\Lambda_p^*(T)) = \mathcal{P}_i(x)^p, \tag{8.5}$$

because then, by the mentioned uniqueness, we will have that $\phi_p(x)$ has entries in $A[[T]]$. The proof of the existence of Λ_p^* given below is a (T)-adic version of the p-adic proof for the existence of Λ_p in Theorem 4.56. For completeness we redo here the argument; the objects below, with a $*$ superscript, roughly correspond to the corresponding objects without a $*$ in the proof of Theorem 4.56. The argument runs as follows. Equation 8.5 is equivalent to the equality

$$V^*(T,s) := \sum_{i=0}^{n}(-1)^i \mathcal{P}_i(d+T)^p s^{n-i} = \sum_{i=0}^{n}(-1)^i \mathcal{P}_i((d+T)^{(p)}\Lambda_p^*(T))s^{n-i},$$

hence to the equality

$$V^*(T,s^p) = \sum_{i=0}^{n}(-1)^i \mathcal{P}_i((d+T)^{(p)}\Lambda_p^*(T))s^{p(n-i)}.$$

To prove the existence of $\Lambda_p^*(T)$ it is enough to construct a sequence $\Lambda_\nu^*(T)$ of diagonal matrices satisfying the following properties:

(1) $\det(s^p \cdot 1_n - (d+T)^{(p)}\Lambda_\nu^*(T)) \equiv V^*(T,s^p) \mod p(T)^{\nu+1}$, $\nu \geq 0$.
(2) $\Lambda_{\nu+1}^*(T) \equiv \Lambda_\nu^*(T) \mod p(T)^{\nu+1}$, $\nu \geq 0$.
(3) $\Lambda_0^* = 1$.

Take $\Lambda_0^* = 1$ as in (3); then (1) is satisfied for $\nu = 0$ because

$$\det(s^p \cdot 1_n - (d+T)^{(p)}) \equiv \det(s^p \cdot 1_n - d) = s^{pn} - 1 \equiv V^*(T,s^p) \mod (T),$$

$$\det(s^p \cdot 1_n - (d+T)^{(p)}) \equiv V^*(T,s^p) \mod p.$$

Assume now $\Lambda_\nu^*(T)$ was constructed and seek $\Lambda_{\nu+1}^*(T)$ of the form $\Lambda_\nu^*(T) - pZ^*(x)$ where $Z^*(x)$ is a diagonal matrix with entries in $(T)^{\nu+1}$ such that (1) holds. Write

$$\begin{aligned}
A^*(T) &= s^p \cdot 1_n - (d+T)^{(p)}\Lambda_\nu^*(T) = (a_{ij}),\\
\det(A^*(T)) &= V^*(T,s^p) + pU^*(T,s^p),\\
U^*(T,s) &= \textstyle\sum_{i=0}^{n-1} U_i^*(T)s^{n-1-i}, \quad U_i^*(T) \in (T)^{\nu+1},\\
Z^* = Z^*(T) &= \mathrm{diag}(z_1^*, ..., z_n^*),\\
B^* &= (d+T)^{(p)}Z^*(T) = (b_{ij}) = (z_j^* x_{ij}^p).
\end{aligned}$$

Then

$$\det(s^p \cdot 1_n - (d+T)^{(p)}\Lambda^*_{\nu+1}(T)) = \det \begin{pmatrix} a_{11}+pb_{11} & a_{12}+pb_{12} & * \\ a_{21}+pb_{21} & a_{22}+pb_{22} & * \\ * & * & * \end{pmatrix}$$

which is $\equiv$ mod $p(T)^{\nu+2}$ to

$$\det(A^*(T)) + pW^*(T, s^p)$$

where

$$W^*(T, s^p) = \det \begin{pmatrix} b_{11} & a_{12} & * \\ b_{21} & a_{22} & * \\ * & * & * \end{pmatrix} + \det \begin{pmatrix} b_{11} & a_{12} & * \\ b_{21} & a_{22} & * \\ * & * & * \end{pmatrix} + \ldots$$

So, for property (1) above, it is enough to find $Z^*(T)$ such that

$$U^*(T, s^p) + W^*(T, s^p) \equiv 0 \mod (T)^{\nu+2}. \tag{8.6}$$

Consider the system

$$\sum_{j=1}^{n} \mathcal{P}_{ij}(x^{(p)})z_j^* = (-1)^i U_i^*(T), \quad i = 0, ..., n-1. \tag{8.7}$$

This system has determinant $\det(\mathcal{P}_{ij}((d+T)^{(p)}))$; the value of this determinant at $T = 0$ equals

$$\det(\mathcal{P}_{ij}(d^{(p)})) = \det(\mathcal{P}_{ij}(d)) = D^{**}(d) \in A^{\times}$$

hence the determinant of the system 8.7 is invertible in $A[[T]]$ so the system is solvable and, by Cramer's rule, the components z_j^* of the solution lie in $(T)^{\nu+1}$. Fix from now on these values of z_j^*. We have $A^*(T) \equiv s^p \cdot 1_n - x^{(p)}$ mod (T) so we have that $W^*(T, s^p)$ is $\equiv$ mod $(T)^{\nu+2}$ to

$$-z_1^* \det \begin{pmatrix} -x_{11}^p & -x_{12}^p & * \\ -x_{21}^p & s^p - x_{22}^p & * \\ * & * & * \end{pmatrix} - z_2^* \det \begin{pmatrix} s^p - x_{11}^p & -x_{12}^p & * \\ -x_{21}^p & -x_{22}^p & * \\ * & * & * \end{pmatrix} - \ldots$$

The latter congruence plus the equalities 8.7 imply the congruence 8.6. This ends our construction and hence our proof. □

Finally we discuss *isospectral Lax connections.* The local Theorem 4.50 immediately implies the following global result:

THEOREM 8.5. *There exists a unique adelic connection $(\delta_p) = (\delta_p^{G^*})$ on G^*, with attached Frobenius lifts $(\phi_p^{G^*})$, such that for all p, the trivial ϕ_{0p}^G on G is $\mathcal{C}$-horizontal with respect to $\phi_p^{G^*}$.*

As in the local case we make the following:

DEFINITION 8.6. The adelic connection (δ_p) on G^* in the above theorem is called the *canonical isospectral Lax connection* on G^*. More generally if the canonical isospectral Lax connection (δ_p) has attached Frobenius lifts (ϕ_p) and, in addition, a family $(\alpha_p(x))$ is given, where, for each p, $\alpha_p \in \mathfrak{g}(\mathcal{O}(G^*)^{\widehat{p}})$ then the adelic connection $(\delta_p^{(\alpha_p)})$ defined by the Frobenius lifts

$$\phi_p^{(\epsilon_p)}(x) = \Phi_p^{(\epsilon_p)}(x) := \epsilon_p(x) \cdot \phi_p(x) \cdot \epsilon_p(x)^{-1}, \quad \epsilon_p(x) = 1 + p\alpha_p(x)$$

will be referred to as the *isospectral Lax connection* on G^* attached to (α_p).

We have an obvious notion of family (α_p) as above being global along P: one simply considers the Definition 8.3 with G^{**} replaced by G^*.

THEOREM 8.7. *Let P be a regular torsion point of $T(A)$, where T is the diagonal torus of $G = GL_n$. Then the canonical isospectral Lax connection (δ_p) on G^* is global along P. In particular (δ_p) has a well defined curvature $(\varphi_{pp'})$ (with respect to P) and this curvature vanishes: $\varphi_{pp'} = 0$ for all p, p'. If, in addition, a family (α_p) global along P is given then the isospectral Lax connection $(\delta_p^{(\alpha_p)})$ on G^* is global along P and hence has a well defined curvature.*

Note that the theorem above says nothing about the vanishing or non-vanishing of the $(1,1)$-curvature of (δ_p) on G^* with respect to (δ_{0p}) on G^*.

PROOF OF THEOREM 8.7. Writing $t = \text{diag}(t_1, ..., t_n)$, the map $\mathcal{C} : T^* \times G \to G^*$ is given by the ring homomorphism

$$A[y, \det(y)^{-1}, D^*(y)^{-1}] \to A[x, t, \det(x)^{-1}, \det(t)^{-1}, D^*(t)^{-1}], \quad y \mapsto x^{-1}tx.$$

Let P be given by $t \mapsto d \in T^*(A)$, $d = \text{diag}(d_1, ..., d_n)$, so $\phi_p(d_i) = d_i^p$. Consider the point $Q \in T^*(A) \times G(A)$ given by $t \mapsto d$, $x \mapsto 1$, so $\mathcal{C}(Q) = P$. By $\mathcal{C}$-horizontality, ϕ_p sends P into P (because ϕ_{0p} sends Q into Q). Set $T = y - d$, $U = x - 1$, $V = t - d$. Then $\mathcal{C}$ induces an injective ring homomorphism (which we view as an inclusion) $A[[T]] \to A[[U, V]]$ given by

$$\begin{aligned} T &\mapsto (1+U)^{-1}(V+d)(1+U) - d \\ &\equiv V + dU - Ud \mod (U, V)^2. \end{aligned}$$

By $\mathcal{C}$-horizontality the endomorphism ϕ_p of $A^{\widehat{p}}[[U, V]]$ sends $A^{\widehat{p}}[[T]]$ into itself. To conclude the proof it is enough to show that the endomorphism ϕ_{0p} of $A[[U, V]]$ sends $A[[T]]$ into itself. (Indeed this shows that (δ_p) is global along P, while the vanishing of the curvature of (δ_p) follows directly from the commutation of the ϕ_{0p}'s on $A[[U, V]]$.) So we are reduced to checking that

$$A^{\widehat{p}}[[T]] \cap A[[U, V]] = A[[T]] \tag{8.8}$$

in $A^{\widehat{p}}[[U, V]]$. Note that

$$\begin{aligned} (V + dU - Ud)_{ij} &= (d_i - d_j)U_{ij} \quad \text{for } i \neq j, \\ (V + dU - Ud)_{ii} &= V_{ii}. \end{aligned}$$

This immediately implies that $A[[U, V]] = A[[T, W]]$ where $W = \{W_1, ..., W_n\}$, $W_i = U_{ii}$. In its turn this immediately implies 8.8. □

8.2. Non-vanishing/vanishing of curvature

An "automatic vanishing" result for isospectral Lax connections is part of Theorem 8.7. Here is a result on isocharacteristic Lax connections.

THEOREM 8.8. *Let $n = 2$, $A = \mathbb{Z}[1/2]$. Then the curvature and $(1,1)$-curvature of the canonical isocharacteristic Lax connection satisfy:*

(1) $\varphi_{pp'}^{(1)} = 0$ *and* $\varphi_{p\bar{p}'}^{(1)'} = 0$ *for all* p, p'.

(2) $\varphi_{p\bar{p}}^{(2)} \neq 0$ *for all* p; *in particular* $\varphi_{p\bar{p}} \neq 0$ *for all* p.

It is not clear at this point what happens in case $n \geq 3$ or whether $\varphi_{pp'}$ vanishes in case $n = 2$.

PROOF OF THEOREM 8.8. The canonical isocharacteristic Lax connection (δ_p) can be explicitly given in case $n = 2$ as follows. Write $x = \begin{pmatrix} a & b \\ c & d \end{pmatrix}$ and assume, as we may, that $P = \begin{pmatrix} 1 & 0 \\ 0 & -1 \end{pmatrix}$. Then D^{**} is, up to sign, $(ad - bc)(a - d)$ and

$$\phi_p \begin{pmatrix} a & b \\ c & d \end{pmatrix} = \begin{pmatrix} u_1 & \frac{b^p}{d^p} u_2 \\ \frac{c^p}{a^p} u_1 & u_2 \end{pmatrix} \tag{8.9}$$

where u_1, u_2 are the two roots, in the ring

$$A[a, b, c, d, \frac{1}{(ad - bc)(a - d)}]^{\widehat{p}}, \tag{8.10}$$

of the equation

$$u^2 - (a + d)^p u + a^p d^p \frac{(ad - bc)^p}{a^p d^p - b^p c^p} = 0,$$

satisfying $u_1 \equiv 1 \bmod p$ and $u_2 \equiv -1 \bmod p$; the coefficients on the matrix 8.9 turn out to be also in the ring 8.10. On the other hand, if $T = x - \begin{pmatrix} 1 & 0 \\ 0 & -1 \end{pmatrix}$, then ϕ_p on $A[[T]]$ acts as

$$\phi_p \begin{pmatrix} T_{11} & T_{12} \\ T_{21} & T_{22} \end{pmatrix} = \begin{pmatrix} U_1 - 1 & \frac{T_{12}^p}{(T_{22}-1)^p} U_2 \\ \frac{T_{21}^p}{(T_{11}+1)^p} U_1 & U_2 + 1 \end{pmatrix}$$

where U_1, U_2 are the corresponding roots in $A[[T]]$ of the equation

$$U^2 - (T_{11} + T_{22})^p U + (T_{11} + 1)^p (T_{22} - 1)^p \frac{((T_{11} + 1)(T_{22} - 1) - T_{12}T_{21})^p}{(T_{11} + 1)^p (T_{22} - 1)^p - T_{12}^p T_{21}^p} = 0.$$

Note that

$$\phi_p(T) \equiv \frac{p}{2}(T_{11} - T_{22}) \cdot \begin{pmatrix} 1 & 0 \\ 0 & -1 \end{pmatrix} \quad \bmod \ (T)^2$$

and hence ϕ_p and $\phi_{p'}$ commute on $A[[T]]/(T)^2$; in the same way one checks that ϕ_p and $\phi_{\overline{p}'}$ commute on $A[[T]]/(T)^2$, which proves assertion (1).

To prove assertion (2) consider two indeterminates s and t and let v_1, v_2 be the two roots in $A[[s, t]]$ of the equation

$$v^2 - s^p v - \left(1 + t + \frac{t^2}{4} - \frac{s^2}{4}\right)^p = 0$$

that are $\equiv 1$ and $-1 \bmod (s, t)$ respectively. We have a commutative diagram

$$\begin{array}{ccc} A[[s,t]] & \xrightarrow{\phi_p} & A[[s,t]] \\ \pi \uparrow & & \uparrow \pi \\ A[[T]] & \xrightarrow{\phi_p} & A[[T]] \end{array}, \tag{8.11}$$

where

$$\pi(T) = \operatorname{diag}\left(\tfrac{s+t}{2}, \tfrac{s-t}{2}\right),$$

$$\phi_p(s) = v_1 + v_2,$$

$$\phi_p(t) = v_1 - v_2 - 2.$$

Note that

$$v_1 \equiv -v_2 \equiv 1 + \frac{p}{2}t + \frac{p^2-p}{8}t^2 - \frac{p}{8}s^2 \mod (s,t)^3,$$

hence

$$\phi_p(s) \equiv 0 \mod (s,t)^3$$

$$\phi_p(t) \equiv pt + \tfrac{p^2-p}{4}t^2 - \tfrac{p}{4}s^2 \mod (s,t)^3.$$

On the other hand we have a commutative diagram as in 8.11 with ϕ_p replaced by $\phi_{\bar{p}}$ where

$$\phi_{\bar{p}}(T) = (1+T)^{(p)} - 1$$

$$\phi_{\bar{p}}(s) = \left(1 + \tfrac{s+t}{2}\right)^p + \left(1 + \tfrac{s-t}{2}\right)^p - 2,$$

$$\phi_{\bar{p}}(t) = \left(1 + \tfrac{s+t}{2}\right)^p - \left(1 + \tfrac{s-t}{2}\right)^p.$$

We have

$$\phi_{\bar{p}}(s) \equiv ps + \tfrac{p^2-p}{4}s^2 + \tfrac{p^2-p}{4}t^2 \mod (s,t)^3,$$

$$\phi_{\bar{p}}(t) \equiv pt + \tfrac{p^2-p}{2}st \mod (s,t)^3.$$

A direct computation gives

$$\phi_{\bar{p}}\phi_p(s) \equiv 0 \mod (s,t)^3,$$

$$\phi_p\phi_{\bar{p}}(s) \equiv \tfrac{p^4-p^2}{4}t^2 \mod (s,t)^3,$$

hence $\phi_{\bar{p}}\phi_p(s) \not\equiv \phi_p\phi_{\bar{p}}(s) \mod (s,t)^3$, which ends the proof of assertion (2). □

CHAPTER 9

Open problems

At this point *arithmetic differential geometry* is far from being a fully developed theory. What are then the next expected steps? And where is the theory headed? A number of puzzles, internal to the theory, as well as suggestive of connections with the rest of number theory, may be showing the right direction. We collect here a sample of such puzzles/problems, both technical and conceptual, that seem to indicate a general common tendency.

9.1. Unifying $\mathfrak{hol}_{\mathbb{Q}}$ and $\Gamma_{\mathbb{Q}}$

One of the main objects of the theory presented in this book is the *holonomy algebra* $\mathfrak{hol}_{\mathbb{Q}}$ of an adelic connection; cf. Definition 3.59. (This is defined only for adelic connections that are global along appropriate subschemes; and as we saw this is the case with real Chern and Levi-Cività connections attached to large classes of metrics, in particular to δ-constant metrics.) We would like to view the Lie algebra $\mathfrak{hol}_{\mathbb{Q}}$ as an infinitesimal analogue of the absolute Galois group $\Gamma_{\mathbb{Q}} = Gal(\mathbb{Q}^a/\mathbb{Q})$.

On the other hand if Hol is the holonomy group of a connection on principal G-bundle over a manifold M in classical differential geometry and if Hol^0 is its connected component then the quotient Hol/Hol^0 is usually referred to as the *monodromy group* of the connection; it is isomorphic to the image Γ_M of the monodromy representation

$$\pi_1(M) \to G \tag{9.1}$$

defined by our connection. (We have ignored here the base points.) The exact sequence

$$1 \to Hol^0 \to Hol \to \Gamma_M \to 1 \tag{9.2}$$

gives rise to a natural homomorphism

$$\Gamma_M \to \text{Out}(Hol^0). \tag{9.3}$$

Now it is classical to see the absolute Galois group $\Gamma_{\mathbb{Q}}$ as an arithmetic analogue of a fundamental group $\pi_1(M)$. Galois representations are then analogous to monodromy representations 9.1 and their images are analogous to the groups Γ_M. Since our Lie algebras $\mathfrak{hol}_{\mathbb{Q}}$ attached to real Chern or Levi-Cività adelic connections are an arithmetic analogue of the Lie algebra of Hol^0 and since we attached Lie algebras to Galois connections it is reasonable to pose the following:

Problem 1. "Unify" our holonomy on GL_n with Galois theory by constructing canonical extensions of Lie algebras

$$0 \to \mathfrak{hol}_1 \to \mathfrak{hol}_2 \to \mathfrak{hol}_3 \to 0 \tag{9.4}$$

where $\mathfrak{hol}_3$ is attached to a Galois connection and $\mathfrak{hol}_1$ is attached to a real Chern or Levi-Cività (adelic) connection on GL_n. Such an extension 9.4 should function as an arithmetic analogue of the extension 9.2 and the induced representation

$$\mathfrak{hol}_3 \to \mathrm{Out}(\mathfrak{hol}_1) \tag{9.5}$$

should be an analogue of the representation 9.3. This construction could involve interesting Galois representations; indeed an extension such as 9.4 could be, in its turn, analogous to the basic exact sequence

$$1 \to \pi_1(X^a) \to \pi_1(X) \to \Gamma_{\mathbb{Q}} \to 1 \tag{9.6}$$

attached to a geometrically connected scheme X over $\mathbb{Q}$ (where $X^a = X \otimes \mathbb{Q}^a$ and we suppressed, again, reference to base points); cf. [**57, 77**]. The representation 9.5 could then be analogous to the representation

$$\Gamma_{\mathbb{Q}} \to \mathrm{Out}(\pi_1(X^a)) \tag{9.7}$$

arising from 9.6. The representations 9.7 play a prominent role in work of Grothendieck, Ihara, Deligne, and many others.

Here are some related problems:

Problem 2. Find links between our curvature of adelic connections and reciprocity (as it appears in arithmetic topology [**104**]); the appearance of the Legendre symbol in our Christoffel symbols may be an indication that such links may exist.

Problem 3. Develop a Galois correspondence for Ehresmann connections; extend the Galois theory from Ehresmann connections to other types of connections, especially real Chern and special linear connections.

Problem 4. Find links between our adelic (flat) connections and Galois representations similar to the classical link between vector bundles with flat connections and monodromy.

Problem 5. Find an arithmetic analogue of Atiyah's theory [**3**] according to which:

(1) the algebraic connections on an algebraic vector bundle over a variety are in bijection with the splittings of the Atiyah extension;
(2) the obstruction to the splitting of the Atiyah extension generates the characteristic ring of the vector bundle.

Problem 6. The curvature of adelic Chern connections led, in our theory, to objects that we called *Chern* $(1,1)$*-forms*; it is tempting to use these forms, and higher versions of them, to define cohomology classes in the style of Chern-Weil theory; the target cohomology groups could be Galois (étale) cohomology groups of appropriate arithmetic objects.

9.2. Unifying "$\partial/\partial p$" and "$\partial/\partial\zeta_{p^\infty}$"

As discussed in the Introduction the theory of p-adic periods of Fontaine and Colmez involves, in particular, a "totally ramified arithmetic calculus" that is perpendicular to our (unramified) arithmetic calculus. In its simplest form this totally ramified calculus morally comes from the fact that the modules of Kähler differentials $\Omega_{\mathbb{Z}[\zeta_{p^n}]}$ are torsion but non-trivial: $d\zeta_{p^n} \neq 0$. It is conceivable that this situation could lead to a "derivation like" operator, "$\partial/\partial\zeta_{p^\infty}$" on an appropriate space on which our p-derivation δ_p also acts. Then one is led to:

Problem 7. Unify our unramified arithmetic calculus (based on p-derivations δ_p, thought of as "$\partial/\partial p$") with the totally ramified arithmetic calculus of Fontaine-Colmez [**64**] (possibly involving the conjectural "derivation like" operators "$\partial/\partial_{p^\infty}$"). One would presumably get, for each p, a partial differential situation with two "directions" and, in particular, one could attempt to study the resulting "curvature." A way to proceed could be to start from the paper [**32**] where the base ring was the ring of power series $R[[q]]$ in the variable q, equipped with the "standard" p-derivation δ_p (extending that of $R = W(\mathbb{F}_p^a)$ and killing q) and with the usual derivation $\delta_q = qd/dq$. Then one could try to specialize the theory in [**32**] via $q \mapsto \zeta_{p^n}$ with $n \to \infty$. Such a specialization does not a priori make sense, of course, and no recipe to give it a meaning seems available at this point. However an indication that this procedure might be meaningful can be found in the formulas giving the "explicit reciprocity laws" of local class field theory due to Shafarevich, Brückner, Vostokov, etc. Cf. [**63**].

9.3. Unifying Sh and GL_n

Problem 8. Relate the arithmetic differential calculus on Shimura varieties Sh (appearing in our book [**35**]) with the arithmetic differential calculus on the classical groups (appearing in our present book). In particular examine the case of

$$Sh = SL_2(\mathbb{Z})\backslash SL_2(\mathbb{R})/SO_2(\mathbb{R}).$$

An indication that such a relation may exist is the δ-algebraic theory of δ-Hodge structures in [**22**] giving δ-algebraic correspondences between classical groups and moduli spaces of abelian varieties; one would have to develop an arithmetic analogue of [**22**].

Here are some related questions:

Problem 9. Find a link between the concept of curvature in the present book and the concept of curvature based on the arithmetic Laplacian in [**34**]. More generally find a common ground between our theory here and [**34**].

Problem 10. The theory in this book is mainly about adelic connections on GL_n possibly compatible with involutions of GL_n. It is natural to explore what happens if one replaces GL_n by a general reductive group.

Problem 11. Explore the arithmetic differential geometry of homogeneous spaces such as "spheres" $S^n := SO_{n+1}/SO_n$.

Problem 12. Develop an arithmetic analogue of the Cassidy-Kolchin theory of differential algebraic groups. Compute the rings of invariant δ-functions for actions of such groups on the varieties naturally appearing in Riemannian geometry. For instance compute the δ-functions on $SO_n\backslash GL_n/\Gamma$ where Γ is a δ-subgroup of GL_n. The δ-invariants of the curvature of the Levi-Cività connection may be playing a role in this problem.

Problem 13. Find interactions between the classical theory of quadratic forms over rings of integers in number fields [**107**] and our theory here (around symmetric matrices that play the role of metrics). Both theories are about the geometry of GL_n/SO_n.

9.4. Further concepts and computations

Problem 14. Decide if the curvature of the real Chern connection attached to a split symmetric $q \in GL_n$ vanishes for $n = 2, 3$. Compute (or prove non-vanishing of) curvature for the correspondence structures attached to real Chern connections on GL_n in case q is symmetric and $n = 2$. Same problem for curvature and $(1,1)$-curvature in case $n \geq 3$.

Problem 15. Prove vanishing/non-vanishing for the mixed curvature of the Levi-Cività connection attached to split q's. Perform analytic continuation of Levi-Cività connection along partition tori and analyze the corresponding mixed curvature and gauge action.

Problem 16. Investigate the existence of correspondence structure for Levi-Cività connection; same question for the canonical isospectral and isocharacteristic Lax connections. In case such correspondence structures exist compute (or prove vanishing/non-vanishing of) their curvatures.

Problem 17. Study the impact of conformal changes of metrics on curvature.

Problem 18. Find the explicit structure of holonomy algebras $\mathfrak{hol}, \mathfrak{hol}^{\Omega}$ attached to real Chern connections.

Problem 19. Investigate the analytic properties of the Dirichlet series arising as our mean and scalar $(1,1)$-curvatures.

Problem 20. Further develop the formalism of curvature in the context of the ring of correspondences on a field.

Problem 21. Develop the theory of Ricci curvature and scalar curvature attached to the Levi-Cività connection. This will require a better understanding of the problem of soldering i.e., the problem of how to view GL_∞ as an analogue of the frame bundle of $Spec\ \mathbb{Z}$.

Problem 22. Develop an arithmetic analogue of connections that are Hamiltonian with respect to the Lie-Poisson bracket. Find an arithmetic analogue of the theory of algebraically completely integrable systems [**2**], along the lines of the arithmetic analogue of the Euler equations for the rigid body discussed in the book; cf. [**48**]. Find a strengthening of the concept of Euler connection by requiring equality instead of congruence in the linearization condition.

Problem 23. Find an arithmetic analogue of Fedosov connections; furthermore find an arithmetic counterpart of deformation quantization. Indeed, the “strong” arithmetic analogue of Fedosov connections introduced in the text was proved not to exist in dimension ≥ 4; what is needed is the “correct weakening” of that concept.

Problem 24. Many of the curvature computations in our theory have the flavor of computations in *algebraic dynamics* (study of semigroups of endomorphisms of algebraic varieties); an instance of this is our use of ideas from [**65**] in the study of adelic Ehresmann connections. The connection with algebraic dynamics deserves further study.

Problem 25. Find an arithmetic analogue of geodesics. A hint could come from [**32**] where “time” was a geometric variable and “space” was represented by arithmetic variables (primes).

Problem 26. Perform curvature computations for complex Chern connections; as a rule they are easier than the corresponding computations for real Chern connections which were presented in the book. In particular complex Chern connections have obvious correspondence structure which is represented by field endomorphisms; and if attached to diagonal matrices they are always global along the diagonal maximal torus so Jordan multiplication ceases to play the role it played in the real case.

9.5. What lies at infinity?

The theory of the present book involves non-Archimedian (finite) adeles only. A natural question is whether there is an Archimedian (infinite) counterpart to our theory. So one can ask, for instance:

Problem 27. Could Borger's positivity approach to study "Witt vectors at infinity" [**15**] interact with our theory? This sounds reasonable in view of the compatibility between our theory and Borger's in the case of a single prime.

Problem 28. Could concepts of *discrete differential geometry* [**10**] play the role of Archimedian counterpart to our theory?

Problem 29. Is there an Archimedian counterpart to our theory that has the flavor of the classical *geometry of numbers* [**49**] or more generally of *Arakelov geometry*?

Problem 30. Could an *arithmetic physics* (in the sense of [**100**]) be developed, with our *arithmetic differential geometry* as background, that would have "real physics" as its Archimedian counterpart? This would offer a picture rather different from that in [**120**], as explained in the Introduction. Indeed the underlying analogy in [**120**] is between p-adic numbers and real numbers whereas the underlying analogy in our arithmetic differential geometry is between p-adic numbers and functions.

Bibliography

[1] D.V. Alekseevskij, V.V. Lychagin, A.M. Vinogradov, *Geometry, I: Basic Ideas and Concepts of Differential Geometry*, Springer, 1991.

[2] M. Adler, P. van Moerbeke, P. Vanhaecke, *Algebraic Integrability, Painlevé Geometry and Lie Algebras* , Springer, 2004.

[3] M. F. Atiyah, *Complex analytic connections in fiber bundles*, Trans. AMS, 85 (1957), 181-207.

[4] M. F. Atiyah, I. G. MacDonald, *Introduction to Commutative Algebra*, Addison-Wesley Series in Mathematics, Westview, 1994.

[5] F. Bayen, M. Flato, C. Fronsdal, A. Lichnerowicz, D. Sternheimer, *Deformation theory and quantization.I*, Ann. Phys., 111 (1978), 61-110.

[6] A. L. Besse, *Einstein Manifolds*, Springer, 1987.

[7] M. Barcau, A. Buium, *Siegel differential modular forms*, International Math. Res. Notices, 2002, No. 28, pp.1459-1503.

[8] M. Barrett, A. Buium, *Curvature on the integers, I*, J. Number Theory, 167 (2016), 481-508.

[9] F. Bayen, M. Flato, C. Fronsdal, A. Lichnerowicz and D. Sternheimer, *Quantum mechanics as a deformation of classical mechanics, Lett. Math. Phys. 1 (1977) 521-530*, and *Deformation theory and quantization*, part I, Ann. of Phys. 111 (1978) 61-110.

[10] A. I. Bobenko, Y. B. Suris, *Discrete Differential Geometry*, Graduate Studies in Math, AMS, 2008.

[11] J. Borger, *The basic geometry of Witt vectors, I*: the affine case, Algebra and Number Theory 5 (2011), no. 2, pp 231-285.

[12] J. Borger, *The basic geometry of Witt vectors, II: Spaces*, Mathematische Annalen 351 (2011), no. 4, pp 877-933.

[13] J. Borger, *Λ-rings and the field with one element*, arXiv:0906.3146.

[14] J. Borger, A. Buium, *Differential forms on arithmetic jet spaces*, Selecta Mathematica 17, 2 (2011), pp. 301-335.

[15] J. Borger, *Witt vectors, semirings, and total positivity*, Absolute Arithmetic and F1-Geometry, K. Thas, editor. EMS Publishing House, 2016.

[16] S. Bosch, W. Lütkebohmert, M. Raynaud, *Néron Models*, Ergebnisse der Mathematik und ihrer Grenzgebiete, 3, 21, Springer, 1990.

[17] J-B. Bost, K. Künnemann, *Hermitian vector bundles and extension groups on arithmetic schemes II. The arithmetic Atiyah class.* Astérisque 327 (2009), 357-420.

[18] A. Buium, *Differential Function Fields and Moduli of Algebraic Varieties*, Lecture Notes in Math. 1226, Springer 1986.

[19] A. Buium, *Intersections in jet spaces and a conjecture of S.Lang*, Annals of Math. 136 (1992) 557-567.

[20] A. Buium, *Geometry of differential polynomial functions I: algebraic groups*, Amer. J. Math. 115, 6 (1993), 1385-1444.

[21] A. Buium, *Geometry of differential polynomial functions II: algebraic curves*, Amer J. Math. 116, 4, (1994), 785-819.

[22] A. Buium, *Geometry of differential polynomial functions III: moduli spaces*, Amer J. Math. 117 (1995), 1-73.

[23] A. Buium, *Differential characters of Abelian varieties over p-adic fields*, Invent. Math., 122 (1995), pp. 309-340.

[24] A. Buium, *Geometry of $p-$jets*, Duke J. Math. 82, 2 (1996), 349-367.

[25] A.Buium, *Differential characters and characteristic polynomial of Frobenius*, Crelle J., 485 (1997), 209-219.

[26] A. Buium, *Differential subgroups of simple algebraic groups over p-adic fields*, Amer. J. Math. 120 (1998), 1277-1287.

[27] A. Buium, *Differential modular forms*, Crelle. J. 520 (2000), 95-167.

[28] A. Buium, *Differential modular forms on Shimura curves,I*, Compositio Math. 139 (2003), 197-237.

[29] A. Buium, *Differential modular forms on Shimura curves,II: Serre operators*, Compositio Math. 140 (2004), 1113-1134.

[30] A. Buium, *Pfaffian equations satisfied by differential modular forms*, Math. Research Letters 11 (2004), 453-466.

[31] A. Buium, *Geometry of Fermat adeles*, Trans. AMS 357 (2004), 901-964.

[32] A. Buium, S. R. Simanca, *Arithmetic partial differential equations I*, Advances in Math. 225 (2010), 689-793.

[33] A. Buium, S. R. Simanca, *Arithmetic partial differential equations II*, Advances in Math., 225 (2010), 1308-1340.

[34] A. Buium, S. R. Simanca, *Arithmetic Laplacians*, Advances in Math. 220 (2009), 246-277.

[35] A. Buium, *Arithmetic Differential Equations*, Math. Surveys and Monographs, 118, American Mathematical Society, Providence, RI, 2005. xxxii+310 pp.

[36] A. Buium, B. Poonen, *Independence of points on elliptic curves arising from special points on modular and Shimura curves, II: local results*, Compositio Math., 145 (2009), 566-602.

[37] A. Buium, A. Saha, *Hecke operators on differential modular forms mod p*, J. Number Theory 132 (2012), 966-997.

[38] A. Buium, A. Saha, *The ring of differential Fourier expansions*, J. of Number Theory 132 (2012), 896-937.

[39] A. Buium, *p-jets of finite algebras, I: p-divisible groups*, Documenta Math. 18 (2013) 943-969.

[40] A. Buium, *p-jets of finite algebras, II: p-typical Witt rings*, Documenta Math. 18 (2013) 971-996.

[41] A. Buium, A. Saha, *The first p-jet space of an elliptic curve: global functions and lifts of Frobenius*, Math. Res. Letters, Vol. 21, Number 04 (2014), 677-689.

[42] A. Buium, Yu. I. Manin, *Arithmetic differential equations of Painlevé VI type*, in: Arithmetic and Geometry, London Mathematical Society Lecture Note Series: 420, L. Dieulefait, G. Faltings, D. R. Heath-Brown, Yu. V. Manin, B. Z. Moroz and J.-P. Wintenberger (eds), Cambridge University Press, 2015, pp. 114-138.

[43] A. Buium, *Differential calculus with integers*, in: Arithmetic and Geometry, London Mathematical Society Lecture Note Series: 420, L. Dieulefait, G. Faltings, D. R. Heath-Brown, Yu. V. Manin, B. Z. Moroz and J.-P. Wintenberger (eds), Cambridge University Press, 2015, pp. 139-187.

[44] A. Buium, T. Dupuy, *Arithmetic differential equations on GL_n, I: differential cocyles*, J.Algebra, 454 (2016), 273-291.

[45] A. Buium, T. Dupuy, *Arithmetic differential equations on GL_n, II: arithmetic Lie-Cartan theory*, Selecta Math. 22, 2, (2016), 447-528.

[46] A. Buium, T. Dupuy, *Arithmetic differential equations on GL_n, III: Galois groups*, Selecta Math. 22, 2, (2016), 529-552.

[47] A. Buium, *Curvature on the integers, II*, J. Number Theory, 167 (2016), 509-545.

[48] A. Buium, E. Previato. *Arithmetic Euler top*, J. Number Theory, 173 (2017) pp. 37-63.

[49] J. W. S. Cassels, *An introduction to the Geometry of Numbers*, Classics in Mathematics, Springer, 1997.

[50] J. W. S. Cassels, *An Introduction to Diophantine Approximation*, Cambridge Tracts in Mathematics and Mathematical Physics, No. 45, CUB, 1957.

[51] P. J. Cassidy, *Differential algebraic groups*, Amer. J. Math. 94 (1972), 891-954.

[52] P. J. Cassidy, *The classification of semisimple differential algebraic groups and the linear semisimple differential algebraic Lie algebras*, J. Algebra (1989), 169-238.

[53] A. Connes, *Non-commutative Geometry*, Academic Press, NY, 1994.

[54] A. Connes, M. Dubois-Violette, *Yang-Mills algebra*, Letters in Mathematical Physics, 61 (2002), 149-158.

[55] A. Connes, C. Consani, *Schemes over $\mathbb{F}_1$ and zeta functions*, Compositio Mathematica, 146 (6), (2010) 1383-1415.

[56] A. Deitmar, *Schemes over* $\mathbb{F}_1$, in: *Number Fields and Function Fields, Two Parallel Worlds*, Progr. Math. 239, Birkhäuser, 2005, 87-100.
[57] P. Deligne, *Le groupe fondamental de la droite projective moins trois points*, Galois groups over $\mathbb{Q}$, MSRI. Publ., 16 (1989), 79-297.
[58] M. Demazure, P. Gabriel, *Groupes Algebriques*, North Holland 1970.
[59] C. Deninger, *Analogies between analysis on foliated spaces and arithmetic geometry*, Groups and analysis, London Math. Soc. Lecture Notes, 354, 174-190, Cambridge Univ. Press 2008.
[60] J. Dieudonné, *La Géométrie des Groupes Classiques*, Springer, 1963.
[61] J. Duistermaat, *On Hessian Riemannian structures*, Asian J. Math., 5 (2001), 79-91.
[62] B.V. Fedosov, *A simple geometrical construction of deformation quantization*, J. Diff. Geom. 40 (1994) 213–238.
[63] I. Fesenko, S. Vostokov, *Local Fields and their Extensions*, Transl. Math. Monogr., vol. 121, Amer. Math. Soc., Providence, RI, 1993.
[64] J-M. Fontaine, *Le corps des périodes p-adiques*, Astérisque 223 (1994), 59-111.
[65] J.K. Fornaes, N. Sibony, *Complex dynamics in higher dimensions*, in: Several Complex Variables, MSRI Publ. 37, Cambridge Univ. Press, Cambridge (1999), 273-296.
[66] I. Gelfand, V. Retakh and M. Shubin, *Fedosov Manifolds*, Advances in Math. 136 (1998) 104–140.
[67] R. Goodman, N. R. Wallach, *Representations and Invariants of the Classical Groups*, Cambridge U. Press, 1998.
[68] V. Guillemin, S. Sternberg, *Symplectic Techniques in Physics*, Cambridge Univ. Press, 1984.
[69] P. Griffiths, J. Harris, *Principles of Algebraic Geometry*, Wiley and Sons, 1978.
[70] S. Haran, *Non-Additive Prolegomena (to any future Arithmetic that will be able to present itself as a Geometry)*, arXiv:0911.3522 [math.NT].
[71] D. Harbater, *Formal patching and adding branch points*, Amer. J. Math., Vol. 115, No. 3 (1993), pp. 487-508.
[72] R. Hartshorne, *Algebraic Geometry*, GTM 52, Springer, 1977.
[73] S. Helgason, *Differential Geometry, Lie Groups, and Symmetric Spaces*, Graduate Studies in Math., Vol. 14, AMS, 2001.
[74] L. Hesselholt, *The big DeRham-Witt complex*, Acta Math. 214 (2015), 135-207.
[75] N. J. Hitchin, G. B. Segal, R. S. Ward, *Integrable Systems*, Oxford GTM, Oxford 1999.
[76] J. E. Humphreys, *Linear Algebraic Groups*, GTM 21, Springer, New York, 1995.
[77] Y. Ihara, *Profinite braid groups, Galois representations and complex multiplication*, Annals of Math. 123 (1986), 3-106.
[78] Y. Ihara, *On Fermat quotient and differentiation of numbers*, RIMS Kokyuroku 810 (1992), 324-341, (In Japanese). English translation by S. Hahn, Univ. of Georgia preprint.
[79] N. Jacobson, *Lie Algebras*, Dover, New York, 1979.
[80] P. Jordan, J. v. Neumann, E. Wigner, *On an algebraic generalization of the quantum mechanical formalism*, Annals of Math., Second Series, Vol. 35, No. 1 (1934), pp. 29-64.
[81] J. Jost, *Geometry and Physics*, Springer, 2009.
[82] J. Jost, *Riemannian Geometry and Geometric Analysis*, Springer, 2010.
[83] A. Joyal, *$\delta-$anneaux et vecteurs de Witt*, C.R. Acad. Sci. Canada, Vol. VII, No. 3, (1985), 177-182.
[84] V. Kac, P. Cheung, *Quantum Calculus*, Springer, 2002.
[85] D. S. Kaliuzhnyi-Verbovetskyi, V. Vinnikov, *Foundations of Noncommutative Function Theory*, Math. Surveys and Monographs 199, American Mathematical Society, Providence, 2014.
[86] M. Kapranov, A. Smirnov, *Cohomology determinants and reciprocity laws: number field case*, unpublished manuscript.
[87] N. Katz, *A conjecture in the arithmetic theory of differential equations*, Bull. Soc. Math. France 110 (2), 203-239.
[88] S. Kobayashi, *Differential Geometry of Complex Vector Bundles*, Princeton University Press, 1987.
[89] S. Kobayashi, K. Nomizu, *Foundations of Differential Geometry*, Vol I and II, Wiley and Sons, New York, 1969.
[90] N. Koblitz, *p-adic Numbers, p-adic Analysis, and Zeta Functions*, Springer, 1984.
[91] E. R. Kolchin, *Differential Algebra and Algebraic Groups*. Pure and Applied Mathematics, Vol. 54. Academic Press, New York-London, 1973. xviii+446 pp.
[92] E. R. Kolchin, *Differential Algebraic Groups*, Academic Press, New York 1985.

[93] N. Kurokawa, H. Ochiai, M. Wakayama, *Absolute derivations and zeta functions*, Documenta Math., Extra Volume Kato (2003), 565-584.

[94] S. Lang, *Algebraic Number Theory*, GTM 110, Springer, 2000.

[95] S. Lang, *Algebra*, GTM 211, Springer, 2002.

[96] A. Lichnerowicz, *Déformations d'algèbres associées à une variété symplectique*, Ann. Inst. Fourier, Grenoble 32 (1982) 157-209.

[97] J. Lubin, *One parameter formal Lie groups over $\mathfrak{p}$-adic integer rings*, Annals of Math. 80 (1964), 464-484.

[98] S. Mac Lane, *Categories for the Working Mathematician*, GTM 5, Springer, 1971.

[99] Yu. I. Manin, *Rational points on algebraic curves over function fields*, Izv. Acad. Nauk USSR, 27 (1963), 1395-1440.

[100] Yu. I. Manin, *Reflections on arithmetical physics*, in: "Conformal Invariance and String Theory," P. Dita and V, Georgescu Editors, Boston, Acadic Press, 1989, 293-303.

[101] Yu. I. Manin, *Sixth Painlevé equation, universal elliptic curve, and mirror of* $\mathbb{P}^2$, Geometry of Differential Equations, A. Khovanskii, A. Varchenko, V. Vassiliev, Editors, Amer. Math. Soc. Transl. (2), vol. 186, (1998), 131-151.

[102] Yu. I. Manin, *Numbers as functions*, arXiv:1312.5160.

[103] H. Matsumura, *Commutative Ring Theory*, Cambridge Studies in Advanced Mathematics, 8. Cambridge University Press, Cambridge, 1989. xiv+320 pp.

[104] M. Morishita, *Knots and Primes*, Springer, 2012.

[105] M. Mulase, *Algebraic theory of the KP equations*, in: Perspectives in Mathematical Physics, R. Penner and S.T. Yau, Editors (1994), 151-218.

[106] R. Narasimhan, *Analysis on Real and Complex Manifolds*, North-Holland, 1968.

[107] O. T. O'Meara, *Introduction to Quadratic Forms*, Springer, 2000.

[108] B. O'Neill, *Semi-Riemannian Geometry*, Academic Press, NY, 1983.

[109] I. B. S. Passi, D. S. Passman, S. K. Sehgal, *Lie solvable group rings*, Can. J. Math., Vol. XXV, No.4, 1973, pp. 748-757.

[110] P. Petersen, *Riemannian Geometry*, GTM 171, Springer, 2006.

[111] M. Raynaud, *Courbes sur une variété abélienne et points de torsion*, Invent. Math. 71 (1983), 207-235.

[112] J. F. Ritt, *Differential Equations from an Algebraic Standpoint*, Colloquium Publications Vol. 14, AMS, 2008.

[113] J. P. Serre, *Local Fields*, Springer, 1995.

[114] M. Singer, M. van der Put, *Galois Theory of Difference Equations*, LNM, Springer 1997.

[115] C. Soulé, *Les variétés sur le corps à un élément.*, Mosc. Math. J. 4 (2004), no. 1, 217-244.

[116] S. Sternberg, *Curvature in Mathematics and Physics*, Dover, New York, 2012.

[117] K. Thas (Editor), *Absolute Arithmetic and $\mathbb{F}_1$-Geometries*, European Math. Soc, 2016.

[118] J. Tits, *Sur les analogues algébriques des groupes semisimples complexes*, Centre Belge Rech. Math., Colloque d'algèbre superieure, (1957), 261-289.

[119] I.Vaisman, *Symplectic curvature tensors*, Monatshefte für Mathematik, 100, 4 (1985), 299-327.

[120] V. S. Vladimirov, I. V. Volovich, E. I. Zelenov, *p-adic Analysis and Mathematical Physics*, World Scientific, 1994.

[121] V. Voevodsky, *Cancellation theorem*, Documenta Mathematica, Extra Volume Suslin (2010) 671-685.

[122] C. Wilkerson, *Lambda rings, binomial domains, and vector bundles over $CP(\infty)$*, Comm. Algebra 10 (1982), 311-328.

[123] K. Yoshida, *Functional Analysis*, Springer, New York, 1995.

[124] O. Zariski, *Collected Papers, volume II*, M. Artin and D. Mumford, eds. MIT Press, 1973.

Index

Selected Published Titles in This Series

222 **Alexandru Buium,** Foundations of Arithmetic Differential Geometry, 2017
215 **Robert Bieri and Ralph Strebel,** On Groups of PL-homeomorphisms of the Real Line, 2016
214 **Jared Speck,** Shock Formation in Small-Data Solutions to 3D Quasilinear Wave Equations, 2016
213 **Harold G. Diamond and Wen-Bin Zhang (Cheung Man Ping),** Beurling Generalized Numbers, 2016
212 **Pandelis Dodos and Vassilis Kanellopoulos,** Ramsey Theory for Product Spaces, 2016
211 **Charlotte Hardouin, Jacques Sauloy, and Michael F. Singer,** Galois Theories of Linear Difference Equations: An Introduction, 2016
210 **Jason P. Bell, Dragos Ghioca, and Thomas J. Tucker,** The Dynamical Mordell–Lang Conjecture, 2016
209 **Steve Y. Oudot,** Persistence Theory: From Quiver Representations to Data Analysis, 2015
208 **Peter S. Ozsváth, András I. Stipsicz, and Zoltán Szabó,** Grid Homology for Knots and Links, 2015
207 **Vladimir I. Bogachev, Nicolai V. Krylov, Michael Röckner, and Stanislav V. Shaposhnikov,** Fokker–Planck–Kolmogorov Equations, 2015
206 **Bennett Chow, Sun-Chin Chu, David Glickenstein, Christine Guenther, James Isenberg, Tom Ivey, Dan Knopf, Peng Lu, Feng Luo, and Lei Ni,** The Ricci Flow: Techniques and Applications: Part IV: Long-Time Solutions and Related Topics, 2015
205 **Pavel Etingof, Shlomo Gelaki, Dmitri Nikshych, and Victor Ostrik,** Tensor Categories, 2015
204 **Victor M. Buchstaber and Taras E. Panov,** Toric Topology, 2015
203 **Donald Yau and Mark W. Johnson,** A Foundation for PROPs, Algebras, and Modules, 2015
202 **Shiri Artstein-Avidan, Apostolos Giannopoulos, and Vitali D. Milman,** Asymptotic Geometric Analysis, Part I, 2015
201 **Christopher L. Douglas, John Francis, André G. Henriques, and Michael A. Hill, Editors,** Topological Modular Forms, 2014
200 **Nikolai Nadirashvili, Vladimir Tkachev, and Serge Vlăduţ,** Nonlinear Elliptic Equations and Nonassociative Algebras, 2014
199 **Dmitry S. Kaliuzhnyi-Verbovetskyi and Victor Vinnikov,** Foundations of Free Noncommutative Function Theory, 2014
198 **Jörg Jahnel,** Brauer Groups, Tamagawa Measures, and Rational Points on Algebraic Varieties, 2014
197 **Richard Evan Schwartz,** The Octagonal PETs, 2014
196 **Silouanos Brazitikos, Apostolos Giannopoulos, Petros Valettas, and Beatrice-Helen Vritsiou,** Geometry of Isotropic Convex Bodies, 2014
195 **Ching-Li Chai, Brian Conrad, and Frans Oort,** Complex Multiplication and Lifting Problems, 2014
194 **Samuel Herrmann, Peter Imkeller, Ilya Pavlyukevich, and Dierk Peithmann,** Stochastic Resonance, 2014
193 **Robert Rumely,** Capacity Theory with Local Rationality, 2013
192 **Messoud Efendiev,** Attractors for Degenerate Parabolic Type Equations, 2013
191 **Grégory Berhuy and Frédérique Oggier,** An Introduction to Central Simple Algebras and Their Applications to Wireless Communication, 2013